# Diagnostics and Prognostics of Engineering Systems:

## Methods and Techniques

Seifedine Kadry
*American University of the Middle East, Kuwait*

| | |
|---|---|
| Managing Director: | Lindsay Johnston |
| Editorial Director: | Joel Gamon |
| Book Production Manager: | Jennifer Romanchak |
| Publishing Systems Analyst: | Adrienne Freeland |
| Development Editor: | Hannah Abelbeck |
| Assistant Acquisitions Editor: | Kayla Wolfe |
| Typesetter: | Deanna Jo Zombro |
| Cover Design: | Nick Newcomer |

Published in the United States of America by
Engineering Science Reference (an imprint of IGI Global)
701 E. Chocolate Avenue
Hershey PA 17033
Tel: 717-533-8845
Fax: 717-533-8661
E-mail: cust@igi-global.com
Web site: http://www.igi-global.com

Library of Congress Cataloging-in-Publication Data

Diagnostics and prognostics of engineering systems: methods and techniques / Seifedine Kadry, editor.
    p. cm.
 Includes bibliographical references and index.
 Summary: "This book provides widespread coverage and discussions on the methods and techniques of diagnosis and prognosis systems, including practical examples to display the method's effectiveness in real-world applications as well as the latest trends and research pertaining to engineering systems"-- Provided by publisher.
 ISBN 978-1-4666-2095-7 (hardcover) -- ISBN 978-1-4666-2096-4 (ebook) -- ISBN 978-1-4666-2097-1 (print & perpetual access) 1. Engineering systems--Testing. I. Kadry, Seifedine, 1977-
 TJ145.D46 2013
 620.001'1--dc23
                              2012017260

British Cataloguing in Publication Data
A Cataloguing in Publication record for this book is available from the British Library.

All work contributed to this book is new, previously-unpublished material. The views expressed in this book are those of the authors, but not necessarily of the publisher.

# Table of Contents

### Section 1
### Fault Tolerant Control

**Chapter 1**
*Zakwan Skaf, University of Manchester, UK.*

### Section 2
### Anomaly/Fault Detection

**Chapter 2**
*Claudia Maria García, Universitat Politècnica de Catalunya UPC, Spain*

### Section 3
### Data Driven Diagnostics

**Chapter 3**
*Veli Lumme, Tampere University of Technology, Finland*

**Section 10**
**Physics Based Diagnostics**

**Section 11**
**Prognostics**

**Section 12**
**Prognostics (Review)**

**Section 13**
**Prognostics (Structure)**

# Detailed Table of Contents

### Section 1
### Fault Tolerant Control

*Zakwan Skaf, University of Manchester, UK.*

A new design of a fault tolerant control (FTC)-based an adaptive, fixed-structure proportional-integral (PI) controller with constraints on the state vector for nonlinear discrete-time system subject to stochastic non-Gaussian disturbance is studied. The objective of the reliable control algorithm scheme is to design a control signal such that the actual probability density function (PDF) of the system is made as close as possible to a desired PDF, and make the tracking performance converge to zero, not only when all components are functional but also in case of admissible faults. A Linear Matrix Inequality (LMI)-based FTC method is presented to ensure that the fault can be estimated and compensated for. A radial basis function (RBF) neural network is used to approximate the output PDF of the system. Thus, the aim of the output PDF control will be a RBF weight control with an adaptive tuning of the basis function parameters. The key issue here is to divide the control horizon into a number of equal time intervals called batches. Within each interval, there are a fixed number of sample points. The design procedure is divided into two main algorithms, within each batch, and between any two adjacent batches. A P-type Iterative Learning Control (ILC) law is employed to tune the parameters of the RBF neural network so that the PDF tracking error decreases along with the batches. Sufficient conditions for the proposed fault tolerance are expressed as LMIs. An analysis of the ILC convergence is carried out. Finally, the effectiveness of the proposed method is demonstrated with an illustrated example.

## Section 2
## Anomaly/Fault Detection

A general methodology for intelligent system monitoring is proposed in this chapter. The methodology combines *degradation hybrid automata* to system degradation tracking and a nonlinear *adaptive model* for model-based diagnosis and prognosis purposes. The principal idea behind this approach is monitoring the plant for any off-nominal system behavior due a wear or degradation. The system degradation is divided in subspaces, from *fully functional*, nominal, or faultless mode to *no functionality* mode, failure. The *degradation hybrid automata*, uses a nonlinear *adaptive model* for continuous flow dynamics and a *system condition* guard to transition between modes. Error Filtering On- line Learning (EFOL) scheme is introduced to design a parametric model and adaptive low in such a way that the unknown part of the adaptive model function is on-line approximated; the on-line approximation is via a Radial Basis Function Neural Network (RBFNN). To validate the proposed methodology, a complete conveyor belt simulator, based on a real system, is designed on *Simulink*; the degradation is characterized using the Paris-Erdogan crack growth function. Once the simulator is designed the measured current, $i$'s, and velocity of the IM, $\omega_m$, are used to modeling the simplified adaptive IM model. EFOL scheme is used to on-line approximate the unknown $T_L$ function. The *simplified adaptive model* estimates the IM velocity,

$\hat{\omega}_m$, as output. $\hat{\omega}_m$ and the measured IM velocity $\omega_m$ are compared to detect any deviation from the nominal system behavior. When the degradation automata detect a system condition change the adaptive model on-line approximate the new $T_L$.

## Section 3
## Data Driven Diagnostics

This chapter discusses the main principles of the creation and use of a classifier in order to predict the interpretation of an unknown data sample. Classification offers the possibility to learn and use learned information received from previous occurrences of various normal and fault modes. This process is continuous and can be generalized to cover the diagnostics of all objects that are substantially of the same type. The effective use of a classifier includes initial training with known data samples, anomaly detection, retraining, and fault detection. With these elements an automated, a continuous learning machine diagnostics system can be developed. The main objective of such a system is to automate various time intensive tasks and allow more time for an expert to interpret unknown anomalies. A secondary objective is to utilize the data collected from previous fault modes to predict the re-occurrence of these faults in a substantially similar machine. It is important to understand the behaviour and functioning of a classifier in the development of software solutions for automated diagnostic methods. Several proven methods that can be used, for instance in software development, are disclosed in this chapter.

## Chapter 4

*Xiaomin Zhao, University of Alberta, Canada*
*Ming J Zuo, University of Alberta, Canada*
*Ramin Moghaddass, University of Alberta, Canada*

Diagnosis of fault levels is an important task in fault diagnosis of rotating machinery. Two or more sensors are usually involved in a condition monitoring system to fully capture the health information on a machine. Generating an indicator that varies monotonically with fault propagation is helpful in diagnosis of fault levels. How to generate such an indicator integrating information from multiple sensors is a challenging problem. This chapter presents two methods to achieve this purpose, following two different ways of integrating information from sensors. The first method treats signals from all sensors together as one multi-dimensional signal, and processes this multi-dimensional signal to generate an indicator. The second method extracts features obtained from each sensor individually, and then combines features from all sensors into a single indicator using a feature fusion technique. These two methods are applied to the diagnosis of the impeller vane trailing edge damage in slurry pumps.

## Chapter 5

*Assia Hakem, Lille 1 University, France*
*Komi Midzodzi Pekpe, Lille 1 University, France*
*Vincent Cocquempot, Lille 1 University, France*

This chapter addresses the problem of Fault Detection and Isolation (FDI) of switching systems where faulty behaviors are represented as faulty modes. The objectives of the online FDI are to identify accurately the current mode and to estimate the switching time. A data-based method is considered here to generate residuals for FDI. Conditions of two important properties, namely discernability between modes and switching detectability, are established. These conditions are different for the two properties. More specifically, it will be shown that a switching occurrence may be detected even if the two considered modes are not discernible. A vehicle rollover prevention example is provided for illustration purpose.

## Section 4
### Data Driven Prognostics

## Chapter 6

*Eric Bechhoefer, NRG Systems, USA*

A prognostic is an estimate of the remaining useful life of a monitored part. While diagnostics alone can support condition based maintenance practices, prognostics facilitates changes to logistics which can greatly reduce cost or increase readiness and availability. A successful prognostic requires four processes: 1) feature extraction of measured data to estimate damage; 2) a threshold for the feature, which, when exceeded, indicates that it is appropriate to perform maintenance; 3) given a future load profile, a model that can estimate the remaining useful life of the component based on the current damage state; and 4) an estimate of the confidence in the prognostic. This chapter outlines a process for data-driven prognostics by: describing appropriate condition indicators (CIs) for gear fault detection; threshold setting for those CIs through fusion into a component health indicator (HI); using a state space process to estimate the remaining useful life given the current component health; and a state estimate to quantify the confidence in the estimate of the remaining useful life.

## Section 5
## Degradation Modeling

### Chapter 7

*Jamie Coble, Pacific Northwest National Laboratory, USA*
*J. Wesley Hines, The University of Tennessee, USA*

The ultimate goal of most prognostic systems is accurate prediction of the remaining useful life of individual systems or components based on their use and performance. Traditionally, individual-based prognostic methods form a measure of degradation which is used to make lifetime estimates. Degradation measures may include sensed measurements, such as temperature or vibration level, or inferred measurements, such as model residuals or physics-based model predictions. Often, it is beneficial to combine several measures of degradation into a single prognostic parameter. Parameter features such as trendability, monotonicity, and prognosability can be used to compare candidate prognostic parameters to determine which is most useful for individual-based prognosis. By quantifying these features for a given parameter, the metrics can be used with any traditional optimization technique to identify an appropriate parameter. This parameter may be used with a parametric extrapolation model to make prognostic estimates for an individual unit. The proposed methods are illustrated with an application to simulated turbofan engine data.

### Chapter 8

*Ramin Moghaddass, University of Alberta, Canada*
*Ming J Zuo, University of Alberta, Canada*
*Xiaomin Zhao, University of Alberta, Canada*

The multi-state reliability analysis has received great attention recently in the domain of reliability and maintenance, specifically for mechanical equipment operating under stress, load, and fatigue conditions. The overall performance of this type of mechanical equipment deteriorates over time, which may result in multi-state health conditions. This deterioration can be represented by a continuous-time degradation process with multiple discrete states. In reality, due to technical problems, directly observing the actual health condition of the equipment may not be possible. In such cases, condition monitoring information may be useful to estimate the actual health condition of the equipment. In this chapter, the authors describe the application of a general stochastic process to multi-state equipment modeling. Also, an unsupervised learning method is presented to estimate the parameters of this stochastic model from condition monitoring data.

### Chapter 9

*Seifedine Kadry, American University of the Middle East, Kuwait*

In this chapter, a new technique is proposed to find the probability density function (pdf) of a stress for a stochastic mechanical system. This technique is based on the combination of the Probabilistic Transformation Method (PTM) and the Finite Element Method (FEM) to obtain the pdf of the response. The PTM has the advantage of evaluating the probability density function pdf of a function with random variable, by multiplying the joint density of the arguments by the Jacobien of the opposite function. Thus, the "exact" pdf can be obtained by using the probabilistic transformation method (PTM) coupled with the deterministic finite elements method (FEM). In the method of the probabilistic transformation, the pdf of the response can be obtained analytically when the pdf of the input random variables is known. An industrial application on a plate perforated with random entries was analyzed followed by a validation of the technique using the simulation of Monte Carlo.

## Chapter 10

*Chao Liu, Tsinghua University, P. R. China*

*Dongxiang Jiang, Tsinghua University, P. R. China*

A transition stage exists during the equipment degradation, which is between the normal condition and the failure condition. The transition stage presents small changes and may not cause significant function loss. However, the transition stage contains the degradation information of the equipment, which is beneficial for the condition classification and prediction in prognostics. The degradation based condition classification and prediction of rotating machinery are studied in this chapter. The normal, abnormal, and failure conditions are defined through anomaly determination of the transition stage. The condition classification methods are analyzed with the degradation conditions. Then the probability of failure occurrence is discussed in the transition stage. Finally, considering the degradation processes in rotating machinery, the condition classification and prediction are carried out with the field data.

## Section 6
### Diagnostics

## Chapter 11

*Omid Geramifard, National University of Singapore, Singapore*

*Jian-Xin Xu, National University of Singapore, Singapore*

*Junhong Zhou, Singapore Institute of Manufacturing Technology, Singapore*

In this chapter, a temporal probabilistic approach based on hidden semi-Markov model is proposed for continuous (real-valued) tool condition monitoring in machinery systems. As an illustrative example, tool wear prediction in CNC-milling machine is conducted using the proposed approach. Results indicate that the additional flexibility provided in the new approach compared to the existing hidden Markov model-based approach improves the performance. 482 features are extracted from 7 signals (three force signals, three vibration signals and acoustic emission) that are acquired for each experiment. After the feature extraction phase, Fisher's discriminant ratio is applied to find the most discriminant features to construct the prediction model. The prediction results are provided for three different cases, i.e. cross-validation, diagnostics, and prognostics. The possibility of incorporating an asymmetric loss function in the proposed approach in order to reflect and consider the cost differences between an under- and over-estimation in tool condition monitoring is also explored and the simulation results are provided.

## Section 7
### Integration of Control and Prognostics

*Teresa Escobet, Universitat Politècnica de Catalunya, Spain*
*Joseba Quevedo, Universitat Politècnica de Catalunya, Spain*
*Vicenç Puig, Universitat Politècnica de Catalunya, Spain*
*Fatiha Nejjari, Universitat Politècnica de Catalunya, Spain*

This chapter proposes the combination of system health monitoring with control and prognosis creating a new paradigm, the health-aware control (HAC) of systems. In this paradigm, the information provided by the prognosis module about the component system health should allow the modification of the controller such that the control objectives will consider the system's health. In this way, the control actions will be generated to fulfill the control objectives, and, at the same time, to extend the life of the system components. HAC control, contrarily to fault-tolerant control (FTC), adjusts the controller even when the system is still in a non-faulty situation. The prognosis module, with the main feature system characteristics provided by condition monitoring, will estimate on-line the component aging for the specific operating conditions. In the non-faulty situation, the control efforts are distributed to the system based on the proposed health indicator. An example is used throughout the chapter to illustrate the ideas and concepts introduced.

## Section 8
### Integrated Prognostics

*David He, The University of Illinois-Chicago, USA*
*Eric Bechhoefer, NRG Systems, USA*
*Jinghua Ma, The University of Illinois-Chicago, USA*
*Junda Zhu, The University of Illinois-Chicago, USA*

In this chapter, a particle filtering based gear prognostics method using a one-dimensional health index for spiral bevel gear subject to pitting failure mode is presented. The presented method effectively addresses the issues in applying particle filtering to mechanical component remaining useful life (RUL) prognostics by integrating a couple of new components into particle filtering: (1) data mining based techniques to effectively define the degradation state transition and measurement functions using a one-dimensional health index obtained by a whitening transform; and (2) an unbiased $l$-step ahead RUL estimator updated with measurement errors. The presented prognostics method is validated using data from a spiral bevel gear case study.

## Section 9
### Life Cycle Cost and Return on Investment for Prognostics and Health Management

**Chapter 14**

*Peter Sandborn, CALCE, University of Maryland, USA*
*Taoufik Jazouli, CALCE, University of Maryland, USA*
*Gilbert Haddad, Schlumberger Technology Center, USA*

The development and demonstration of innovative prognostics and health management (PHM) technology is necessary but not sufficient for widespread adoption of PHM concepts within systems. Without the ability to create viable business cases for the insertion of the new technology into systems and associated management processes, PHM will remain a novelty that is not widely disseminated. This chapter addresses two key capabilities necessary for supporting business cases for the inclusion and optimization of PHM within systems. First, the chapter describes the construction of life-cycle cost models that enable return on investment estimations for the inclusion of PHM within systems and the valuation of maintenance options. Second, the chapter addresses the support of availability-centric requirements (e.g., availability contracts) for critical systems that incorporate PHM, and the resulting value that can be realized. Examples associated with avionics, wind turbines, and wind farms are provided.

### Section 10
### Physics Based Diagnostics

**Chapter 15**

*Chengqing Yuan, Reliability Engineering Institute, Wuhan University of Technology, China*
*Xinping Yan, Reliability Engineering Institute, Wuhan University of Technology, China*
*Zhixiong Li, Reliability Engineering Institute, Wuhan University of Technology, China*
*Yuelei Zhang, Reliability Engineering Institute, Wuhan University of Technology, China*
*Chenxing Sheng, Reliability Engineering Institute, Wuhan University of Technology, China*
*Jiangbin Zhao, Reliability Engineering Institute, Wuhan University of Technology, China*

Marine power machinery parts are key equipments in ships. Ships always work in rigorous conditions such as offshore, heavy load, et cetera. Therefore, the failures in marine power machinery would badly threaten the safety of voyages. Keeping marine power machineries running reliably is the guarantee of voyage safety. For the condition monitoring and fault diagnosis of marine power machinery system, this study established the systemic condition identification approach for the tribo-system of marine power machinery and developed integrated diagnosis method by combining on-line and off-line ways for marine power machinery. Lastly, the remote fault diagnosis system was developed for practical application in marine power machinery, which consists of monitoring system in the ship, diagnosis system in laboratory centre, and maintenance management & maintenance decision support system.

# Section 11
## Prognostics

**Chapter 16**

*Giulio Gola, Institute for Energy Technology, & IO-center for Integrated Operations, Norway*
*Bent H. Nystad, Institute for Energy Technology*

Oil and gas industries are constantly aiming at improving the efficiency of their operations. In this respect, focus is on the development of technology, methods, and work processes related to equipment condition and performance monitoring in order to achieve the highest standards in terms of safety and productivity. To this aim, a key issue is represented by maintenance optimization of critical structures, systems, and components. A way towards this goal is offered by Condition-Based Maintenance (CBM) strategies. CBM aims at regulating maintenance scheduling based on data analyses and system condition monitoring and bears the potential advantage of obtaining relevant cost savings and improved operational safety and availability. A critical aspect of CBM is its integration with condition monitoring technologies for handling a wide range of information sources and eventually making optimal decisions on when and what to repair. In this chapter, a CBM case study concerning choke valves utilized in Norwegian offshore oil and gas platforms is proposed and investigated. The objective is to define a procedure for optimizing maintenance of choke valves by on-line monitoring their condition and determining their Remaining Useful Life (RUL). Choke valves undergo erosion caused by sand grains transported by the oil-water-gas mixture extracted from the well. Erosion is a critical problem which can affect the correct valve functioning, resulting in revenue losses and cause environmental hazards.

# Section 12
## Prognostics (Review)

**Chapter 17**

*E. Zio, Ecole Centrale Paris, France & Politecnico di Milano, Italy*

Prognostics and health management (PHM) is a field of research and application which aims at making use of past, present, and future information on the environmental, operational, and usage conditions of an equipment in order to detect its degradation, diagnose its faults, and predict and proactively manage its failures. This chapter reviews the state of knowledge on the methods for PHM, placing these in context with the different information and data which may be available for performing the task and identifying the current challenges and open issues which must be addressed for achieving reliable deployment in practice. The focus is predominantly on the prognostic part of PHM, which addresses the prediction of equipment failure occurrence and associated residual useful life (RUL).

## Section 13
## Prognostics (Structure)

**Chapter 18**

Structure Reliability and Response Prognostics under Uncertainty Using Bayesian Analysis and
Analytical Approximations ...................................................................................................... 358

*Xuefei Guan, Clarkson University, USA*
*Jingjing He, Clarkson University, USA*
*Ratneshwar Jha, Clarkson University, USA*
*Yongming Liu, Clarkson University, USA*

This study presents an efficient method for system reliability and response prognostics based on Bayesian analysis and analytical approximations. Uncertainties are explicitly included using probabilistic modeling. Usage and health monitoring information is used to perform the Bayesian updating. To improve the computational efficiency, an analytical computation procedure is proposed and formulated to avoid time-consuming simulations in classical methods. Two realistic problems are presented for demonstrations. One is a composite beam reliability analysis, and the other is the structural frame dynamic property estimation with sensor measurement data. The overall efficiency and accuracy of the proposed method is compared with the traditional simulation-based method.

**Chapter 19**

Fatigue Damage Prognostics and Life Prediction with Dynamic Response Reconstruction Using
Indirect Sensor Measurements ................................................................................................ 376

*Jingjing He, Clarkson University, USA*
*Xuefei Guan, Clarkson University, USA*
*Yongming Liu, Clarkson University, USA*

This study presents a general methodology for fatigue damage prognostics and life prediction integrating the structural health monitoring system. A new method for structure response reconstruction of critical locations using measurements from remote sensors is developed. The method is based on the empirical mode decomposition with intermittency criteria and transformation equations derived from finite element modeling. Dynamic responses measured from usage monitoring system or sensors at available locations are decomposed into modal responses directly in time domain. Transformation equations based on finite element modeling are used to extrapolate the modal responses from the measured locations to critical locations where direct sensor measurements are not available. The mode superposition method is employed to obtain dynamic responses at critical locations for fatigue crack propagation analysis. Fatigue analysis and life prediction can be performed given reconstructed responses at the critical location. The method is demonstrated using a multi degree-of-freedom cantilever beam problem.

# Foreword

Arguably, structural health monitoring and damage prognosis has been practiced (perhaps without knowing it as such) since the beginning of humankind. Early hunters invariably listened to the sound their bowstrings made when bent and released to determine whether the bow-and-arrow mechanism was properly working; their very livelihood depended upon it! Today, our own very livelihoods still depend on similar assessments. Each one of us, often subconsciously, expects that the transportation systems, civil infrastructure, machines—one could say *everything*—that we use in daily life will perform in the way it was designed to give us maximum satisfactory usage and maximum safety. I fully expect (and truly hope!) that the very chair I am sitting upon— my favorite chair in my house— to write this foreword has been assessed for damage in some manner and deemed safe for me to sit. Borrowing an idea from the early hunters, I knocked on all four wooden legs of the chair, and the acoustic response sounded reasonably similar in each leg; no leg had a significantly different emission to my (rather untrained, admittedly) ear. Needless to say, I probably wouldn't have survived those early hunting days.

Fortunately, the fields of structural health monitoring and prognostics have advanced a long way. Qualitatively speaking, *structural health monitoring* (SHM) is the general process of making an assessment, based on appropriate analyses of in-situ measured data, about the current ability of a structural component or system to perform its intended design function(s) successfully. *Damage prognosis* (DP) or *prognostics* extends this process by considering how the SHM assessment, when combined with probabilistic future loading and failure mode models with relevant sources of uncertainty adequately quantified, may be used to forecast remaining useful life (RUL) or similar performance-level metrics in a way that facilitates efficient life cycle management of the component or system. A successful SHM/DP strategy may enable significant system ownership cost reduction through maintenance optimization ("change the right engine rear bearing at next scheduled stop"), performance maximization during operation ("set cruise throttle down 10% to avoid reduced flutter onset boundary"), minimization of unscheduled downtime ("only change the right engine rear bearing, nothing else"), and/or enable significant life safety advantage through catastrophic failure mitigation. Quite important outcomes, indeed: I would challenge any structural/system owner to deny that at least some of these goals are all central to optimal life cycle management.

It is unlikely that only one book or collection of works could completely cover the enormous breadth that SHM/DP encompasses, especially when considering the application domain before it: just about *everything*. Nor should anyone ever expect something so comprehensive, as it could quite possibly fill thousands upon thousands of pages. This book, *Diagnostics and Prognostics of Engineering Systems: Methods and Techniques,* edited by Professor Seifedine Kadry, addresses the timely need to aggregate a number of key techniques and approaches used within a well-defined, accessible application area, that of industrial machinery. The collected articles consider the highly interdisciplinary scope of SHM/

DP, as any such SHM/DP strategy broadly must include in-situ data acquisition, feature extraction from the measured data, statistical modeling and classification of those features, and predictive modeling to assist in making risk-informed life cycle decisions, all within whatever economic, regulatory, and other constraints that are inevitably present. The chapters draw from computer science, mechanical engineering, industrial engineering, information technology, control engineering, statistics and decision sciences, and even policy/management in a successful effort to bring the state-of-the-art in how SHM/DP is being developed, evaluated, and deployed in industrial machine practice applications. There is quite a dearth of collected works in SHM/DP, and the appearance of this book couldn't be more timely in the current economic climate, where structural owner/operators are demanding more and more useful life from their systems, often well beyond initial design lifetimes. Just as importantly, the book truly spans the international community, with contributions from thirteen different countries, embracing the "global citizen" concept that has increasingly defined the modern engineer or technical practitioner.

So, whether you are a researcher, practitioner, technical manager, or you just want to learn about the incredibly broad and exciting field of SHM/DP, I invite you to sit back in your own favorite chair (sorry, mine is still taken) and open this book. I am happy to report that a second acoustic test on my chair still reveals no statistically significant (at least, based on my poor hearing) signature change, so I will comfortably read along with you.

*Michael Todd*
*University of California San Diego, USA*

**Michael Todd** *received his B.S.E. (1992), M.S. (1993), and Ph.D. (1996) from Duke University's Department of Mechanical Engineering and Materials Science, where he was an NSF Graduate Research Fellow. In 1996, he began as an A.S.E.E. post-doctoral fellow, then a staff research engineer (1998), and finally Section Head (2000) at the United States Naval Research Laboratory (NRL) in the Fibre Optic Smart Structures Section. In 2003, he joined the Structural Engineering Department at the University of California San Diego, where he currently serves as Professor and Vice Chair. To date, he has published over 75 journal articles, 170 other proceedings, 5 book chapters, and has 4 patents. His research interests are in applying nonlinear time series techniques to structural health monitoring (SHM) applications, adapting Bayesian inference frames for optimal decision-making in SHM, developing novel ultrasonic interrogation strategies for aerospace structural assessment, optimizing sensor networks for various SHM-rooted performance measures, developing RF-based sensing systems for structural assessment, creating real-time shape reconstruction strategies for highly flexible aerospace and naval structural systems based on limited data sets, creating rapid assessment checks for validation of satellite systems, designing and testing fibre optic measurement systems for civil and naval structural applications, and modelling noise propagation in fibre optic measurement systems. He serves on the editorial board of "Structural Health Monitoring: An International Journal."*

# Preface

In human health care, a medical analysis is made, based on the measurements of some parameters related to health conditions; the examination of the collected measurements aims at detecting anomalies, diagnosing illnesses and predicting their evolution. By analogy, technical procedures of health management are used to capture the functional state of industrial equipment from historical recordings of measurable parameters.

Today, most maintenance actions are carried out by either the preventive or the corrective approach. The preventive approach has fixed maintenance intervals in order to prevent components, sub-systems or systems to degrade. Corrective maintenance is performed after an obvious fault or breakdown has occurred. Both approaches have shown to be costly in many applications due to lost production, cost of keeping spare parts, quality deficiencies, et cetera.

Basically, predictive maintenance or Condition Based Maintenance (CBM) differs from preventive maintenance by basing maintenance need on the actual condition of the machine rather than on some preset schedule. As the preventive maintenance is time-based and activities such as changing lubricant are based on time, like calendar time or equipment run time. For example, most people change the engine oil in their car/jeep at every 3000 to 5000 KMs vehicles traveled. No concern is given to the actual condition and performance capability of the oil. This methodology would be analogous to a preventive maintenance task. If on the other hand, the operator of the car discounted the vehicle run time and had the oil analyzed at some periodicity to determine its actual condition and lubrication properties, he/she may be able to extend the oil change until the vehicle had traveled 10000 KMs. This is the fundamental difference between predictive maintenance and preventive maintenance, whereby predictive maintenance is used to define needed maintenance task based on quantified material/equipment condition.

The objective of CBM is to maintain the correct equipment at the correct time. CBM is based on using real-time Prognostics and Health Management (PHM) data to prioritize and optimize maintenance resources. By observing the state of the system (condition monitoring), the system will determine its health, and act only when maintenance is actually necessary thus minimizing the remaining useful life of the equipment that is thrown away. Using CBM, maintenance personnel are able to decide when the right time to perform maintenance is. Ideally CBM will allow the maintenance personnel to do only the right things, minimizing spare parts cost, system downtime, and time spent on maintenance.

The area of intelligent maintenance and diagnostic and prognostic–enabled CBM of machinery is a vital one for today's complex systems in industry, aerospace vehicles, military and merchant ships, the automotive industry, and elsewhere. The industrial and military communities are concerned about critical system and component reliability and availability. The goals are both to maximize equipment up time and to minimize maintenance and operating costs. As manning levels are reduced and equipment becomes more complex, intelligent maintenance schemes must replace the old prescheduled and labor intensive planned maintenance systems to ensure that equipment continues to function. Increased demands on machinery place growing importance on keeping all equipment in service to accommodate mission-critical usage. While fault detection and fault isolation effectiveness with very low false alarm rates continue to improve on these new applications, prognosis requirements are even more ambitious and present very significant challenges to system design teams. These prognostic challenges have been addressed aggressively for mechanical systems for some time but are only recently being explored fully for electronics systems.

Prognosis is one of the more challenging aspects of the modern prognostic and health management (PHM) system. It also has the potential to be the most beneficial in terms of both reduced operational and support cost and life-cycle total ownership cost and improved safety of many types of machinery and complex systems. The evolution of diagnostic monitoring of complex systems has led to the recognition that predictive prognosis is both desired and technically possible.

This book is a collection of chapters that could completely cover the huge coverage that structural health monitoring (SHM) and prognostic health management (PHM) encompasses. The chapters draw from different science fields like mechanical and industrial engineering, information technology, and control engineering in a fruitful effort to bring the state of the art in how SHM and PHM is being established, evaluated, and deployed in industrial machine practice applications.

**Chapter 1**, "Iterative Fault Tolerant Control for General Discrete-Time Stochastic Systems Using Output Probability Density Estimation," presents an ILC-based FTC method for the shape control of the output PDFs for general stochastic systems with non-Gaussian variable using a generalized fixed-structure PI controller with constraints on the state vector that results from the application of square root PDF modeling. The whole control horizon is divided into a number of batches. Within each batch, the state-constrained generalized PI controller is used to shape the output PDF using an LMI approach. Between any two adjacent batches, the parameters of the RBF basis functions are tuned. A P-Type ILC is applied to achieve such tuning between batches, where a sufficient condition for the ILC convergence has been established.

**Chapter 2**, "Intelligent System Monitoring: On-Line Learning and System Condition State," proposes a new intelligent monitoring method for system condition representation. It uses 5 consecutive steps, 1) data acquisition; which in step 3), diagnosis, is compared with a simplified adaptive mode, 2); the degradation hybrid automata flow among degraded modes tracking the system condition, state 4); and, finally EOL or/and RUL forecasting is computed on prognosis step, 5). This chapter addresses an approach for intelligent system monitoring based on a simplified adaptive model in which steps 2) and 4) are studied being out of scope the diagnosis and prognosis part.

**Chapter 3**, "Principles of Classification," shows that it is possible to develop a continuous learning diagnostic system using neural networks. The system will significantly save an analyst's time and effort in anomaly and fault detection, which would otherwise require extensive work. There are several ways to enhance the presentation of data on a classifier in order to predict the interpretation of a new data sample. The confidence level of prediction can be estimated using simple terms.

**Chapter 4**, "Generating Indicators for Diagnosis of Fault Levels by Integrating Information from Two or More Sensors," presents two methods of generating indicators for fault levels by integrating information from possible sensors. The first method regards signals from two sensors and different health conditions as one multivariate signal. Multivariate empirical mode decomposition is adopted to decompose the multivariate signal into a set of IMFs. The fault-sensitive IMF is chosen by a criterion based on mutual information. Then a full spectra based indicator is obtained. The indicator generated by method 1 reveals the characteristics of planar vibration motions. The second method extracts features from each individual sensor, uses global fuzzy preference approximation quality to select features having better monotonic relevance with fault levels, and utilizes PCA to combine information in selected features into a single indicator. The generated indicator makes use of information among different sensors and features, and outperforms each individual feature. This method is general and can work for multiple sensors. The generated indicator, however, because of the linear transformation induced by PCA, doesn't keep the physical meaning of the original selected features.

**Chapter 5**, "Fault Detection and Isolation for Switching Systems using a Parameter-Free Method," proposes a data-based method for switching detection and mode recognition. Both processes can be achieved successfully in both methods when the system switches between discernible modes. Discernibility between modes requires a necessary condition related to the Markov parameters of the mode. Although when the system switches between non discernible modes, the switching is detected, which means that the non-discernibility condition does not imply the non-switching detectability is not a sufficient condition for a non-switching detectability.

**Chapter 6**, "Data Driven Prognostics for Rotating Machinery," outlines a process for data-driven prognostics by: describing appropriate condition indicators (CIs) for gear fault detection; threshold setting for those CIs through fusion into a component health indicator (HI); using a state space process to estimate the remaining useful life given the current component health; and a state estimate to quantify the confidence in the estimate of the remaining useful life.

**Chapter 7**, "Identifying Suitable Degradation Parameters for Individual-Based Prognostics," presents a set of metrics that characterize the suitability of a prognostic parameter. Parameter features such as monotonicity, prognosability, and trendability can be used to compare candidate prognostic parameters to determine which is most useful for individual-based prognosis. Monotonicity characterizes the underlying positive or negative trend of the parameter, which addresses the common assumption that physical systems do not self-heal. Prognosability gives a measure of the variance in the critical failure value of a degradation parameter for a population of systems or components, which improves confidence in the estimate of failure. Finally, trendability indicates the degree to which the developed degradation parameters of a population of systems have the same underlying shape and can be described by the same functional form. These three intuitive metrics can be formalized to give a quantitative measure of prognostic parameter suitableness. The combination of the three measures and the suitability can then be used as a fitness function to optimize the development of a prognostic parameter.

**Chapter 8**, "Modeling Multi-State Equipment Degradation with Non-Homogeneous Continuous-Time Hidden Semi-Markov Process," presents a general stochastic model using nonhomogeneous continuous-time hidden semi-Markov process (NHCTHSMP) to model the degradation process and the observation process of a piece of multi-state equipment with unobservable states. The detailed mathematical structure for the NHCTHSMP associated with the multi-state equipment was described. Important measures of a NHCTHSMP based on the associated kernel function and the transition rate function were illustrated. Finally an estimation method was presented which can be used to estimate the unknown parameters of a NHCTHSMP using condition monitoring information. A simple numerical example was provided to describe the application of NHCTHSMP in modeling the degradation process and the observation process of multi-state equipment with unobservable states

**Chapter 9**, "Stochastic Fatigue of a Mechanical System Using Random Transformation Technique," presents a new technique to find the probability density function (pdf) of a stress for a stochastic mechanical system. This technique is based on the combination of the Probabilistic Transformation Method

(PTM) and the Finite Element Method (FEM) to obtain the pdf of the response. The new technique is verified with 10000 Monte-Carlo simulations.

**Chapter 10**, "Degradation Based Condition Classification and Prediction in Rotating Machinery Prognostics," analyzes the degradation based condition classification and prediction approaches in prognostics. The normal, abnormal, and failure conditions are defined through anomaly determination of the transition stage. The condition classification methods are analyzed with the degradation conditions. Then the probability of failure occurrence is discussed in the transition stage. Finally, considering the degradation processes in rotating machinery, the condition classification and prediction are carried out with the field data.

**Chapter 11**, "A Temporal Probabilistic Approach for Continuous Tool Condition Monitoring," introduces a hidden semi-Markov model based approach for continuous diagnosis and prognosis. Also, a computationally efficient version of forward-backward algorithm for application of HSMM in continuous health condition monitoring is described. Based on the simplified forward-backward algorithm, diagnostics and prognostics procedures are defined. A comparative study is conducted between the suggested HSMM-based approach and the existing HMM-based approach. Performances of the two approaches are compared in three cases i.e. cross-validation, diagnostics and prognostics. Based on the experimental results, HSMM-based approach outperforms the HMM-based approach in both diagnostics and prognostics. Prognosis ability of the suggested HSMM-based approach is tested in case III. Interestingly, the error rate of the HSMM-based approach predicting 10 time steps ahead is less than the acquired one step ahead average prognosis error rate from the HMM-based approach which indicates how powerful HSMM is compared to HMM in capturing the underlying temporal information.

**Chapter 12**, "Combining Health Monitoring and Control," proposes the combination of system health monitoring with control and prognosis by the introduction of the HAC paradigm. In this paradigm, the information provided by the prognosis module about the component system health should allow modifying the controller such that the control objectives will consider the system health. In this way, the control actions will be generated to fulfill the control objectives but at the same time to extend the life of the system components. HAC control contrarily to FTC adjusts the controller even when the system is still in non-faulty situation. The prognosis module will estimate on-line the component aging for the specific operating conditions. In the non-faulty situation, the control efforts are distributed to the system based on the proposed health indicator. An example has been used along the chapter to illustrate the ideas and concepts introduced.

**Chapter 13**, "A Particle Filtering Based Approach for Gear Prognostics," presents a particle filtering based gear prognostics method using a one-dimensional health index for spiral bevel gear subject to pitting failure mode. The presented method effectively addresses the issues in applying particle filtering to mechanical component remaining useful life prognostics by integrating a couple of new components into particle filtering: (1) data mining based techniques to effectively define the degradation state transition and measurement functions using a one-dimensional health index obtained by a whitening transform; (2) an unbiased l-step ahead RUL estimator updated with measurement errors. The presented prognostics method is validated using data from a spiral bevel gear case study. The validation results have shown the effectiveness of the presented method.

**Chapter 14**, "Supporting Business Cases for PHM: Return on Investment and Availability Impacts," addresses two key capabilities necessary for supporting business cases for the inclusion and optimization of PHM within systems. First the chapter describes the construction of life-cycle cost models that enable return on investment estimations for the inclusion of PHM within systems and the valuation of maintenance options. Second, the authors address the support of availability-centric requirements (e.g., availability contracts) for critical systems that incorporate PHM; and the resulting value that can be realized. Examples associated with avionics, wind turbines, and wind farms are provided.

**Chapter 15**, "Remote Fault Diagnosis System for Marine Power Machinery System," reports a new knowledge based remote diagnosis system in the application of condition monitoring and fault diagnosis (CMFD) for marine power machinery systems. The constructed two lever diagnosis system integrates the performance parameters, lubricant oil analysis, vibration, and instantaneous speed analysis to make the remote diagnosis system of marine power machinery systems feasible and available. The gear pump test shows that the proposed system is competent for fault detection. The proposed knowledge based remote diagnosis system has been proven to be feasible in engineering practice, and efficient for failure detection for diesel engines.

**Chapter 16**, "Prognostics and Health Management of Choke Valves Subject to Erosion: A Diagnostic-Prognostic Frame for Optimal Maintenance Scheduling," analyses a practical case study concerning erosion in choke valves used in oil industries with the aim of defining a diagnostic-prognostic frame for optimizing maintenance scheduling of such components. Two objectives have been identified: 1) the development of a condition monitoring system capable of providing reliable calculations of the erosion state based on collected measurements of physical parameters related to the choke erosion and 2) the development of a prognostic system to accurately estimate the remaining useful life of the choke. An empirical, model-based approach has been used to fulfill the diagnostic bjective of providing reliable calculations of the erosion state, whereas a statistical method based on the gamma probability distribution has been adopted to reach the prognostic goal of accurately estimating the remaining useful life of the choke.

**Chapter 17**, "Prognostics and Health Management of Industrial Equipment," reviews the state of knowledge on the methods for PHM, placing these in context with the different information and data which may be available for performing the task and identifying the current challenges and open issues which must be addressed for achieving reliable deployment in practice. The focus is predominantly on the prognostic part of PHM, which addresses the prediction of equipment failure occurrence and associated residual useful life (RUL).

**Chapter 18**, "Structure Reliability and Response Prognostics under Uncertainty Using Bayesian Analysis and Analytical Approximations," develops an efficient analytical Bayesian method for reliability and system response updating. The method is capable of incorporating additional information such as inspection data to reduce uncertainties and improve the estimation accuracy. One major difference between the proposed work and the traditional approach is that the proposed method performs all the calculations including Bayesian updating without using MC or MCMC simulations. A twenty-variable numerical example and a structural scale problem are presented for demonstration. Comparisons are made with traditional simulation-based methods to investigate the accuracy and efficiency.

**Chapter 19**, "Fatigue Damage Prognostics and Life Prediction with Dynamic Response Reconstruction Using Indirect Sensor Measurements," proposes a new methodology for fatigue prognosis integrating usage monitoring system. EMD method is employed to decompose the signal into a series of IMFs with specific filtering process. Those IMFs, which represented the displacement for each mode, are used to extrapolate the dynamic response at critical spot. It should be noticed that the mode shape information is required and can be obtained from classical finite element analysis. The fatigue crack growth prognosis is performed after the extrapolation process using a time-derivative model. Based on the current study, several conclusions are drawn: (1) The presented study provides a concurrent fatigue crack prognosis, which can be used for on-line fatigue life prediction, and (2) The numerical study demonstrates the proposed reconstruction method can effectively identify the dynamic responses for the critical spot where direct sensor measures are unavailable.

## WHO AND HOW TO READ THIS BOOK

This book has three groups of people as its potential audience, (i) undergraduate students and postgraduate students conducting research in the areas of system diagnostic and prognostic; (ii) researchers at universities and other institutions working in these fields; and (iii) practitioners in the Research and Development departments of industrial settings . This book differs from other books that have comprehensive case study and real data from industrial settings. The book can be used as an advanced reference for a course taught at the postgraduate level in industrial engineering, electrical engineering, mechanical engineering, manufacturing intelligence, and industrial electronics.

*Seifedine Kadry*
*American University of the Middle East, Kuwait*

# Acknowledgment

I have taken efforts in this project. However, it would not have been possible without the kind support and help of many people. I would like to extend my sincere thanks to all of them.

I would like first to thank all contributing authors for their effort in both writing and reviewing chapters. The completion of this book would have been impossible without their effort and commitment. Thank you all for making this pledge fruitful and useful to system diagnostic and prognostic community.

Second, I would like to thank all IGI Global publishing team members who assisted me during this process. In particular, I would like to thank Hannah Abelbeck who assisted me administratively during all stages of this project.

My gratitude should go also to my teachers, especially Professor Alaa Chateauneuf, Professor Khaled El-Tawil, Professor Ahmad Jammal, Professor Khaled Smaily, and Mohamed Smaily.

Finally, yet importantly, I would like to express my heartfelt thanks to my beloved family for their blessings, and my colleagues for their help and wishes for the successful completion of this project.

*Seifedine Kadry*
*American University of the Middle East, Kuwait*

# Section 1
# Fault Tolerant Control

# Chapter 1
# Iterative Fault Tolerant Control for General Discrete-Time Stochastic Systems Using Output Probability Density Estimation

**Zakwan Skaf**
*University of Manchester, UK.*

## ABSTRACT

*A new design of a fault tolerant control (FTC)-based an adaptive, fixed-structure proportional-integral (PI) controller with constraints on the state vector for nonlinear discrete-time system subject to stochastic non-Gaussian disturbance is studied. The objective of the reliable control algorithm scheme is to design a control signal such that the actual probability density function (PDF) of the system is made as close as possible to a desired PDF, and make the tracking performance converge to zero, not only when all components are functional but also in case of admissible faults. A Linear Matrix Inequality (LMI)-based FTC method is presented to ensure that the fault can be estimated and compensated for. A radial basis function (RBF) neural network is used to approximate the output PDF of the system. Thus, the aim of the output PDF control will be a RBF weight control with an adaptive tuning of the basis function parameters. The key issue here is to divide the control horizon into a number of equal time intervals called batches. Within each interval, there are a fixed number of sample points. The design procedure is divided into two main algorithms, within each batch, and between any two adjacent batches. A P-type Iterative Learning Control (ILC) law is employed to tune the parameters of the RBF neural network so that the PDF tracking error decreases along with the batches. Sufficient conditions for the proposed fault tolerance are expressed as LMIs. An analysis of the ILC convergence is carried out. Finally, the effectiveness of the proposed method is demonstrated with an illustrated example.*

DOI: 10.4018/978-1-4666-2095-7.ch001

## 1. INTRODUCTION

Under the assumption that the random variables or the noise in the stochastic system are subject to Gaussian processes, the following approaches have been widely applied in theoretical studies: minimum variance control (Astrom, 1970), whose purpose is minimizing the variations in the controlled system outputs or tracking errors, linear optimal control (Anderson, 1971), linear quadratic martingale control (Solo, 1990), and stochastic control for systems with Markovian jump parameters (Xia, Shi, Liu, and Rees, 2006). In all these methods, the targets are the mean and variance of the output. However, this assumption may not hold in some applications. For example, many variables in the paper-making systems do not obey Gaussian distributions (Wang, 2000, Wang,1999). Therefore, a new measure of randomness, called the PDF control, should be employed for general stochastic systems with non-Gaussian variables (Wang, 2000). In PDF control problems, the control objective is to design a control signal so that the PDF shape of the output variable follows a desired distribution.

There are many stochastic systems in practice whose outputs are the PDF of the system output (Wang, 2000) rather than the actual output values. For such cases, the measured output PDFs can be used as an output for the feedback control. Such types of stochastic systems are called Stochastic Distribution Control (SDC) systems (Wang, 2000). Practical examples of SDC systems in industrial applications include: Molecular weight distribution control (Crowley and Choi, 1998, Shibasaki, Araki, Nagahata, and Ueda, 2005), combustion flame distribution processes (Sun, Yue, and Wang, 2006, Wang, Afshar, and Wang, 2008), particle size distribution control in polymerisation and powder processing industries (Dunbar and Hickey, 2000, Shi, El-Farra, Li, Mhaskar, and Christo□des, 2006), and the wet-end of paper-making (Wang, 2000). A well-known example of a SDC system in practice is the 2D paper web solid distribution

in paper-making processes (Wang, 2000). For instance, the PDF distribution of the grammage of the finished sheet can be measured online via digital cameras. Such images from these digital cameras can be processed and used as a feedback signal to the control system (Wang, 2000).

SDC was originally developed by Professor Hong Wang in 1996, when he considered a number of challenging paper machine modelling and control problems (Wang, 2000). The process and the control were presented in a PDF form. As such, the purpose of the controller design was to obtain the PDF of the controller so that the closed-loop PDF would follow the pre-specified PDF. Since then, rapid developments have been made and introduced in different control applications (A. Wang, H. Wang, and Guo, 2009).

The most existing PDF control approaches are based on the B-spline model. Moreover, multi-layer perception (MLP) neural network models have been applied to the shape control for the output PDFs (Wang, 2003). Recently, a RBF neural network has been used to approximate the output PDF of the system (Wang and Afshar, 2009). In this work, we have used RBFs instead of B-Splines which help generalize the output PDF expression and overcome the complexity, limitations, and constrains with B-spline-based functional approximations.

Many effective fault detection and diagnosis (FDD) strategies have been developed by researchers in the last several decades to cover various types of faulty systems (Patton, Frank, and Clark, 1989, Isermann, 1993, Wang and Lin, 2000). For stochastic systems, many significant schemes have been introduced and applied to practical processes successfully. In general, the following approaches have been widely applied and developed for this problem: filter-or observer-based approaches (Wang and Lin, 2000, Zhang, Guo, and Wang, 2006), identification-based approaches (Isermann and Balle, 1996), and static approaches based likelihood approaches (Karny, Nagy, and Novovicova, 2002). The filter based

FDD approaches have been presented as an effective technique for Gaussian variables in stochastic systems. However, in many practical processes, non-Gaussian variables exist in many stochastic systems. Therefore, for non-Gaussian stochastic systems, a new FDD approach has been established by using output distribution function for general stochastic systems in (Wang and Lin, 2000), where the dynamical system was supposed to be a precise linear model and the design algorithm required some technical conditions that were hard to verify. That work was motivated by the retention system of the paper making process, where the system output is replaced by the measured output PDFs to generate the residual of the filter (Wang and Lin, 2000). The residual signal is calculated via the use of either the weighted integration or the integration of the square of the difference between the measured and the estimated PDFs. This method was the first attempt focusing on the application of the PDF model. However, there was a criticism that the used linear B-spline model cannot guarantee that the output PDF of the model is positive (Zhang et al., 2006, Guo and Wang, 2005). Subsequently, an improved design approach has been applied for the general stochastic system by using a square root B-spline model and nonlinear filter design (Guo and Wang, 2005).

Due to the high demand for reliability and safe operation, many FTC methods were developed in the past four decades, which have the capability of detecting the occurrence of faults and maintaining the performance of the system in the presence of faults at a prescribed level (Iserman, 2006). In most cases, the literatures on the FTC algorithms for stochastic systems have been presented under the assumption that the random variables or the noise in the stochastic system are subject to Gaussian distribution (Srichander and Walker, 1993). In (Yao, A. Wang, and H. Wang, 2008), a nonlinear adaptive observer-based fault diagnosis alorithm has been presented for the SDC systems that are based on the rational square-root B-spline approximation model. When faults occur in the

system, the controller was redesigned. This method is suffering from the complexity of modelling of stochastic distribution control. As such, there is a need to develop FTC methods that can be applied to general stochastic systems subject to arbitrary variables distribution.

## 2. ILC-BASED PDF CONTROL

To begin this section, a brief history of some significant developments in the field of ILC is presented. The term Iterative Learning Control was first presented in (Arimoto, Kawamura, and Miyazaki, 1984) and discussed in more detail in (Arimoto, 1990). Traditional iterative learning controllers have been developed over the past decade for nonlinear systems with nonlinearities satisfying the global Lipschitz continuous condition. The application of ILC to nonlinear systems proved to be a good way of improving the performance of these systems, which has led the development of different methods dealing with nonlinear dynamics (Chiang and Liu, 994). Examples of systems that operate in a repetitive manner include robot arm manipulators and chemical batch processes. In each of these tasks, the system is required to perform the same action over and over again.

By using information from previous repetitions, a suitable control action can be found iteratively. A classical control law is of the form:

$$u_k(i) = u_{k-1}(i) + \lambda J_{k-1}(i) \qquad (1)$$

where $i$ represents the time instant which satisfies $0 \le i \le m$. $m$ is the total number of time samples within a batch. $J(i)$ stands for a function that is related to the tracking performance index, and $\lambda$ is a learning rate gain, which is chosen so that the iterative control law is convergent.

In the stochastic control area, the ILC has been applied successfully to nonlinear stochastic systems (Chen and Fang, 2004). Some effort has been made to apply ILC in the design of the output PDF in order to control its shape (Wang and Afshar, 2009). As shown in Figure 1, the control horizon has been divided into a number of equal time-domain intervals called batches indexed by $k = (1, 2, ...)$, and these batches are specified by $\left[(k-1)(N + \Delta N), \ (k)(N + \Delta N)\right]$. Within each batch there are a fixed number of sample points $N$ which is considered as the batch length and should be large enough so that the system reaches the steady state within each batch; $\Delta N$ is the time period between adjacent batches. The design procedure is divided into two main algorithms, within each batch and between any two adjacent batches. Within batches, fixed basis functions are used to generate the required control. Then, the control law is implemented in the stochastic system in such a way that the closed-loop system is stable. Meanwhile, between adjacent batches the RBF parameters are updated to ensure the measured output PDF is closer to the desired PDF within the next batch.

## 3. PROBLEM FORMULATION

In this section, this approach is different from the results in (Guo and Wang, 2005), a discrete-

time RBF neural networks square root model is introduced to approximate the output PDFs and then formulate a discrete-time nonlinear model for the weighting vectors.

Consider $\sigma_1 = \sigma_2 = \sigma_3 = 0.2$ as the input of a discrete-time dynamic stochastic system,

$$A = \begin{bmatrix} -0.45 & 0.03 \\ 0.1 & -0.28 \end{bmatrix}, \quad B = \begin{bmatrix} 0.45 & 0.01 \\ 0.01 & -0.86 \end{bmatrix},$$

as the output, and

$$G = \begin{bmatrix} 0.02 & 0 \\ 0 & 0.01 \end{bmatrix}, \quad D = E = \begin{bmatrix} 1 & 0 \\ 0 & 1 \end{bmatrix}$$

as the fault. At sample time k,

$$g(V(t)) = \begin{bmatrix} 0 \\ \sqrt{v_1^2 + v_2^2} \end{bmatrix}$$

can be described by its PDF $V_1(0) = [0.001, \ 0.001]^T$, which is defined by

$$P(a \le y(k) < \xi, u(k)) = \int_a^\xi \gamma(y, u(k), F(k)) dy$$

*Figure 1. The ILC-based output PDF control scheme*

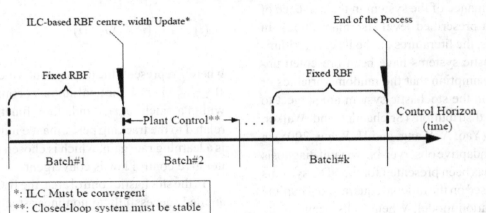

where

$$\Lambda_\sigma = \begin{bmatrix} -0.71 & 0 & 0 \\ 0 & -0.71 & 0 \\ 0 & 0 & -0.71 \end{bmatrix} \begin{bmatrix} 0.005 \times j \\ 0.005 \times j \\ 0.005 \times j \end{bmatrix}$$

denotes the probability of output variable $j = 0, 1, ...., 20$ lying between $U_1$ and $U_2$ when the control is applied to the system. It is assumed that

$$U_1 = \begin{bmatrix} 0.1 & 0 \\ 0 & 0.1 \end{bmatrix}, \quad U_2 = \begin{bmatrix} 1 & 0 \\ 0 & 1 \end{bmatrix},$$

in known and the PDF is measurable. The well-known RBF neural networks can be used to approximate the square root of the output PDF as follows (Wang, 2000).

$$\sqrt{\gamma(y, u(k), F(k))} = R(y)V(k) + r_n(y)h(V(k)) + \omega(y, u(k), F(k)) \tag{2}$$

where $\sigma_{g1} = \sigma_{g2} = \sigma_{g3} = 0.1$ is the output measured PDF.

$$R(y) = [r_1(y), r_2(y), ......, r_{n-1}(y)]$$
$$V(k) = [v_1(k), v_2(k), ......, v_{n-1}(k)]^T$$
$$\Lambda_4 = \Lambda_2^T \Lambda_2 - \Lambda_3 \Lambda_1, \Lambda_1 = \int_a^b R^T(y)R(y)dy$$
$$\Lambda_2 = \int_a^b R^T(y)r_n(y)dy, \Lambda_3 = \int_a^b r_n^2(y)dy \tag{3}$$

and

$$h(V(k)) = \frac{1}{\Lambda_3}\left(-\Lambda_2 V(k) + \sqrt{V^T(k)\Lambda_4 V(k)}\right) \tag{4}$$

F(k) is supposed to be an actuator fault to be diagnosed and compensated. Term $\omega(y, u(k), F(k))$ represents the model uncertainties or the error term on the approximation of PDFs. In addition, $\omega(y, u(k), F(k))$ must satisfy the following condition (Wang, 2000):

$$|\omega(y, u(k), F(k))| \leq \delta$$

where $\delta > 0$ is a known positive constant. In (2), $R(y)$ and $V(k)$ are the activation function and weight element corresponding to RBF neural network used for PDF modelling, respectively. Similar to (Wang and Afshar, 2006, Wang, Afshar, and Yue, 2006), the RBF activation functions are chosen as Gaussian shapes and expressed as follows:

$$r_l(y) = exp\left(-\frac{(y_j - \mu_l)^2}{2\sigma_l^2}\right) \tag{5}$$

where $\mu_l$, $\sigma_l$ are the centres and widths of the RBF basis functions, respectively. In (4), the nonlinear function $h(V(k))$ should satisfy the following Lipschitz condition for any $V_1(k)$, $V_2(k)$ and a known matrix $U_1$:

$$\|h(V_1(k)) - h(V_2(k))\| \leq \|U_1(V_1(k) - V_2(k))\| \tag{6}$$

## 3.1 Nonlinear Dynamic Weight Model

In many cases, the dynamic relation between the input and the output PDFs can be transformed into dynamic relation between the control input and the weights of the RBFs neural network approximation to the output PDFs. In this section, the following discrete-time nonlinear weighting model will be used

$$x(k+1) = Ax(k) + Bu(k) + Gg(x(k)) + DF(k)$$
$$V(k) = Ex(k)$$
$$(7)$$

where $x(k) \in R^n$ is the state vector, and $u(k) \in R^r$ is the measurable input vector. Moreover, $A$, $B$, $C$, $D$, and $E$ represent the identified coefficient matrices of the weight system with suitable dimensions. $g(x(k))$ is a nonlinear vector function that stands for the nonlinear dynamics of the model, and is supposed to satisfy $g(0) = 0$ and the following Lipschitz condition.

$$\left\| g(x_1(k)) - g(x_2(k)) \right\| \le \left\| U_2(x_1(k) - x_2(k)) \right\|$$
$$(8)$$

for any $x_1(k)$ and $x_2(k)$, where $U_2$ is a known matrix. $F(k)$ is an actuator fault to be estimated and rejected. With model (7), Equation (2) can be written as a nonlinear function of $x(k)$ as follows:

$$\sqrt{\gamma(y, u(k), F(k))}$$
$$= R(y)Ex(k) + r_n(y)h(Ex(k)) + \omega(y, u(k), F(k))$$
$$(9)$$

Different from the models considered in (Guo and Wang, 2005), the proposed discrete-time square root RBFNN model is more practical and better suited to digital control.

## 4. FAULT DETECTION

In order to detect the fault based on the changes of PDFs, the following nonlinear observer is considered:

$$\hat{x}(k+1) = A\hat{x}(k) + Bu(k) + Gg(\hat{x}(k)) + L\varepsilon(k)$$
$$\varepsilon(t) = \int_a^b \mu(y)(\sqrt{\gamma(y, u(k, F))} - \sqrt{\hat{\gamma}(y, u(k))})dy$$
$$\sqrt{\hat{\gamma}(y, u(k))} = R(y)E\hat{x}(k) + h(E\hat{x}(k))r_n(y)$$
$$(10)$$

where $\hat{x}(k) \in R^n$ is the estimated state and $L \in R^{n \times p}$ is the filter gain to be determined. Residual $\varepsilon(k)$ is formulated as an integral of the difference between the measured PDFs and the estimated ones, where $\mu(y) \in R^{p \times 1}$ is a prespecified weighting vector.

Denoting the estimation error as $e(k) = x(k) - \hat{x}(k)$, the dynamic of the estimation error will be expressed as

$$e(k+1) = (A - L\Gamma_1)e(k) + [Gg(x(k))$$
$$- Gg(\hat{x}(k))] - L\Gamma_2[h(Ex(k))$$
$$- h(E\hat{x}(k))] - L\Delta(k) + DF(k) \quad (11)$$

where

$$\Gamma_1 = \int_a^b \mu(y)R(y)Edy$$
$$\Gamma_2 = \int_a^b \mu(y)r_n(y)dy \qquad (12)$$
$$\Delta(k) = \int_a^b \mu(y)\omega(y, u(k))dy$$

It can be seen that

$$\varepsilon(k) = \Gamma_1 e(k) + \Gamma_2[h(Ex(k))$$
$$- h(E\hat{x}(k))] + \Delta(k) \qquad (13)$$

From $|\omega(y, u(k), F)| \le \delta$, it can be verified that

$$\left\| \Delta(k) \right\| = \left\| \int_a^b \mu(y)\omega(y, u(k))dy \right\| \le \tilde{\delta}$$

where $\tilde{\delta} = \delta \left\| \int_a^b \mu(y)dy \right\|$.

**Theorem 1 :** *For the parameter* $\lambda_i > 0 (i = 1, 2)$, *if there exist matrices* $P > 0$, *and* $R$ *satisfying*

$$\Psi =$$

$$\begin{bmatrix} M_1 & 0 & 0 & A^T P - \Gamma_1^T R^T \\ 0 & -\lambda_2^2 I & 0 & G^T P \\ 0 & 0 & -\lambda_1^2 I & \Gamma_2^T R^T \\ PA - R\Gamma_1 & PG & -R\Gamma_2 & -P \end{bmatrix} < 0 \tag{14}$$

$$M_1 = -P + \lambda_2^2 U_2^T U_2 + \lambda_1^2 E^T U_1^T U_1 E$$

*then in the absence of fault, the error dynamic system with gain* $L = P^{-1}R$ *is stable and the error satisfies* $\lim_{k \to \infty} e(k) = 0$

**Proof:** For this purpose, the following Lyapunov function is considered.

$$\Phi(k) = e^T(k)Pe(k)$$
$$+ \lambda_1^2 \sum_{i=1}^{k-1} \left[ \left\| U_1 Ee(i) \right\|^2 - \left\| h(Ex(i)) - h(E\hat{x}(i)) \right\|^2 \right]$$
$$+ \lambda_2^2 \sum_{i=1}^{k-1} \left[ \left\| U_2 e(i) \right\|^2 - \left\| g(x(i)) - g(\hat{x}(i)) \right\|^2 \right] \tag{15}$$

In the absence of $F(k)$, along with (11) it can be verified that

$$\Delta\Phi = \Phi(k+1) - \Phi(k)$$
$$= e^T(k+1)Pe(k+1) - e^T(k)Pe(k)$$
$$+ \lambda_1^2 \left[ \left\| U_1 Ee(k) \right\|^2 - \left\| h(Ex(k)) - h(E\hat{x}(k)) \right\|^2 \right]$$
$$+ \lambda_2^2 \left[ \left\| U_2 e(k) \right\|^2 - \left\| g(x(k)) - g(\hat{x}(k)) \right\|^2 \right]$$
$$= S_k^T \Psi_1 S_k + 2 S_k^T \begin{bmatrix} -(A - L\Gamma_1)^T PL \\ -G^T PL \\ \Gamma_2^T L^T PL \end{bmatrix}$$
$$\Delta(k) + \Delta^T(k) L^T PL \Delta(k) < 0 \tag{16}$$

where

$$\Psi_1 =$$
$$\begin{bmatrix} \Psi_2 & (A - L\Gamma_1)^T PG & -(A - L\Gamma_1)^T PL\Gamma_2 \\ * & -\lambda_2^2 I + G^T PG & -G^T PL\Gamma_2 \\ * & * & -\lambda_1^2 I + \Gamma_2^T L^T PL\Gamma_2 \end{bmatrix} \tag{17}$$

$$S_k^T = \left[ e^T(k), (g(x(k)) - g(\hat{x}(k))^T, \right.$$
$$\left. (h(Ex(k)) - h(E\hat{x}(k)))^T \right]$$
$$\Psi_2 = (A - L\Gamma_1)^T P(A - L\Gamma_1) + M_1$$

Denote $R = PL$, then it can be seen that

$$\Delta\Phi = S_k^T \Psi_1 S_k + 2 S_k^T$$
$$\begin{bmatrix} -(A - L\Gamma_1)^T R \\ -G^T R \\ \Gamma_2^T R^T P^{-1} R \end{bmatrix} \tag{18}$$
$$\Delta(k) + \Delta^T(k) R^T P^{-1} R \Delta(k) < 0$$

By using the Schur complement formula, (17) can be further reduced to

$$\Psi_5 = \begin{bmatrix} \Psi_3 & \Psi_4^T \\ \Psi_4 & -P \end{bmatrix} < 0 \tag{19}$$

where

$$\Psi_3 =$$
$$\begin{bmatrix} -P + \lambda_2^2 U_2^T U_2 + \lambda_1^2 E^T U_1^T U_1 E & 0 & 0 \\ * & -\lambda_2^2 I & 0 \\ * & * & -\lambda_1^2 I \end{bmatrix} \tag{20}$$

and

$$\Psi_4 = \begin{bmatrix} PA - R\Gamma_1 & PG & -R\Gamma_2 \end{bmatrix} \tag{21}$$

which is equivalent to (14). If (14) holds, a positive scalar exists $\rho$ so that $\Psi \leq -\rho I$. Thus, it can be seen that

$$
\begin{aligned}
\Delta \Phi &\leq -\rho \left\| S_k \right\|^2 - 2 \left\| \tilde{A}^T R \right\| \left\| \Delta(k) \right\| \left\| S_k \right\| \\
&\quad + \left\| \Delta(k) \right\|^2 \left\| R^T P^{-1} R \right\| \\
&\leq -\rho \| S_k \|^2 - 2\tilde{\delta} \| \tilde{A}^T R \| \| S_k \| \\
&\quad + \tilde{\delta}^2 \| R^T P^{-1} R \|
\end{aligned} \tag{22}
$$

where

$$
\tilde{A} = \begin{bmatrix} A - L\Gamma_1 & PG & -R\Gamma_2 \end{bmatrix} \tag{23}
$$

it can be shown that

$$
\left\| S_k \right\| \geq \tilde{\delta} \rho^{-1} \left( \left\| \tilde{A}^T R \right\| + \sqrt{ \left\| \tilde{A}^T R \right\|^2 + \rho \left\| R^T P^{-1} R \right\| } \right) \tag{24}
$$

which implies

$$
\begin{aligned}
& \left\| S_k \right\| \leq max \\
& \left\{ \left\| S_k(1) \right\|, \tilde{\delta} \rho^{-1} \left( \left\| \tilde{A}^T R \right\|^2 + \sqrt{ \left\| \tilde{A}^T R \right\|^2 + \rho \left\| R^T P^{-1} R \right\| } \right) \right\}
\end{aligned} \tag{25}
$$

By a similar proof to Theorem 1, it can get that the error system in (11) is asymptotically stable under $\Psi < 0$. Because $\lim_{k \to \infty} e(k) = 0$, Equations (6) and (8) guarantee that $\lim_{k \to \infty} \left( h(Ex(k)) - h(E\hat{x}(k)) \right) = 0$ and $\lim_{k \to \infty} \left( g(x(k)) - g(\hat{x}(k)) \right) = 0$, which implies that $\lim_{k \to \infty} \Delta \Phi(k) = 0$.

Theorem 1 provides a sufficient criterion for the stability of the error system in the absence of the fault, which presents a necessary condition for fault detection. In order to detect $F$, the following theorem should be considered.

**Theorem 2:** *For the parameter* $\lambda_i > 0 (i = 1, 2)$, *if there exist matrices* $P > 0$, *and* $R$ *satisfying (14), then fault* $F$ *can be detected by the following criterion*

$$
\left\| \varepsilon(k) \right\| > \tilde{\vartheta} = \tilde{\vartheta}_0 \left( \left\| \Gamma_1 \right\| + \left\| \Gamma_2 \right\| \left\| U_1 E \right\| \right) + \tilde{\delta} \tag{26}
$$

*which means that (26) implies* $F \neq 0$

Once fault is detected, it needs to be estimated, which follows an adaptive fault diagnosis algorithm in the next section.

## 5. FAULT DIAGNOSIS

Once the fault is detected, the fault value must be estimated. For this purpose, the following observer is considered:

$$
\hat{x}(k+1) = A\hat{x}(k) + Bu(k) + Gg(\hat{x}(k)) + L\varepsilon(k) + D\hat{F}(k)
$$

$$
\begin{aligned}
\sqrt{\hat{\gamma}(y, u(k))} &= R(y)E\hat{x}(k) + r_n(y)h(E\hat{x}(k)) \\
\hat{F}(k+1) &= -\Upsilon_1 \hat{F}(k) + \Upsilon_2 \varepsilon(k)
\end{aligned} \tag{27}
$$

where $\hat{F}(k)$ is the estimation of $F(k)$. $\Upsilon_1 \left( -I < \Upsilon_1 < I \right)$ and $\Upsilon_2$ are the learning operators to be determined together with $L$ by the diagnosis algorithm in (27). Denoting $\bar{F}(k) = F(k) - \hat{F}(k)$ and $e(k) = x(k) - \hat{x}(k)$. The dynamic of the estimation error will be expressed as

$$
\begin{aligned}
e(k+1) = {} & (A - L\Gamma_1)e(k) + [Gg(x(k)) \\
& - Gg(\hat{x}(k))] - L\Gamma_2[h(Ex(k)) \\
& - h(E\hat{x}(k))] - L\Delta(k) + D\bar{F}(k)
\end{aligned} \tag{28}
$$

and

$$\bar{F}(k+1) = F(k+1) - \hat{F}(k-1)$$
$$= F(k+1) + \Upsilon_1\hat{F}(k) - \Upsilon_2\varepsilon(k)$$
$$= F(k+1) - \Upsilon_1 F(k) + \Upsilon_1\bar{F}(k) - \Upsilon_2\varepsilon(k)$$
$$= \Delta F(k) + \Upsilon_1\bar{F}(k) - \Upsilon_2\Gamma_1 e(k) - \Upsilon_2\Delta(k)$$
$$- \Upsilon_2\Gamma_2[h(Ex(k)) - h(E\hat{x}(k))]$$

(29)

where $\Delta F(k) = F(k+1) - \Upsilon_1 F(k)$

**Theorem 3:** *For the parameter* $\lambda_i > 0 (i = 1, 2)$, *if there exist matrices* $P > 0$, $R$, *and* $\Upsilon_i > 0 (i = 1, 2)$ *satisfying*

$$\bar{\Psi} =$$
$$\begin{bmatrix} \bar{\Psi}_1 & 0 & M_2 & M_3 & (A - L\Gamma_1)^T P \\ * & -\lambda_2^2 I & 0 & 0 & G^T P \\ * & * & M_4 & M_5 & \Gamma_2^T R^T \\ * & * & * & M_6 & D^T P \\ * & * & * & * & -P \end{bmatrix} < 0$$

(30)

where

$$\bar{\Psi}_1 = -P + \lambda_2^2 U_2^T U_2$$
$$+ \lambda_1^2 E^T U_1^T U_1 E + \Gamma_1^T \Upsilon_2^T \Upsilon_2 \Gamma_1$$
$$M_2 = \Gamma_1^T \Upsilon_2^T \Upsilon_2 \Gamma_2, M_3 = \Gamma_1^T \Upsilon_2^T \Upsilon_1$$
$$M_4 = -\lambda_1^2 I + \Gamma_2^T \Upsilon_2^T \Upsilon_2 \Gamma_2, M_5 = -\Gamma_2^T \Upsilon_2^T \Upsilon_1$$
$$M_6 = -I + \Upsilon_1^T \Upsilon_1$$

*then the filtering gain* $L = P^{-1}R$, *the error dynamic system is stable and the error satisfies*

$$\|\bar{S}_k\| \leq$$
$$max\left\{\|\bar{S}_k(1)\|, \rho_1^{-1}\left(\|\Xi_1\| + \sqrt{\|\Xi_1\|^2 + \rho_1\|\Xi_2\|}\right)\right\}$$

(31)

**Proof:** For this purpose, the following Lyapunov function is considered.

$$\Phi_1(k) = e^T(k)Pe(k) + \bar{F}^T(k)\bar{F}(k)$$
$$+ \lambda_1^2 \sum_{i=1}^{k-1}\left[\|U_1 Ee(i)\|^2 - \|h(Ex(i)) - h(E\hat{x}(i))\|^2\right]$$
$$+ \lambda_2^2 \sum_{i=1}^{k-1}\left[\|U_2 e(i)\|^2 - \|g(x(i)) - g(\hat{x}(i))\|^2\right]$$

(32)

It can be verfied that

$$\Delta\Phi_1 = \Phi_1(k+1) - \Phi_1(k)$$
$$= e^T(k+1)Pe(k+1) + \bar{F}^T(k+1)\bar{F}(k+1)$$
$$- e^T(k)Pe(k) - \bar{F}^T(k)\bar{F}(k)$$
$$+ \lambda_1^2\left[\|U_1 Ee(k)\|^2 - \|h(Ex(k)) - h(E\hat{x}(k))\|^2\right]$$
$$+ \lambda_2^2\left[\|U_2 e(k)\|^2 - \|g(x(k)) - g(\hat{x}(k))\|^2\right]$$
$$= \bar{S}_k^T\bar{\Psi}_2\bar{S}_k + \Delta F^T(k)\Delta F(k) - 2\Delta F^T(k)\Upsilon_2\Delta(k)$$
$$+ \Delta^T(k)\left(L^T PL + \Upsilon_2^T\Upsilon_2\right)\Delta(k)$$
$$- 2\bar{S}_k^T\left[\begin{bmatrix}(A - L\Gamma_1)^T PL \\ G^T PL \\ -\Gamma_2^T L^T PL \\ D^T PL\end{bmatrix} + \begin{bmatrix}-\Gamma_1^T\Upsilon_2^T \\ 0 \\ -\Gamma_2^T\Upsilon_2^T \\ \Upsilon_1^T\end{bmatrix}\Upsilon_2\right]$$
$$\Delta(k) + 2\bar{S}_k^T\begin{bmatrix}-\Gamma_1^T\Upsilon_2^T \\ 0 \\ -\Gamma_2^T\Upsilon_2^T \\ \Upsilon_1^T\end{bmatrix}\Delta F(k) < 0$$

(33)

where

$$\bar{\Psi}_2 = \begin{bmatrix} \bar{\Psi}_3 & (A - L\Gamma_1)^T PG & \bar{\Psi}_4 & \bar{\Psi}_5 \\ * & -\lambda_2^2 I + G^T PG & -G^T PL\Gamma_2 & G^T PD \\ * & * & \bar{\Psi}_6 & N_1 \\ * & * & * & N_2 \end{bmatrix}$$

$$(34)$$

$$N_1 = -\Gamma_2^T L^T PD - M_5$$
$$N_2 = D^T PD + M_6$$

$$\bar{\Psi}_3 = (A - L\Gamma_1)^T P(A - L\Gamma_1) - P + \lambda_2^2 U_2^T U_2$$
$$\quad + \lambda_1^2 E^T U_1^T U_1 E + \Gamma_1^T \Upsilon_2^T \Upsilon_2 \Gamma_1$$
$$\bar{\Psi}_4 = -(A - L\Gamma_1)^T PL\Gamma_2 + \Gamma_1^T \Upsilon_2^T \Upsilon_2 \Gamma_2$$
$$\bar{\Psi}_5 = (A - L\Gamma_1)^T PD - \Gamma_1^T \Upsilon_2^T \Upsilon_1$$
$$\bar{\Psi}_6 = -\lambda_1^2 I + \Gamma_2^T L^T PL\Gamma_2 + \Gamma_2^T \Upsilon_2^T \Upsilon_2 \Gamma_2$$

Denote $R = PL$, then it can be seen that

$$\Delta \Phi_1 = \bar{S}_k^T \bar{\Psi}_2 \bar{S}_k + \Delta F^T(k) \Delta F(k)$$
$$-2\Delta F^T(k) \Upsilon_2 \Delta(k) + \Delta^T(k)\left(R^T P^{-1} R + \Upsilon_2^T \Upsilon_2\right) \Delta(k)$$
$$-2\bar{S}_k^T \left\{ \begin{bmatrix} (A - L\Gamma_1)^T R \\ G^T R \\ -\Gamma_2^T L^T R \\ D^T R \end{bmatrix} + \begin{bmatrix} -\Gamma_1^T \Upsilon_2^T \\ 0 \\ -\Gamma_2^T \Upsilon_2^T \\ \Upsilon_1^T \end{bmatrix} \Upsilon_2 \right\}$$
$$\Delta(k) + 2\bar{S}_k^T \begin{bmatrix} -\Gamma_1^T \Upsilon_2^T \\ 0 \\ -\Gamma_2^T \Upsilon_2^T \\ \Upsilon_1^T \end{bmatrix} \Delta F(k) < 0$$

$$(35)$$

By using the Schur complement formula, (34) can be further reduced to

$$\bar{\Psi}_9 = \begin{bmatrix} \bar{\Psi}_7 & \bar{\Psi}_7^T \\ \bar{\Psi}_8 & -P \end{bmatrix} < 0$$

$$(36)$$

where

$$\bar{\Psi}_7 = \begin{bmatrix} \bar{\Psi}_1 & 0 & \Gamma_1^T \Upsilon_2^T \Upsilon_2 \Gamma_2 & \Gamma_1^T \Upsilon_2^T \Upsilon_1 \\ * & -\lambda_2^2 I & 0 & 0 \\ * & * & -\lambda_1^2 I + \Gamma_2^T \Upsilon_2^T \Upsilon_2 \Gamma_2 & -\Gamma_2^T \Upsilon_2^T \Upsilon_1 \\ * & * & * & -I + \Upsilon_1^T \Upsilon_1 \end{bmatrix}$$

$$(37)$$

and

$$\bar{\Psi}_8 = \begin{bmatrix} PA - R\Gamma_1 & PG & -R\Gamma_2 & PD \end{bmatrix} \quad (38)$$

which is equivalent to (30). If (30) holds, a positive scalar exists $\rho_1$ so that $\bar{\Psi} \leq -\rho_1 I$. Thus, it can be seen that

$$\Delta \Phi_1$$
$$\leq -\rho_1 \|S_k\|^2 - 2\left\{ \|\tilde{A}^T R\| + \|\tilde{T}^T\| \|\Upsilon_2\| \right\} \|\Delta(k)\| \|S_k\|$$
$$-2\|\Delta F(k)\| \|\Upsilon_2\| \|\Delta(k)\| + 2\|\tilde{T}^T\| \|\Delta F(k)\| \|S_k\|$$
$$+\|\Delta(k)\|^2 \left\{ \|R^T P^{-1} R\| + \|\Upsilon_2\|^2 \right\} + \|\Delta F(k)\|^2$$
$$\leq -\rho \|S_k\|^2 - 2\|\Xi_1\| \|S_k\| + \Xi_2$$

$$(39)$$

where

$$\tilde{A} = \begin{bmatrix} A - L\Gamma_1 & PG & -R\Gamma_2 & PD \end{bmatrix}$$

$$\tilde{T} = \begin{bmatrix} -\Upsilon_2 \Gamma_1 & 0 & -\Upsilon_2 \Gamma_2 & \Upsilon_2 \end{bmatrix}$$

$$\Xi_1 = \tilde{\delta} \left\{ \|\tilde{A}^T R\| + \|\tilde{T}^T\| \|\Upsilon_2\| \right\} - \|\tilde{T}^T\| \|\Delta F(k)\|$$

and

$$\Xi_2 = \left\|\Delta F(k)\right\|^2 - 2\left\|\Delta F(k)\right\|\left\|\Upsilon_2\right\|\tilde{\delta}$$
$$+ \tilde{\delta}^2\left\{\left\|R^T P^{-1} R\right\| + \left\|\Upsilon_2\right\|^2\right\}$$

It can be shown that

$$\left\|\bar{S}_k\right\| \geq \rho_1^{-1}\left(\left\|\Xi_1\right\| + \sqrt{\left\|\Xi_1\right\|^2 + \rho_1\left\|\Xi_2\right\|}\right) \quad (40)$$

which implies

$$\left\|\bar{S}_k\right\| \leq$$
$$max\left\{\left\|\bar{S}_k(1)\right\|, \rho_1^{-1}\left(\left\|\Xi_1\right\| + \sqrt{\left\|\Xi_1\right\|^2 + \rho_1\left\|\Xi_2\right\|}\right)\right\} \quad (41)$$

This means that the error system in (28) is asymptotically stable under $\bar{\bar{\Psi}} < 0$.

# 6. FAULT TOLERANT CONTROL

## 6.1. Problem Formulation

Similar to (Wang, et al., 2006), consider $u_k(i) \in R^r$ is the input of a discrete-time dynamic stochastic system at the $i^{th}$ time instant within the $k^{th}$ batch and $y \in [a,b]$ is the output. At sample time $k$, $y$ can be described by its PDF, $\gamma_k(y, u_k(i))$. Assuming that $[a,b]$ is known and the probability density function is continuous and bounded within each iteration, then the well-known RBF neural networks can be used to approximate the square root of the output PDF as

$$\sqrt{\gamma(y, u_k(i))} = \sum_{l=1}^n \nu_{l,k}\left(u_k(i)\right) r_{l,k}\left(y\right) \quad (42)$$

where $\gamma(y, u_k(i))$ is the output PDF measured at the $i^{th}$ time instant within the $k^{th}$ batch. Also

$\nu_{l,k}(i)$ is the $l^{th}$ weight element of the RBF neural network in the $i^{th}$ $(i = 1,2,....,m)$ sample time within the $k^{th}$ batch, and $r_{l,k}(y)$ denotes the $l^{th}$ $(l = 1,2,....,n)$ RBF activation function within $k^{th}$ batch. Assume $n$ and $k$ represent the number of RBFs and the batch length, respectively. The RBF activation functions are expressed as follows (Wang and Afshar, 2009).

$$r_{l,k}(y) = exp\left(-\frac{(y_j - \mu_{l,k})^2}{2\sigma_{l,k}^2}\right) \quad (43)$$

where $\mu_{l,k}$, $\sigma_{l,k}$ are the centres and widths of the RBF basis functions within the $k^{th}$ batch, respectively.

Different from (Guo and Wang, 2006), the output PDF described in (42) can be re-written as the following vector form:

$$\sqrt{\gamma(y, u_k(i))} = \begin{bmatrix} R_k(y) & r_{n,k}(y) \end{bmatrix}\begin{bmatrix} V_k(i) \\ \nu_{n,k}(i) \end{bmatrix} \quad (44)$$

where

$$R_k(y) = [r_{1,k}(y), r_{2,k}(y),....,r_{n-1,k}(y)]$$

$$V_k(i) = [\nu_{l,k}(i), \nu_{2,k}(i),....,\nu_{n-1,k}(i)]^T.$$

Since, $\gamma(y, u_k(i))$ is a probability density function, it must satisfy the following integral constraint (Wang, 2000).

$$\int_a^b (\gamma(y, u_k(i)))dy = \int_a^b (\sqrt{\gamma(y, u_k(i))})^2 dy = 1 \quad (45)$$

By substituting $\gamma(y, u_k(i))$ from (44) and solving the equation for $\nu_{n,k}(i)$ similar to (Wang,

2000), it can be shown that the following state constraint shall be satisfied within each batch to guarantee that the measured $\gamma(y, u_k(i))$ is a probability density function (Wang, 2000).

$$V_k^T(i)Q_{ab,k}V_k(i) \leq 1 \tag{46}$$

where

$$Q_{ab,k} = b_{1,k} - b_{3,k}^{-1}b_{2,k}^T b_{2,k}$$

$$b_{1,k} = \int_a^b R_k^T(y)R_k(y)dy$$

$$b_{2,k} = \int_a^b r_{n,k}(y)R_k(y)dy$$

$$b_{3,k} = \int_a^b r_{n,k}^2(y)dy$$

It has been proven in (Wang, 2000) that $Q_{ab,k}$ is always positive definite. When (5) holds, it can be seen that $\nu_{n,k}$ can be represented as a known nonlinear function of $V_k$ called $h(V(k))$. Thus the output PDF described in (1) can be re-written as follows:

$$\sqrt{\gamma(y, u_k(i))} = R_k(y)V_k(i) + r_{n,k}(y)h(V(k)) \tag{47}$$

In (47), the nonlinear function $h(V(k))$ should satisfy the following Lipschitz condition

$$\left\| h(V_{1,k}(i)) - h(V_{2,k}(i)) \right\| \leq \left\| \bar{U}_1(V_{1,k}(i) - V_{2,k}(i)) \right\| \tag{48}$$

where $\bar{U}_1$ is a known matrix.

Thus the dynamic model between the output PDF and the RBF neural network weight vectors in the presence of the actuator fault will be established as follows:

$$V_k(i+1) = A_k V_k(i) + B_k u_k(i) + Gg(V_k(i)) + DF_k(i)$$
$$\sqrt{\gamma(y, u_k(i))} = R_k(y)V_k(i) + r_{n,k}(y)h(V_k(i)) \tag{49}$$

Similar to (Guo and Wang, 2006), the nonlinear dynamics of the model in (49) is supposed to satisfy the following Lipschitz condition.

$$\left\| g(V_{1,k}(i)) - g(V_{2,k}(i)) \right\| \leq \left\| \bar{U}_2(V_{1,k}(i) - V_{2,k}(i)) \right\| \tag{50}$$

where $\bar{U}_2$ is a known matrix.

## 6.2. Controller Design

A generalized PI controller with tuneable coefficients is considered an adaptive controller in this work as follows

$$\xi_k(i) = \xi_k(i-1) + T_s e_k(i-1)$$

$$u_k(i) = K_{P,k}.e_k(i) + K_{I,k}\xi_k(i) \tag{51}$$

where $e_k(i) = V_g - V_k(i)$ represents the dynamical weight tracking error and $T_s$ is the sampling time. Substituting (51) in (49) yields the following closed-loop system for the weight control loop with the $k^{th}$ batch:

$$M_k(i+1) = \bar{A}_k M_k(i) + \bar{B}_k V_g + \bar{G}g(M_k(i)) + \bar{D}F_k(i) \tag{52}$$

where

$$M_k(i) = \begin{bmatrix} V_k(i) \\ \xi_k(i) \end{bmatrix}, \quad \bar{A}_k = \begin{bmatrix} A_k - B_k K_{P,k} & B_k K_{I,k} \\ -T_s I & I \end{bmatrix}$$

$$\bar{B}_k = \begin{bmatrix} B_k K_{P,k} \\ \\ T_s I \end{bmatrix}, \quad \bar{G} = \begin{bmatrix} G & 0 \\ \\ 0 & 0 \end{bmatrix},$$

$$\bar{D} = \begin{bmatrix} D \\ \\ 0 \end{bmatrix}, \quad g(M_k(i)) = \begin{bmatrix} g(V_k(i)) \\ \\ 0 \end{bmatrix}$$

Denote $\tilde{A}_k = \begin{bmatrix} A_k & 0 \\ -T_s I & I \end{bmatrix}, \quad \tilde{B}_k = \begin{bmatrix} B_k \\ 0 \end{bmatrix}.$

The following theorem represents the solvability conditions of the general PI controller.

**Theorem 4:** *Within the $k^{th}$ batch, for the parameter $\lambda$, if there exist matrices $P > 0$, and $R$ satisfying the following LMIs for any initial condition $M(0)$ satisfying constraint (46)*

$$\tilde{\Psi}_k =$$

$$\begin{bmatrix} -P_k & 0 & 0 & 0 & 0 & N_1^T & N_2^T \\ 0 & -\lambda^2 I & 0 & 0 & 0 & \bar{G}^T & 0 \\ 0 & 0 & M_6 & \Upsilon_1^T & N_3^T & \bar{D}^T & 0 \\ 0 & 0 & \Upsilon_1 & I & -\Upsilon_2^T & 0 & 0 \\ 0 & 0 & N_3 & -\Upsilon_2 & N_4 & 0 & 0 \\ N_1 & \bar{G} & \bar{D} & 0 & 0 & -P_k & 0 \\ N_2 & 0 & 0 & 0 & 0 & 0 & -I \end{bmatrix} < 0 \quad (53)$$

and

$$2\tilde{\alpha} \left\| Q_{ab} \right\| \left\| \bar{B}_k V_g \right\|^2 \leq \lambda_{min}(P_k) \quad (54)$$

where

$$N_1 = \tilde{A}_k P_k + \tilde{B}_k R, \quad N_2 = \lambda \bar{U}_2 P_k$$
$$N_3 = -\Upsilon_2^T \Upsilon_1, \quad N_4 = \Upsilon_2^T \Upsilon_2$$

*then, the closed loop system is stable with $\lim_{k\to\infty} e(i) = 0$, and the controller parameters can be calculated by using*

$$\begin{bmatrix} K_P & K_I \end{bmatrix} = R P^{-1}$$

**Proof:** For this purpose, the following Lyapunov function is considered.

$$\Phi_3(i) = M_k^T(i) P^{-1} M_k(i) + \bar{F}_k^T(i) \bar{F}_k(i) + \lambda^2 \sum_{j=1}^{i-1} \left[ \left\| \bar{U}_2 M_k(i) \right\|^2 - \left\| g(M_k(i)) \right\|^2 \right] \quad (55)$$

where $\bar{F}(k) = F(k) - \hat{F}(k)$

It can be verfied that

$$\Delta\Phi_3 = \Phi_3(M_k(i+1), i+1) - \Phi_3(M_k(i), i)$$
$$= \tilde{M}_k^T(i) \tilde{\Psi}_{1,k} \tilde{M}_k(i) + 2\tilde{M}_k(i) \tilde{N}_k^T P_k^{-1} \bar{B}_k V_g + V_g^T \bar{B}^T P^{-1} \bar{B} V_g \quad (56)$$

where

$$\tilde{\Psi}_{1,k} = \begin{bmatrix} \tilde{N}_{k1} & Q_1^T & Q_2^T & 0 & 0 \\ Q_1 & Q_3 & Q_4^T & 0 & 0 \\ Q_2 & Q_4 & \tilde{N}_{k,2} & \Upsilon_1^T & -\Upsilon_1^T \Upsilon_2 \\ 0 & 0 & \Upsilon_1 & I & -\Upsilon_2^T \\ 0 & 0 & -\Upsilon_2^T \Upsilon_1 & -\Upsilon_2 & \Upsilon_2^T \Upsilon_2 \end{bmatrix}$$

$$\tilde{N}_k = [\bar{A}_k^T, \bar{G}_k^T, \bar{D}_k^T, 0, 0]$$
$$\tilde{N}_{k,1} = \bar{A}_k^T P_k^{-1} \bar{A}_k - P_k^{-1} + \lambda^2 \bar{U}_2^T \bar{U}_2$$
$$\tilde{N}_{k,2} = \bar{D}_k^T P_k^{-1} \bar{D}_k - I + \Upsilon_1^T \Upsilon_1$$
$$Q_1 = \bar{G}^T P_k^{-T} \bar{A}_k$$
$$Q_2 = \bar{D}_k^T P_k^{-T} \bar{A}_k$$
$$Q_3 = \bar{G}^T P_k^{-1} \bar{G} - \lambda^2 I$$
$$Q_4 = \bar{D}_k^T P_k^{-T} \bar{G}$$

and

$$\tilde{M}_k^T(i) = [M_k^T(i), g^T(M_k(i)), \bar{F}_k^T, \Delta F^T(k), \varepsilon(k)]$$

By using the well-know Schur complement formula, (56) can be as follows:

$$\tilde{\Psi}_{2,k} =$$
$$\begin{bmatrix} Q_5 & 0 & 0 & 0 & 0 & \bar{A}_k^T \\ 0 & -\lambda^2 I & 0 & 0 & 0 & \bar{G}^T \\ 0 & 0 & M_6 & \Upsilon_1^T & N_3^T & \bar{D}^T \\ 0 & 0 & \Upsilon_1 & I & -\Upsilon_2^T & 0 \\ 0 & 0 & N_3 & -\Upsilon_2 & N_4 & 0 \\ \bar{A}_k & \bar{G} & \bar{D} & 0 & 0 & -P_k \end{bmatrix} < 0$$

(57)

where $Q_5 = -P_k^{-1} + \lambda^2 \bar{U}_2^T \bar{U}_2$.

By pre-multiplying $\tilde{\Psi}_{2,k}$ by $diag\left(P_k^T, I, I, I, I, I\right)$ and post multiplying it by $diag\left(P_k, I, I, I, I, I\right)$, and applying the well-know Schur complement formula, the condition for stability will be as follows:

$$\tilde{\Psi}_{3,k}$$
$$\begin{bmatrix} -P_k & 0 & 0 & 0 & 0 & Q_6^T & Q_7^T \\ 0 & -\lambda^2 I & 0 & 0 & 0 & \bar{G}^T & 0 \\ 0 & 0 & M_6 & \Upsilon_1^T & N_3^T & \bar{D}^T & 0 \\ 0 & 0 & \Upsilon_1 & I & -\Upsilon_2^T & 0 & 0 \\ 0 & 0 & N_3 & -\Upsilon_2 & N_4 & 0 & 0 \\ Q_6 & \bar{G} & \bar{D} & 0 & 0 & -P_k & 0 \\ Q_7 & 0 & 0 & 0 & 0 & 0 & -I \end{bmatrix} < 0$$

(58)

where $Q_6 = \bar{A}_k P_k$ and $Q_7 = \lambda \bar{U}_2 P_k$.

By substituting matrices $\tilde{A}$ and $\tilde{B}$ into $\tilde{\Psi}_3$, $\tilde{\Psi}$ can be obtained.

If (53) holds, a positive scalar $\tilde{\alpha}$ exists so that $\tilde{\Psi} \le -\tilde{\alpha} I$. Along with (52) it can be verified that

$$\Delta \Phi_3 \le -\tilde{\alpha} \left\| \tilde{M}_k \right\|^2 + 2 \left\| \tilde{M}_k \right\| \left\| \tilde{N}_k^T P^{-1} \bar{B} V_g \right\|$$
$$+ \left\| V_g \right\|^2 \left\| \bar{B}^T P^{-1} \bar{B} \right\|$$

(59)

It is obvious that the right-hand side of inequality is a second order polynomial with respect to $\left\| \tilde{M}_k \right\|$. Denote

$$\tilde{\sigma} = V_g^T \bar{B}^T P^{-1} \bar{B} V_g$$

(60)

Thus, it can be shown that $\Delta \Phi_3 \le 0$ holds if

$$\left\| \tilde{M}_k \right\| \ge$$
$$\tilde{\alpha}^{-1} \left( \left\| \tilde{N}_k^T P^{-1} \bar{B} V_g \right\| + \sqrt{\left\| \tilde{N}_k^T P^{-1} \bar{B} V_g \right\|^2 + \tilde{\alpha} \tilde{\sigma}} \right)$$

(61)

which implies

$$\left\| \tilde{M}_k \right\| \le max$$
$$\left\{ \left\| \tilde{M}_k(1) \right\|, \tilde{\alpha}^{-1} \left( \left\| \tilde{N}_k^T P^{-1} \bar{B} V_g \right\| + \sqrt{\left\| \tilde{N}_k^T P^{-1} \bar{B} V_g \right\|^2 + \tilde{\alpha} \tilde{\sigma}} \right) \right\}$$

(62)

This confirm that the closed-loop system is bounded and internally stable.

Based on (52) and (55) it can be shown that

$$\Delta \Phi_3 = \Phi_3(M_k(i+1), i+1) - \Phi_3(M_k(i), i)$$
$$= \tilde{M}_k^T(i)(\bar{A}_k^T P_k^{-1} \bar{S}_k - P_k^{-1}) \tilde{M}_k(i)$$
$$+ 2\tilde{M}_k^T(i) \tilde{N}_k^T P_k^{-1} \bar{B}_k V_g + V_g^T \bar{B}^T P^{-1} \bar{B} V_g$$

(63)

It can be verified that $(\bar{A}_k^T P_k^{-1} \bar{S}_k - P_k^{-1}) < -\tilde{\alpha}^{-1} I$ as long as $(\bar{A}_k^T P_k^{-1} \bar{S}_k - P_k^{-1}) < 0$ is guaranteed by (53) Thus it can be seen that

$$\Delta\Phi_3 \leq -\tilde{\alpha}^{-1}\left\|\tilde{M}_k(i)\right\|^2 + 2\tilde{M}_k(i)\tilde{N}_k^T P_k^{-1}\bar{B}_k V_g$$
$$+ V_g^T \bar{B}^T P^{-1}\bar{B}V_g$$
$$\leq -\tilde{\alpha}^{-1}\left\|\tilde{M}_k(i)\right\|^2 - (P_k^{-1/2}\tilde{N}_k^T\tilde{M}_k(i)$$
$$- P_k^{-1/2}\bar{B}_k V_g)^T(P_k^{-1/2}\tilde{N}_k^T\tilde{M}_k(i) - P_k^{-1/2}\bar{B}_k V_g)$$
$$+ 2V_g^T\tilde{N}_k^T P_k^{-1}\bar{B}_k V_g$$
$$\leq -\tilde{\alpha}^{-1}\left\|\tilde{M}_k(i)\right\|^2 + 2\lambda_{max}(P^{-1})\left\|\bar{B}_k V_g\right\|^2$$

$$(64)$$

Denote $\tilde{\beta} = 2\lambda_{max}(P^{-1})\left\|\bar{B}_k V_g\right\|^2$.

Then it can be shown that $\Delta\Phi_3 < 0$ hold if $\left\|\tilde{M}_k(i)\right\|^2 > \tilde{\alpha}\tilde{\beta}$. From the constraint (46), it can be seen that

$$V_k^T(i)Q_{ab}V_k(i) < \left\|V_k(i)\right\|^2\left\|Q_{ab}\right\|$$
$$\leq \left\|\tilde{M}_k(i)\right\|^2\left\|Q_{ab}\right\| \qquad (65)$$
$$\leq \tilde{\alpha}\tilde{\beta}\left\|Q_{ab}\right\| \leq 1$$

Thus, the constraint in (54) can be guaranteed by the obtained results from (64) and (65). To discuss the system tracking performance, suppose that $\tilde{\varphi}_1(i)$ and $\tilde{\varphi}_2(i)$ are two trajectories of the nonlinear closed-loop system (52) corresponding to fixed initial conditions and fault, and $V_g$ is the input. Denote the error between the two trajectories as $\chi(i) = \tilde{\varphi}_1(i) - \tilde{\varphi}_2(i)$ with $\chi(1) = 0$. Then, the dynamic of $\chi(i+1)$ can be presented as follows.

$$\chi(i+1) = \bar{A}_k\chi(i) + \bar{G}\left[g(\tilde{\varphi}_1) - g(\tilde{\varphi}_2)\right] \qquad (66)$$

The following Lyapunov function will be considered

$$\Phi_4\left(\chi(i), \tilde{\varphi}_1(i), \tilde{\varphi}_2(i), i\right) = \chi^T(i)P^{-1}\chi(i)$$
$$+ \lambda^2\sum_{j=1}^{i-1}\left[\left\|U\chi(i)\right\|^2 - \left\|g(\tilde{\varphi}_1(i)) - g(\tilde{\varphi}_2(i))\right\|^2\right]$$

$$(67)$$

It can be verified that

$$\Delta\Phi_4 = \tilde{\chi}^T(i)\tilde{\Psi}_{4,k}\tilde{\chi}(i) < -\tilde{\alpha}\left\|\tilde{\chi}(i)\right\| \qquad (68)$$

where

$$\tilde{\chi}(i) = [\chi^T(i), g(\tilde{\varphi}_1) - g(\tilde{\varphi}_2)]$$

$$\tilde{\Psi}_{4,k} =$$
$$\begin{bmatrix} \bar{A}_k^T P_k^{-1}\bar{A}_k - P_k^{-1} + \lambda^2\bar{U}_2^T\bar{U}_2 & \bar{A}_k^T P_k^{-1}\bar{G}_k \\ \bar{G}_k^T P_k^{-T}\bar{A}_k & \bar{G}_k^T P_k^{-1}\bar{G}_k - \lambda^2 I \end{bmatrix}$$

This means that the closed loop system is exponentially stable around $\chi = 0$ neighborhood. Thus, the tracking performance of the system has been satisfied.

## 7. TUNING OF RADIAL BASIS FUNCTION

Similar to (Wang et al., 2006), the following P-type ILC law will be used to tune the basis function parameters (RBF centres and widths) between any two batches

$$\mu_{l,k} = \mu_{l,k-1} + \Lambda_\mu E_{k-1}\sigma_{l,k} = \sigma_{l,k-1} + \Lambda_\sigma E_{k-1}$$

$$(69)$$

where the performance indices of the $(k-1)^{th}$ batch will be as follows.

$$E_{k-1} = [J_{k-1}(1), J_{k-1}(2), \ldots, J_{k-1}(m)]^T$$

where $m$ represents the total number of time instants within a batch. $J_{k-1}(i)$ is the performance at the $i^{th}$ sampling instant of the $(k-1)^{th}$ batch, it can be expressed as follows.

$$J_{k-1}(i) = \int_a^b \left( \sqrt{\gamma\left(y, u_k(i)\right)} - \sqrt{g(y)} \right)^2 dy$$

In addition, the learning parameters in (29) are defined as

$$\begin{aligned} \Lambda_\mu &= \alpha_\mu [\lambda_1, \lambda_2, ..., \lambda_m] \\ \Lambda_\sigma &= \alpha_\sigma [\bar{\lambda}_1, \bar{\lambda}_2, ..., \bar{\lambda}_m] \end{aligned} \qquad (70)$$

where $\lambda, \bar{\lambda}$ are the learning elements, and $\alpha_\mu, \alpha_\sigma$ are the learning rates to be determined.

## 8. CONVERGENCE ANALYSIS

Similar to (Wang et al., 2006), the learning vectors in (70) should be selected carefully to ensure the convergence of the ILC-based tuning algorithm between batches. Therefore, the closed loop performance should satisfy the following condition between batches:

$$\frac{F_k}{F_{k-1}} = \frac{\sum\limits_{i=1}^m J_k(i)}{\sum\limits_{i=1}^m J_{k-1}(i)} \leq 1 \qquad (71)$$

where

$$F_k = \sum_{i=1}^m J_k(i) \qquad (72)$$

where $F_k$ is the measure of the overall closed loop performance within the $k^{th}$ batch. Since $J_k(i)$ is non-negative, it can be verified that

$$\Delta F_k = F_k - F_{k-1} \leq 0 \qquad (73)$$

The conditions of convergence have been discussed in (Wang et al., 2006), which can be summarized as follows:

$$\sum_{i=1}^m \int_a^b \left[ \left( \sqrt{\gamma_{k-1,i}(y)} - \sqrt{g(y)} \right) \Delta\sqrt{\gamma_{k-1,i}(y)} \right] dy \leq 0 \qquad (74)$$

together with

$$\Delta R_{l,k-1}(y) = \frac{y_j - \mu_{k-1}}{\sigma_{k-1}^2} R_{l,k-1}(y)\Lambda_\mu E_{k-1} + \frac{\left(y_j - \mu_{k-1}\right)^2}{\sigma_{k-1}^3} R_{l,k-1}(y)\Lambda_\mu E_{k-1} \qquad (75)$$

and

$$\begin{aligned} \Delta\sqrt{\gamma_{k-1,i}(y)} &= \sqrt{\gamma_{k-1,i}(y)} - \sqrt{\gamma_{k-2,i}(y)} \\ &= \sum_{l=1}^n V_l(i)\Delta R_{l,k-1}(y) \end{aligned} \qquad (76)$$

where

$$\begin{aligned} \Delta R_{l,k-1}(y) &= R_{l,k-1}(y) - R_{l,k-2}(y) \\ \Delta\mu_k &= \mu_k - \mu_{k-1} \\ \Delta\sigma_k &= \sigma_k - \sigma_{k-1} \end{aligned} \qquad (77)$$

## 9. AN ILLUSTRATED EXAMPLE

In this section, a simulation study of the proposed method will be described. First, the system model and RBF neural network components are introduced, and then the performance of the FTC control law will be investigated.

For a stochastic system with non-Gaussian process, it is supposed that the output PDF can be formulated by using three-layer neural network with three radial basis activation functions as shown in Figure 2 with the following initial conditions over its definition interval [a,b]:

$$y \in [0,2],$$

$$\mu_1 = 0.5, \mu_2 = 1.0, \mu_3 = 1.5$$

$$\sigma_1 = \sigma_2 = \sigma_3 = 0.2$$

This would mean that the output PDF of the stochastic system is described as:

$$\sqrt{\gamma(y, u(k), F(k))}$$
$$= R(y)V(k) + r_n(y)h(V(k)) + \omega(y, u(k), F(k)) \tag{78}$$

where

$$R(y) = [r_1(y), r_2(y)]$$

and

$$V(t) = [v_1(t), v_2]^T$$

The weight vector behaves dynamically as described in (7) with the following parameters:

$$A = \begin{bmatrix} -0.45 & 0.03 \\ 0.1 & -0.28 \end{bmatrix}, \quad B = \begin{bmatrix} 0.45 & 0.01 \\ 0.01 & -0.86 \end{bmatrix},$$
$$G = \begin{bmatrix} 0.02 & 0 \\ 0 & 0.01 \end{bmatrix}, \quad D = E = \begin{bmatrix} 1 & 0 \\ 0 & 1 \end{bmatrix}.$$

The nonlinear function was chosen as follows:

$$g(V(t)) = \begin{bmatrix} 0 \\ \sqrt{v_1^2 + v_2^2} \end{bmatrix}.$$

The initial value of the weight vector is set as $V_1(0) = [0.001, 0.001]^T$. In addition, the matrices $U_1$ and $U_2$ were chosen as follows.

$$U_1 = \begin{bmatrix} 0.1 & 0 \\ 0 & 0.1 \end{bmatrix}, \quad U_2 = \begin{bmatrix} 1 & 0 \\ 0 & 1 \end{bmatrix}.$$

Each batch is divided into 200 samples, i.e., m = 200, and the total number of batches, i.e. k, is set to 20. Also, the RBF basis functions parameters of the desired output PDF are as follows:

*Figure 2. Initial distribution of RBFs*

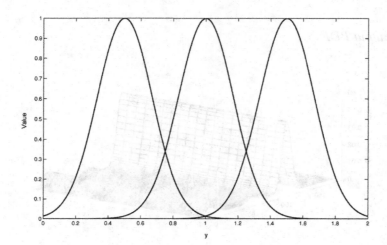

$\mu_{g1} = 0.2, \mu_{g2} = 0.8, \mu_{g3} = 1.3$

$\sigma_{g1} = \sigma_{g2} = \sigma_{g3} = 0.1$

Moreover, the desired dynamical weights are set as $V_g = [0.09; \ 0.06]$. With the above parameters, the 3-D plot of the desired PDF shape is shown in Figure 3. Assume that the modelling error satisfies $|\ \omega(y, \ u(k), \ F)\ |\leq 0.002$. The bound of modelling error satisfies $\tilde{\delta} = 0.0008$ for $\mu(y) = 1$. From (3), it can be computed that

$$\Lambda_1 = \begin{bmatrix} 0.3010 & 0.0347 \\ 0.0347 & 0.3010 \end{bmatrix},$$
$$\Lambda_2 = \begin{bmatrix} 0.0001 & 0.0350 \end{bmatrix},$$
$$\Lambda_3 = 0.3413$$

Also, from (12), it can be seen that

$$\Gamma_1 = \begin{bmatrix} 0.0389 & 0.0348 \end{bmatrix}, \ \Gamma_2 = 0.4225$$

To demonstrate the effectiveness of the proposed algorithm, the fault is chosen to be a constant signal as $F(t) = 0.8$, and it is supposed to commence at $T = 2s$.

Firstly, we consider the fault detection problem. Using Theorem 1 with $\lambda_1 = \lambda_2 = 1$ it can be calculated that

$$P = \begin{bmatrix} 1.5941 & 0.1479 \\ 0.1479 & 95.5499 \end{bmatrix},$$
$$R = \begin{bmatrix} 14.9186 \\ 50.1447 \end{bmatrix}, \ L = \begin{bmatrix} 9.3113 \\ 0.5104 \end{bmatrix}$$

Next, the fault diagnosis problem is considered for the above system and fault. Using Theorem 3, the following results can be obtained:

$$P = \begin{bmatrix} 2.272 & -0.023 \\ -0.023 & 2.305 \end{bmatrix},$$
$$R = \begin{bmatrix} -3.956 \\ 0.920 \end{bmatrix}, \ L = \begin{bmatrix} -1.737 \\ 0.382 \end{bmatrix}$$

$$\Upsilon_1 = 0.97, \ \Upsilon_2 = 1.3$$

By applying the nonlinear fault isolation filter, Figure 4 shows that such a filter can effectively diagnose the actuator fault.

The difference of the PDF tracking within the second batch between the measured and desired PDF is shown in Figure 5. The parameters of controller in this batch are

*Figure 3. Desired output PDF*

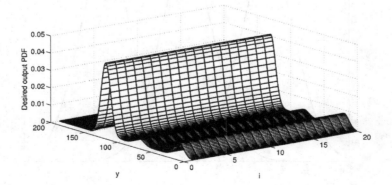

*Figure 4. Fault and its estimation under the filter*

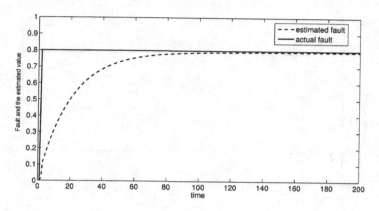

$$K_P = 1.0e + 003 \begin{bmatrix} -5.5224 & -3.5340 \\ 7.4317 & 4,4533 \end{bmatrix},$$

$$K_I = 1.0e + 002 \begin{bmatrix} 0.0394 & 0.0394 \\ 0.4860 & 0.4860 \end{bmatrix}$$

After the final batch (k=10), the coefficients of PI controller are as follows.

$$K_P = 1.0e + 002 \begin{bmatrix} 1.7588 & -2.7559 \\ -3.0794 & 4,8251 \end{bmatrix},$$

$$K_I = \begin{bmatrix} 0.1876 & 0.1876 \\ -0.3285 & -0.3285 \end{bmatrix}$$

As a result, the corresponding of the weight control loop with the last batch of operation can be described in Figure 6. Figure 6 clearly demonstrates the effect of the fault-tolerant action on maintaining the correct value for the controlled weights. Moreover, it reflects the effectiveness of LMI feasbility results.

The 3D mesh plot of the output PDF in the last batch of operation is shown in Figure 7. In addition, the PDF tracking performance within the last batch of operation is shown in Figure 8.

Finally, the trend of the ILC performance function along the batches implies the effectiveness of the proposed algorithm as shown in Figure 9.

*Figure 5. 2D plot of measured and desired PDF, k=2*

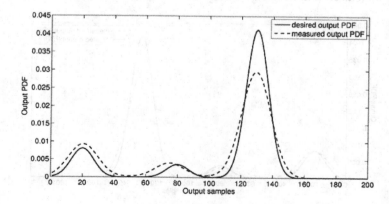

*Figure 6. 2D RBF neural network weights under FTC*

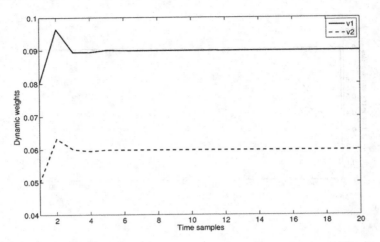

*Figure 7. 3D mesh plot of the measured output PDF after the fault*

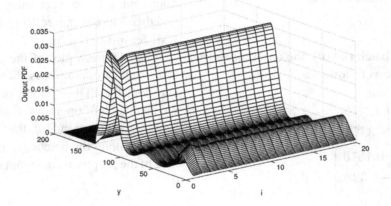

*Figure 8. 2D plot of measured and desired PDF, k=10*

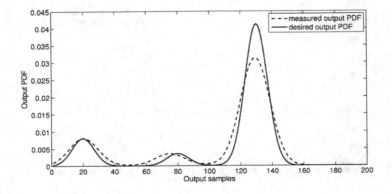

*Figure 9. Performance function of the ILC*

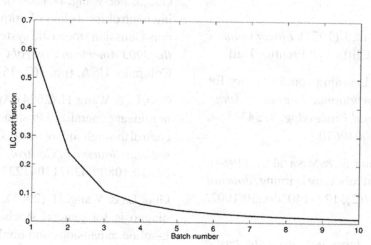

## 10. CONCLUSION

An ILC-based FTC method is presented, for the shape control of the output PDFs for general stochastic systems with non-Gaussian variable using a generalized fixed-structure PI controller with constraints on the state vector that results from the application of square-root PDF modelling. The whole control horizon is divided into a number of batches. Within each batch, the state-constrained generalized PI controller is used to shape the output PDF using an LMI approach. Between any two adjacent batches, the parameters of the RBF basis functions are tuned. A P-Type ILC is applied to achieve such tuning between batches, where a sufficient condition for the ILC convergence has been established.

After a fault happens, the fault estimation information is introduced into the controller re-configuration to retain the stability and tracking convergence of the post-fault system, leading to the FTC of the whole stochastic nonlinear system. The simulation of the illustrated example demonstrates the use of the control algorithm and shows that a satisfactory estimation and control performance is established.

The FTC framework in this chapter developed the ILC-based SDC with the application of RBFNNs instead of a fixed rational square-root B-spline expansion, which has the advantage of more general output PDF approximation in terms of tuneable RBF parameters. The application of ILC decreases the complexity of the stochastic model through the use of tuneable basis functions. Moreover, a fixed-structure PI controller is easy to implement in practical process control problems.

As a future work of this work, an investigation could be carried out into the FTC performance-based PID/PI when the stochastic discrete-time system is subject to a time delay and parametric uncertainty. The performance could be compared to other FTC framework-based conventional controllers. Similarly, an investigation of the ILC FTC performance-based PI when the non-Gaussian singular stochastic discrete-time system is subject to a time delay and parametric uncertainty could be carried out. Once again, the performance could be compared with other ILC FTC framework-based conventional controllers.

# REFERENCES

Anderson, B., & Moore, J. (1971). *Linear optimal control*. Englewood Cliffs, NJ: Prentice Hall.

Arimoto, S. (1990). Learning control theory for robotic motion. *International Journal of Adaptive Control and Signal Processing, 4*, 543–564. doi:10.1002/acs.4480040610

Arimoto, S., Kawamura, S., & Miyazaki, F. (1984). Bettering operation of robots by learning. *Journal of Robotic Systems, 1*(2), 123–140. doi:10.1002/rob.4620010203

Astrom, K. (1970). *Introduction to stochastic control theory*. New York, NY: Academic Press.

Chen, H.-F., & Fang, H.-T. (2004). Output tracking for nonlinear stochastic systems by iterative learning control. *IEEE Transactions on Automatic Control, 49*(4), 583–588. doi:10.1109/TAC.2004.825613

Chien, C.-J., & Liu, J.-S. (1994). *A P-type iterative learning controller for robust output tracking of nonlinear time-varying systems*. American Control Conference.

Crowley, J. T., & Choi, Y. K. (1998). Experimental studies on optimal molecular weight distribution control in a batch-free radical polymerization process. *Chemical Engineering Science, 53*(15), 2769–2790. doi:10.1016/S0009-2509(98)00095-5

Dunbar, C. A., & Hickey, A. J. (2000). Evaluation of probability density functions to approximate particle size distributions of representative pharmaceutical aerosols. *Journal of Aerosol Science, 31*(7), 813–831. doi:10.1016/S0021-8502(99)00557-1

Guo, L., & Wang, H. (2003). Pseudo-PID tracking control for a class of output PDFs of general non-Gaussian stochastic systems. *Proceedings of the 2003 American Control Conference,* Denver, Colorado, USA, (pp. 362–367).

Guo, L., & Wang, H. (2004). Applying constrained nonlinear generalized PI strategy to PDF tracking control through square root b-spline models. *International Journal of Control, 77*(17), 1481–1492. doi:10.1080/00207170412331326972

Guo, L., & Wang, H. (2005). Fault detection and diagnosis for general stochastic systems using b-spline expansions and nonlinear filters. *IEEE Transactions on Circuits and Systems, 52*(8), 1644–1652. doi:10.1109/TCSI.2005.851686

Iserman, R. (2006). *Fault-diagnosis systems: An introduction from fault detection to fault tolerance.* Berlin, Germany: Springer-Verlag.

Isermann, R. (1993). Fault diagnosis of machines via parameter estimation and knowledge processing, tutorial paper. *Automatica, 29*(4), 815–836. doi:10.1016/0005-1098(93)90088-B

Isermann, R., & Balle, P. (1996). Trends in the application of model based fault detection and diagnosis of technical process. *Proceedings of the IFAC World Congress,* (pp. 1–12).

Karny, M., Nagy, I., & Novovicova, J. (2002). Mixed-data multi-modelling for fault detection and isolation. *International Journal of Adaptive Control and Signal Processing, 16*, 61–83. doi:10.1002/acs.672

Patton, R., Frank, P., & Clark, R. (1989). *Fault diagnosis in dynamic systems: Theory and application.* Englewood Cliffs, NJ: Prentice Hall.

Sawyer, S., & Tapia, A. (2005). The sociotechnical nature of mobile computing work: Evidence from a study of policing in the United States. *International Journal of Technology and Human Interaction, 1*(3), 1–14. doi:10.4018/jthi.2005070101

Shi, D., El-Farra, N. H., Mhaskar, M., & Li, P., & Christofides, P. D. (2006). Predictive control of particle size distribution in particulate processes. *Chemical Engineering Science, 61*(1), 268–28. doi:10.1016/j.ces.2004.12.059

Shibasaki, Y., Araki, T., Nagahata, R., & Ueda, M. (2005). Control of molecular weight distribution in polycondensation polymers 2- poly(ether ke-tone) synthesis. *European Polymer Journal, 41*(10), 2428–2433. doi:10.1016/j.eurpolymj.2005.05.001

Solo, V. (1990). Stochastic adaptive control and martingale limit theory. *IEEE Transactions on Automatic Control, 35*(1), 66–71. doi:10.1109/9.45146

Srichander, R., & Walker, K. B. (1993). Stochastic stability analysis for continuous-time fault tolerant control systems. *International Journal of Control, 57*, 433–452. doi:10.1080/00207179308934397

Sun, X., Yue, H., & Wang, H. (2006). Modelling and control of the flame temperature distribution using probability density function shaping. *Transactions of the Institute of Measurement and Control, 28*(5), 401–428. doi:10.1177/0142331206073124

Wang, A., Afshar, P., & Wang, H. (2008). Complex stochastic systems modelling and control via iterative machine learning. *Neurocomputing, 71*(13-15), 2685–2692. doi:10.1016/j.neucom.2007.06.018

Wang, A., Wang, H., & Guo, L. (2009). Recent advances on stochastic distribution control: Probability density function control. *Control and Decision Conference, CCDC'09* (pp. xxxv–xli).

Wang, H. (1999). Robust control of the output probability density functions for multivariable stochastic systems with guaranteed stability. *IEEE Transactions on Automatic Control, 44*(11), 2103–2107. doi:10.1109/9.802925

Wang, H. (2000). *Bounded dynamic stochastic systems: Modelling and control.* London, UK: Springer-Verlag.

Wang, H. (2003). Multivariable output probability density function control for non-Gaussian stochastic systems using simple MLP neural networks. *Proceedings of the IFAC International Conference on Intelligent Control Systems and Signal Processing,* Algarve, Portugal, (pp. 84– 89).

Wang, H., & Afshar, P. (2006). Radial basis function based iterative learning control for stochastic distribution systems. *Proceedings of the IEEE International Symposium on Intelligent Control,* (pp. 100–105).

Wang, H., & Afshar, P. (2009). ILC-based fixed-structure controller design for output PDF shaping in stochastic systems using LMI techniques. *IEEE Transactions on Automatic Control, 54*(4), 760–773. doi:10.1109/TAC.2009.2014934

Wang, H., Afshar, P., & Yue, H. (2006). ILC-based generalised PI control for output PDF of stochastic systems using LMI and RBF neural net-works. *Proceedings of the IEEE Conference on Decision and Control,* (pp. 5048–5053).

Wang, H., & Lin, W. (2000). Applying observer based FDI techniques to detect faults in dynamic and bounded stochastic distributions. *International Journal of Control, 73*, 1424–1436. doi:10.1080/002071700445433

Xia, Y., Shi, P., Liu, G., & Rees, D. (2006). Sliding mode control for stochastic jump systems with time-delay. *The Sixth World Congress on Intelligent Control and Automation, WCICA 2006,* Vol. 1, (pp. 354–358).

Yao, L.-N., Wang, A., & Wang, H. (2008). Fault detection, diagnosis and tolerant control for non-Gaussian stochastic distribution systems using a rational square-root approximation model. *International Journal of Modelling. Identification and Control, 3*(2), 162–172. doi:10.1504/IJMIC.2008.019355

Zhang, Y. M., Guo, L., & Wang, H. (2006). Filter-based fault detection and diagnosis using output PDFs for stochastic systems with time delays. *International Journal of Adaptive Control and Signal Processing, 20*(4), 175–194. doi:10.1002/acs.894

Zhao, F. (Ed.). (2006). *Maximize business profits through e-partnerships.* Hershey, PA: IRM Press.

# Section 2
# Anomaly/Fault Detection

# Chapter 2
# Intelligent System Monitoring:
## On-Line Learning and System Condition State

**Claudia Maria García**
*Universitat Politècnica de Catalunya UPC, Spain*

## ABSTRACT

*A general methodology for intelligent system monitoring is proposed in this chapter. The methodology combines degradation hybrid automata to system degradation tracking and a nonlinear adaptive model for model-based diagnosis and prognosis purposes. The principal idea behind this approach is monitoring the plant for any off-nominal system behavior due a wear or degradation. The system degradation is divided in subspaces, from fully functional, nominal, or faultless mode to no functionality mode, failure. The degradation hybrid automata, uses a nonlinear adaptive model for continuous flow dynamics and a system condition guard to transition between modes. Error Filtering On- line Learning (EFOL) scheme is introduced to design a parametric model and adaptive low in such a way that the unknown part of the adaptive model function is on-line approximated; the on-line approximation is via a Radial Basis Function Neural Network (RBFNN). To validate the proposed methodology, a complete conveyor belt simulator, based on a real system, is designed on Simulink; the degradation is characterized using the Paris-Erdogan crack growth function. Once the simulator is designed the measured current, $i$'s, and velocity of the IM, $\omega_m$, are used to modeling the simplified adaptive IM model. EFOL scheme is used to on-line approximate the unknown $T_L$ function. The simplified adaptive model estimates the IM velocity, $\hat{\omega}_m$, as output. $\hat{\omega}_m$ and the measured IM velocity $\omega_m$ are compared to detect any deviation from the nominal system behavior. When the degradation automata detect a system condition change the adaptive model on-line approximate the new $T_L$.*

DOI: 10.4018/978-1-4666-2095-7.ch002

# INTRODUCTION

The Prognosis and Health Management (PHM) field is predicated on four fundamental notions (Uckun, Goebel, & Lucas, 2008): first, all electromechanical systems ages as a function of use, passage of time and environmental conditions; second, component aging and damage accumulation is a monotonic process, and it's shown at the physical or chemical composition of the component; third, signs of aging are detectable prior to over fatal or total failure; and finally, it is possible to correlate ageing signs with a component or system model.

Based on this four predicates, Figure 1 shows a proposed intelligent monitoring system bloc diagram. The diagnosis and prognosis is based on an adaptive system model which is on-line approximated to follow the system degradation. The system monitoring is due by hybrid automata, which change mode depending on the system condition stage:

**Step 1:** Data acquisition: this procedure is used to obtain the different signals available in the system. It covers the range of normal operation to fault and degradation behavior of the plant.

**Step 2:** Simplified adaptive model: From mathematical viewpoint the selection of a function approximator provides a way for parametrizing unknown parts of the physical model; designing on-line learning algorithms and parameter adaptive laws.

**Step 3:** Diagnosis: Allows comparing the acquired data and the physic model to detect any off-nominal system situation.

**Step 4:** System monitoring: It is based in the proposed degradation hybrid automata; allowing system condition tracking.

**Step 5:** Prognosis: based on actual data and the physic model, prognosis techniques are capable of forecast the future system behavior.

The main area of interest in this chapter is the impact and potential benefit of use nonlinear parametric models to approximate unknown functions in such a way that the descriptive physical model can be on-line adaptable. The use of on-line learning techniques allows the use of the *degradation hybrid automata* for system condition monitoring.

*Figure 1. Intelligent monitoring system bloc diagram*

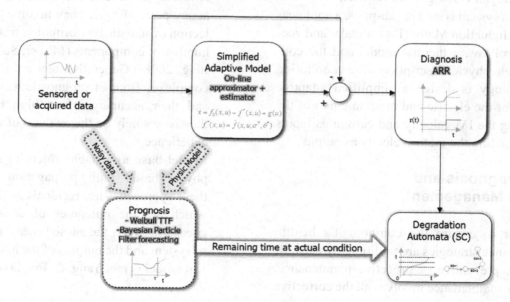

*Figure 2. Conveyor belt system*

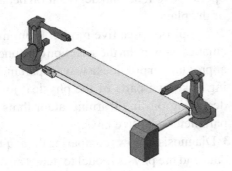

The unknown function approximation is done by using a Radial Basis Function Neural Network (RBFNN). Diagnosis and Prognosis are out of scope, centering this work in the degradation automata for intelligent monitoring, *step 4*, and parametric estimator to adapt the model under degradation, *step 2*.

The methodology is exemplified via a complete conveyor belt system designed in *Simulink* and based on a real system. The *Simulink* conveyor belt process is employ to carry out fatigue or wear simulations; the process degradation is introduced by two friction parameters, $\theta_{static}, \theta_{kinetic} \in \Theta$, the static and kinetic degradation friction respectively, which are time varying and follow the Paris-Erdogan crack growth function. The conveyor belt system is split in sub-parts: a nonlinear nominal Induction Motor (IM), a scalar and vector control law, a thermal model and the conveyor belt physic description. The monitoring methodology is design a simplified adaptive model, for the electric and mechanic part of the IM, using the IM velocity and current as inputs and estimating the motor velocity as output.

## The Prognosis and Health Management

In order to ensure the component's health, maintenance strategies are traditionally used by, reactive, preventive or proactive maintenance. Reactive maintenance involves all the corrective actions performed as a result of system failure, to restore the specified system condition as the main objective. Preventive maintenance is based on a schedule of planned maintenance actions that aims to "prevent" commonly failures. On the other hand, proactive maintenance consists on an intended set of measures to maximize the operational availability and safety of the system. It reacts when maintenance conditions fail, but also anticipate the non satisfaction of the specified condition to avoid the failure.

Condition Based Maintenance (CBM) is characterized as a proactive maintenance strategy that uses diagnosis and prognosis to determine system's health (Luo, Pattipati, Qiao, & Chigusa, 2008). Prognosis in CBM involves prediction of system degradation based on the analysis of monitored data. Based on the current condition, its main porpoise is to assess, whether the process needs maintenance tasks, and if a maintenance process is needed, determining when the maintenance actions should be executed (Tran, Yang, & Tan, 2009).

From engineering viewpoint, the deployment of prognosis process can come from three types of approaches:

- **Experience-base approach:** use statistical reliability to predict probability of failure at any point of time. They involve the collection of statistical information from large number of components (Muller, Suhner, & Iung, 2008). Generally, they are the least complexes from of prognostic techniques and their accuracy is not high because they base solely on the analysis of the past experience.
- **Model-base approach:** refers to specific physics-based fault propagation model, these approaches use residuals as features, which are the outcomes of consistency checks between the sensed measurements of system and the outputs of the mathematical model (Tran, Yang, & Tan, 2009).

- **Data-driven or artificial intelligent approach:** They utilize and require large amount of historical failure data to build a prognostic model that learns the system behavior (Bolander, Qiu, Eklund, Hindle, & Rosenfeld, 2009). When a hybrid neuro-fuzzy approaches techniques appears are possible to complete the big amount of data with the expert knowledge of the system.

The goal is to predict End Of Life (EOL) or Remaining Useful Life (RUL) at a given time, $\tau_R$, using the system observations, where:

$$RUL(\tau_R) \triangleq EOL(\tau_R) - \tau_R$$

*Data-driven* or *experience* based approaches are more commonly used in industry, several statistical techniques are used on PHM as Weibull Time To Failure (TTF) (Candy, 2009) or Survival and Bayes Analysis (Klein & Moeschberger, 1997); however *model* based approaches presents a more accurate health forecasting and in the last years is more often find model based prognosis examples, as (Gucik-Derigny, Outbib, & Ouladsine, 2009) or (Orchard & Vachtsevanos, 2007) where a particle filtering-based diagnosis and prognosis on turbine engine is presented. *Particle filtering* or *Kalman filtering* data-driven prognosis techniques creates more accurate predictor by means of physical models.

The use of model-based approaches is not a trivial task; it deals with severe uncertainties, system/process and measures noise, disturbances and unknown parameters/functions. The use of prognosis filtering techniques such as *Kalman, Smoothing* or *Particle* filters introduce a solution for measurement and process noise; in addition, employs system measurements and a physical model representation to predict $n$ steps ahead of output.

## Problem Formulation

In general, we define a system model as

$$\dot{x}(t) = f(t, x(t), \theta(t), u(t), v(t))$$
$$y(t) = h(t, x(t), \theta(t), u(t), n(t)) \tag{1}$$

where $t \in \mathbb{R}$ is a continuous time variable, $x(t) \in \mathbb{R}^{n_x}$ is the state vector, assumed to be available for measurement, $\theta(t) \in \mathbb{R}^{n_\theta}$ is the parameter vector, $u(t) \in \mathbb{R}^{n_u}$ is the input vector, $v(t) \in \mathbb{R}^{n_v}$ is the process noise vector, $f$ is the state equation, $y(t) \in \mathbb{R}^{n_y}$ is the measured output vector, $n(t) \in \mathbb{R}^{n_n}$ is the measurement noise vector and $h$ is the output equation. The parameter $\theta(t)$ evolves in an unknown way.

In most application the vector field $f$ is partially known. The known part of $f$, usually referred to as the *nominal* model, is divided either by analytical methods or identification methods. Therefore, it is assumed that $f$ can be decomposed as

$$f(x, u) = f_0(x, u) + f^*(x, u),$$

where $f_0$ represents the known system dynamics and $f^*$ represents the discrepancy between the actual dynamics $f$ and the nominal dynamics $f_0$; $f^*$ contains all the uncertain and $f_0$ contains the remaining (known) information.

The nonlinear system (1) can be rewritten as

$$\dot{x}(t) = f_0(x, u) + \hat{f}(x, u; \theta^*, \sigma^*) + e_f(x, u), \tag{2}$$

where $\hat{f}$ is an approximating function of the $f^*$, and $(\theta^*, \sigma^*)$ is a set of optimal parameters that minimize a suitable cost function between $f^*$ and the adaptive approximator $\hat{f}$ for all $(x, u)$ belonging to a compact set $\mathcal{D} \subset (\mathbb{R}^n \times \mathbb{R}^m)$. The error $e_f$ defines as

$$e_f(x,u) = f^*(x,u) - \hat{f}(x,u;\theta^*,\sigma^*),$$

represents the Minimum Functional Approximation Error (MFAE), which is the minimum possible deviation between the unknown function $f^*$ and the adaptive approximator $\hat{f}$ in the $\infty - norm$ sense over the compact set $\mathcal{D}$ :

$$(\theta^*,\sigma^*) =$$

$$\arg \min_{(\theta,\sigma)\in\mathbb{R}^{q_\theta}\times\mathbb{R}^{q_\sigma}} \left\{ \sup_{(x,u)\in\mathcal{D}} \left\| f^*(x,u) - \hat{f}(x,u;\theta,\sigma) \right\| \right\}$$

In general, increasing the number of adjustable parameters in the adaptive approximator reduces the MFAE. If $\dot{x}$ is available for measurement then from (2) the parameter estimation problem becomes a static nonlinear approximation of the general form

$$\bar{\chi} = \hat{f}(x,u;\theta^*,\sigma^*) + e_f(x,u), \tag{3}$$

where $\bar{\chi} = \dot{x} - f_0(x,u)$ is measurable variable, $e_f$ is the minimum functional approximation error (or noise term) and $(\theta^*,\sigma^*)$ are the unknown parameter vector to be estimated.

## System Degradation Monitoring

The performance degradation of mechanical systems is mainly due to component wear, abrasion and fatigue during the operation process; *degradation* can be described as a change on the process expected behavior or the incapability of perform a required or design function, operation or job. The *system degradation* is typically seen as a slowly increasing incipient fault. Therefore, an incipient fault become a failure when system is unable of perform the preferred or designed task. From system degradation fault apparition to the total failure exists a certain period of time, called the RUL of the system. PHM techniques

deal with forecast the EOL or TTF of components or systems in order to ensure functionality and avoid failures. The proposed intelligent monitoring methodology, describes the *system life* as a *degradation hybrid system*. Hybrid Systems (HS) comprise continuous state evolution within a mode and discrete transitions from one mode to another; either controlled or autonomous. For each mode of operation, the system dynamics is governed by a different continuous behavioral model. Control signals, may transition the system from current mode to a different operating mode – this is called 'controlled' transition. Other transitions are 'autonomous' because they are governed by the values of internal state variables. The mathematical representation of HS is the hybrid automata.

In an hybrid automata, at every mode $q$ is assumed that the continuous dynamics are described by a parameterized system of the form $\dot{x} = f_q(x(t),\theta(t))$, where the systems behavior depends on the fault parameter $\theta \in \Theta$, therefore a finite set of subspaces $\Theta_d \in \mathbb{R}^d$ representing the possible *degradation system* levels $\theta_d \in \Theta_d$, $q_d \in Q_d$ is used. This set of faults parameters is denoted by $\Theta_d = U_{q\in Q}\Theta_q$. Is assumed that the set of modes of the HS is partitioned as $Q = Q_N \cup Q_F$, where $Q_N$ and $Q_F$ are the set of normal or nominal modes and faulty modes respectively. At this present work fault region $\Theta_d$ is split in 5 subspaces, $d = \{1..5\}$ following the *grey-scale health index* (Kalgren, Byington, P.E., M.J., & Watson, 2006), from *fully functional* mode $q_0$, (faultless=0) to *no functionality* mode $q_4$,(failure=1); where $q_0 \in Q_N$ and $q_{1..4} \in Q_F$, Table 1.

The set of transitions or failure events $\Sigma_F$ labels the transitions to faulty modes, has information about the continuous dynamics for faulty modes is available then a flow condition is associated with this modes, the System Condition (SC).

*Table 1. Grey-scale health index representation (Kalgren, Byington, P.E., M.J., & Watson, 2006)*

| Operational Capability | Maintenance Action | Logistics Action |
|---|---|---|
| Fully Functional | No Maintenance Required | No Logistic Changes |
| Functional with degraded Performance | Maintenance at Convenience | Trigger Opportunistic Logistic Sparing |
| Reduced Functionality | Schedule Maintenance Now | On-Demand Logistic Sparing |
| Functionality Severely Impinged | Remove from Service ASAP | Logistic Emergency Sparing |
| No functionality | Remove from Service Now | Logistic reflect unit out of Service |

## CONVEYOR BELT PROCESS SIMULATION

Although induction machines are reliable, they are subjected to some failures; the understanding degradation behavior of IM is crucial for the adoption and application of health management methodologies on systems and industrial processes. Conveyor belt systems are subjected to velocity changes along the working route and a big amount of start-stops cycles, this operating service causes on the IM a fatigue and degradation due to thermal and mechanical stresses. As aging test are destructives and substantially expensive a conveyor belt simulator (in *Simulink*) based on a real system is presented (Figure 2).

### The Induction Motor and Control

Two kind of induction motor model are presented on this chapter, the scalar and vector control base models depending of the kind of controller; firstly is presented an easy model assuming the steady-state simplified circuit on the IM (Chapman, 2005), and adding some corrections for low

velocity scalar control. The second model is based on the full equivalent dynamic IM circuit and using a vector control, the so called Field Oriented Controllers (FOC) (Ong, 1998).

The induction motor dynamics is complicated due to the physical properties of the IM constituent materials, characteristics, geometric construction, the electromagnetic fluxes rotation and the inter-dependence of machine operation velocity and load points. Therefore, along the literature and the induction machine standards a simplified model based on an internal electric circuit (Figure 3) is commonly used, not only for machine test purpose also for control discipline.

The dynamic model of three-phase induction motor assumes the following main hypothesis:

- The stator is a hollow cylinder containing a concentric solid cylinder representing the rotor.
- The air gap is narrow, with radial constant length and delimited by the smooth cylinders surfaces, rotor and stator.
- The windings are dimensionless radial surfaces and the slots are not considered.

*Figure 3. Simplified IM per phase circuit at steady-state*

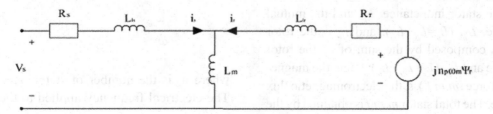

*Box 1. RST for stator and rotor equations*

$$\theta_s(\alpha, t) = N_s \left[ i_{sR}(t) \cdot \cos(\alpha) + i_{sS}(t) \cdot \cos(\alpha + \frac{2\pi}{3}) + i_{sT}(t) \cdot \cos(\alpha + \frac{4\pi}{3}) \right]$$

$$\psi_s(t) = c \left[ a^0 \psi_{sR}(t) + a^1 \psi_{sS}(t) + a^2 \psi_{sT}(t) \right]$$

$$\psi_r(t) = c \left[ a^0 \psi_{rR}(t) + a^1 \psi_{rS}(t) + a^2 \psi_{rT}(t) \right]$$

- In both, stator and rotor, the windings are sets of three, symmetrical with individual harmonic winding distribution and uniform spatial arrangement relative or symmetric to $120°$ ( $\frac{2}{3} \pi$ radians).

- The windings are star connected with neutral and rotor winding are short circuit.

- Both, rotor and stator are enough axially long to neglect edge effects.

- Both, rotor and stator are completely laminated with infinite magnetic permeability material. Therefore, magnetic saturation effects and heating losses due to eddy currents in the iron are neglected.

$$\begin{cases} u_s(t) = R_s i_s(t) + \dot{\psi}_s(t) \\ 0 = R_r i_r(t) + \dot{\psi}_r(t) - j n_p \omega_m \psi_r(t) \end{cases} \quad (4)$$

where, sub index '$s$' and '$r$' mean stator and rotor respectively, $R$ is resistance, $i$ is current and $\psi$ is electromagnetic flux:

$$\begin{cases} \psi_s = L_s i_s(t) + L_m i_r(t) \\ \psi_r = L_r i_r(t) + L_m i_s(t) \end{cases}$$

where, $L_s$ is the stator reactance, composed by the sum of the stator inductance, $L_{ls}$, and the mutual inductance, $L_m$, ($L_s = L_{ls} + L_m$); and $L_r$ is the rotor reactance, composed by the sum of $L_{lr}$ the rotor inductance and $L_m$, ($L_r = L_{lr} + L_m$); being the magnetomotive force (*m.m.f.*), $\varepsilon$, the electromagnetic flux derivative. The total stator *m.m.f* is obtained by the superposition of the three windings; remember that the IM is composed by three windings spread $120°$, composing then a six equation system simplified in (4) (RST for stator and rotor equations) where $a^0 = 1$, $a^1 = e^{j\frac{2\pi}{3}}$ and $a^2 = e^{j\frac{4\pi}{3}}$; $c$ is the transformation constant and $N$ is the number of turns of wire in the coil. (see Box 1)

The nominal region of the IM is from the synchronism condition (slip, $s=0$, and torque, $T_e=0$) to the situation of maximum torque ($s_{MAX}$, $T_{eMAX}$); the region in which the applied torque increase leads to an almost proportional increase of the per phase current consumption, slip and rotor magnetic field; the power factor or $cos(\theta)$ is approximately 1 and the rotor reactance, $L_r$, is negligible. The region of low velocity is from maximum torque condition ($s_{MAX}$, $T_{eMAX}$) to idle condition (slip, $s=1$, and motor velocity, $\omega_m=0$); region in which the applied torque increases and the power factor and slip decrease, this is the transient region; the motor is accelerating or decelerating (Chapman, 2005). The maximum torque cannot be exceeded and is usually between two to three times the nominal or full load torque.

The asynchronous motor slip or *unit difference* in velocity between the mechanical frequency, $\omega_m$, and electric rotor $\omega/n_p$, is defined as:

$$s = 1 - \frac{n_p \omega_m}{\omega}$$

where $n_p$ is the number of stator pairs of pole. The electrical frequency applied to the stator is represented by the Greek letter $\omega$, in radians per second, and $f$, in hertz. The rotor rotation velocity in revolutions per minute, $n$, is:

$$n = \frac{30}{\pi} \omega_m = \frac{60f}{n_p}(1-s) \qquad (5)$$

Note that the slip is one at start-up and zero at synchronous velocity.

## Scalar Control Motor Modelling

Scalar control is based on the steady-state circuit of the induction machine. The rotating motor velocity is approaching to synchronism when the balance slip is less than the corresponding to the maximum torque ($0 < s_{eq} < s_{max}$) the motor is at steady-state, at the nominal region; the rotor circuit can be simplified neglecting $L_{lr}$ ( $L_{lr} \approx 0$ ), and treating rotor effects as a *varying resistance* supplied by a constant-voltage source, $\varepsilon_r$ (magnetomotive force), the equivalent rotor resistance from this point of view is $R_r + R_{load} = \dfrac{R_r}{s}$ and the simplified circuit:

$$\begin{cases} u_s(t) = R_s i_s(t) + \dot{\psi}_s(t) \\ \varepsilon_r = \dfrac{R_r}{s} i_r(t) \end{cases}$$

With the previous rotor circuit simplification is seen that the rotor energy consumed per unit time is in function of the varying resistance $\dfrac{R_r}{s}$ :

$$P = 3 \frac{R_r}{s} \left| i_r \right|^2$$

where $i_r$ is assumed the effective rotor current. Part of the power is dissipated in the form of electrical losses in the resistance $R_r$; the difference is the

mechanical power supplied to the motor shaft. The above expression can be written:

$$P = cR_r \left| i_r \right|^2 + c\left( \frac{1-s}{s} \right) R_r \left| i_r \right|^2,$$

it follows that the mechanical power delivered is:

$$P_m = c\left( \frac{1-s}{s} \right) R_r \left| i_r \right|^2,$$

using the (5) is:

$$P_m = c \cdot n_p \frac{\omega_m}{\omega_s} R_r \left| i_r \right|^2$$

The torque as a function of the rotor current is obviously:

$$T_e = c \cdot n_p \frac{1}{\omega_s} \frac{R_r}{s} \left| i_r \right|^2 \qquad (6)$$

The motor velocity control is done by the manipulation of the electromagnetic torque. If $s$ is small, the rotor velocity can be controlled by varying the synchronous velocity, $fs$. To maintain constant flux, $u_s$ should be proportionally changed with $fs$. On induction motor steady-state electromagnetic torque is proportional to the slip frequency $f_{slip}$. Therefore, the PI output, rotor velocity loop, is used to fix the reference rate for the slip frequency, $f_{slip}^*$, which is equivalent to set a reference torque, (7). The reference for the power frequency, $f_m^*$, is calculated by adding $f_{slip}^*$, (generated by the PI velocity) and the rotation frequency of the rotor (measured) (8).

$$f_{slip} = \left( K_P + \frac{K_I}{s} \right) \cdot (\omega_{ref}^* - \omega_m) \qquad (7)$$

33

$$f_m^* = f_m + f_{slip} \qquad (8)$$

where $f_m = \omega_m \cdot n_p$, and the control law is:

$$u_s^* = \frac{f_m^*}{\omega_0} \cdot u_{s0} \qquad (9)$$

where $\omega_0 = 2\pi f_0$ with $f_0$ as the rated base frequency and $u_{s0} = \dfrac{u_{l0}}{\sqrt{3}}$ with $U_{l0}$ as the line to line voltage. With the angel and modulus of $\theta_e$ the $u_s^*$ control law can be described in *ab*, polar coordinates (see appendix), $\theta_e = \int 2\pi f_m^* dt$

The basic V/F strategy supposes the rotor resistance *Rs* negligible; at low velocity this is an invalid hypothesis, to correct such a problem, a compensation expression is applied to the control law (9); which consists in adding to the applied stator voltage a term of the voltage drop in *Rs*:

$$\left| u_s^* \right| = \left| \frac{f_m^*}{\omega_0} + \frac{R_s}{j\omega_0 L_s} \cdot \frac{1 + j\dfrac{1}{\varpi} \cdot \dfrac{s}{s_p}}{1 + j\dfrac{s}{s_p}} \right| \cdot u_{s0} \qquad (10)$$

where $s_p = \dfrac{R_r}{2\pi f_0 \varpi L_r}$ and σ is the total leakage factor, $\varpi = 1 - \dfrac{L_m^{\,2}}{L_s L_r}$;

## Vector Control Motor Modelling

Accurate knowledge of induction motor model and its parameters is critical when field oriented techniques are used. The induction motor parameters vary with the operation conditions, as happens with all electric motors. The inductances tend to saturate at high flux and the resistances trend to increase as heating and skin effect. To deal with the change in the IM parameters in Marino,

Tomei, & Verrelli (2010) and Wang and Wang (1998), among others, are presented a nonlinear adaptive approach based in feedback linearization and the adaptation of some IM parameters. Also in the IM branch of research are found some works in Fault Diagnosis (FD), as in Espinoza-Trejo, Campos-Delgado, & Loredo-Flores (2010) where a FD scheme strategy based in FOC is presented. Furthermore other researcher has been focused in a more optimal control strategy (Wasynczuk, et al., 1998), where are presented a maximum torque per ampere control law as well as a maximum eficiency and maximum power factor in combination with FOC. In Aissa and Eddine (2009) a series iron losses model for induction motor is presented and a power losses minimization control strategy is used for FOC control.

The IM model in (4) is reduced to four equations by using the park or *dq* (for *direct* and *quadrature*) transformation (Appendix), it is based on a set of assumptions, among others symetrical three phase machine and neglected saturation. The symbols used and their meaning are listed at the end, in the Appendix.

From now let be the state variables, $x = (i_{sd}, i_{sq}, \psi_{rd}, \rho, \omega_m)$, stator *dq* currents, rotor *d* flux, $\rho$ the reference frame for the rotating field model, and $\omega_m$ the motor velocity, consider a IM described by (see also Box 2):

$$\dot{x} = f(x) + g(x)u \qquad (11)$$

where $L_m$, $L_r$, $L_s$ are the mutual, rotor and stator inductances; $n_p$, $B$, $J$, $c$ and $T_L$ are the pare of poles, the viscous coefficient, the motor inertia, the coordinates transformation factor and the load torque respectively. And the rotor fluxe, $\psi_{rd}' = \dfrac{\psi_{rd}}{L_m}$ and $\psi_{rq}' = \dfrac{\psi_{rq}}{L_m}$, $T_s$ and $T_r$ are the stator and rotor time constant respectively, and $\varpi$ is the total leakage factor, $\varpi = 1 - \dfrac{L_m^{\,2}}{L_s L_r}$.

*Box 2.*

$$
f(x) = \begin{pmatrix}
-(\frac{1}{\varpi T_s} + \frac{1-\varpi}{\varpi T_r})i_{sd} + n_p\omega_m i_{sq} + \frac{1}{T_r}\frac{i_{sq}^2}{\psi_{rd}'} + \frac{1-\varpi}{\varpi T_r}\psi_{rd}' \\
-n_p\omega_m i_{sd} + \frac{1}{T_r}\frac{i_{sq}}{\psi_{rd}'}i_{sd} - (\frac{1}{\varpi T_s} + \frac{1-\varpi}{\varpi T_r})i_{sq} - \frac{1-\varpi}{\varpi}n_p\omega_m\psi_{rd}' \\
\frac{1}{T_r}i_{sd} - \frac{1}{T_r}\psi_{rd}' \\
n_p\omega_m + \frac{1}{T_r}\frac{i_{sq}}{\psi_{rd}'} \\
c\frac{np}{J}\frac{L_m}{L_r}(\psi_{dr}i_{qs}) - \frac{T_L}{J} - \frac{B}{J}\omega_m
\end{pmatrix}
\qquad
g(x) = \begin{pmatrix}
\frac{1}{\varpi L_s} & 0 \\
0 & \frac{1}{\varpi L_s} \\
0 & 0 \\
0 & 0 \\
0 & 0
\end{pmatrix}
$$

A vector control (FOC) is employed for the IM drive in order to satisfy high-performance velocity control. This control strategy has been widely studied previously in the literature, by now as a classical control technique, (Bose, 2002), (Bocker & Mathapati, 2007). The basic principle of this technique consists in holding the rotor flux magnitude at a constat value $\psi^*$. In this way, a lineal relationship is obtained between control variables and velocity. Once the rotor flux has been regulated, the rotor velocity is asymtotically decoupled from the magnetic flux. Hence the vector control objective is as follows:

$$\lim_{t\to\infty}\left|\omega_m(t) - \omega_{ref}(t)\right| = 0.$$
$$\lim_{t\to\infty}\left|\psi_{dr}(t) - \psi^*\right| = 0 \quad \& \quad \lim_{t\to\infty}\psi_{qr}(t) = 0.$$

These control objectives can be satisfied by using the next control feedback, as showed in (Espinoza-Trejo, Campos-Delgado, & Loredo-Flores, 2010), (Marino, Tomei, & Verrelli, 2010).

From the two first equations of the IM model (11) is defined the feedback state control:

$$u_{sd} = \varpi L_s[-n_p\omega_m i_{sq} - \frac{1}{T_r}\frac{i_{sq}^2}{\psi_{rd}'} - \frac{1-\varpi}{\varpi T_r}\psi_{rd}' + v_d]$$

$$u_{sq} = \varpi L_s[n_p\omega_m i_{sd} - \frac{1}{T_r}\frac{i_{sq}}{\psi_{rd}'}i_{sd} + \frac{1-\varpi}{\varpi}n_p\omega_m\psi_{rd}' + v_q]$$

$v_d$ can be indepently controlled via a PI:

$$v_d^{pi} = k_{p1}^d(i_{ds}^* - i_{ds}) + k_{i1}^d\int_0^t(i_{ds}^*(\tau) - i_{ds}(\tau))d\tau$$

$$i_{ds}^* = k_{p2}^d(\psi^* - \psi_{dr}) + k_{i2}^d\int_0^t(\psi^*(\tau) - \psi_{dr}(\tau))d\tau$$

Meanwhile the input control $v_q$ is expressed as:

$$v_q^{pi} = k_{p1}^q(T_e^* - T_e) + k_{i1}^q\int_0^t(T_e^*(\tau) - T_e(\tau))d\tau$$

$$T_e^* = k_{p2}^q(\omega^* - \omega_m) + k_{i2}^q\int_0^t(\omega^*(\tau) - \omega_m(\tau))d\tau$$

$$T_e = c \cdot np\frac{L_m}{L_r}(\psi_{dr}i_{qs})$$

Then the flux amplitude dynamics are linear and can be independently controlled by $v_d^{pi}$, and the stator currents present a first-order lineal description. When the rotor flux $\psi_{rd}$ reaches its reference $\psi^*$, then the dynamics of the velocity $\omega_m$ are linear, and can be independently controlled by $v_q^{pi}$. Lastly

$$k_{p1}^d, \quad k_{p2}^d, \quad k_{i1}^d, \quad k_{i2}^d, \quad k_{p1}^q, \quad k_{p2}^q, \quad k_{i1}^q, \quad k_{i2}^q$$

must be tuned in order to guarantee stability and tracking performance.

When the reference velocity, $\omega^*$, is constant $i_{qs}$ is too and $\dfrac{d\psi_{rd}}{dt} = 0$ (consequently $\psi_{rd} = L_m i_{sd}$), constituting a steady-state solution.

The stator current amplitude is defined as the peak AC current. In terms of *dq* variables (Wasynczuk, et al., 1998):

$$\left| i_s \right| = \sqrt{i_{qs}^2 + i_{ds}^2}$$

Operation at maximum torque per ampere (MTA) is achieved when at a given torque and velocity, the slip frequency is adjusted so that the stator current amplitude is minimized. An expression for the slip frequency which minimizes the stator current amplitude is discussed in Wasynczuk (1998) by noting that to maximize the product of $i_{qs}$ and $i_{ds}$ subject to $\left| i_s \right|$, $i_{qs}$ should be equal to $i_{ds}$.

The forth equation in (11) shows the operating condition:

$$\frac{d\rho}{dt} = n_p \omega_m + \frac{1}{T_r} \frac{i_{sq}}{\psi'_{rd}} = n_p \omega_m + \omega_{si} \tag{12}$$

where, $\omega_{si}$ is the slip frequency. Let first determine the constant vectors for the flux and current reference $(\psi^*, 0) \mapsto \psi_{rq} = 0, \left| \psi^* \right| = \psi_{rd}$ and $(i_{sd}^*, i_{sq}^*) \mapsto i_{sd} = i_{sq} = i_{MTA}$, then (12) can be rewrited as:

$$\dot{\rho}^* = n_p \omega^* + \frac{1}{T_r} \frac{i_{MTA}}{i_{MTA}} \tag{13}$$

And, from (13) and the third equation in (11), the reference flux, $\psi^*$, for MTA strategy is:

$$\psi^{*2} = \frac{T_e^* L_r}{c \cdot n_p} \tag{14}$$

## Thermal Model

The increase of the temperature in the IM is interpreted as a first order model. It depends on the generated thermal power and two typical temperatures: the ambient temperature, that supposes the initial or starting point, $T_{st}$, and the equilibrium temperature, which as well depends on the thermal power and the time constant $\tau$:

$$mc \frac{dT}{dt} = P - \frac{T - T_{st}}{R} \tag{15}$$

$$P_e = P_u + P$$

$$u_{sd} i_{sd} + u_{sq} i_{sq} = T_L \cdot \omega_m + P \tag{16}$$

where, $P$ is the thermal power associated to the power losses; $P_e$ is the input power associated to the supply motor voltage and current, and $P_u$ is the output power associated with the torque proportionated by the IM; $mc$ is the thermal capacity, with $R$ as the thermal resistance defined by, $\alpha$, the thermal conductivity, $\alpha = \dfrac{1}{R}$; $\tau$ is the process heating time constant, defined as $\tau = R \times mc$.

$P$ and $R$ values are obtained through $mc$, $T$ and $\tau$, from the real system stator heating curve.

The proposed thermal model is based on the determination of one unique and global value for the thermal resistance, $R$, and the constant of time, $\tau$, at continuous work, allowing the estimation of the thermal power lost by the IM and the equilibrium temperature.

## Conveyor Belt

The motor shaft is directly connected to one conveyor pulley, and then, the conveyor belt movement can be described as:

$$V_m - V_{belt} = V \tag{17}$$

$$F_m = \left(V \cdot D_P + \frac{d}{dt} V \cdot R_S\right) - V_{belt} \cdot R_W - F_{friction}$$

$$a_{belt} = \frac{F_m}{m_{belt}} \tag{18}$$

$$T_L = \omega_m \cdot R_O + F_m \cdot R_P \tag{19}$$

where, $V_m$, $V_{belt}$ and $V$ are the motor, belt and the resulting lineal velocity, respectively. With $D_P$ the damping factor, $R_S$ pulley stiffness, $R_w$ and $m_{belt}$ the conveyor belt load and mass; and, $R_O$ and $R_P$ the dynamic viscous coefficient and the pulley radius, the reader is referred to the Nomenclature (Appendix) for further information.

## Degradation Simulation

The most common incipient fault in a motor is a friction fault. Friction between surfaces trends to resist motion; however, friction forces changes with velocity and trends to be greater stationary, this result in motion which alternative *sticks* and *slips* as force balance requires. This *stiction* phenomenon is common in many mechanical systems also found as *stick-slip* event. Aging deterioration may cause motor stress, increasing friction forces, with the consequent increase on the load torque; this kind of *deterioration friction* is mod-

eled by introducing a fault parameter $\theta \in \mathbb{R}^{n_\theta}$ to simulate the motor friction effects, where the value of $\theta$ is increased. The nominal condition with no motor friction implies $\theta=0$, $\theta_0$. Mathematically, stress $(\sigma)$, is expressed as follows, where, $F$ is force (load) and $a$ is the material area:

$$\Delta\sigma = \lim \frac{\Delta F}{\Delta a}$$

Deformation and degradation of the IM insulation material caused by the increase of stress can be characterized as Paris-Erdogan crack growth rate equation; where $F$ curve is sigmoid, (see Figure 4); Existing two stress limits, the lower limit to $\Delta\sigma_{th}$, below which no crack growth take place and the upper limit, $\Delta\sigma_{CR}$, which crack growth is explosive. These limits divide the insulation material deterioration in three zones, 1) near threshold region, 2) striation growth; and 3) crack growth instability (Wei, 2010).

At the conveyor belt process the induction motor shaft is attached to one belt pulley, on the pulley is the belt; the real belt velocity depends upon the transmitted velocity from the motor shaft to the pulley and the load friction forces; expressing the force at the pulley as $F_{in}$ and the belt force as $F_{belt}$:

$$m_{belt} a_{belt} = F_m$$

$$m_{belt} a_{belt} = F_{in} - F_{belt} - F_{friction} \tag{20}$$

The friction force is more complex, however:

$$F_{friction} = \begin{cases} \text{sgn}(V_{belt})\mu F_n, & |F_{stationary}| > \mu F_n \\ F_{stationary}, & otherwise, V_{belt} = 0 \end{cases} \tag{21}$$

where $\mu$ is the coefficient of friction, $\mu(V_{belt})$, $F_n$ is the normal force and $F_{stationary}$ is the instantaneous force such $V_{belt}=0$.

*Figure 4. Deterioration friction forces under stress following Paris-Erdogan postulate*

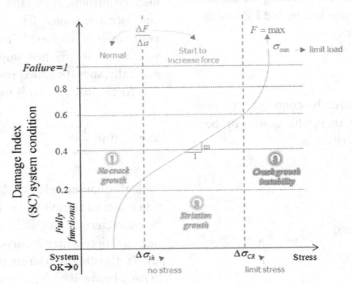

In many applications the friction capacity is described by its static and kinetic magnitudes. This approach is used in the present model:

$$\mu F_n = \begin{cases} \mu_{static} F_n = F_{static}, & V_{belt} = 0 \\ \mu_{kinetic} F_n = F_{sliding}, & V_{belt} \neq 0 \end{cases} \quad (22)$$

The following logic determines $F_{stationary}$; whenever the velocity is nonzero, an impulsive force would be needed to make it zero instantaneously. This always exceeds the capacity, $F_{sliding}$, so the latter magnitude is used. When the velocity is already zero, however, $F_{stationary}$ is the force which maintains this condition by making the acceleration zero.

$$F_{stationary} = F_{in} - F_{belt}$$
$$= F_{sum}$$

The friction force can thus be expressed as:

$$F_{friction} = \begin{cases} sgn(V_{belt})F_{sliding}, & V_{belt} \neq 0 \\ F_{sum}, & V_{belt} = 0, |F_{sum}| < F_{static} \\ sgn(F_{sum})F_{static}, & V_{belt} = 0, |F_{sum}| \geq F_{static} \end{cases}$$

The system deterioration is incorporated by $\theta_{static}, \theta_{kinetic} \in \Theta$, varying terms, forcing $F_{static}$ and $F_{sliding}$ to follow the sigmoid crack growth Paris-Erdogan function, Figure 4.

$$\mu F_n = \begin{cases} (\mu_{static} + \theta_{static})F_n = F_{static}, & V_{belt} = 0 \\ (\mu_{kinetic} + \theta_{kinetic})F_n = F_{sliding}, & V_{belt} \neq 0 \end{cases}$$

## ON-LINE LEARNING ADAPTIVE MODELLING

In Farrell and Polycarpou (2005) is described a design methodology for on-line approximate unknown functions for dynamic systems, creating on this way adaptive models to be used on adaptive control theory. One of the approaches is the Error Filtering On-line Learning (EFOL).

Following, EFOL is used to design a parametric model to on-line approximate the nonlinear

*Figure 5. Simplified circuit stator current as model inputs*

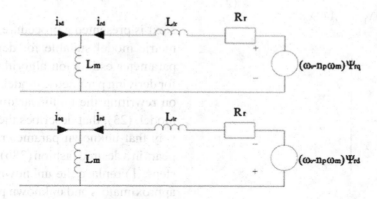

unknown load torque function, $T_L$. $T_L$ is approximated based on Radial Basis Function Neural Network, RBFNN, trained on-line using two inputs, the electromagnetic estimated torque, $\hat{T}_e$, and the acquired angular motor velocity, $\omega_m$. A simplified adaptive model of the IM based on the electric circuit, Figure 5, is presented, which on-line adapt the $T_L$ function to compute an estimation of the motor velocity, $\hat{\omega}_m$.

## Induction Motor Adaptive Model

The proposed simplified model is shown at Figure 5, where the model inputs are directly the $dq$ stator currents. At (23) is presented the model equations.

$$\frac{d\hat{\psi}_{rd}}{dt} = -\frac{R_r}{L_r}\hat{\psi}_{rd} + \hat{\psi}_{rq}(\omega_{si} - np\hat{\omega}_m) + \frac{L_m R_r}{L_r}i_{sd}$$

$$\frac{d\hat{\psi}_{rq}}{dt} = -\frac{R_r}{L_r}\hat{\psi}_{rq} - \hat{\psi}_{rd}(\omega_{si} - np\hat{\omega}_m) + \frac{L_m R_r}{L_r}i_{sq}$$

$$\frac{d\hat{\omega}_m}{dt} = c\frac{np}{J}\frac{L_m}{L_r}(\hat{\psi}_{dr}i_{qs} - \hat{\psi}_{qr}i_{ds}) - \frac{T_L}{J} - \frac{B}{J}\hat{\omega}_m$$

$$(23)$$

with the estimated electromagnetic torque, $\hat{T}_e$:

$$\hat{T}_e = c \cdot np \cdot \frac{L_m}{L_r}(\hat{\psi}_{dr}i_{qs} - \hat{\psi}_{qr}i_{ds}) \qquad (24)$$

From now let the input variables of the adaptive model be the $dq$ stator currents and the state variables:

$$x = (\hat{\psi}_{rd}, \hat{\psi}_{rq}, \hat{\omega}_m) \qquad (25)$$

the unknown function $T_L$ and unknown parameter $R_r$ be the deviation from their nominal value $T_{LN}$ and $R_{rN}$ of load torque $T_L$ and rotor resistance $R_r$:

$$\chi = (\chi_1, \chi_2)^T = (T_L - T_{LN}; R_r - R_{rN})^T \qquad (26)$$

$T_L$ is typically unknown because is the summatory of all opposing forces (or load) to the rotor movement, i.e. in this case of study is considered as the opposing force carried by the conveyor belt system in opposition to the IM $T_e$; if the system is wear, under fatigue it will increase (more opposing force); causing the IM velocity to decrease with a consequence reduction of the stator flux thus will increase the stator current origin an augment of the electromagnetic torque $T_e$; resulting in enhance of temperature. Due to rotor heating $R_r$ my vary ±50% around it nominal value. Replacing (26) in system (23) with the state variables (25) and replacing $\omega_{si}$ by the operating condition (13) the IM model can be rewritten in compact form as:

$$\dot{x} = f(x) + g_1(x)i_{sd}$$
$$+ g_2(x)i_{sq} + f_1(x)\chi_1 + f_2(x)\chi_2 \quad (27)$$

$$f(x) = \begin{pmatrix} -\dfrac{R_{rN}}{L_r}\hat{\psi}_{rd} \\[2mm] -\dfrac{R_{rN}}{L_r}\hat{\psi}_{rq} \\[2mm] -\dfrac{T_{LN}}{J} - \dfrac{B}{J}\hat{\omega}_m \end{pmatrix}$$

$$g_1(x) = \begin{pmatrix} \dfrac{L_m R_{rN}}{L_r} \\[2mm] 0 \\[2mm] -c\,\dfrac{n_p}{J}\dfrac{L_m}{L_r}\hat{\psi}_{rq} \end{pmatrix}$$

$$g_2(x) = \begin{pmatrix} \dfrac{L_m R_{rN}}{L_r}\dfrac{\hat{\psi}_{rq}}{\hat{\psi}_{rd}} \\[2mm] \dfrac{L_m R_{rN}}{L_r}\dfrac{1}{\hat{\psi}_{rd}^2} + \dfrac{L_m R_{rN}}{L_r} \\[2mm] c\,\dfrac{n_p}{J}\dfrac{L_m}{L_r}\hat{\psi}_{rq} \end{pmatrix}$$

$$f_1(x) = \begin{pmatrix} 0 \\ 0 \\ -1/J \end{pmatrix}$$

$$f_2(x) = \begin{pmatrix} -\dfrac{1}{Lr}\hat{\psi}_{rd} + -\dfrac{L_m}{Lr}\dfrac{i_{sq}}{\hat{\psi}_{rd}}\hat{\psi}_{rq} + \dfrac{L_m}{Lr}i_{sd} \\[2mm] -\dfrac{1}{Lr}\hat{\psi}_{rq} + -\dfrac{L_m}{Lr}\dfrac{i_{sq}}{\hat{\psi}_{rd}^2} + \dfrac{L_m}{Lr}i_{sq} \\[2mm] 0 \end{pmatrix}$$

## Load Torque Adaptive Approximator

Next is presented a procedure for design a parametric model suitable for develop an adaptive parameter estimation algorithm. The procedure for deriving parametric models basically consists on rewriting the nonlinear differential equation model, (23), that describes the system, in such a way that unknown parameters or functions appears in a desired fashion (28b). There are two key steps: 1) replace the unknown nonlinearities by approximators and unknown parameters by their estimates; ensuring that all available information of the plant knowledge is used; and 2) avoid the use of differentiators and facilitate the derivation of convenient parametric models; filtering techniques are employed, where certain signals are passed through a stable filter.

Let's state the load torque function as:

$$T_L = f_L^*(\hat{T}_e, \omega_m; \theta^*, \sigma^*)$$

where $f_L^*$ is an unknown function with unknown parameters $\theta^*$ and $\sigma^*$, and depends on the estimated electromagnetic torque, $\hat{T}_e$, and the motor velocity, $\omega_m$, (x's); then the third equation of (23) can be expressed as:

$$J\dot{\omega}_m = \hat{T}_e - f_L^*(T_e, \omega_m; \theta^*, \sigma^*) - B\omega_m \quad (28)$$

(28) can be divided into

$$\chi_1 = -J\dot{\omega}_m + \hat{T}_e - B\omega_m, \quad (28a)$$

the measurable variable, and:

$$\hat{\chi}_1 = \hat{f}_L(\hat{T}_e, \omega_m; \theta^*, \sigma^*), \quad (28b)$$

the Linearly Parameterized Approximator (LPA) of the unknown $T_L$. For LPA, $\sigma$ is selected a priory, and therefore the approximation function $\hat{f}_L$

can be written as $\hat{f}_L(\hat{T}_e, \omega_m : \theta) = \theta^{*T}\phi(\hat{T}_e, \omega_m)$ the on-line $T_L$ approximation is due by a Radial Basis Function (RBF), where $z(t)$ is the output of the RBFNN:

$$\hat{f}_L(x) = z(t) = \sum_{i=1}^{N} \theta_i G_i(\|x_i - \sigma_i\|)$$

$G_i(\cdot)$ is the network hidden layer where the RBF is applied, and, $x_i$ are the network inputs signals; $\theta$ and $\sigma$, are the weights and centers respectively. Centers, $\sigma$ is known parameters, pre-established between the $x_i$ region inputs, $\hat{T}_e$ and $\omega_m$; then $\hat{f}_L$ is LPA. The computation of the $T_L$ on-line approximator is not a difficult task but some factors may be taken in mind, depending on the number of RBF network inputs and nodes; the number of parameters required for localized approximators grows exponentially with the number of dimensions $D$, known as *curse of dimensionality*, in other words, as much inputs your RBF network has more neurons it will need; $d = \dim(D)$, $D$ is partitioned into $\bar{N}$ divisions, centers or neurons, then there will be $N = \bar{N}^d$ total partitions. This exponentially increases the RBF based approximator, the increase of $N$ with $d$ is a problem if either the computation time or memory requirements of approximator become too large.

The evaluation of $\hat{f}_L$ for general approximator requires calculation of the $N$ elements of $\phi(x's)$; A Gaussian RBF monotonically decreases with distance from the center. Gaussian like RBF are local (give a significant response only in neighborhood near to the center). The two dimensions Gaussian function are:

$$g(x) = e^{-\left(\frac{\left(\hat{T}_e - c_{T_e}\right)^2}{\sigma_{T_e}^2} + \frac{\left(\omega_m - c_{\omega_m}\right)^2}{\sigma_{\omega_m}^2}\right)} = \phi_j$$

$\sigma_{Te,com}$ must be chosen according to the center distance and can be different for $\hat{T}_e$ and $\omega_m$; e.g. for a domain $D_{Te}$ if the center distances are a unit $\sigma$ can be chosen as 0.5, then $Gaussian(c1_{Te})$ and $Gaussian(c2_{Te})$ will cross at 50% approximately.

When the number of neurons is too big the computation time is also too big, one easy way to solve this problem is *refresh* only the neighbor's weights of the input point, *lattice based approximators*, see recommendations.

Next the Error Filtering On-line Learning scheme is presented, EFOL can be used for LPA and Nonlinear Parameterized Approximators (NPA). The on-line learning model will generate a training signal $e_1(t)$ that will be used to approximate the unknown nonlinearities in the system. The on-line learning model consists of the adaptive approximator augmented by identifier dynamics. The identified dynamics are used to incorporate any a priori knowledge into the identification design and to filter some of the signals to avoid the use of differentiations and decrease the effects of noise.

By filtering (28a) and (28b) with stable first-order filter:

$$\chi_1(t) = \frac{\lambda}{s + \lambda}[\hat{f}_L(z(t); \theta^*)] + \delta(t), \qquad (29)$$

and the measurable variable using the fact that $\dot{\omega}_m(t) = s[\omega_m(t)]$ is computed as:

$$\chi_1(t) = -\frac{\lambda s}{s + \lambda}[J\omega_m(t)] + \frac{\lambda}{s + \lambda}[\hat{T}_e(t) - B\omega_m(t)] \rightarrow$$
$$\chi_1(t) = -\lambda J\omega_m(t)$$
$$+ \frac{\lambda}{s + \lambda}[+\lambda J\omega_m(t) + \hat{T}_e(t) - B\omega_m(t)],$$

$$(30)$$

where $\delta(t)$ is the filtered Minimum Function Approximation Error (MFAE), $e_f$. The MFAE value is typically small compared with the estimation error, $e_1(t)$, then it can be neglected; and the filter $\dfrac{\lambda}{s+\lambda}$ is a strictly positive real (SPR).

The EFOL is expressed as:

$$\hat{\chi}_1(t) = \frac{\lambda}{s+\lambda}[\hat{f}_L(z(t);\hat{\theta})] \tag{31}$$

Therefore, the estimator is obtained by replacing the unknown parameters $\theta^*$ by their parameter estimates $\hat{\theta}(t)$,

$$\chi_1(t) = \frac{\lambda}{s+\lambda}[\theta^{*T}\varphi(z(t))] \Rightarrow \hat{\chi}_1(t)$$
$$= \frac{\lambda}{s+\lambda}[\hat{\theta}^T\varphi(z(t))] \tag{32}$$

and the estimation error $e_1(t)$, which will be used in the update of the parametric estimates, is given by:

$$e_1(t) = \hat{\chi}_1(t) - \chi_1(t), \tag{33}$$

with $\hat{\chi}_1(t)$ as the approximator and $\chi_1(t)$ the known measured information.

*Figure 6. EFOL scheme in Simulink*

The architecture of EFOL scheme is depicted as *Simulink* block diagram in Figure 6, the inputs of EFOL are $\hat{T}_e$ and $\omega_m$; recall that $\hat{T}_e = c \cdot np \cdot \dfrac{L_m}{L_r}(\hat{\psi}_{dr}i_{qs} - \hat{\psi}_{qr}i_{ds})$ depends on the plant inputs $u(t) = (i_{qs}, i_{ds})$ and the measurable state vector $x(t) = (\hat{\psi}_{rd}, \hat{\psi}_{rq})$. The output estimation error $e_1(t)$, used in the update of parameter estimates $\hat{\theta}(t)$, can be regarded as the output of the EFOL model.

EFOL model consists on two components, 1) the adaptive approximator done with the RBFNN; and 2) the rest of parts, referred to as the estimator, which contains the filters and a priori known nonlinearities $f_0$.

To understand why EFOL is referred to as *'error filtering'* scheme, retake (29), (31),

$$e_1(t) = \frac{\lambda}{s+\lambda}[\hat{f}_L(z(t);\hat{\theta}(t)) - \hat{f}_L(z(t);\theta^*)] - \delta(t)$$
$$= \frac{\lambda}{s+\lambda}[\hat{f}_L(z(t);\hat{\theta}(t)) - \hat{f}_L(z(t);\theta^*) - e_f(z(t))]$$
$$= \frac{\lambda}{s+\lambda}[\hat{f}_L(z(t);\hat{\theta}(t)) - f_L^*(z(t))].$$

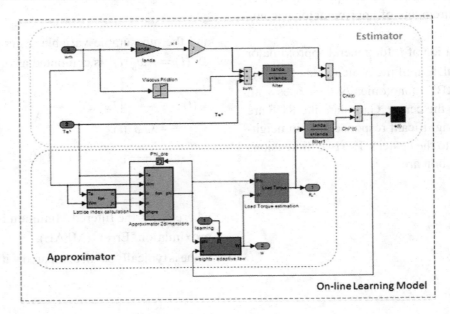

Therefore, $e_1(t)$ is equal to the filtered version of the approximation error $\hat{f}_L(z(t); \hat{\theta}(t)) - f_L^*(z(t))$ at time $t$; thus the term *'error filtering'*. In general the estimation error $e_1(t)$ follows the approximation error signal $\hat{f}_L(z(t); \hat{\theta}(t)) - f_L^*(z(t))$ with some decay dynamics that depend on the value of $\lambda$. It is easy to see that a large value of $\lambda$ the closer the estimation error, $e_1(t)$, will follow the approximation error, $e_f$. On the other hand, in the presence of measurement noise, a large value of $\lambda$ will allow to have a greater effect on the approximation parameters. In Figure 6 $\lambda$ multiplies the measurement $\omega_m$. The reader is refered to Farrell & Polycarpou (2005), chapter 4 for a full methodology understearing.

For the approximation adaptive law derivation, the *Lyapunov synthesis method* is applied with the next standard quadratic candidate function of the output error $e_1$, (33), and the parameter estimation error $\tilde{\theta}$, $\tilde{\theta} = \hat{\theta} - \theta^*$, which implies $\dot{e}_1 = -\lambda e_1 + \lambda \tilde{\theta}^T \varphi(z)$:

$$V(e_1, \tilde{\theta}) = \frac{\mu}{2\lambda} e_1^2 + \tilde{\theta}^T \Gamma^{-1} \tilde{\theta} \qquad (34)$$

with $\mu = 2$ (for simplification) and $\Gamma$ a positive constant to be defined. By taking the time derivative and using the fact that $\theta^*$ is constant implying $\dot{\tilde{\theta}} = \dot{\hat{\theta}}$ we obtain:

$$\dot{V} = -2e_1^2 + \tilde{\theta}^T \Gamma^{-1}(\dot{\hat{\theta}} + \Gamma \varphi(z) e_1) \qquad (35)$$

To obtain desirable stability and convergence, for the IM model, is needed the derivative of $V$ to be at least negative semi-definite. The first term of (35) is negative, while the second term is indefinite; as the sign of the RBFNN weights, $\theta$, are unknown is not possible to force the term $\tilde{\theta}$ to be negative, the best that can be done is force it to zero. If now we define the estimation law as:

$$\dot{\hat{\theta}} = -\Gamma \varphi(z(t)) e_1(t) \qquad (36)$$

Which defines the parameter estimation dynamics, then (35) yields to:

$$\dot{V} = -2e_1^2 \qquad (37)$$

This guarantees that $e_1(t)$ and therefore $\tilde{\theta}(t)$ are bounded, moreover since $\theta^*$ is constant, $\hat{\theta}(t)$ is also uniformly bounded ($\hat{\theta}(t) \in L_\infty$). Also since $\dot{V} \le 0$ and by definition $V(t) \ge 0$, following Lyapunov theory it implies that $V(t)$ converge at some value in a finite time, then $e_1(t)$ is an $L_2$ signal. Assuming that the RBFNN is also bounded, the dynamic error, $\dot{e}_1 = -\lambda e_1 + \lambda k_L \tilde{\theta}^T \varphi(z)$, is also bounded. Since the requirements for Barbălat's Lemma are satisfied;

$$\lim_{t \to \infty} e_1(t) = 0 \qquad (38)$$

moreover, since (38), the convergence of $e$, (37), and the boundedness of the RBFNN, it can be concluded that:

$$\lim_{t \to \infty} \dot{\hat{\theta}}(t) = \lim_{t \to \infty} \dot{\tilde{\theta}}(t) = 0 \qquad (39)$$

showing that the rate of change of the parameters estimation approaches to zero, meaning:

$$\lim_{t \to \infty} \left| \hat{T}_L(t) - T_L^* \right| = 0 \qquad (40)$$

where $\hat{T}_L = \hat{f}_L(\hat{T}_e, \omega_m : \hat{\theta}) + T_{LN}$.

## Rotor Resistance Parameter Estimation

The estimation of the rotor resistance, is developed by NEMA standards (2006) which describes the physical behavior of the rotor resistance governed by the IM temperature, as:

$$\frac{T + 234.5}{R_{r2}} = \frac{T_{st} + 234.5}{R_{rN}} \qquad (41)$$

Remembering that $T_{st}$ is the ambient temperature and $R_{rN}$ the nominal rotor resistance, while $T$ is the actual temperature and $R_{r2}$ is the associated rotor resistance to $T$ temperature. If $T_{st}$, the ambient temperature and $R_{rN}$, the rotor resistance are known, the actual rotor resistance $R_{r2} = \chi_2$, is calculable monitoring the motor temperature, where $\hat{R}_r = \chi_2(t) + R_{rN}$.

$$\chi_2 = \frac{(T + 234.5) \times R_{rN}}{T_{st} + 234.5} \qquad (42)$$

Estimating the rotor resistance, $\hat{R}_r$, the monitoring adaptive model follows the warm-up transient state of the IM.

## DEGRADATION AUTOMATA AND SYSTEM CONDITION STATE

Aging deterioration may slow down the motor causing an increasing stead-state error between the reference value and the actual angular velocity achieved by the IM. In, this case the velocity of the motor is smaller than the nominal value, and therefore, the model (23) will be different than the measured value $\omega_m$.

$$r(t) = \omega_m(t) - \hat{\omega}_m(t) \qquad (43)$$

(43) may be used to detect faults, $r(t)$ is zero for all $t$ in the absence of fault and it becomes nonzero upon occurrence of a system fault (degradation in this case); $r(t)$ is said to be a residual that allows detection of faults on the system; while residu-

als are zero in ideal situations, in practice, this is seldom the case (Gertler, 1988). Then usually are computed some residual bounds or thresholds whereas inside is considered a nominal behavior and outside an off-nominal situation, or fault.

The system degradation is monitored by hybrid automata, which interleave continuous evolution segments and discrete transitions. Continuous evolution corresponds to the progress of time and modifies the continuous state $x$ according to the flow condition of mode $q$. Discrete or model transitions change the mode $q$ and are assumed to be instantaneous. Autonomous mode transitions are labeled by guard condition $G$ which is logical predicates over the continuous state space X. If the continuous state $x$ satisfies the guard condition, then the system transition to a new mode. In general, fault in the autonomous transitions are represented by parameterized guard condition of the form $G(x, \theta_d) \subset x \times \Theta_d$ where $\theta_d \in \Theta_d$ is the *degradation fault parameter;* and $\theta_{d0}$ describes the faultless system. Therefore, a set of sub-spaces $\theta_d \in \Theta_{d=5}$ are used, representing the fault hypothesis $\theta_d \in \Theta_d$, $d \in E$.

The degradation hybrid automata is descrived as:

$$M = (Q, X, \Sigma, Init, E, f, G) \qquad (44)$$

where:

- $Q$ is the set of discrete states or modes $\{q_0, q_1, q_2, q_3, q_4\}$, from: *fully functional*, $q_0$, to *no functionality*, $q_4$, Table 1;
- X as the set of continuos system dynamics, $\{\hat{\psi}_{rd}, \hat{\psi}_{rq}, \hat{\omega}_m\}$;
- $\Sigma$ is a finit set of transitions or events, $\{SC_1, SC_2, SC_3, SC_4\}$ corresponding to the damage index presented in figure 3.
- $Init \subseteq Q \times X$ is the set of initial conditions;

- $E \subseteq Q \times \Sigma \times Q$ is the set of controlled and autonomous discrete transitions;
- $f : \mathbb{R} \times Q \times X \times \Theta \to X$ is the flow condition, descrivied by the set of trajectories showed at (39) and $\Theta : Q \times X$ is the system degradation;
- And, $G : \Theta \times E \to 2^X$ is a partial function that associate by a guard condition, on this case $g_i : g_i\big(r(t) \leq \mathrm{SC}_i\big)$; meaning that when (43) violates $\mathrm{SC}_i$ with $i\ \{1..4\}$, a tranistion occurs, where $\mathrm{SC}_{i\{1..4\}} \in (0,1) \to 1 = failure$, Figure 3. SC acts has a threshold of the adaptive model, $\hat{\omega}_m$, when deviates from measured $\omega_m$.

The domain of the system degradation X is described in Table 1, and the continuous dynamics (23) flows from one mode to another. When transition occurs $\hat{f}_L(z(t);\hat{\theta})$ is on-line adapted in orther to fullfil (43); on-line estimate the new $T_L$ function at the new degraded mode $q_i$, from *fully functional* ($i$=0), no degradation, to ($i$=4) *no functionality*. See Figure 8 for an example.

Figure 7, illustrates the system condition state as hybrid automata, where:

- *Fully functional mode* refers to the electro-mechanical conveyor belt system with no discernible damage.
- A *degraded performance* or *reduced functionality mode*, are the condition system state that exhibits signs of wear and tear due to use or time, the system continues to serve its primary purpose.
- *Functionality severely impinged mode* is one whose intended function has ceased to perform.
- *No functionality mode* or failure is the loose of system functionality.
- *Maintenance tasks* are those that revert the faulted system to *fully functional mode*, the maintenance tasks depends on the *system condition mode*, goes from replacement of components to repairs.

Figure 8 presents a simulation of the full methodology previously presented, where simulation starts once the IM is at thermal steady-state, (~11600s). $\theta_{static}, \theta_{kinetic} \in \Theta$ are increasing following the sigmoid crack growth Paris-Erdogan function, Figure 2. The monitoring degradation mode changes from *fully functional* to *degraded performance*. In this example, assume that the IM testing is under stressful conditions, so that each second of testing is equivalent to 1 minute of

*Figure 7. Degradation hybrid automata*

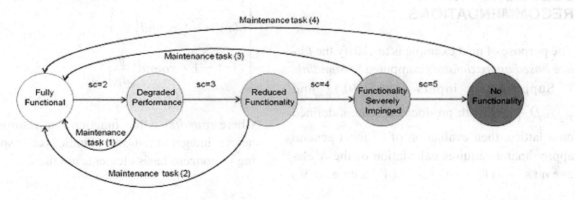

*Figure 8. Conveyor belt intelligent monitoring, IM velocity, learning, Estimated $T_L$, and RBFNN weights*

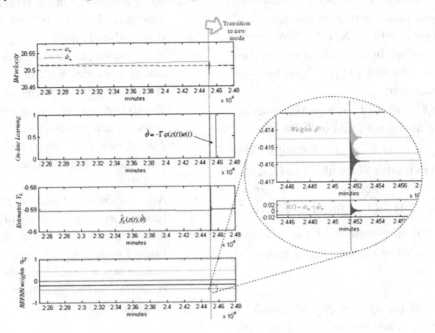

actual use in the field. So that, at the graphics the view time starts at 22500 minutes (375h).

The conveyor belt is under degradation and $\omega_m(t) \neq \hat{\omega}_m(t)$ implying that $r(t) > SC_1$(the residual cross the guard), and the degradation monitoring automata change mode, from $q_0$ to $q_1$; at the new mode, the learning law is activated and $\hat{f}_L$ is on-line estimated, once the residual, $r(t)$, is recovered the learning stops and the automata remains at mode $q_1$ since $r(t)$ cross $SC_2$.

## SOLUTIONS AND RECOMMENDATIONS

The purpose of next example is to clarify the *lattice based approximators* computed in *Simulink*:

Suppose two inputs $x_1 \in D_1\{0,1\}$ and $x_2 \in D_2\{0,1\}$ with pre-located centers defined on a lattice, then evaluation of $\hat{f}$ for a general approximator requires calculation of the $N$ elements of $\phi(x)$, (Figure 9)

$c_m = c_{i,j} = \left( (i-1)\cdot dx, (j-1)\cdot dy \right)$, for $i = 1, ..., n_x$ and $j = 1, ..., n_y$ where $N = n_x n_y$, $m = i + n_x \times (j-1)$, $dx = \dfrac{1}{n_x - 1}$ and

$dy = \dfrac{1}{n_y - 1}$. On that way, Figure 9 presents the lattice diagram with $x$'s as nodal centers $m$ and $*$ indicates an evaluating point, $* = (x, y)$. The appropriate elements of $\phi$ and $\theta$, without directly calculate all $\phi(x)$ can be found by algorithm such as:

$$\begin{cases} i_c(x) = 1 + round\left(\dfrac{x}{dx}\right) \\ j_c(x) = 1 + round\left(\dfrac{y}{dy}\right) \end{cases}$$

where *round(z)* is the function that returns the nearest integer to $z$. the set of indices corresponding to nonzero basis elements are then:

*Figure 9. Example: lattice diagram*

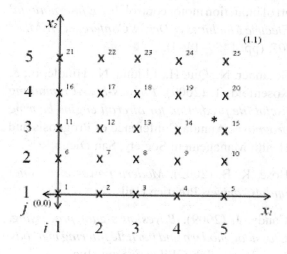

$$(i_c - 1, j_c + 1) \quad (i_c, j_c + 1) \quad (i_c + 1, j_c + 1)$$
$$(i_c - 1, j_c) \quad\quad (i_c, j_c) \quad\quad (i_c + 1, j_c)$$
$$(i_c - 1, j_c - 1) \quad (i_c, j_c - 1) \quad (i_c + 1, j_c - 1)$$

At evaluation point indicated by *, $(ic,jc)=(4,2)$, $m=14$, and the nodal addresses of nonzero basis elements are $\{8,9,10,13,14,15,18,19,20\}$.

On that way the learning algorithm doesn't need update the 25 neurons every step-time just 12 each time, if input data domain $x \notin D\{0,1\}$ the easy way is standardize the data to $\{0,1\}$.

## FUTURE RESEARCH DIRECTIONS

For instance, the approximation of $T_L$ can be easily used to control the IM, furthermore under the *monitoring degraded automata* presented on this chapter, the conveyor belt system can be controlled depending on the *system condition state*, reconfiguring and relaxing the system specifications while the system is wear or damage. Furthermore, the *remaining system life* is divided in subspaces, modes, then the prognosis methods as Particle, Kalman, etc. filters techniques; can be developed to forecast the nearest system condition

transition and estimate the remaining time at the actual degradation mode.

Many advances on diagnosis and prognosis, especially on model-based approaches, have been done recently. Nevertheless, there are some open issues that need to be solved. The first issue concerns the computational complexity of model-based monitoring. Methods that work for a certain number of physic equations become useless when the number is scarcely increased, and the system representation becomes hard to compute and study. An alternative to deal with a high computational complexity is to investigate distributed monitoring techniques. Such techniques are based on the *divide and conquer* principle, transforming a hard problem into several feasible sub-problems. One possibility is to design the conveyor belt simulator as a hybrid system, and split the full belt working cycle in several modes. Departing from idle situation *transition* to different modes like *accelerating mode* where the angular velocity $\omega_m$, is in ramp-up; a *steady-state mode*, where $\omega_m$ remains constant; and, a *decelerating mode*, with the angular velocity behavior as ramp-down, on such a way that the dynamic representation of system will be easier to compute and study.

## CONCLUSION

A new intelligent monitoring method for system condition representation is proposed. It uses 5 consecutive steps, 1) *data acquisition*; which in step 3), *diagnosis*, is compared with a *simplified adaptive mode*, 2); the *degradation hybrid automata* flow among degraded modes tracking the *system condition*, state 4); and, finally EOL or/and RUL forecasting is computed on *prognosis* step, 5). This chapter addresses an approach for intelligent system monitoring based on a simplified adaptive model in which steps 2) and 4) are studied being out of scope the diagnosis and prognosis part.

A general *degradation hybrid automata* to system condition tracking is proposed, it combines a nonlinear adaptive model for continuous dynamics and a system condition guard to transition among degradation modes. EFOL scheme is used to design a parametric model and adaptive low in such a way that the unknown nonlinear systems model part is on-line approximated. Adaptive model output is compared with the measured acquired data to detect any deviation from the nominal system behavior. Once degradation is detected there is a mode change to new degradation stage and the adaptive model is on-line approximate via a RBFNN to estimate the new system dynamics.

The methodology is suitable and validated for the conveyor belt example, a conveyor belt simulator in *Simulink*, based on a real system, is fully developed, where a scalar and vector control strategies are presented. The conveyor belt actuator is modeled as IM *T*-circuit in combination with a thermal model; the degradation is characterized due the Paris-Erdogan crack growth function. Once the simulator is designed the measured current and velocity of the IM are used to modeling the simplified adaptive IM model. EFOL scheme is used to on-line approximate the unknown $T_L$ function. The *simplified adaptive model* estimates the IM velocity, $\hat{\omega}_m$, as output. $\hat{\omega}_m$ and the measured IM velocity $\omega_m$ are compared to detect system degradation. The degradation automata detect a system condition change and the adaptive model on-line approximate the new $T_L$.

## REFERENCES

Aissa, K., & Eddine, K. D. (2009). Vector control using series iron loss model of induction, motors and power loss minimization. *World Academy of Science. Engineering and Technology, 52,* 142–148.

Bocker, J., & Mathapati, S. (2007). State of the art of induction motor control. *IEEE International Electric Machines & Drives Conference, IEMDC '07,* (pp. 1459–1464).

Bolander, N., Qiu, H., Eklund, N., Hindle, E., & Rosenfeld, T. (2009). *Physics-based remaining useful life prediction for aircraft engine bearing prognosis.* Annula Conference of Prognosis and Health Management Society, San Diego.

Bose, K. B. (2002). *Modern power electronics and AC drives.* Prentice Hall.

Candy, J. (2009). *Bayesian signal processing: Classical, modern and particle filtering methods.* New Jersey: John Wiley & Sons, Inc.

Chapman, S. (2005). *Electric machinery fundamentals* (4th ed.). McGraw Hill.

Espinoza-Trejo, D., Campos-Delgado, D., & Loredo-Flores, A. (2010). *A novel fault diagnosis scheme for FOC induction motor drives by using variable structure observers. Industrial Electronics* (pp. 2601–2606). ISIE.

Farrell, J., & Polycarpou, M. (2005). *Adaptive approximation based control, unifying neural, fuzzy and traditional adaptive approximation approaches.* Wiley-Interscience.

Gertler, J. J. (1988). *Survey of model-based failure detection and isolation in complex plants* (pp. 3–11). IEEE Control Systems. doi:10.1109/37.9163

Gucik-Derigny, D., Outbib, R., & Ouladsine, M. (2009). Estimation of damage behaviour for model-based prognostic. *7th IFAC Symposium on Fault Detection and Safety of Technical Processes,* (pp. 1444-1449). Barcelona.

Kalgren, P., Byington, C., & Watson, M. (2006). *Defining PHM, a lexical evolution of maintenance and logistics* (pp. 353–358). Anaheim, California: IEEE AUTOTESTCON. doi:10.1109/AUTEST.2006.283685

Klein, J., & Moeschberger, M. L. (1997). *Survival analysis: Techniques for censored and truncated data*. New York, NY: Springer.

Luo, J., Pattipati, K., Qiao, L., & Chigusa, S. (2008). Model-based prognostic techniques applied to a suspension system. *IEEE Transactions on Systems, Man, and Cybernetics. Part A, Systems and Humans*, 38(5), 1156–1168. doi:10.1109/TSMCA.2008.2001055

Marino, R., Tomei, P., & Verrelli, C. (2010). *Induction motor control design*. London, UK: Springer.

Muller, A., Suhner, M., & Iung, B. (2008). Formalisation of a new prognosis model for supporting proactive maintenance implementation on industrial system. *Reliability Engineering & System Safety*, 93(2), 234–253. doi:10.1016/j.ress.2006.12.004

Ong, C.-M. (1998). *Dynamic simulations of electric machinery using MATLAB Simulink*. Prentice Hall.

Orchard, E., & Vachtsevanos, G. (2007). *A particle filtering-based framework for real-time fault diagnosis and failure prognosis in a turbine engine*. 15th Mediterranean Conference on Control & Automation, Athens, Greece.

Standards. (2006). *Electric motors and generators MG 1-2006*.

Tran, V., Yang, B., & Tan, A. (2009). Multi-step ahead direct prediction for machine condition prognosis using regression trees and neuro-fuzzy systems. *Expert Systems with Applications*, 36, 9378–9387. doi:10.1016/j.eswa.2009.01.007

Uckun, S., Goebel, K., & Lucas, P. (2008). Standardizing research methods for prognostics. *PHM 2008, International Conference on Prognostics and Health Management*, (pp. 1-10).

Wang, W.-J., & Wang, C.-C. (1998). A new composite adaptive speed controller for induction motor based on feedback linearization. *IEEE Transactions on Energy Conversion*, 1–6. doi:10.1109/60.658196

Wasynczuk, O., Sudhoff, S. D., Corzine, K. A., Tichenor, J. L., Krause, P. C., & Hamen, I. G. (1998). A maximum torque per ampere control strategy for induction motor drivers. *IEEE Transactions on Energy Conversion*, 13(2), 163–169. doi:10.1109/60.678980

Wei, P. R. (2010). *Fracture mechanics: Integration of mechanics, materials science & chemistry*. Cambridge, UK: Cambridge University press. doi:10.1017/CBO9780511806865

## ADDITIONAL READING

Bayoudh, M., Travé-Massuyès, L., & Olive, X. (2006). *Hybrid system diagnosability by abstracting faulty continuous dynamics*. 17th International Principles of Diagnosis Workshop.

Beebe, R. S. (2004). *Predictive maintenance of pumps using condition monitoring*. Elsevier - Technology and Engineering.

Bemporad, A. (2006). *Master course tutorial: Model predictive control of hybrid systems*. Barcelona, Spain: Terrassa.

Bemporad, A., & Morari, M. (1999). Control of systems integrating logic, dynamics, and constraints. *Automatica*, 35(3), 407–427. doi:10.1016/S0005-1098(98)00178-2

Blanusa, B. (2010). *New trends in technologies: Devices, computer, communication and industrial systems*. InTech.

Brown, R. B. (1963). *Smoothing forecast and prediction*. Englewood Cliffs, NJ: Prentice-Hall.

Chaohai, Z., Zongyuan, M., & Qijie, Z. (1994). On-line incipient fault detection of induction motors using artificial neural networks. *Proceedings of the IEEE International Conference on Industrial Technology*, (pp. 458 - 462).

Chow, M. Y. (1997). *Methodologies of using neural network and fuzzy logic technologies for motor incipient fault detection*. Singapore: World Scientific. doi:10.1142/9789812819383

Cupertino, F., Giordano, V., Mininno, E., & Salvatore, L. (2005). Application of supervised and unsupervised neural networks for broken rotor bar detection in induction motors. *2005 IEEE International Conference on Electric Machines and Drives*, (pp. 1895-1901).

Elsaadawi, A., Kalas, A., & Fawzi, M. (2008). Development of an expert system to fault diagnosis of three phase induction motor drive system. *12th International Middle-East Power System Conference, MEPCON 2008*, (pp. 497 - 502).

Ferrari, R., Parisini, T., & Polycarpou, M. (2008). A robust fault detection and isolation scheme for a class of uncertain input-output discrete-time nonlinear systems. *American Control Conference, 2008*, (pp. 2804 - 2809). Seattle, WA.

García, C., Escobet, T., & Quevedo, J. (2010). PHM techniques for condition-based maintenance based on hybrid system model representation. *Annual Conference of the Prognostics and Health Management Society*, (pp. 1-8).

Garcia, C., & Quevedo, J. (2011). An accelerated aging and degradation characterization of V/F controlled induction motor. *ICAT 2011 XXIII International Symposium on Information, Communication and Automation Technologies*, (p. 7). Bosnia & Herzegovina, Sarajevo.

Gebraeel, N., Lawley, M., Liu, R., & Parmeshwaran, V. (2004). Residual life predictions from vibration-based degradation signals: A neural network approach. *IEEE Transactions on Industrial Electronics*, 694–700. doi:10.1109/TIE.2004.824875

Heng, A., Tan, A., Mathewa, J., Montgomery, N., Banjevic, D., & Jardine, A. (2009). Intelligent condition-based prediction of machinery reliability. *Mechanical Systems and Signal Processing, 23*, 1600–1614. doi:10.1016/j.ymssp.2008.12.006

Katipamula, S., & Brambley, M. R. (2005). *Methods for fault detection, diagnostics, and prognostics for building systems-A review, Part I* (pp. 3–25). HVAC&R RESEARCH. doi:10.1080/10789669.2005.10391123

Korbicz, J., Koscielny, J. M., & Kowalczuk, Z. (2004). *Fault diagnosis: Models, artificial intelligence, applications*. Springer.

Krstic, M., Kanellakopoulos, I., & Kokotovic, P. (1995). *Nonlinear and adaptive control design*. John Wiley & Sons, Inc.

Lee, B.-S., & Krishnan, R. (1998). Adaptive stator resistance compensator for high performance direct torque controlled induction motor drives. *Thirty-Third IAS Annual Meeting, The 1998 IEEE Industry Applications Conference*, (pp. 423-430). St. Louis, MO, USA.

Lee, J., Ni, J., Djurdjanovic, D., Qui, H., & Liao, H. (2006). Intelligent prognostics tools and e-maintenance. *Computers in Industry, 57*, 476–489. doi:10.1016/j.compind.2006.02.014

Lygeros, J. (2004). *Lecture notes on hybrid systems*.

Ma, Z. (2009). A new life system approach to the prognostic and health management (PHM) with survival analysis, dynamic hybrid fault models, evolutionary game theory, and three-layer survivability analysis. *2009 IEEE Aerospace Conference,* (pp. 1-20). Big Sky, MT.

Martins, J., Pires, V., & Pires, A. (2007). Unsupervised neural-network-based algorithm for an on-line diagnosis of three-phase induction motor stator fault. *IEEE Transactions on Industrial Electronics,* 259–264. doi:10.1109/TIE.2006.888790

Melero, M. G., Cabanas, M. F., Rojas, C., Orcajo, G. A., Cano, J. M., & Solares, J. (2003). Study of an induction motor working under stator winding inter-turn short circuit condition. *IEEE International Symposium on Diagnostics for Electrical Machines, Power Electronics and Drivers SDEMPED,* (pp. 24-26). Atlanta, USA.

Mobley, R. K. (2002). *An introduction to predictive maintenance.* USA: Elsevier Science.

Mouna, B. H., Lassaâd, S., & Mustapha, M. (2010). A robust adaptive control algorithm for sensorless induction motor drives. *ICGST-ACSE Journal, 10*(1), 61–69.

NEMA. (2007). *Electric motors and generators (Revision) MG 1-2007.*

Polycarpou, M., Ellinas, G., Kyriakides, E., & Panayiotou, C. (2010). *Intelligent health monitoring of critical infrastructure systems* (pp. 18–20). IEEE Computer Society. doi:10.1109/COMPENG.2010.46

Polycarpou, M. N., & Helmicki, A. J. (1995). Automated fault detection and accomodation: A learning system approach. *IEEE Transactions on Systems, Man, and Cybernetics, 25*(11), 1447–1458. doi:10.1109/21.467710

Quang, N., & Dittrich, J.-A. (2008). *Vector control of three-phase AC machines.* Springer.

Schoen, R., Lin, B., Habetler, T., Schlag, J., & Farag, S. (1995). An unsupervised, on-line system for induction motor fault detection using stator current monitoring. *IEEE Transactions on Industry Applications,* 1280–1286. doi:10.1109/28.475698

Sheppard, J., Wilmering, T., & Kaufman, M. (2008). *IEEE standards for prognostics and health management.* Salt Lake City: IEEE AUTOTESTCON.

Shing, J., & Jang, R. (1993). ANFIS: Adaptive-network-based fuzzy inference system. *IEEE Transactions on Systems, Man, and Cybernetics, 23*(3), 665–685. doi:10.1109/21.256541

Stone, G., Boulter, E. A., Culbert, I., & Dhirani, H. (2004). *Electrical insulation for rotating machines.* IEEE Press Series on Power Engineering.

Su, H., & Chong, K. T. (2007). Induction machine condition monitoring using neural network modeling. *IEEE Transactions on Industrial Electronics,* 241–249. doi:10.1109/TIE.2006.888786

Tavner, P. (2008). Review of condition monitoring of rotating electrical machines. *IET Electric Power Applications, 2*(4), 215–247. doi:10.1049/iet-epa:20070280

Tavner, P., & Penman, J. (1987). *Condition monitoring of electrical machines.* Letchworth, UK: Research Studies Press and John Wiley & Sons.

Tian, X., Cao, Y. P., & Chen, S. (2011). Process fault prognosis using a fuzzy-adaptive unscented Kalman predictor. *International Journal of Adaptive Control and Signal Processing,* 813–830. doi:10.1002/acs.1243

Vas, P. (1996). *Parameter estimation, conidtion monitoring and diagnosis of electric machines.* Oxford, UK: Clarendon Press.

Vas, P. (1999). *Artificial-intelligent-based electrical machines and drives. Application of fuzzy, neural, fuzzy-neural and genetic-algorithm-based techniques*. Oxford, UK: Oxford Univ. Press.

Vemuri, A. T., & Polycarpou, M. M. (1997). Neural-network-based robust fault diagnosis in robotic systems. *IEEE Transactions on Neural Networks*, *8*(6), 1410–1420. doi:10.1109/72.641464

Verghese, G. C., & Sanders, S. R. (1988). Observers for flux estimation in induction machines. *IEEE Transactions on Industrial Electronics*, *35*, 85–95. doi:10.1109/41.3067

Vichare, N., & Pecht, M. (2006). Prognostics and health management of electronics. *IEEE Transactions on Components and Packaging Technologies*, 222–229. doi:10.1109/TCAPT.2006.870387

Yu, M., Wang, D., Luo, M., & Huang, L. (2011). *Prognosis of hybrid systems with multiple incipient faults: Augmented global analytical redundancy relations approach* (pp. 540–551). Systems and Humans, IEEE Transactions on Systems, Man and Cybernetics, Part A. doi:10.1109/TSMCA.2010.2076396

Zhang, H., Kang, R., & Pecht, M. (2009). A hybrid prognostic and health management approach for condition-based maintenance. *IEEE International Conference on Industrial Engineering and Engineering Management, IEEM 2009*, (pp. 1165-1169). Hong Kong.

## KEY TERMS AND DEFINITIONS

**Analytical Redundancy Relations (ARR):** Are equations that are deduced from an analytical model and which solely involve measured variables.

**Condition Based Maintenance (CBM):** Is a maintenance equipment based on an estimation of it condition and maintenance logistics; only executes repairs when objective evidence indicates the need for such action.

**Detect:** Recognize that a monitored or modeled parameter(s) has departed its normal operating envelope or threshold.

**Diagnose:** Identify, localize and determine the severity of an evolving fault condition.

**Diagnosis:** In our application domain, the disease is evolving failure in the equipment and the observation of signs and symptoms is accomplished by test, measurement, and reasoning on the results.

**Incipient Fault:** The earliest stage of a condition change, when it's just beginning to come into being or become apparent that will ultimately progress to functional failure.

**On-Line Learning:** The learning is into operation.

**Prognosis:** Reliability and accurately forecast the remaining useful life or risk to complete a planned mission.

**Reamining Useful Life:** The remaining time forecasted until failure.

**System Monitoring:** Device or program used to monitor, observe and check a system over a period of time.

# APPENDIX

*Table A1. Conveyor belt data*

| | | |
|---|---|---|
| $R_s$ | stator resistance | (14 Ω) |
| $R_r$ | rotor resistance | (13Ω) |
| $i_s$ | stator current | |
| $i_r$ | rotor current | |
| $\psi_r$ | rotor flux linkage | |
| $n_p$ | number of pole pairs | 3 |
| $\omega_m$ | motor angular velocity | (900rpm) rated |
| $\omega_r = n_p \omega_m$ | rotor angular velocity | |
| $u$ | voltage input | |
| $\rho$ | angle of rotation | |
| $Lm$ | mutual inductance | (0.45 H) |
| $L_s = L_{ls} + L_m$ | stator inductance | (0.452 H) |
| $L_r = L_{lr} + L_m$ | rotor inductance | (0.452 H) |
| $J$ | rotor inertia | (0.025 Kgm²) |
| $B$ | rotor viscous coefficient | (0.1 Nms/rad) |
| $T_L$ | load torque | (rated 1.72 Nm) |
| $T_e$ | electromagnetic torque | (rated *power 250W*) |
| $T$ | temperature | |
| $T_{st}$ | ambient temperature | (20°C) |
| $V_{belt}$ | belt lineal velocity | |
| $D_P$ | damping factor | (100e³ Ns/m) |
| $R_p$ | roller radius | (0.01275 m) |
| $R_w$ | conveyor belt load | (0.80 Ns/m) |
| $m_{belt}$ | belt mass | (24 kg) |
| $R_O$ | dynamic viscous coef. | (0.01 Nms/rad) |
| $R_S$ | roller stiffness | (105e⁻⁶ N/m) |
| $mc$ | thermal capacity | (1500 J/K) |
| $R$ | thermal resistance | (0.8 K/W) |

$$\begin{bmatrix} X_d \\ X_q \end{bmatrix} = \begin{bmatrix} \cos \rho & \sin \rho \\ -\sin \rho & \cos \rho \end{bmatrix} \cdot \begin{bmatrix} X_a \\ X_b \end{bmatrix}$$

$$\begin{bmatrix} X_a \\ X_b \end{bmatrix} = c \cdot \begin{bmatrix} 1 & -\dfrac{1}{2} & -\dfrac{1}{2} \\ 0 & \sqrt{\dfrac{3}{2}} & -\sqrt{\dfrac{3}{2}} \end{bmatrix} \cdot \begin{bmatrix} X_R \\ X_S \\ X_T \end{bmatrix}$$

# Section 3
# Data Driven Diagnostics

# Chapter 3
# Principles of Classification

**Veli Lumme**
*Tampere University of Technology, Finland*

## ABSTRACT

*This chapter discusses the main principles of the creation and use of a classifier in order to predict the interpretation of an unknown data sample. Classification offers the possibility to learn and use learned information received from previous occurrences of various normal and fault modes. This process is continuous and can be generalized to cover the diagnostics of all objects that are substantially of the same type. The effective use of a classifier includes initial training with known data samples, anomaly detection, retraining, and fault detection. With these elements an automated, a continuous learning machine diagnostics system can be developed. The main objective of such a system is to automate various time intensive tasks and allow more time for an expert to interpret unknown anomalies. A secondary objective is to utilize the data collected from previous fault modes to predict the re-occurrence of these faults in a substantially similar machine. It is important to understand the behaviour and functioning of a classifier in the development of software solutions for automated diagnostic methods. Several proven methods that can be used, for instance in software development, are disclosed in this chapter.*

## INTRODUCTION

Diagnostics of condition related data collected from machines requires the understanding of the machine's operation, fault modes and symptoms. Symptoms are commonly understood as human perceptions, or more often as changes in descriptor values that are descriptive of various fault modes.

For a particular fault mode, some symptoms need to be present, some might be and others should not be in order to reach a specific conclusion.

A single descriptor value may change without any indication of machine deterioration. This brings up two concerns. First, it may take a significant amount of an analyst's time and effort to detect and verify a single anomaly. Second, the

DOI: 10.4018/978-1-4666-2095-7.ch003

change in a descriptor value often is not linearly related to the severity of the fault. These problems are emphasized in most diagnostics systems that rely on the trending of single descriptors only.

Diagnostics is mainly based on an analyst's knowledge and experience of various fault modes in various machines. The diagnostic process usually begins, when an anomaly has been observed. Systems have been developed to detect anomalies automatically starting more than fifteen years ago (Milne, 1996). These systems typically lack the ability to adapt to variations in measured or calculated parameters that are caused by changes in the environment or operational parameters of the machine. In many cases the systems rely on known rules that need to be created and updated.

In condition monitoring, classification can be understood as the prediction of a label for any new data based on training data collected earlier. In the case of a single or only a few descriptors, classification is trivial and can easily be done manually or using simple arithmetic. When we are studying the machine condition, there are numerous descriptors related to various types of machines and fault modes. In such a multivariate environment, more effective tools are required (Lumme, 2011).

## BACKGROUND

A symptom is a perception made by means of descriptors, which may indicate the presence of one or more faults with a certain probability. If a descriptor value remains stable, there is no a symptom and therefore no indication of a fault. A descriptor as such is therefore not a good indicator of a machine condition, but a symptom will more likely react to a change in descriptor magnitude.

Descriptors indicating the operation or condition of a machine are derived from the measurement data. These descriptors may be either directly measured quantities or extracted properties, derived or calculated from these quantities,

characteristics such as vibration, rotation speed, temperature, pressure and so on. A characteristic vector, i.e. syndrome descriptive of the current operational state of the machine concerned, is formed from these symptoms. From characteristic vectors describing the same operational state, classes are formed so that in each class all characteristics have an allowed range of variation. The classes describing different operational states of the machines are stored to form a database.

To examine the operation of an individual object, a characteristic vector descriptive of its current operational state is formed and the vector is compared to the database, which contains stored classes. If a vector being examined corresponds to a class describing a fault mode, an alarm is activated. Similarly, an alarm is activated, if no class corresponding to this vector can be found in the database, because in this case the object being examined is in a completely new and unknown operational state, which has to be immediately analysed to establish whether it is a normal and acceptable state or a new fault mode.

## CLASSIFICATION

Classification is the process of finding a set of models that describe and distinguish data classes or concepts in order to use the model to predict the class of objects whose class label is unknown. The derived model is based on the analysis of a set of training data (Han & Kamber, 2001, p. 24).

For presentation purposes all figures in this chapter show data in two dimensions. In reality the presentation shall be extended to a multi-dimensional hyperspace. The number of dimensions represents the number of descriptors or symptoms used. Depending on the object under investigation and on the potential fault modes, the number of symptoms might be more than twenty and can even be over one hundred in some cases.

Various classifiers are available for prediction. These include Bayesian, statistical, fuzzy logic and neural network based classifiers, but their differences are not discussed in this chapter. The general principles of classification apply to all of them more or less. There are, however, two general categories of classifiers: supervised and unsupervised. A supervised learning algorithm requires that the output, also referred to as a label or an interpretation, of the training data is known. The creation of classes is based both on the similarity of the training data samples and their output. In unsupervised learning the classes are created entirely based on the similarity of the data samples. Kohonen (2001) explains the principles of both types of classifiers in details using self-organizing maps and learning vector quantification as examples.

## Training of Classifier

The training algorithm attempts to organize all similar data samples into the same class. The similarity is typically measured by the Euclidean distance between the data samples. During the training, the data samples adjust the weight vector coordinates through an iterative process towards the weight centre of all data samples used for training of a particular class. See Figure 1.

*Figure 1. Weight centre of data samples*

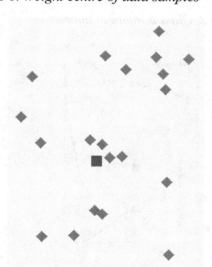

After training the classifier consists of a set of classes. Each class has a characteristic weight vector, which will be used to search for the best matching class for any new data. A classifier that was trained using supervised learning will have all classes labelled. For a classifier trained with unsupervised learning the classes should be labelled manually. Typically in the initial training phase all class labels should present a normal mode only, unless a known fault mode has been included in the training data set. In that case, the classes that were trained using these data samples should be labelled accordingly.

Once a set of classes has been created, the label of any new data can be predicted by locating the best matching class. Figure 2 shows an example on a set of ten classes with their weight vectors. The lines mark the approximate borders between the classes. Any data vector within borders will have the shortest distance to the same weight centre and therefore will have the same label. Clearly the presentation is much more complicated in a multi-dimensional hyperspace, where the hyperplanes are used instead of border lines.

Figure 2 suggests that for any new data sample, a best matching class can always be found by locating a respective weight centre with a minimum distance to the data sample. However, one

*Figure 2. Borders between classes*

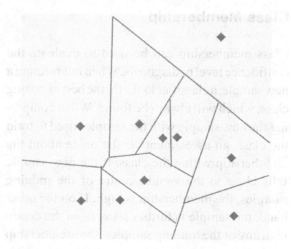

should keep in mind that the new data sample might still localize at a distance, which is rather far from the weight centre of the best matching class. Such a data sample does not belong to a data set used to train the class or any other known class.

Figure 3 presents an example of a class that has been trained using the twenty data sets shown. The data samples with a short distance to the weight centre have a high similarity. A contour has been drawn through the furthest data vectors. Such a contour would probably be an excellent class border in a hyperspace, but is difficult to implement. Another solution for a class border is the use of a rectangle in hyperspace. The shape and size of such a rectangle shall be defined by the minimum and maximum symptom values in the data set. As an easy solution one could use the hypersphere with the radius at the maximum distance of all training data samples and centred at the weight vector. Any new data vector falling within the hypersphere or close to it would have the same class label. Any data vector clearly outside the hypersphere represents an anomaly meaning that similar data has not yet been presented to the classifier. Data vectors close to the surface of the hypersphere might belong to the same class at some confidence level or are anomalies. A closer analysis at the data should reveal, if the data should have the same label.

## Class Membership

Class membership can be used to evaluate the confidence level of diagnosis. When interpreting a new sample, a classifier looks for the best matching class, which will always be found. When comparing the new sample with the samples used to train the class, an assessment can be made about the membership within the class. If the new sample falls close to the weight centre of the training samples, the membership is high. If on the other hand, the sample is further away from the centre than any of the training samples, the membership

is low. Taken that the classifier has been accurately labelled, the confidence level of diagnosis would then be high or low consequentially.

Membership functions are typically used in fuzzy logic, but let's define a useful function for classification purposes. We may take a linear approach by defining that we have a maximum (1.0) membership (confidence of diagnosis) at the centre and a pre-defined (for instance 0.5) membership at the maximum distance. This can be expressed as follows:

$$y = \max(1 - c\frac{s}{s_{max}}; 0) \qquad (1)$$

where $y$ is the membership value, $c$ is the membership constant, $s$ is the new sample's Euclidean distance from the median and $s_{max}$ is the Euclidean distance of furthest training sample from the weight centre.

When $s$ is zero, i.e. the sample is at the centre, the membership value will be 1. When $s$ is $s_{max}$, the function will return 1-c as the membership value. For any distances greater than $s_{max}/c$, the membership will be zero. A value of 0.5 is recommended for the membership constant. This will result in

*Figure 3. Class borders to anomalies*

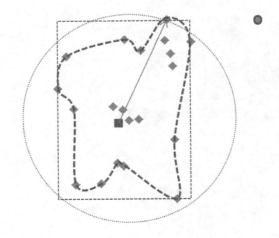

a 0.5 membership at $s_{max}$ and zero membership at distances longer than two times $s_{max}$.

A linear model is not ideal, because it results in relatively high membership values at long distances from the centre and low membership values quite close to the centre. Instead a sigmoid function expressed in the following equations is more appropriate.

$$y\ (t) = \frac{1}{1 + e^{-t}} \tag{2}$$

$$t = c \times (1 - \frac{s}{s_{max}}) \tag{3}$$

where $c$ is a skew factor, $s$ the distance from median and the $s_{max}$ the maximum distance from centre.

A skew factor of 5 would yield a membership function given in Figure 4. The function gives a membership value of 0.99 at the centre, 0.5 at the maximum distance and 0.01 at two times the maximum distance.

## Multivariate Classifier

The previous figures used two-dimensional presentations of data for simplicity. Figure 5 highlights the complexity of a classifier in three dimensions.

Additional dimensions are difficult to show on a two-dimensional paper. The numerous spheres represent the various classes created through training. The radiuses of the spheres represent the distances of the furthest training data from the sphere's centre point. Some of them form clusters and might intersect. However, a hyperplane can always be found, which separates the two close classes from each other. On the other hand, the closely neighbouring classes likely represent data with the small variations only. In reality the fault modes will show in a low membership, i.e. in a high novelty rate, to any class labelled as normal, if the descriptors are correctly selected.

Clusters define a group of classes with close resemblance. Empty spaces between the clusters indicate still "unknown classes" that have not been trained with any data samples. These clusters might be difficult to distinguish in a low dimensional presentation.

## Class Label

For machine diagnostic purposes the class label should be as exact and descriptive as possible. For all classes that are related to the normal mode without any early indications of beginning fault modes, the simple label "normal" or no label may suffice, but for other classes the label should include both

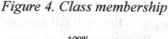

*Figure 4. Class membership*

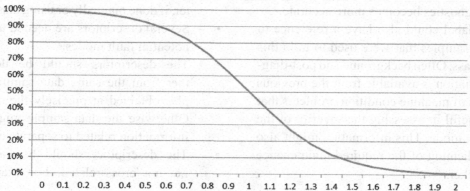

*Figure 5. Three-dimensional classifier*

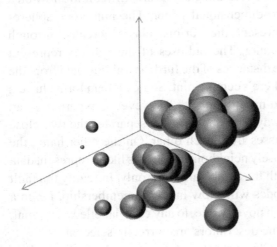

the fault modes and the severity of the fault. One should keep in mind that the characteristics of a data sample may be caused by more than one fault at a time. Condition monitoring typically reveals first a primary fault, for instance an imbalance that causes a secondary fault, such as a bearing failure. Symptoms of both fault modes appear in the data.

The fault severity as a part of the class label can be expressed in several ways. Fuzzy definitions, such as "moderate", "serious" and "extreme" are often considered practical, but in many cases prognosis of the remaining operation time is more informative. The prognosis should include the confidence level, for instance "the machine can be safely operated 30 days at 80 per cent confidence". As more data becomes available, the confidence level of prognostics becomes more accurate.

A class label could also have a reference to earlier data samples that were used to train this particular class. Often background and post-diagnosis information is available from the previous occurrences of machine condition problems that might be useful in assessing the severity of the current fault mode. This information might also guide the planning of the required maintenance actions quicker.

## DATA PRE-PROCESSING

There are several general aspects that need to be considered in data analysis. These include missing or erroneous data, missing feature values, irrelevant data, random values, outliers etc. The data analysis may lead to wrong conclusions, if these aspects are not recognized and handled properly. An additional concern is the presentation of features in two or more different units. In condition monitoring applications, we may wish to analyse vibration velocity and acceleration values or vibration and process values simultaneously. Some feature values may vary, for instance between 0.1 and 5, and others between 20 and 100. Without data pre-processing the lower values will probably lose their significance in the analysis. In order take advantage of all feature values the data should be normalized.

One of the goals of classification is to generalize the data collected from different objects. This would allow usage of the data related to a specific fault in a single machine in order to benefit the other machines in comparable conditions. Machines are individual and the feature values may not be the same, even if the conditions are equivalent. Use of symptom values instead of descriptor values will solve this problem. The following matters should be considered, when selecting the descriptors:

- A descriptor should be sensitive to change, when machine state deteriorates. If this is not the case, the prediction will fail no matter, if done manually or using classification.
- Several descriptors are needed to cover all potential fault modes.
- The descriptors should be derived either from the same data sample or samples obtained at precisely the same time. Otherwise the data sample might contain information related to separate events.
- The descriptors should be independent, i.e. not cross-related. For example, two

highly correlated descriptors have as much information as just a single, independent descriptor.

- The descriptor definitions may need to be overlapped. A non-significant variation of a parameter value should not cause a major increase or decrease in a descriptor value

- Some descriptors should be used as explanatory variables only. For instance the running speed or load are not descriptors of machine's condition, but can be used to explain the variations in descriptor values.

In an ideal situation a complete set of data including all faults in various progressions for all machines is available. In this way the system could be trained to identify all potential operation and fault modes. In practice this is impossible and especially data related to fault modes will typically be missing in the beginning. Also it is likely that all the normal modes cannot be experienced during the initial training period. Due to the ability to retrain the system, this is really not a major problem. Whenever data that was initially missing is tested with the system, it will appear as an anomaly and as such attention will be drawn to it. The generalization of data and classes can be used so that each machine does not have to experience all potential failure modes in order for the faults to be identified.

Errors in the data might have been generated during the data acquisition and collection or symptom extraction. The use of erroneous data is difficult to handle. In fact the classifier as such has no means to detect errors in the data. The errors may, however, cause anomaly detection and would be investigated more closely. Data values should be justified during the pre-processing. Any abnormal set of data should be discarded from further processing. In a condition monitoring application the changes in data values between successive measurements are typically small. Even during a fault progression, the feature values change slowly depending on the measurement interval. If a significant change or deviation is detected, a re-measurement could perhaps be made. If the change is not permanent, the data should be discarded.

Some of the descriptor values might be missing from the data set because of various reasons, such as a faulty sensor or cable. It is not always desirable to discard the whole data set because of this. These descriptor values must not be replaced by zero, which is a definite value. This would result in errors during data analysis. A Not a Number value (NaN) can safely be used, if a descriptor value is not available. Later during the normalisation process a descriptor value NaN can be replaced with a symptom value zero, which defines that this particular descriptor value has not changed in this data sample. In other words, a mean value is assumed instead of the missing value.

Sometimes a collected value appears to be random. This could happen because of a measurement error or a sudden distortion in the environment. An outlier value deviates significantly from the successive values taken in comparable situations in the same position. If the value is not permanent, it should be re-taken or replaced with a NaN value. Novelty detection can be a result of an outlier, when several of the values deviate from the average values.

The values may be imprecise for several reasons. In the tested data, some of the values suffered from poor dynamic range. This caused low amplitude values to fluctuate relatively strongly from the average value. A small change in the actual value causes a significant change in the collected value. This may result in unexpected classification results depending on the type of normalization.

Too precise values may also induce a problem. In many applications the variations in data values are random and caused by noise. If the conditions are otherwise stable, the only variation in the data might be a consequence of an external factor. Because of this, the classifier might try to build classes based on randomness.

Some of the derived descriptor values may be extremely low, close to zero. In such a case the only variation is random and does not depend on changes in the machine operation or condition. Such values should not be used as an input to the classifier. It is recommended all values below a specified threshold value are rounded to zero.

In some cases the collected data may include samples that are not representative of to the current condition of a machine. In particular this happens at low speeds. The system can handle irrelevant data, but neurons (classes) in the classifier will be unnecessarily occupied.

The purpose of normalisation is to isolate statistical error in measured data by making it commensurable. Normalization refers to the division of multiple sets of data by a common variable in order to negate that variable's effect on the data, thus allowing underlying characteristics of the data sets to be compared. This allows data on different *scales* to be compared, by bringing them to a common scale.

It has an effect on several aspects:

- Values in different units
- Values with different ranges of change
- Symptom creation
- Generalization

Berthold & Hand (2003) advise to usage of the standard score method with principal component analysis. In statistics, a standard score indicates, how many standard deviations an observation is above or below the mean. It is a dimensionless quantity derived by subtracting the population mean from an individual raw score and then dividing the difference by the population standard deviation. The data can be normalized with a standard score by carrying out the following steps.

Centre each variable by subtracting the mean of each variable to give

$$\dot{x}_j = x_j - \bar{x} \tag{4}$$

Divide each element by its standard deviation

$$z_{ij} = \frac{x_{ij}}{s_i} \tag{5}$$

The method in fact makes data commensurable by being dimensionless and equalizing the ranges of changes. With the centring of the variables the original feature values $x_{ij}$ turn to symptom values $z_{ij}$.

When converting feature values to symptom values one should consider, which data population should be used to calculate the mean and standard deviation of each variable. A logical solution would be to use the initial data taken, when the machine was running in a normal mode before any changes caused by developing faults. For convenience it is much easier to use the exact training population when calculating the mean and standard deviation for each variable.

In a practical application, the normalization of data needs to be performed before the classifier is applied. The offset (mean) and scale (standard deviation) shall be saved before the classifier is created. After the training process, the weight vectors, minima and maxima of each variable can be restored to the original values, if required. They can also be used as normalized values, in which case the new data samples should be normalized accordingly before using the classifier for prediction. Upon retraining of the classifier the entire normalization process has to be repeated with a new population of data.

## CONTINUOUS LEARNING

### General

The following sections present the various operations, where classification is used as a part of continuous learning. In this particular case more than a hundred various vibration descriptor values were taken from a gearbox every 90 minutes.

A self-organizing map (SOM), which is an unsupervised learning algorithm, can be used to create the classifier. SOM creates a two-dimensional presentation of multi-dimensional data as shown in the following illustrations. For the creation of classifiers in this chapter a special variation of SOM, namely a tree-structured self-organizing map (TS-SOM) was used (Lensu, 2002). TS-SOM is a hierarchical classifier, where each upper level class has four descendants as illustrated in Figure 6. The number of layers actually defines the number of available classes to the power of four, i.e. 1, 4, 16, 64, 256… The required number of classes depends mainly on the divergence of the data. In a condition monitoring application there should be enough classes to correlate with the number of various fault modes and severities. In general TS-SOM offers faster processing due to the smaller amount of hierarchical searches for the best matching class. This is accomplished by locating first the best of four descendants on the second highest layer. The next search is aimed at the descendants of the previous class and so on until the lowest layer has been reached.

A self-organizing map provides the following information for each of the trained classes:

- Cclass reference (number, label)
- Number of samples used to train the class
- Euclidean distance of the furthest sample from the weight centre
- Weight centre as a set of symptom weights
- Minimum values in the data set for each symptom
- Maximum values in the data set for each symptom

## Initial Training

Before the initial training, a reasonable number of data samples should be collected and the descriptors derived from the data. The volume of the data

should preferably include data from the various operational modes of the machine, but data can also be added through a retraining process, if such data later becomes available. Depending on the machine's operational cycles the collection of data may take from a few hours to several weeks or months. For statistical reasons the number of data samples should exceed 100. Typically, data from a machine running in a fault mode is not available for initial training. This can also be added later through retraining.

The map in Figure 7 was created bases on a set of 2736 data samples from a machine in good condition. A five layer classifier with 256 classes was selected. Before the training operation, all data samples were normalized using the standard score method presented earlier. Due to the initialization algorithm the map tends to be organized so that the classes with high weight vector norms are located in the top left corner, while those with low weight vector norms are in the opposite corner. The weight vectors in adjacent classes are in the same order of magnitude.

Some of the classes in the map have not learned from a single data sample. These classes have no value in classification and should be excluded in the search for the best matching class. They have

*Figure 6. Tree-structured self-organizing map*

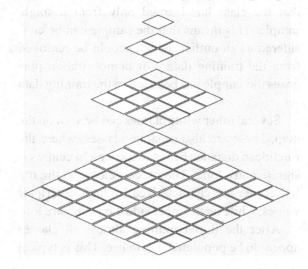

*Figure 7. Map after initial training*

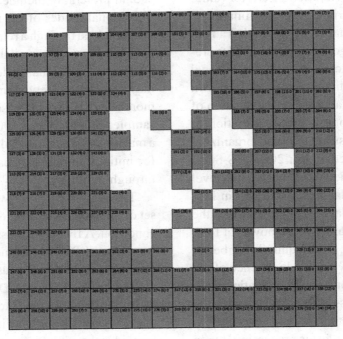

been marked with a white background on the map. It is, however, worth noting that these classes typically separate classes that deviate significantly from their neighbours. For instance it can be seen that the class in the top left corner is totally isolated, surrounded by non-learned classes. One can conclude that the class represents a significant deviation in the weight centre in comparison with the other classes. It can also be seen that the class has learned only from a single sample. This means that the sample can be considered as an outlier, which should be removed from the training data. For demonstration purposes the sample has been left in the training data set.

Several other white classes can be seen on the map. These are also separate classes, where the Euclidean distance between the weight centres is significantly long. The classes closer to the top left corner are characterized by high descriptor values, while on the other side the values are low.

After the initial training almost all classes appear to be populated, i.e. in use. This is typical even if the initial training data is homogenous. As the classes are always used effectively, it may look like there is no space left for any retraining of anomalies or known fault modes. The classifier will, however, adapt to the data volume, when new data is appended to the training data. This means that a smaller number of classes will be used to represent slight deviations between adjacent data samples. One might say that the classifier becomes less accurate. On the other hand, in most cases the symptoms should have been selected carefully so that any changes in the machine condition would cause a major change in the symptoms.

## Prediction

While classification algorithm creates the classification rules, prediction can be viewed as the construction and use of a model to assess the class of an unlabelled sample or to assess the value or value ranges of an attribute that a given sample is likely to have (Han & Kamber, 2001).

After the initial training the classifier is ready to be used for prediction of any new samples. Before the actual prediction the symptom values should be normalized using the same scale and offset as with the initial training data, or the weight factors of all classes should be de-normalized. During prediction, the Euclidean distance between the data sample and the weight centre of each class is calculated, and the one with the minimum distance shall be selected as the best matching class.

After the initial training the prediction basically returns information, whether the tested samples have known labels or if they are anomalies. This information can be expressed in a simple method, for instance using traffic light signals, where green indicates that the object is in a known good condition, yellow that an anomaly has been detected or red that a known fault condition has occurred. The prediction could first be performed on the training data, which should result in an equal amount of prediction hits and training samples in the classes and no anomalies. If this is not the case, the classifier might be inaccurate.

While the borders between classes are defined by the closest distances, a border with an anomaly is a fuzzy definition. If class membership is high, the tested sample belongs to the same class and will have the same label. Low membership represents an anomaly. A class border has to be drawn somewhere. If incorrectly set, an under-diagnosis or over-diagnosis may happen. Over-diagnosis

occurs when a machine fault is diagnosed correctly, but the diagnosis is irrelevant. A correct diagnosis may be irrelevant because the severity of the fault is not significant and requires no immediate actions. Under-diagnosis occurs as a failure to recognize or correctly diagnose a fault or condition.

Figure 8 shows the so called anomaly index against time for a certain data block. Anomaly index is inversely proportional to the class membership. At the class centre it receives value 0 and at the class border value 1.0. The first set of the data until 31.12.2008 has been collected on a gearbox, when it was known to operate in a good condition. The anomaly index for this data set is obviously maximum 1.0 for all training samples. The second set of data starting 01.01.2009 was collected on the same gearbox and this unlabeled data was used for prediction. Starting about 20.01.2009 high anomaly index values can be detected indicating that the classifier has not been trained with any data samples similar to these.

When an anomaly is detected, it should be interpreted with human expertise. This may involve analysis of raw data, comparison of syndromes or additional data to be collected. The interpretation does not need to be completed at the time of detection and not even necessarily before retraining. The samples related to the anomalies should in any case be labelled for reference so that the respective class on a new

*Figure 8. Anomaly detection*

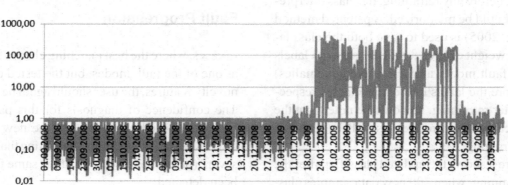

map after retraining can be found and labelled accordingly. Usually it is, however, not necessary to label classes representing normal modes, as normal the mode is considered the default mode. There might be times, where we would prefer a certain normal mode to be detected. In such a case the respective class can also be labelled.

Data classified from an abnormal condition represents a novelty and therefore should to be retained for two reasons. If the data was collected, when an object has experienced a previously unknown normal mode, new similar samples will no longer be diagnosed as novelties. In the case of a known fault mode, any new data from the various severities of the fault is valuable in prediction of the fault progression.

## Retraining

Retraining enables a continuous learning process. Retraining is necessary, whenever new data is appended to (or removed from) the training data. This happens, when new anomalies caused by new normal or new fault modes are encountered. Retraining is also required, when the training parameters are changed.

Before retraining all data samples should be normalized in the same way as the initial training. In this case it is recommended that the whole population, including the appended data samples, be used in calculating of the mean and standard deviation, which will be changed as compared with the ones used in the initial training.

Also, before any retraining, the class interpretations should be memorized. A patented method (Lumme, 2005) is used to save both the class labels and weight centres for all classes with labels (known fault modes and unresolved anomalies) and restore the labels to the new classes respectively after retraining. It should be noted that the class locations and references will be changed upon retraining.

Figure 9 shows the updated classifier after the first retraining, when 200 new data samples clas-

sified as anomalies, were appended to the initial data. It can be seen that the map has changed in shape. This is a result of the classifier's ability to adapt to new data volumes and types of data.

## Fault Detection

A classifier will not be able to detect fault modes until it has been trained with data containing descriptors related to a specific fault mode. This is, in fact, similar to human expertise. A machine analyst cannot diagnose faults that he has not experienced nor has any knowledge about. While an analyst can receive knowledge through education, a classifier will gain knowledge through retraining supported by human expertise. Once an anomaly has been detected and the respective class in the re-trained classifier has been labelled as a fault mode (without anomalies), any new data finding its best match in this class will be automatically interpreted accordingly. The confidence of diagnosis is very highly dependent on the accuracy and correctness of the original labelling accomplished by the human analyst. In an ideal case, the analyst would give a precise description of the fault mode and severity including the remaining safe operational time. This interpretation can be based on post-diagnosis, when the problem has been studied in details and uncompromised evidence has been provided to support the diagnosis. For any new sample classified in this group of data samples, the membership within the class will indicate the confidence of diagnosis.

## Fault Progression

In cases, where the best matching class is labelled as one of the fault modes, but the tested data has novelty features, the user should again be alerted. The confidence of diagnosis for this particular fault mode might be poor, but the new sample might still have descriptors suggesting that a more severe (or less severe) mode of the same fault has been detected.

*Figure 9. Map after retraining*

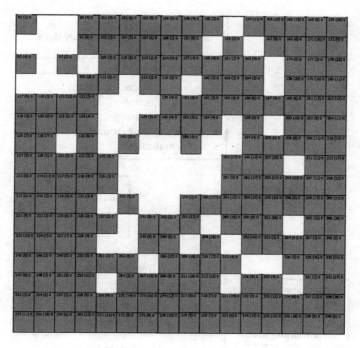

When the fault progresses in most cases to a more severe mode, the symptoms will also change. It is very important to understand that some symptoms may increase and others may decrease in value. Also, in the beginning there might be only a primary fault present, but due to the excessive load caused by it a secondary fault with its own characteristic symptoms may develop. For these reasons, the syndromes should be studied instead of single symptoms. An early warning of fault progression is received as an appearance of anomalies, as was the case for the first instance of the early fault. If not retrained with new data, the fault progression will be detected as a lower membership within the class.

On the user's initiative, retraining can be started in order to differentiate between the various severity modes of the fault. When the fault progression has ceased, possibly because of maintenance actions or a failure, the full history of the fault progression can be displayed as a trajectory on a map. This path allows the prediction, of what will probably happen next, when a new instance of a similar fault is first detected.

## Summary

Figure 10 summarizes the training process. The procedure is slightly different for retraining, where the class labels and weights shall be memorized before the training process and retraced after the process. The system shall allow the user to label the classes at any time, except between the memorization and retracing processes. Class labelling may happen typically after any retraining process, but can wait until the diagnosis and prognosis have been confirmed.

The prediction procedure is summarized in Figure 11. Data samples can be tested one by one or in a batch. After the data pre-processing the proper classifier shall be retrieved from the database. There might be several classifiers available for each object. It is recommended that descriptors that are related to different kinds of characteristics or different timestamps have their own classifiers. For instance a classifier with vibration, temperature and oil descriptors mixed might fail, because the parameters may have been collected at different spans.

*Figure 10. Training procedure*

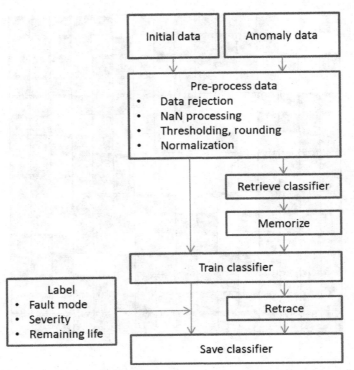

After finding the best matching class it is important to test first, if its label refers to a fault and only after that if the data is an anomaly. In fact it is not necessary to test separately, if a data sample classified as a fault also is anomaly. Each fault occurrence will always have variations mainly because of various degrees of fault severity. All data related to fault modes is valuable and will increase the confidence level of the class label. However, before proceeding with retraining, both the anomalies and fault detections should be verified by human expertise. This may include for instance outlier removal.

All data samples that were classified as normal can be discarded. Typically there is enough training data for classes labelled as the normal mode and there is no further use for more of this kind of data. In fact this brings a major benefit in optimizing the data to be saved. Many current condition monitoring systems keep collecting and trending data that stays stable and has no use in prediction of machine condition.

## CONFIDENCE OF PREDICTION

There are several points that should be taken into consideration, when the confidence of a prediction is evaluated. This is equally important, when using a neural classifier for failure mode and severity assessment.

First, it is a common assumption that the probability of a failure at a given time is proportional to the magnitude of descriptors. In many cases this might be true, but more often the fault progression can be seen as the change in the symptom distribution, i.e. in the syndrome. Depending on the fault mode some symptom values may or may not increase, stay steady or decrease in magnitude. Therefore the evaluation of the probability of a failure cannot rely on the magnitude of a single symptom only.

Additionally, the relationship between the severity of a fault and the symptom magnitude is often non-linear. Usually this relationship is not known. Even, if the failure has happened before

*Figure 11. Prediction procedure*

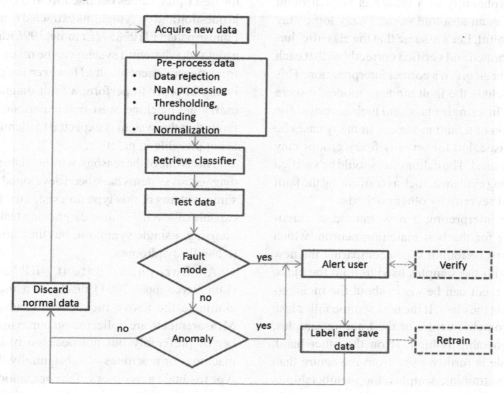

and the symptoms were recorded in detail, the next occurrence of the same failure mode might not give the same symptom values. It is therefore often considered satisfactory, if we can estimate the probability of a failure mode diagnosis being accurate at a certain time (ISO 13379).

Some additional uncertainty is added, when we take into consideration the ability of the vibration analyst to diagnose the symptoms and severity of the fault modes. The symptoms might have been detectable, but the analyst could not interpret them. He could perhaps diagnose the fault mode, but not the severity of the fault. If the fault has been misdiagnosed once and was not verified, it could be misinterpreted again.

Another problem is introduced, when two or more fault modes appear simultaneously. Some of the symptoms might be a result of a single root cause, such as imbalance or misalignment, while other symptoms might be related to the consequential fault modes, such as a bearing defect.

A classifier relies on the weight vectors that have been defined using data samples with syndromes. In order for a classifier to give a plain text diagnosis and severity as an output, it needs to be calibrated. This means that all classes, which were trained using data samples from a known condition, should be made identifiable. Typically the training samples used to train a single class would have the same interpretation. If not, the symptom extraction might have been imperfect. In other words, the symptom extraction chosen could not differentiate between two (or more) fault modes and should be improved. If this is not possible, the interpretation of a new sample falling into this particular class, would mean that in some probability the new sample represents any, or a combination of all possible fault modes.

The probability of a failure is very difficult to define as an absolute value. Fuzzy logic may here be useful. Let's assume that the classifier has been calibrated and verified correctly so that each class has been given a correct interpretation. This should include the fault mode or modes, if there are many in a single class, and their severity. The description of a fault mode can in many cases be deterministic, but for severity fuzzy groups may have to be used. The calibration should be verified using strong evidence, such as confirming the fault mode and severity by other methods.

When interpreting a new sample, a classifier looks for the best matching neuron, which will always be found. When comparing the new sample with the samples used to train the class, an assessment can be made about the membership within the class. If the new sample falls close to the geometric centre of the training samples, the membership is high. If on the other hand, the sample is further away from the centre than any of the training samples, the membership is low. Taken that the classifier has been accurately labelled, the confidence of diagnosis would then be high or low consequentially.

## GENERALIZATION

The prediction of the condition of an individual machine is difficult, because the reference data are generally deficient. The amount of existing measured data is scant especially at the initial phase and such data are not available in all operational states of the machine. Therefore, the detection of anomalies cannot be based on historical data, but general information has to be utilized instead. Consequently, the anomaly detection is uncertain and inaccurate.

The absence of empiric data is a special disadvantage in the diagnosis of faults. In practice, fault diagnosis is based on known symptom rules, which have been widely published. In many cases, the rules are of a general nature and are not based on the descriptor values obtained from the machine in question and on symptoms extracted from them. Finnish patent FI102857 (Lumme, 1996) discloses a method, whereby a system can be made to learn from measurement results. However, the problem is that in order to perform a fault diagnosis, an individual machine must first experience all the faults that the system is expected to identify. This is not possible in practice.

This is one of the reasons, why no viable remote diagnostics systems have been developed. Even if some systems of this type do exist, they are only capable of solving simple diagnosing tasks based on using a single symptom, but they are unable to handle syndromes.

A European patent EP1292812 (Lumme&Seppä, 2001) discloses a method to eliminate the above mentioned disadvantages. Measurements are collected on a maximal number of preferably but not necessarily identical machines or machines of substantially the same type to obtain descriptors of the operation and the condition of the machines. As a result any fault mode detected on a single object can be generalized to the diagnostics of a similar fault mode in any other object of the same type.

An essential point about the use and functionality of the patented method is that the database should be as large as possible and contain characteristic vectors descriptive of different operational states of machines in question in as large an area of application as possible. This method has proven effective in the anomaly and fault detection in a gearbox application with a large number of descriptors (Lumme, 2011).

Figure 12 presents a classifier map that has been created based on data collected from several similar objects. Small circles have been used to show the class labels, i.e. fault modes. The numbering in the boxes denotes various cases. A case is typically initiated, when an anomaly is detected. It consists of one or more events during which the data records were collected. A case concept typically contains all information related to a specific anomaly and fault detection process. The case usually includes data also from events after the problem was solved.

## FUTURE RESEARCH DIRECTIONS

The main research work is aimed at developing the user interface. A classifier provides much information as discussed earlier in this chapter. However, a normal user is typically interested only in simple terms, if the object is currently in a normal mode, a fault mode or is experiencing an anomaly. This can be easily displayed for instance with a traffic light indicator. If historical data is required, a weekly view might be more informative. For both indicators, a green colour marks a normal mode, red a known fault and yellow an anomaly.

In many cases, this might not be adequate for an analyst. A trajectory view shows the classes the data has progressed through to reach the current class. This view is extremely useful in the prediction of which class the next samples will be classified to, if such knowledge already exists in the classifier database. For analysis purposes a

class view showing the distribution of symptom weights is useful. This may be extended to a trend view, which shows the progress of the weight norm within the class. The view can be used in the prediction of when the data will fall outside the class borders.

Other views may become interesting. For a classifier map with hundreds of classes it would be practical to cluster some of the classes to form a common class. A map view should also visualize all classes representing various fault modes.

## CONCLUSION

This chapter has shown that it is possible to develop a continuous learning diagnostic system using neural networks. The system will significantly save an analyst's time and effort in anomaly and fault detection, which would otherwise require extensive work. There are several ways to enhance the presentation of data on a classifier in order to predict the interpretation of a new data sample. The confidence level of prediction can be estimated using simple terms.

*Figure 12. Map trained with data collected from various objects*

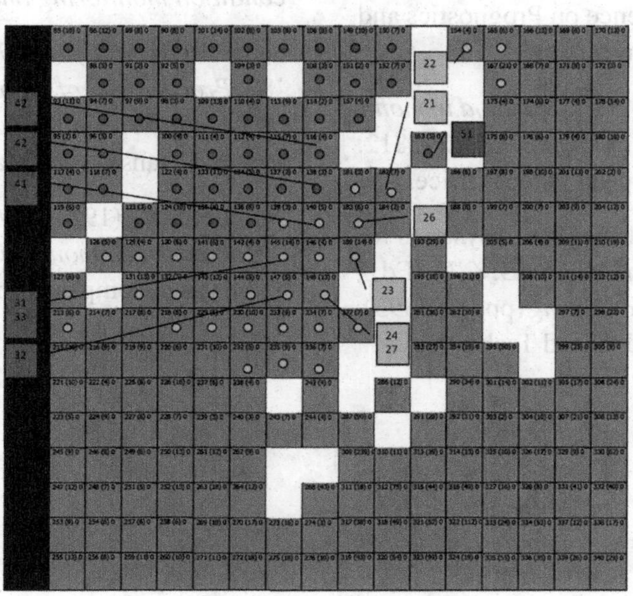

# REFERENCES

Berthold, M., & Hand, J. H. (Eds.). (2003). *Intelligent data analysis*. Berlin, Germany: Springer-Verlag. doi:10.1007/978-3-540-48625-1

Han, J., & Kamber, M. (2001). *Data mining: Concepts and techniques*. London, UK: Academic Press.

ISO 13379. (2003). *Condition monitoring and diagnostics of machines — General guidelines on data interpretation and diagnostic techniques*.

Kohonen, T. (2001). *Self-organizing maps*. Berlin, Germany: Springer-Verlag.

Lensu, A. (2002). *Computationally intelligent methods for qualitative data analysis*. Jyväskylä, Finland: Jyväskylä University Printing House.

Lumme, V. (1996). *Self learning method in the condition monitoring and diagnostics of rotating machines*. FI102857, Helsinki, Finland: Finnish Patent Office

Lumme, V. (2005). *Method in handing of samples*. FI115486. Helsinki, Finland: Finnish Patent Office

Lumme, V. (2011). *Diagnosis of multi-descriptor condition monitoring data*. Paper presented at the International Conference on Prognostics and Heath Management, Denver, CO.

Lumme, V., & Seppä, J. (2001). *Method in monitoring the condition of machines*. EP1292812. Munich, Germany: European Patent Office.

Milne, R. (1996). Knowledge based systems for maintenance management. In Rao, B. K. N. (Ed.), *Handbook of condition monitoring* (pp. 377–393). Oxford, UK: Elsevier Advanced Technology.

# ADDITIONAL READING

ISO 10816-1, *Mechanical vibration — Evaluation of machine vibration by measurements on non-rotating parts — Part 1: General guidelines*

ISO 13373-1, *Condition monitoring and diagnostics of machines — Vibration condition monitoring — Part 1: General procedures*

ISO 13374-1, *Condition monitoring and diagnostics of machines — Data processing, communication and presentation — Part 1: General guidelines*

ISO 13380, *Condition monitoring and diagnostics of machines — General guidelines on using performance parameters*

ISO 13381, *Condition monitoring and diagnostics of machines — Prognostics — Part 1: General guidelines*

ISO 18436-1, *Condition monitoring and diagnostics of machines — Requirements for training and certification of personnel — Part 1: Requirements for certifying bodies and the certification process*

ISO 18436-2, *Condition monitoring and diagnostics of machines — Requirements for training and certification of personnel — Part 2: Vibration condition monitoring and diagnostics*

Jones, M. H. (Ed.). (1987). Condition monitoring '87. *Proceedings of an International Conference on Condition Monitoring* held at University College of Swansea, 31st March – 3rd April, 1987.

Mitchell, J. S. (1981). *An introduction to machinery analysis and monitoring*. Tulsa, OK: PennWell Publishing Company. doi:10.1115/1.3256342

## KEY TERMS AND DEFINITIONS

**Classification:** A machine learning procedure in which individual items are placed into groups based on quantitative information on one or more characteristics inherent in the items and based on a training set of previously labeled items.

**Classifier:** A function that maps sets of input attributes to tagged classes.

**Fault:** Condition of a component that occurs when one of its components or assemblies degrades or exhibits abnormal behavior, which may lead to the failure of the machine.

**Hyperspace:** A Euclidean space of dimension greater than three.

**Neuron:** Any of the numerous types of specialized cell in the brain or other nervous system that transmit and process neural signals. The nodes of artificial neural networks are also called neurons.

**Normalization:** Refers to the division of multiple sets of data by a common variable in order to negate that variable's effect on the data, thus allowing underlying characteristics of the data sets to be compared.

**Outlier:** An observation that is numerically distant from the rest of the data.

**Symptom:** A perception, made by means of human observations and measurements (descriptors), which may indicate the presence of one or more faults with a certain probability.

**Syndrome:** A group of signs or symptoms that collectively indicate or characterize an abnormal condition.

# Chapter 4
# Generating Indicators for Diagnosis of Fault Levels by Integrating Information from Two or More Sensors

**Xiaomin Zhao**
*University of Alberta, Canada*

**Ming J Zuo**
*University of Alberta, Canada*

**Ramin Moghaddass**
*University of Alberta, Canada*

## ABSTRACT

*Diagnosis of fault levels is an important task in fault diagnosis of rotating machinery. Two or more sensors are usually involved in a condition monitoring system to fully capture the health information on a machine. Generating an indicator that varies monotonically with fault propagation is helpful in diagnosis of fault levels. How to generate such an indicator integrating information from multiple sensors is a challenging problem. This chapter presents two methods to achieve this purpose, following two different ways of integrating information from sensors. The first method treats signals from all sensors together as one multi-dimensional signal, and processes this multi-dimensional signal to generate an indicator. The second method extracts features obtained from each sensor individually, and then combines features from all sensors into a single indicator using a feature fusion technique. These two methods are applied to the diagnosis of the impeller vane trailing edge damage in slurry pumps.*

DOI: 10.4018/978-1-4666-2095-7.ch004

## INTRODUCTION

Fault diagnosis is of prime importance to the safe operation of rotating machinery in various industries such as mining, power, and aerospace. It provides information on the health condition of a machine, based on which preventive maintenance or other actions can be conducted to avoid consequences of severe damage or failure. There are mainly three tasks in fault diagnosis (Bocaniala & Palade, 2006): fault detection, fault isolation and fault identification. The first task (fault detection) is to determine whether a fault occurs or not. The second task (fault isolation) is to determine the location of a fault and to specify the fault type. The third task (*fault identification*) is to estimate the severity of a fault. The first two tasks are also called fault detection and isolation (FDI). They are the first steps in fault diagnosis and have been widely studied in the literature (Bocaniala & Palade, 2006; Jardine et al., 2006). The third task provides detailed information on a fault and gains more and more attention. For the convenience of description, fault identification is referred as the *diagnosis of fault levels* in this chapter.

One important characteristic of fault levels is the inherent *ordinal information* among different levels. For example, "a severe fault" is worse than "a medium fault", and even worse than "a slight fault". This makes the diagnosis of fault levels more complicated than the diagnosis of fault types (Lei & Zuo, 2009b; Zhao et al., 2011b). Having ordinal information is an important characteristic of fault levels. Thus keeping the ordinal information is a necessity for the diagnosis of fault levels. One way to achieve this is to generate an indicator reflecting the fault propagation. If such an indicator is obtained, the fault propagation can be tracked by monitoring the value of this indicator and the fault level can be easily estimated by comparing the value of this indicator with pre-determined thresholds. How to generate such an indicator is a challenging problem.

Vibration analysis is widely used for fault diagnosis, in which various features can be obtained to express the health information on machines. The indicator that can be used for the diagnosis of fault levels must signify the propagation of a fault. In another word, the indicator must monotonically vary with the fault levels. This requirement makes many traditional features unqualified as an effective indicator. Previous researches have shown that different features are sensitive to different stages of fault propagation. For instance, kurtosis, crest factor and impulse factor are sensitive to the existence of sharp peaks (impact faults), especially at the initial stage of a fault. But they will decrease to the value of a normal case as the damage grows (Tandon & Choudhury, 1999). Li and Limmer (2000) studied some statistical features' correlations with gear tooth wear, including RMS, FM0, FM4 and NA4 (Sait & Sharaf-Eldeen, 2011), and found that these features showed little change during most of the gear life.

In order to obtain an indicator for fault levels, various techniques have been used, which can be generally classified into two groups. The first group resorts to an assessment model which is obtained through machine learning techniques such as neural network (Jay, 1996) and hidden Markov model (Ocak et al., 2007). An indicator is generated using the assessment model. Li and Limmer (2000) built a linear auto-regression model based on normal vibration data, and then applied this model to vibration data over the gear's life. The distance (prediction error) between the model output value and the monitored vibration data was used as an indicator for gear damage levels. Qiu et al. (2003) employed Self Organizing Map (SOM) to build a model for bearing health assessment. An indicator evaluating the distance between the new input data and best matching unit of the SOM was extracted. Pan et al. (2010) built an assessment model based on training samples including normal and failure data using fuzzy c-means. The subjection of tested data to the normal state was extracted as an indicator. Ocak et al.

(2007) utilized a hidden Markov model (HMM) to train a model based on normal bearing data. It was found that the probabilities of the normal bearing HMM kept decreasing as the bearing damage propagated. The indicators generated above are based on the distance (Li & Limmer, 2000; Pan et al., 2010; Qiu et al., 2003) or the probability (Ocak et al., 2007) of new vibration data to the model built by machine learning techniques. These indicators cannot give a physical insight to the machine generating the vibration signals. This chapter will not focus on the first group.

The second group uses advanced signal processing techniques and generates an indicator that provides an insight into the physical phenomena of the machines, e.g. vibration amplitude (Bachschmid et al., 2010). Bachschmid et al. (2010) studied the dynamic response of heavy, horizontal axis, turbogenerator units. They found that "the 2X components that are excited in resonance corresponding to the first natural frequencies of the different shafts during a run-down transient of the unit" were reliable indicators for a developing crack. Yesilyurt (2006) used wavelet transform and found a consistent trend between the mean frequency of the scalograms and the wear progression of tools in milling. Loutridis (2006) used Wigner-Ville distribution, wavelet transform and empirical mode decomposition (EMD) and found that these techniques are capable of producing a reliable indicator (i.e. the instantaneous energy of vibration) representing the propagation of gear crack. Feng et al. (2010) applied the regularization dimension theory to the Fourier spectra of gearbox vibration signals and found that the regulation dimension increase monotonically with the gear fault severity. The indicators generated by this approach provide an insight to the physical phenomena inside the machine. This chapter focuses on the techniques in the second group. The contents of this chapter are based on Zhao et al. (2011a) and Zhao et al. (2011c).

## BACKGROUND

Depending upon different machines and fault types, vibration in more than one direction may become significant. Moreover, the location of a fault and its structural path greatly influence the quality of a measured vibration signal. For this reason, a set of vibration sensors at different directions/locations are usually involved in the fault diagnosis and condition monitoring process. For example, Tse et al. (2001) employed a set of accelerometers for bearing fault diagnosis. The above mentioned methods of the second group in the last section can handle one-dimensional signal only, i.e. the vibration signal from one sensor only. Different sensors measure the health information from different perspectives. The indicator representing fault propagation, especially for complex machines, needs to make full use of the information provided by different sensors. Thus, how to efficiently extract an indicator from two or more sensors using proper signal processing techniques becomes a key issue.

Fourier spectrum is a commonly used technique in vibration analysis. Conventional Fourier spectrum (also called half spectrum) deals with one-dimensional signals only. In conventional Fourier spectrum, the positive and negative frequency components are mirror images of each other (complex conjugate); hence only half is analyzed. Full spectrum (Goldman & Muszynska, 1999; Lee & Han, 1998) is capable of handling a two-dimensional signal measured from two orthogonal sensors, i.e. a planar motion. In a full spectrum, the positive and negative frequency components are not mirror images of each other anymore. The positive frequency component represents a motion in the direction of the planar motion; the negative frequency component represents a motion in the opposite direction of the planar motion. Thus the full spectrum reveals, not only the amplitude of a frequency component (as does the half spectrum), but also the directivity of a frequency component with respect to the planar rotational direction. This gives full spectrum great potential in the condition monitoring of rotating machinery (Patel & Darpe, 2009; Wu & Meng, 2006). Patel and Darpe (2008) used full spectrum to detect the crack depth of a rotor and found that the positive 2X (twice the rotor rotation frequency) frequency component becomes stronger with the increase of crack depth.

Besides full spectrum, statistical-based methods such as principle component analysis (PCA) and independent component analysis (ICA) are also widely used for multi-dimensional signals. Widodo and Yang (2007) applied PCA, ICA and their kernel extensions to features from six sensors and diagnosed six fault types of motors. Zhang et al. (2006) utilized principal component (PC) representations of features from two sensors to monitor a double-suction pump. Cempel (2003) applied singular vector decomposition (SVD) to a set of features, and then proposed an indicator based on the singular values to track the health condition of a diesel engine.

The above reported methods of integrating information from different sensors fall into two ways. The first way regards the signals from different sensors as one multi-dimensional signal, uses signal processing techniques that are capable of handling multi-dimensional signals to analyze this multi-dimensional signal. The work by Patel and Darpe (2008) using full spectrum directly on a two-dimensional signal (i.e. two signals measured from two orthogonal sensors respectively) follows this way. This way of integrating information from different sensors is often used when the physical pattern due to a fault (e.g. the characteristic of a planar vibration) is known. A proper signal processing technique (e.g. full spectrum) is applied to capture this pattern. One key issue when applying full spectrum for this purpose is the selection of the sensitive frequency component, because not all frequency components are sensitive to fault levels.

The second way regards signals from different sensors as a set of one-dimensional signals, applies signal processing techniques to each individual sensor for feature extraction, and then combines the features from all sensors. The work by Cempel (2003) applying SVD to features calculated from each individual sensor follows this way. However, when using this way to generate an indicator for fault levels, the selection of sensitive features is an issue to be addressed. Because not all the features have positive contributions to the indicator generation, especially considering that the indicator needs to show a monotonic trend with the fault levels.

This chapter presents two methods of generating indicators, following each of the two ways of utilizing information from two or more sensors. It is worth mentioning that indicators which represent the damage propagation could be different for different target machines. Expert knowledge on the specific machine is often required in searching for these indictors. The target machine of this chapter is a centrifugal slurry pump. The first method was reported in Zhao et al. (2011a), in which a sensitive frequency component was selected using mutual information and empirical mode decomposition. The second method was reported in Zhao et al. (2011c), in which sensitive features were selected by rough sets.

## GENERATING AN INDICATOR FOR DIAGNOSIS OF FAULT LEVELS

In this section, two indicator generation methods are presented first, and then are applied to diagnose fault levels of impellers in centrifugal slurry pumps.

### Method I: Process Signals from Two Sensors Together

The idea of this method is to integrate information from two sensors by processing signals from two sensors together using full spectrum. Full spectrum is capable of revealing the directivity and the energy of each spectral component in a planar vibration motion. However, not all the spectral components are sensitive to faults. Selecting the sensitive (fault-affected) spectral components can help diagnose fault levels more efficiently. Fixed band pass filtering can be used to choose spectral components in a certain frequency range. However, prior knowledge on the sensitive frequency ranges is required before processing the vibration data. Moreover, the noise contamination inevitably contained during vibration data measurements makes the fixed band pass filtering less efficient. If noise is mixed with the interested frequency range of a true signal, then the fixed band pass filtering process is ineffective.

*Empirical mode decomposition* (EMD) can be used as an adaptive filter (Flandrin et al., 2004). EMD is a self-adaptive signal processing technology, requiring no prior knowledge on the data. EMD decomposes a raw signal into a set of complete and almost orthogonal components called intrinsic mode functions (IMFs). IMFs represent the natural oscillatory modes embedded in the raw signal. Each IMF covers a certain frequency range. The IMFs work as the basis functions which are determined by the raw signal rather than by pre-determined functions. EMD has been widely used for fault diagnosis of rotating machinery (Fan & Zuo, 2008; Lei et al., 2010). However, standard EMD has the limitation in that it works only for one-dimensional signals (i.e. real-valued signals). When dealing with data from multiple sensors, standard EMD needs to decompose signals from each sensor individually. However, because of the local and self-adaptive nature of the standard EMD, the decomposition results of signals from multiple sources may not match in either the number or the frequency content (Looney & Mandic, 2009). For illustration purposes, here we take two signals ($x$ and $y$) as an example. If $x$ and $y$ are decomposed by the standard EMD, separately, signal $x$ may result in 10 IMFs, whereas signal $y$ may result in 9 IMFs. This is an unmatched problem in terms of the number of IMF. The $i^{th}$ IMF of signal $x$ may be in

the frequency range of 50-100 Hz, whereas the $i^{th}$ IMF of signal $y$ may be in a different frequency range, e.g. 25-50 Hz. This is an unmatched problem in terms of the frequency content of an IMF. The unmatched problems make the generation of an indicator for comparing signals at different fault levels difficult or even impossible. To overcome the unmatched problems, Rehman and Mandic (2010b) extended the standard EMD to the multivariate EMD. *Multivariate EMD* is able to find common oscillatory modes within signals from multiple sources.

It is worth mentioning that when comparing signals from different health conditions, signals from not only all sensors but also all different health conditions need to be combined together into a multi-dimensional signal and then decomposed using the multivariate EMD. Otherwise, the unmatched problem will occur among signals from different health conditions. Therefore, even though only two sensors are involved in a full spectrum, a multi-dimensional signal is generated to be decomposed by the multivariate EMD. In the following, the concepts of multivariate EMD (Rehman & Mandic, 2010b) and full spectrum (Goldman & Muszynska, 1999) are introduced first, and then method I (Zhao et al., 2011a) is presented.

## Multivariate Empirical Mode Decomposition (Multivariate EMD)

Standard EMD decomposes a one-dimensional signal, $x(t)$, into a set of IMFs, $c_i(t)$, and a residual signal, $r(t)$, so that

$$x(t) = \sum_{i=1}^{N} c_i(t) + r(t). \qquad (1)$$

The IMFs are defined so as to have the symmetric upper and lower envelopes with the number of zero crossings and the number of extrema differing at most by one. To extract IMFs, a sifting process is employed, which is described below (Rehman & Mandic, 2010b).

**Step 1:** Find the locations of all the extrema of $x(t)$.

**Step 2:** Interpolate among all the minima to obtain the lower signal envelope, $e_{\min}(t)$. Interpolate among all the maxima to obtain the upper signal envelope, $e_{\max}(t)$.

**Step 3:** Compute the local mean, $m(t) = [e_{\min}(t) + e_{\max}(t)] / 2$.

**Step 4:** Subtract the mean from $x(t)$ to obtain the "oscillatory mode", $s(t) = x(t) - m(t)$.

**Step 5:** If $s(t)$ satisfies the stopping criterion, then define $c_1(t) = s(t)$ as the first IMF; otherwise, set new $x(t) = s(t)$ and repeat the process from Step 1.

The same procedure is applied iteratively to the residue, $r(t) = x(t) - c_1(t)$, to extract other IMFs. The standard stopping criterion terminates the shifting process when the defined conditions for an IMF is met for a certain consecutive times (Huang et al., 2003).

Standard EMD considers only one-dimensional signals. Thus the local mean can be calculated by averaging the upper and lower envelopes which are obtained by interpolating among the local maxima and minima, respectively. However, for multi-dimensional signals, the local maxima and minima cannot be defined directly, and the notion of "oscillatory modes" defining an IMF is rather confusing. To solve these problems, multi-dimensional envelopes are firstly generated by taking a signal's projections along different directions, and then the average of these multi-dimensional envelopes are taken as the local mean (Rehman & Mandic, 2010a, 2010b; Rilling et al., 2007). This calculation of local mean can be considered as an approximation of the integral of all envelopes along the multiple projection directions in $n$-dimensional space. The accuracy of this

approximation depends on the uniformity of the chosen direction vectors. Thus, how to choose a suitable set of direction vectors for projection becomes the main issue. For bivariate signals, points can be uniformly selected along a unit 1-sphere (i.e. a circle of radius 1), where each point represents a direction vector (Rilling et al., 2007). For trivariate signals, points can be selected uniformly on a unit 2-sphere (i.e. a ball of radius 1) (Rehman & Mandic, 2010a). For a more general case (a $n$-dimensional signal), low-discrepancy sequences can be used to generate a set of uniformly distributed points on a unit ($n$-1)-sphere (Rehman & Mandic, 2010b). "Low-discrepancy sequences are also called quasi-random or subrandom sequences, due to their common use as a replacement of uniformly distributed random numbers. Roughly speaking, the discrepancy of a sequence is low if the number of points in the sequence falling into an arbitrary set B is close to proportional to the measure of B, as would happen on average in the case of a uniform distribution" (Wikipedia, 2008).

Let $X(t) = [x_1(t), x_2(t), ..., x_n(t)]$ be an $n$-dimensional signal and $V^k = \{v_1^k, v_2^k, ..., v_n^k\}$ denote the $k^{th}$ direction vector in a direction set, $V$. The procedure for multivariate EMD is outlined as follows (Rehman & Mandic, 2010b).

**Step 1:** Choose a suitable set of direction vectors, $V$.

**Step 2:** Calculate the $k^{th}$ projection, $p^k(t)$, of the input signal $X(t)$ along the $k^{th}$ direction vector, $V^k$, for each $k$ (i.e. $k = 1, 2, ..., l$ where $l$ is the total number of direction vectors in $V$).

**Step 3:** Find the time instants, $t_i^k$, corresponding to the maxima of the projected signal $p^k(t)$ for each $k$.

**Step 4:** Interpolate $[t_i^k, X(t_i^k)]$ to obtain multivariate envelopes, $E^k(t)$, for each $k$.

**Step 5:** The mean is estimated by

$$M(t) = \frac{1}{l} \sum_{k=1}^{l} E^k(t) \qquad (2)$$

**Step 6:** Calculate $D(t) = X(t) - M(t)$. If $D(t)$ fulfills the stopping criterion for a multivariate IMF, then assign $D(t)$ as an IMF and apply the above procedures from Step 2 to $X(t)$-$D(t)$ to extract the next IMF; otherwise, apply it to $D(t)$.

The stopping criterion for multivariate IMFs is similar to that in standard EMD. The difference is that the condition on the difference between the number of extrema and the number of zero crossings is not imposed.

## Full Spectrum

Conventional Fourier spectrum (also called half spectrum) deals with one-dimensional data (i.e. data from only one coordinate direction, either X-coordinate or Y-coordinate). In conventional Fourier spectrum, the negative frequency component is the complex conjugate of the positive frequency component. The positive and negative parts of the spectra are mirror images of each other; hence, only the positive half needs to be analyzed. However, individual Fourier spectra obtained from X and Y motions separately are unable to reveal any phase correlation between the X and Y motions. *Full spectrum* (Goldman & Muszynska, 1999) overcomes this limitation by utilizing vibration data measured from two orthogonal directions together (i.e. X and Y) for a planar vibration motion.

Figure 1 shows the procedure for obtaining a full spectrum from half spectra of X and Y signals. The FFT module has two inputs: one is called the direct buffer and the other is called the quadrature buffer. Simultaneously sampled X signal and Y signal are put into the direct part and the quadrature part of the FFT module, respectively. Results are then subjected to another transform from X to Y

and from Y to X to get the direct output and the quadrature output. Consequently, the positive and the negative frequency halves of this FFT are not mirror images any more. In the right half of the full spectrum plot, the amplitudes of the forward components (also known as positive frequency components) are shown. In the left half, the amplitudes of the backward components (also known as negative frequency components) are shown. It is worth noting that an elliptical orbit in one rotating direction is made up of two contra-rotating circular orbits of different amplitudes. The full spectrum considers the rotating directivity, and therefore generates a two-sided spectrum for this elliptical rotation. The full spectrum reveals not only the amplitude but also the directional nature (i.e. either forward or backward) of each frequency component with respect to the rotation direction. This attribute is very important for the fault diagnosis of rotating machines. There have been studies where researchers have proposed full spectrum-based features for fault diagnosis of rotating machine (Lee & Han, 1998; Patel & Darpe, 2009; Wu & Meng, 2006).

## Indicator Generation Method I (for Two Sensors Only)

If full spectrum is applied directly to the raw signals, many spectral components would be generated and it is hard to focus on the most sensitive one. In order to choose a fault-sensitive spectral component for comparison, a sensitive IMF should be selected. To select the sensitive IMF, criteria based on the correlation coefficient between an IMF and its raw signal were used in (Lei & Zuo, 2009a; Lin et al., 2009; Z. K. Peng et al., 2005). Note that the correlation coefficient reflects only linear relationships. To account for nonlinear relationships as well, Zhao et al. (2011a) proposed a selection criterion based on mutual information (Kraskov et al., 2004). This criterion takes into account two kinds of mutual information: that between the $n^{th}$ IMF and its raw

*Figure 1. Procedure of obtaining a full spectrum (Goldman & Muszynska, 1999)*

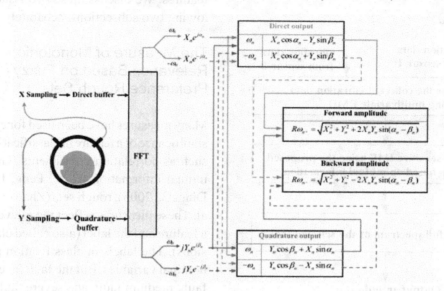

signal, and that between the $n^{th}$ IMF of a signal with certain health condition and the $n^{th}$ IMFs of signals with different health conditions.

Suppose that two signals, $x_{nor}(t)$ and $y_{nor}(t)$, collected from two orthogonal directions under the normal operation, are denoted by a two-dimensional signal, $x_{nor}(t)\vec{i} + y_{nor}(t)\vec{j}$; and two signals, $x(t)$ and $y(t)$, collected from the same two orthogonal directions under a fault condition, are denoted by $x(t)\vec{i} + y(t)\vec{j}$. Let $cn_{xnor}(t)$, $cn_{ynor}(t)$, $cn_x(t)$, and $cn_y(t)$ be the $n^{th}$ IMF of $x_{nor}(t)$, $y_{nor}(t)$, $x(t)$ and $y(t)$, respectively, obtained by multivariate EMD. The criterion for selecting a sensitive IMF is described below (Zhao et al., 2011a).

1. Calculate the mutual information, $a_n$, between the $n^{th}$ IMF of the normal signal, $cn_{xnor}(t)\vec{i} + cn_{ynor}(t)\vec{j}$, and the signal itself, $x_{nor}(t)\vec{i} + y_{nor}(t)\vec{j}$.

2. Calculate the mutual information, $b_n$, between the $n^{th}$ IMF of the fault signal, $cn_x(t)\vec{i} + cn_y(t)\vec{j}$, and the signal itself, $x(t)\vec{i} + y(t)\vec{j}$.

3. Calculate the mutual information, $e_n$, between $cn_{xnor}(t)\vec{i} + cn_{ynor}(t)\vec{j}$ and $cn_x(t)\vec{i} + cn_y(t)\vec{j}$.

4. Calculate the sensitivity factor, $\lambda_n$, for the $n^{th}$ IMF by

$$\lambda_n = \frac{a_n + b_n}{2} - e_n \qquad (3)$$

In Equation (3), the first part, $(a_n + b_n)/2$, represents the average mutual information between the $n^{th}$ IMFs and the raw signals; the second part, $e_n$, represents the mutual information between the $n^{th}$ IMF of the normal signal and the $n^{th}$ IMF of the signal under different health conditions (i.e. the fault signal in this example). To ensure that an IMF is informative enough to represent the original signal, the first part of Equation (3) is expected to be high; to enable the easy detection of the fault, the second part is expected to be low. The higher the $\lambda_n$ value is, the more sensitive the IMF is. The IMF having the highest value of $\lambda_n$ is selected as the most sensitive IMF to be further investigated.

*Figure 2. Flow chart of method I for indicator generation*

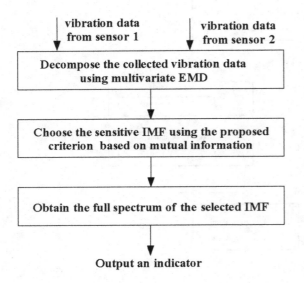

After the sensitive IMF is selected, the full spectrum of this IMF can be obtained. Then a proper full spectral indicator is extracted for fault diagnosis. A flow chart of the method (Zhao et al., 2011a) is given in Figure 2.

## Method II: Process Signals from Each Sensor Individually

The idea of this method is to extract features from each individual sensor first, and then combine features from all sensors together. The first step (feature extraction) can be achieved by reported signal processing techniques from the time-domain, the frequency-domain and the time-frequency domain. In the second step, an indicator that has monotonic trend with the fault levels is generated. How to combine the health information (features) from all sensors into an indicator that represents the health condition (i.e. exhibit monotonic trend) is the focus of this method. To address this issue, we need 1) a measure to select features exhibiting better monotonic relevance to the fault level; and 2) a strategy to combine

features. We discuss these two issues in the following two subsections, separately.

## The Measure of Monotonic Relevance Based on Fuzzy Preference Rough Sets

Many measures have been used for evaluating the significance of a feature in classification problems, such as correlation coefficients (Guyon, 2008), mutual information (H. C. Peng, Long, F., and Ding, C., 2005), rough sets (Zhao et al., 2010a) et al. These measures reflect the relevance between a feature and the label (also called classes or decisions). The labels in classification problems are nominal variables. But the fault levels (e.g. slight fault, medium fault, and severe fault) are ordinal variables, as stated in the introduction part of this chapter. Features selected for generating an indicator are expected to carry the ordinal information among different fault levels. Thus the evaluation value of a feature is based on the ability of this feature in expressing the ordinal information. In another word, the measure evaluates the monotonic relevance between the feature and the fault level. Most existing measures, however, do not consider the monotonic relation.

*Rough set* has been proved to be an effective tool in selecting important features. Traditional rough sets consider only the equivalence relation which is suitable for nominal variables. In order to consider the preference relation (i.e. monotonic relation), Greco et al. (2002) introduced dominance rough sets to measure the preference relation qualitatively. Hu et al. (2010) extended dominance rough sets and proposed a fuzzy preference rough set model which can reflect the monotonic relevance between two variables quantitatively. In this work, the fuzzy preference rough set model is used for the evaluation of monotonic relevance. The general ideas on dominance rough sets and fuzzy preference rough sets are briefly stated here. The materials that follow in this section are based on (Hu et al., 2010).

The theory of rough sets deals with the approximation of an arbitrary subset of a universe (*U*) by the lower and upper approximations. Let <*U*, *C* ∪ D> be a decision table, where *U* is a non-empty set of finite samples (the universe), *C* is a set of features describing the characteristics of each sample, and *D* is the decision describing the label (e.g. the fault level) of each sample. In dominance rough sets, without loss of generality, we assume that elements in *D* has a preference relation, $d_1 \leq d_2 \leq ... \leq d_p$. For $\forall x, h \in U$ and $a \in C$, let $f(x, a)$ and $f(h, a)$ be the values of feature *a* for sample *x* and sample *h*, respectively. The following sets are associated:

$$[x]_a^\geq = \{h \mid f(h,a) \geq f(x,a)\}, \tag{4}$$

$$[x]_a^\leq = \{h \mid f(h,a) \leq f(x,a)\}. \tag{5}$$

The first set, $[x]_a^\geq$, consists of samples that are not worse than sample *x* with respect to *a*. The second set, $[x]_a^\leq$, consists of samples that are not better than sample *x* with respect to *a*. As to decision *D*, let $d_k^\geq = \bigcup_{l \geq k} d_l$ and $d_k^\leq = \bigcup_{l \leq k} d_l$. The lower approximations of $d_k^\geq$ and $d_k^\leq$ with respect to *a* are defined with the upward lower approximation (Equation (6)) and downward lower approximation (Equation (7)).

$$\underline{a}_{d_k^\geq}^\geq = \{x : [x]_a^\geq \subseteq d_k^\geq\}, \tag{6}$$

$$\underline{a}_{d_k^\leq}^\leq = \{x : [x]_a^\leq \subseteq d_k^\leq\}. \tag{7}$$

Dominance rough sets could not reveal the preference relation quantitatively. It provides only a qualitative measure. To overcome this, Hu et al. (2010) proposed *fuzzy preference based rough sets* by replacing dominance relations with fuzzy preference relations.

The fuzzy preference relation is represented by an $n \times n$ matrix $R = (r_{ij})_{n \times n}$, where $r_{ij}$ is the preference degree of sample $x_i$ over sample $x_j$. In this chapter, we use $r_{ij} = 1$ to represent that $x_i$ is absolutely preferred to $x_j$, $r_{ij} = 0.5$ to represent that there is no difference between $x_i$ and $x_j$, $r_{ij} > 0.5$ to represent that $x_i$ is more likely to be preferred to $x_j$, and $r_{ij} < 0.5$ to represent that $x_i$ is less likely to be preferred to $x_j$. Let $f(x, a)$ be the value of feature *a* for sample *x*. The upward and downward fuzzy preference relations between samples $x_i$ and $x_j$ can be computed by

$$r_{ij}^> = \frac{1}{1 + e^{-s(f(x_i,a) - f(x_j,a))}},$$

and

$$r_{ij}^< = \frac{1}{1 + e^{s(f(x_i,a) - f(x_j,a))}} \tag{8}$$

where *s* is a positive parameter which can be used to adjust the shape of the function and it is determined by specific applications, $r_{ij}^>$ represents how much $x_i$ is larger than $x_j$, and $r_{ij}^<$ represents how much $x_i$ is smaller than $x_j$. Thus fuzzy preference relations reflect not only the fact that sample $x_i$ is larger/smaller than $x_j$ (qualitatively), but also how much $x_i$ is larger/smaller than $x_j$ (quantitatively).

With the upward fuzzy preference relation ($R^> = [r_{ij}^>]$) and the downward fuzzy preference relation ($R^< = [r_{ij}^<]$) induced by feature *a*, the memberships of a sample *x* to the lower approximations of $d_i^\geq$ and $d_i^\leq$ are defined with the upward fuzzy lower approximation (Equation (9)) and the downward fuzzy lower approximation (Equation (10)).

$$\underline{R}^{>}_{d^{\geq}_k(x)} = \inf_{h \in U} \max\{1 - R^{>}(h, x), d^{\geq}_k(h)\}, \qquad (9)$$

$$\underline{R}^{<}_{d^{\leq}_k(x)} = \inf_{h \in U} \max\{1 - R^{<}(h, x), d^{\leq}_k(h)\}. \qquad (10)$$

The approximation quality, also called dependency, is defined as the ratio between the union of the lower approximations and the universe. The approximation qualities of $D$ with respect to $a$ are then defined with:

- *Upward fuzzy preference approximation quality* (upward FPAQ):

$$r^{>}_a(D) = \frac{\sum_k \sum_{x \in d_k} \underline{R}^{>}_{d^{\geq}_k(x)}}{\sum_k |d^{\geq}_k|} \qquad (11)$$

- *Downward fuzzy preference approximation quality* (downward FPAQ)

$$r^{<}_a(D) = \frac{\sum_k \sum_{x \in d_k} \underline{R}^{<}_{d^{\leq}_k(x)}}{\sum_k |d^{\leq}_k|} \qquad (12)$$

- *Global fuzzy preference approximation quality* (global FPAQ)

$$r_a(D) = \frac{\sum_k \sum_{x \in d_k} \underline{R}^{<}_{d^{\leq}_k(x)} + \sum_k \sum_{x \in d_k} \underline{R}^{>}_{d^{\geq}_k(x)}}{\sum_k |d^{\leq}_k| + \sum_k |d^{\geq}_k|} \qquad (13)$$

where $|d^{\geq}_k|$ and $|d^{\leq}_k|$ are the numbers of samples with decisions dominating and dominated by $d_k$, respectively. The fuzzy preference approximation qualities measure the preference relation (monotonic relation) between feature $a$ and decision $D$. The upward FPAQ reflects the preference in the sense that when the feature of a sample ($h$) is not worse than the feature of a sample ($x$), the decision of $h$ would not be worse than the decision of $x$. The downward FPAQ reflects the preference in the sense that when the feature of $h$ is not better than the feature of $x$, the decision of $h$ would not be better than the decision of $x$. The global FPAQ averages the upward and downward information, and is used as a measure of the *monotonic relevance*

between a feature and decision (fault levels). The higher the value of global FPAQ is, the more significant the feature is.

## Feature Fusion Using PCA

The global FPAQ defined in fuzzy preference based rough sets can help select features having better monotonic relations with the fault levels. Each of the features may contain complementary information on machinery conditions. To make the final indicator more robust, feature fusion technology is needed for combining information in different features. Eigenvalue/Eigenvector analysis has been widely used as a feature fusion method in condition monitoring (Natke & Cempel, 2001; Pires et al., 2010; Turhan-Sayan, 2005). Pires et al. (2010) found a severity index for stator winding fault, and rotor broken bars could be detected from the obtained eigenvalues. Turhan-Sayan (2005) used *Principal Component Analysis* (PCA) for feature fusion and obtained a single indicator that can effectively represent the electromagnetic target of concern. PCA is a simple eigenvector-based multivariate analysis method. It transforms a number of possibly correlated variables into a number of uncorrelated variables called principal components. The calculation of PCA is stated as follows. The materials described below are from Jolliffe (2002).

Let $A$ be a zero-mean matrix $m{\times}n$, where each of the $m$ rows represents a sample and each of the $n$ columns stands for a feature, the singular value decomposition on $A$ is

$$A = U\Sigma V^{*}, \qquad (14)$$

where $U$ is an $m{\times}m$ unitary matrix, $\Sigma$ is an $m{\times}n$ rectangular diagonal matrix with nonnegative real numbers on the diagonal, $V^{*}$ (the conjugate transpose of $V$) is an $n{\times}n$ unitary matrix. The diagonal entries of $\Sigma$ are known as the singular values of $A$. The maximum singular value is $\Sigma_{11}$ followed by $\Sigma_{22}$ and $\Sigma_{33}$, so on and so forth. The columns of $U$ (respectively $V$) are called the left (respectively right) singular vectors of $A$ which are actually the eigenvalues of $AA^T$ (respectively $A^TA$). The non-zero singular values of $A$ are the square roots of the non-zero eigenvalues of $AA^T$ or $A^TA$. Each eigenvalue is proportional to the portion of the variance (more correctly of the sum of the squared distances of the samples from their multidimensional mean) that is correlated with each eigenvector.

The PCA transformation is given by:

$$P = A V. \qquad (15)$$

Since $V$ are unitary matrixes, each column of $P$ is simply a rotation of the corresponding column of $A$. The $i^{th}$ column of $P$ represents the $i^{th}$ principal component of $A$, corresponding to the $i^{th}$ eigenvalue of $AA^T$ or $A^TA$.

Natke and Cempel (2001) found that the non-zero eigenvalues, ordered by their magnitudes, can be regarded as fault ranking indices which measure the fault intensity. The first principal component corresponds to the largest eigenvalue, and therefore contains most information on damage conditions. In this chapter, we consider different fault levels of the same fault type. The first principal component is used as a single indicator representing the fault levels.

## Indicator Generation Method II (for Multiple Sensors)

A method of generating an indicator for fault levels using both fuzzy preference based rough sets and PCA is presented by Zhao et al. (2011c). Let $D$ be the fault levels, $C$ be the set of features, $S$ be the set of selected features, $I_s$ be the indicator generated by $S$, and $e_s$ represent the monotonic relevance between the indicator and $D$ (fault levels). The steps are described as follows.

1. Extract features from a signal measured from an individual sensor using proper signal processing techniques. Features from all sensors are put together and stored in set $C = [c_1, c_2, ... c_m]$, where $m$ is the total number of features from all sensors.

2. Employ Equation (13) to evaluate the monotonic relevance between each feature and the fault level ($D$). Results are saved as $E = [e_1, e_2, ... e_m]$, where $e_i$ is the monotonic relevance between feature $c_i$ and $D$.

3. Set $S = \square$. Find $c_k$ such that $e_k = \max_i(e_i)$, put $c_k$ into $S$ (i.e. $S = [S, c_k]$), delete it from $C$ (i.e. $C = C - c_k$) and Set $I_s = c_k$ and $e_s = e_k$.

4. For each feature, $c_i$, in $C$, generate a temporary feature set $T_i = [S, c_i]$, where $i = 1, 2, ..., p$ and $p$ is the current total number of features in $C$.

5. Compute $I_{tempi}$ ($I_{tempi}$ is the first principal component of $T_i$).

6. Use Equation (13) to calculate $e_{tempi}$, the monotonic relevance between $I_{tempi}$ and $D$.

7. Find $T_k$ that corresponds to the highest monotonic relevance, i.e. $e_{tempk} = \max_i(e_{tempi})$.

8. If $e_{tempk} > e_s$, then let $S = T_k$, $C = C - c_k$, $e_s = e_{tempk}$, $I_s = I_{tempi}$ and go to Step (4); Otherwise, go to Step (9).

9. Output the indicator $I_s$.

## Application to Impeller Fault Diagnosis in Pumps

The two proposed methods are tested with experimental data on a centrifugal slurry pump. Centrifugal pumps are widely used for moving slurries in applications such as the oil sands and mining. As slurries contain abrasive and erosive solid particles, the slurry pump impellers are subjected to harsh impingement which may cause wear on impeller vanes. Khalid and Sapuan (2007) fabricated a wear testing rig for centrifugal slurry pump impellers. They found that wear occurred at both the region near the centre (vane leading edge) of the impeller and the region at the rim (vane trailing edge) and that the vane trailing edge of an impeller encounters more wear compared to the vane leading edge. In this chapter, we study the wear at *impeller vane trailing edge*.

## Experimental Data

An experimental test rig was set up at Reliability Research Lab at University of Alberta. Details on the experimental system refer to Patel and Zuo (2010). To track the damage propagation at the vane trailing edge, four damage levels are considered: no damage (level 0), slight damage (level 1), medium damage (level 2), and severe damage (level 3). Each of the five vanes of an impeller was shortened, from the vane trailing edge side, by 0%, 22%, 26% and 29.3% in length, respectively. For each damage level, vibration data were collected at a sampling frequency of 5 KHz with the pump running at 2400 RPM at the best efficiency point (BEP) flow rate. One tri-axial accelerometer (A1) was installed in this test rig. The accelerometer A1 measures vibration from three directions X, Y and Z. We use A1X, A1Y and A1Z to denote the sensor measured from each of the three directions. The sensors' locations and directions are labeled in Figure 3.

In pumps, hydrodynamic forces are often the major sources of vibration (Gulich, 2010). It is believed that the vane trailing edge damage influences the velocity distribution near the cutwater (where sensors are located as shown in Figure 3), affects hydrodynamic forces and therefore vibrations on the casing. To validate this, computational fluid dynamic (CFD) simulation were conducted to calculate the flow field for each damage condition by Zhao et al. (2010b). Figure 4 shows the relative velocity fields near the cutwater under the four damage conditions. The liquid flows out from the impeller into the volute, experiences a rotation of a certain degree in the volute, and enters to the outlet. Thus near the cutwater area, part of the flow is directed into the volute (forward direction), and the rest is directed to the outlet (backward direction). For the no damage case (Figure 4), the velocity vectors are well directed towards the outlet and the volute. This directivity is distorted when the damage occurs at the vane trailing edge. The degree of distortion increases as the level of damage increases. At the severe damage case, the flow has an obviously weak trend in the forward direction. In the following space, we use methods I and II to find indicators for tracking the damage propagation.

*Figure 3. Sensors' locations and directions (Patel & Zuo, 2010)*

(a) Vibration sensor location

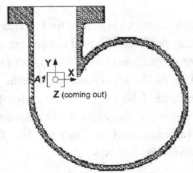

(b) Schematic for direction convention

*Figure 4. Zoomed view of the relative velocity fields near the cutwater area (Zhao et al., 2011a)*

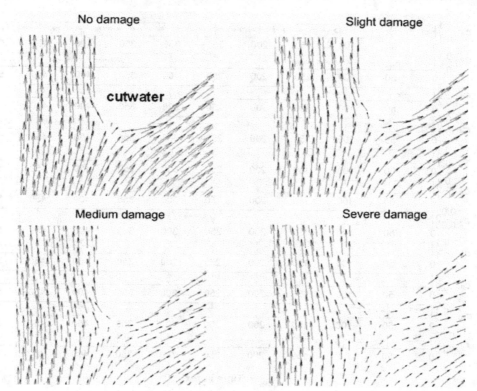

## Results of Method I

The flow pattern revealed in Figure 4 is in the plane of A1X and A1Y. So here in method I, we analyze signals from A1X and A1Y to capture the change of the flow pattern shown in Figure 4. Method I processes signals from A1X and A1Y together using the multivariate EMD and full spectrum. We use $x_0$, $x_1$, $x_2$, and $x_3$ (respectively $y_0$, $y_1$, $y_2$, and $y_3$) to denote the signal measured by A1X (respectively A1Y) for the no damage, slight damage, medium damage, and severe damage cases, respectively.

First, the eight raw signals (i.e. $x_0$, $y_0$, $x_1$, $y_1$, $x_2$, $y_2$, $x_3$, and $y_3$) are combined into one eight-dimensional signal and decomposed together by the multivariate EMD. The first ten IMFs are obtained for each signal, and the rest IMFs having small amplitudes are put into the residual. As examples, Figure 5 shows the decomposition

results (IMFs) of $y_0$ and $y_3$. It can be seen that the $i^{th}$ IMF of $y_0$ and the $i^{th}$ IMF of $y_3$ have the same frequency content (This can also be seen in the full spectrum of the 4th IMF which will be shown in Figure6). This is an important advantage of the multivariate EMD over the standard EMD, because the latter cannot ensure the match of frequency contents in the $i^{th}$ IMF of multiple signals (Looney & Mandic, 2009; Zhao et al., 2011a).

Then, the criterion based on mutual information is used to select the most sensitive IMF from the ten IMFs. The sensitive factor of the $n^{th}$ IMF is calculated using the averaged mutual information between the $n^{th}$ IMF and the raw signal subtracted by the averaged mutual information between the $n^{th}$ IMF of one health condition and that of three other health conditions. The sensitive factors for the ten IMFs are (in order): 0.09, 0.16, 0.19, 0.24, 0.12, 0.21, 0.05, -0.11, -0.35 and -0.77.

*Figure 5. Decomposition results for $y_0$ (a) and $y_3$(b) using multivariate EMD (Zhao et al., 2011a)*

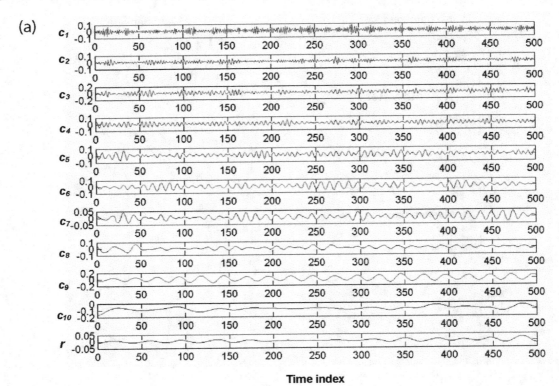

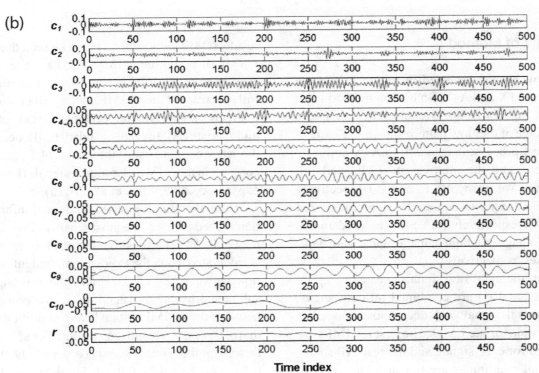

*Figure 6. Full spectra of the 4ᵗʰ IMFs of the signal at each of the four damage levels (Zhao et al., 2011a)*

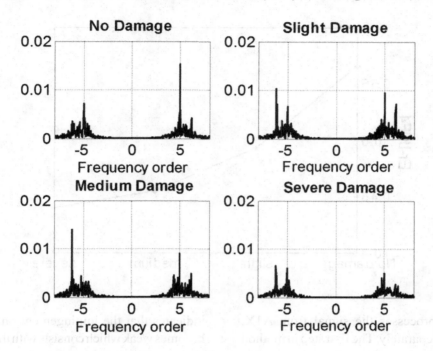

The 4ᵗʰ IMF has the highest value (0.24) and is chosen as the most sensitive IMF.

Finally, the full spectrum of the 4ᵗʰ IMF for each damage level is calculated and compared. The rotation direction of the impeller is clockwise (Figure 3). To make the forward direction be the impeller's rotation direction, the 4ᵗʰ IMF of signal $x_i$ (where $i$=0, 1, 2, 3 stands for no damage, slight damage, medium damage and severe damage) and the 4ᵗʰ IMF of signal $y_i$ ($i$=0, 1, 2, 3) are used as the quadrature part and the direct part to calculate the full spectrum for each damage level. The results are shown in Figure 6. It can be seen from Figure 6 that at the no damage condition, the forward components are dominant; and as the health condition gets worse, the portion of the forward components becomes smaller. An indicator defined in Equation (15) is extracted to express the information on forward components, where $a(f)$ stands for the amplitude at a frequency $f$. The change of this indicator with the damage levels is plotted in Figure 7. It can be seen from Figure 7 that the indicator monotonically decreases

as the damage level increases. This is consistent with the flow pattern shown in Figure 6, that is, the flow in the forward direction becomes weak as the damage grows. Results show that method I can capture the change of a planar vibration motion. One disadvantage of method I, however, is that it utilizes information from two sensors only.

$$Er = \frac{\sqrt{\sum_{f>0} a(f)^2}}{\sqrt{\sum_{f>0} a(f)^2} + \sqrt{\sum_{f<0} a(f)^2}} \quad (15)$$

## Results of Method II

Method II doesn't have the limitation on the number of sensors, and can work for one, two or more sensors. Now we apply method II to the experimental data measured from three sensors (A1X, A1Y and A1Z).

*Figure 7. Indicator generated by method I (Zhao et al., 2011a)*

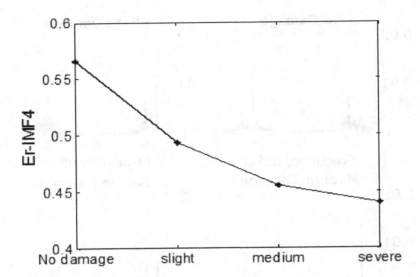

Method II processes the signal from A1X, A1Y and A1Z, separately. The first step in method II is feature extraction from measured signals. Conventional Fourier spectrum is employed for this purpose. For centrifugal pumps, the pump rotating frequency (1X), its 2nd harmonic (2X), vane passing frequency (5X, because impellers used in these experiments have five vanes) and its 2nd harmonic (10X) carry useful information on pump conditions (Gulich, 2010; Zhao et al., 2010a). Therefore, the amplitudes at 1X, 2X, 5X and 10X were extracted from Fourier spectra of A1X signal and A1Y signal, respectively. Totally, there were twelve features extracted for each damage condition.

Not all the twelve features reflect the pump health condition (i.e. damage level) monotonically. Equation (13) is used to measure the monotonic relevance between a feature and the damage levels. The parameter $s$ of the preference relations (Equation (8)) is set to be 25. Using method II, two features are finally selected for the indicator generation. They are the amplitudes at 5X from A1X and A1Y, as shown in Figure 8 (a). It can be seen that the two amplitudes generally show a decreasing trend with the damage level. This

indicates that the impingement on the cutwater becomes weak which consists with the flow pattern shown in Figure 4. However, the amplitude at 5X from A1X cannot clearly distinguish the medium and the severe damage levels, and the amplitude at 5X from A1Y cannot clearly distinguish the no damage and the slight damage levels. The generated indicator is shown in Figure 8 (b). The indicator is generated by applying PCA to these two features, it can be seen that different fault levels can now be clearly distinguished and the monotonic trend is also clearer. This verifies the effectiveness of method II.

The sensitive features selected in Method II are the amplitudes at 5X (the vane passing frequency) from sensor A1X and A1Y. This is consistent with the results from Method I, in which the sensitive IMF selected is centered at 5X (Figure 6). This can be explained as follows. The local flow field near the cutwater is subjected to significant variations each time an impeller vane passes it, so the vane passing frequency is the characteristic frequency for impeller vane trailing edge damage.

Now, we discuss the impact of additional sensors included in the indicator generation process. The impact of additional sensor depends on how

*Figure 8. (a) Two features selected by method I; (b) indicator generated by method I*

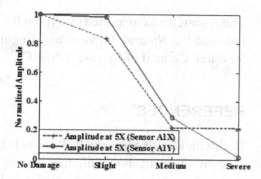
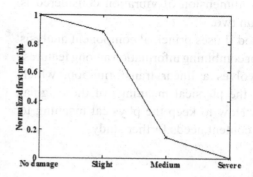

much additional useful information the sensor provides; this is affected by the sensor's location and direction. If the sensor captures valuable information on the health condition of a machine, then the performance (prediction accuracy) of the generated indicator would increase; otherwise, the performance of the indicator would not increase. In the pump example, sensor A1X and A1Y provide valuable information on pump's health condition. So adding each of these two sensors would improve the performance of the generated indicator, as can be seen from Figure 8. The monotonic trend in Figure 8 (b) is clearer than that in Figure 8 (a). However, sensor A1Z, which is in the direction perpendicular to the plane of the main flow, provides little information on the pump's health condition. So the features from this sensor did not get selected for indicator generation. Thus the performance of the generated indicator using sensor A1X, A1Y and A1Z remained the same as the indicator generated using sensor A1X and A1Y only. Theoretically, the additional sensors would not decrease the performance of the indicator, because those irrelevant sensors won't be selected in the feature selection process anyway. However, they will add computational burden to feature selection process. From this sense, it is always helpful to include only informative sensors.

The computational efficiency of the two methods is problem-dependent. In method I, the process of multivariate empirical mode decomposition is time-consuming. Increase of the number of sensors and the number of damage levels will increase the dimension of the multivariate signals and thus increase the computational time. In method II, the process of feature selection is time-consuming. Increase of the number of the sensors and/or the number of features extracted from each sensor will increase the computational time. As of the pump example, the total computational time on an AMD Opteron™ Processor 246 (2.00 GHz) needed for Method I was 6140 seconds; and for Method II it was 189 seconds because the number of features extracted is not very large (totally 12 features).

In this application, both method I and method II successfully extract an indicator for impeller trailing edge damage levels. Method I is capable of revealing the characteristic of a planar vibration motion; however it uses information for two sensors only. Method II can integrate information from multiple sensors. However, it uses PCA to fuse information from multiple sensors. PCA involves linear transformation, thus the original meaning of the selected features cannot be well kept in the generated indicator.

## FUTURE RESEARCH DIRECTIONS

In method I, full spectrum is employed to analyze two orthogonal vibration signals. Full spectrum is suitable only for analyzing planar vibration motions. Further studies need to be conducted when the dimension of vibration considered is larger than two.

Method II uses principal component analysis (PCA) for combining information among features. PCA involves a linear transformation, which destroys the physical meanings of the original features. How to keep the physical meaning to its large content needs further study.

## CONCLUSION

In this chapter, two methods of generating indicators for fault levels by integrating information from possible sensors are presented. The first method regards signals from two sensors and different health conditions as one multivariate signal. Multivariate empirical mode decomposition is adopted to decompose the multivariate signal into a set of IMFs. The fault-sensitive IMF is chosen by a criterion based on mutual information. Then a full spectra based indicator is obtained. The indicator generated by method I reveals the characteristics of planar vibration motions.

The second method extracts features from each individual sensor, uses global fuzzy preference approximation quality to select features having better monotonic relevance with fault levels, and utilizes PCA to combine information in selected features into a single indicator. The generated indicator makes use of information among different sensors and features, and outperforms each individual feature. This method is general and can work for multiple sensors. The generated indicator, however, because of the linear transformation induced by PCA, doesn't keep the physical meaning of the original selected features.

These two methods are applied to track impeller vane trailing edge damage. The experimental results validate the effectiveness of the two methods.

## ACKNOWLEDGMENT

This research was supported by Syncrude Canada Ltd. and the Natural Sciences and Engineering Research Council of Canada (NSERC).

## REFERENCES

Bachschmid, N., Pennacchi, P., & Tanzi, E. (2010). A sensitivity analysis of vibrations in cracked turbogenerator units versus crack position and depth. *Mechanical Systems and Signal Processing, 24*(3), 844–859. doi:10.1016/j.ymssp.2009.10.001

Bocaniala, C. D., & Palade, V. (2006). Computational intelligence methodologies in fault diagnosis: Review and state of the art computational intelligence in fault diagnosis. In Palade, V., Jain, L., & Bocaniala, C. D. (Eds.), *Computational intelligence in fault diagnosis* (pp. 1–36). London, UK: Springer. doi:10.1007/978-1-84628-631-5_1

Cempel, C. (2003). Multidimensional condition monitoring of mechanical system in operation. *Mechanical Systems and Signal Processing, 17*(6), 1291–1303. doi:10.1006/mssp.2002.1573

Fan, X., & Zuo, M. J. (2008). Machine fault feature extraction based on intrinsic mode functions. *Measurement Science & Technology, 19*(4), 1–13. doi:10.1088/0957-0233/19/4/045105

Feng, Z. P., Zuo, M. J., & Chu, F. L. (2010). Application of regularization dimension to gear damage assessment. *Mechanical Systems and Signal Processing, 24*(4), 1081–1098. doi:10.1016/j.ymssp.2009.08.006

Flandrin, P., Rilling, G., & Goncalves, P. (2004). Empirical mode decomposition as a filter bank. *IEEE Signal Processing Letters, 11*(2), 112–114. doi:10.1109/LSP.2003.821662

Goldman, P., & Muszynska, A. (1999). Application of full spectrum to rotating machinery diagnostics. *Orbit,* First Quarter, 17-21.

Greco, S., Inuiguchi, M., & Slowiński, R. (2002). In Alpigini, J., Peters, J., Skowron, A., & Zhong, N. (Eds.), *Dominance-based rough set approach using possibility and necessity measures rough sets and current trends in computing* (pp. 84–85). Berlin, Germany: Springer.

Gulich, J. F. (2010). *Centrifugal pumps.* New York, NY: Spinger. doi:10.1007/978-3-642-12824-0

Guyon, I. (2008). Practical feature selection: From correlation to causality. In *Mining massive data sets for security*. IOS Press.

Hu, Q., Yu, D., & Guo, M. (2010). Fuzzy preference based rough sets. *Information Science, 180*(10), 2003–2022. doi:10.1016/j.ins.2010.01.015

Huang, N. E., Wu, M.-L. C., Long, S. R., Shen, S. S. P., Qu, W., Gloersen, P., & Fan, K. L. (2003). A confidence limit for the empirical mode decomposition and Hilbert spectral analysis. *Proceedings of the Royal Society of London. Series A: Mathematical, Physical and Engineering Sciences, 459*(2037), 2317-2345.

Jardine, A. K. S., Lin, D., & Banjevic, D. (2006). A review on machinery diagnostics and prognostics implementing condition-based maintenance. *Mechanical Systems and Signal Processing, 20*(7), 1483–1510. doi:10.1016/j.ymssp.2005.09.012

Jay, L. (1996). Measurement of machine performance degradation using a neural network model. *Computers in Industry, 30*(3), 193–209. doi:10.1016/0166-3615(96)00013-9

Jolliffe, I. T. (2002). *Principal component analysis.* New York, NY: Springer-Verlag.

Khalid, Y. A., & Sapuan, S. M. (2007). Wear analysis of centrifugal slurry pump impellers. *Industrial Lubrication and Tribology, 59*(1), 18–28. doi:10.1108/00368790710723106

Kraskov, A., Stogbauer, H., & Grassberger, P. (2004). Estimating mutual information. *Physical Review E: Statistical, Nonlinear, and Soft Matter Physics, 69*(6), 1–16. doi:10.1103/PhysRevE.69.066138

Lee, C. W., & Han, Y. S. (1998). The directional Wigner distribution and its applications. *Journal of Sound and Vibration, 216*(4), 585–600. doi:10.1006/jsvi.1998.1715

Lei, Y., & Zuo, M. J. (2009a). Fault diagnosis of rotating machinery using an improved HHT based on EEMD and sensitive IMFs. *Measurement Science & Technology, 20*(12), 1–12. doi:10.1088/0957-0233/20/12/125701

Lei, Y., & Zuo, M. J. (2009b). Gear crack level identification based on weighted K nearest neighbor classification algorithm. *Mechanical Systems and Signal Processing, 23*(5), 1535–1547. doi:10.1016/j.ymssp.2009.01.009

Lei, Y., Zuo, M. J., He, Z. J., & Zi, Y. Y. (2010). A multidimensional hybrid intelligent method for gear fault diagnosis. *Expert Systems with Applications, 37*(2), 1419–1430. doi:10.1016/j.eswa.2009.06.060

Li, C. J., & Limmer, J. D. (2000). Model-based condition index for tracking gear wear and fatigue damage. *Wear, 241*(1), 26–32. doi:10.1016/S0043-1648(00)00356-2

Lin, C. C., Liu, P. L., & Yeh, P. L. (2009). Application of empirical mode decomposition in the impact-echo test. *NDT & E International, 42*(7), 589–598. doi:10.1016/j.ndteint.2009.03.003

Looney, D., & Mandic, D. P. (2009). Multiscale image fusion using complex extensions of EMD. *IEEE Transactions on Signal Processing, 57*(4), 1626–1630. doi:10.1109/TSP.2008.2011836

Loutridis, S. J. (2006). Instantaneous energy density as a feature for gear fault detection. *Mechanical Systems and Signal Processing, 20*(5), 1239–1253. doi:10.1016/j.ymssp.2004.12.001

Natke, H. G., & Cempel, C. (2001). The symptom observation matrix for monitoring and diagnostics. *Journal of Sound and Vibration, 248*(4), 597–620. doi:10.1006/jsvi.2001.3800

Ocak, H., Loparo, K. A., & Discenzo, F. M. (2007). Online tracking of bearing wear using wavelet packet decomposition and probabilistic modeling: A method for bearing prognostics. *Journal of Sound and Vibration, 302*(4-5), 951–961. doi:10.1016/j.jsv.2007.01.001

Pan, Y., Chen, J., & Li, X. (2010). Bearing performance degradation assessment based on lifting wavelet packet decomposition and fuzzy c-means. *Mechanical Systems and Signal Processing, 24*(2), 559–566. doi:10.1016/j.ymssp.2009.07.012

Patel, T. H., & Darpe, A. K. (2008). Vibration response of a cracked rotor in presence of rotor–stator rub. *Journal of Sound and Vibration, 317*(3-5), 841–865. doi:10.1016/j.jsv.2008.03.032

Patel, T. H., & Darpe, A. K. (2009). Use of full spectrum cascade for rotor rub identification. *Advances in Vibration Engineering, 8*(2), 139–151.

Patel, T. H., & Zuo, M. J. (2010). *The pump loop, its modifications, and experiments conducted during phase II of the slurry pump project* (p. 25). Edmonton, Canada: University of Alberta.

Peng, H. C., Long, F., & Ding, C. (2005). Feature selection based on mutual information: criteria of max-dependency, max-relevance, and min-redundancy. *IEEE Transactions on Pattern Analysis and Machine Intelligence, 27*(8), 1226–1238. doi:10.1109/TPAMI.2005.159

Peng, Z. K., Tse, P. W., & Chu, F. L. (2005). An improved Hilbert-Huang transform and its application in vibration signal analysis. *Journal of Sound and Vibration, 286*(1-2), 187–205. doi:10.1016/j.jsv.2004.10.005

Pires, V. F., Martins, J. F., & Pires, A. J. (2010). Eigenvector/eigenvalue analysis of a 3D current referential fault detection and diagnosis of an induction motor. *Energy Conversion and Management, 51*(5), 901–907. doi:10.1016/j.enconman.2009.11.028

Qiu, H., Lee, J., Lin, J., & Yu, G. (2003). Robust performance degradation assessment methods for enhanced rolling element bearing prognostics. *Advanced Engineering Informatics, 17*(3-4), 127–140. doi:10.1016/j.aei.2004.08.001

Rehman, N., & Mandic, D. P. (2010a). Empirical mode decomposition for trivariate signals. *IEEE Transactions on Signal Processing, 58*(3), 1059–1068. doi:10.1109/TSP.2009.2033730

Rehman, N., & Mandic, D. P. (2010b). Multivariate empirical mode decomposition. *Proceedings of the Royal Society A: Mathematical, Physical and Engineering Science, 466*(2117), 1291-1302.

Rilling, G., Flandrin, P., Goncalves, P., & Lilly, J. M. (2007). Bivariate empirical mode decomposition. *IEEE Signal Processing Letters, 14*(12), 936–939. doi:10.1109/LSP.2007.904710

Sait, A. S., & Sharaf-Eldeen, Y. I. (2011). *A review of gearbox condition monitoring based on vibration analysis techniques diagnostics and prognostics*, Vol. 5. Paper presented at the Rotating Machinery, Structural Health Monitoring, Shock and Vibration.

Tandon, N., & Choudhury, A. (1999). A review of vibration and acoustic measurement methods for the detection of defects in rolling element bearings. *Tribology International, 32*(8), 469–480. doi:10.1016/S0301-679X(99)00077-8

Tse, P. W., Peng, Y. H., & Yam, R. (2001). Wavelet analysis and envelope detection for rolling element bearing fault diagnosis---Their effectiveness and flexibilities. *Journal of Vibration and Acoustics, 123*(3), 303–310. doi:10.1115/1.1379745

Turhan-Sayan, G. (2005). Real time electromagnetic target classification using a novel feature extraction technique with PCA-based fusion. *IEEE Transactions on Antennas and Propagation, 53*(2), 766–776. doi:10.1109/TAP.2004.841326

Widodo, A., & Yang, B.-S. (2007). Application of nonlinear feature extraction and support vector machines for fault diagnosis of induction motors. *Expert Systems with Applications, 33*(1), 241–250. doi:10.1016/j.eswa.2006.04.020

Wikipedia. (2008). *Low-discrepancy sequence.* Retrieved from http://en.wikipedia.org/wiki/Low-discrepancy_sequence

Wu, F. Q., & Meng, G. (2006). Compound rub malfunctions feature extraction based on full-spectrum cascade analysis and SVM. *Mechanical Systems and Signal Processing, 20*(8), 2007–2021. doi:10.1016/j.ymssp.2005.10.004

Yesilyurt, I., & Ozturk, H. (2006). Tool condition monitoring in milling using vibration analysis. *International Journal of Production Research, 45*(4), 1013–1028. doi:10.1080/00207540600677781

Zhang, S., Hodkiewicz, M., Ma, L., & Mathew, J. (2006). *Machinery condition prognosis using multivariate analysis engineering asset management* (pp. 847–854). London, UK: Springer.

Zhao, X., Hu, Q., Lei, Y., & Zuo, M. J. (2010a). Vibration-based fault diagnosis of slurry pump impellers using neighbourhood rough set models. *Proceedings of the Institution of Mechanical Engineers. Part C, Journal of Mechanical Engineering Science, 224*(4), 995–1006. doi:10.1243/09544062JMES1777

Zhao, X., Patel, T. H., Sahoo, A., & Zuo, M. J. (2010b). *Numerical simulation of slurry pumps with leading or trailing edge damage on the impeller* (p. 34). Edmonton, Canada: University of Alberta.

Zhao, X., Patel, T. H., & Zuo, M. J. (2011a). Multivariate EMD and full spectrum based condition monitoring for rotating machinery. *Mechanical Systems and Signal Processing.*

Zhao, X., Zuo, M. J., & Liu, Z. (2011b). *Diagnosis of pitting damage levels of planet gears based on ordinal ranking.* Paper presented at the IEEE International Conference on Prognostics and Health management, Denver, U.S.

Zhao, X., Zuo, M. J., & Patel, T. H. (2011c). (Manuscript submitted for publication). Generating an indicator for pump impeller damage using half and full spectra, fuzzy preference based rough sets, and PCA. *Measurement Science & Technology.*

## ADDITIONAL READING

Baccianella, S., Esuli, A., & Sebastiani, F. (2010). *Feature selection for ordinal regression.* Paper presented at the 2010 ACM Symposium on Applied Computing, New York.

Gaudette, L., & Japkowicz, N. (2009). Evaluation methods for ordinal classification. In Gao, Y., & Japkowicz, N. (Eds.), *Advances in artificial intelligence* (pp. 207–210). Berlin, Germany: Springer. doi:10.1007/978-3-642-01818-3_25

Geng, X., Liu, T. Y., Qin, T., & Li, H. (2007). *Feature selection for ranking.* Paper presented at the 30th Annual International ACM SIGIR Conference on Research and Development in Information Retrieval, Amsterdam.

Hastie, R. T., & Friedman, J. (2009). *The elements of statistical learning: Data mining, inference, and prediction.* Springer. doi:10.1007/BF02985802

Jardine, A. K. S., Lin, D., & Banjevic, D. (2006). A review on machinery diagnostics and prognostics implementing condition-based maintenance. *Mechanical Systems and Signal Processing, 20*(7), 1483–1510. doi:10.1016/j.ymssp.2005.09.012

Lei, Y., & Zuo, M. J. (2009). Gear crack level identification based on weighted K nearest neighbor classification algorithm. *Mechanical Systems and Signal Processing, 23*(5), 1535–1547. doi:10.1016/j.ymssp.2009.01.009

Lei, Y., Zuo, M. J., He, Z. J., & Zi, Y. Y. (2010). A multidimensional hybrid intelligent method for gear fault diagnosis. *Expert Systems with Applications, 37*(2), 1419–1430. doi:10.1016/j.eswa.2009.06.060

Lin, H.-T. (2008). *From ordinal ranking to binary classification. Doctor of Philosophy, California Institute of Technology*. Pasadena: Unites States.

Loutridis, S. J. (2004). Damage detection in gear systems using empirical mode decomposition. *Engineering Structures, 26*(12), 1833–1841. doi:10.1016/j.engstruct.2004.07.007

Martin, H. (1992). Detection of gear damage by statistical vibration analysis. *Proceedings of the Institute of Mechanical Engineers, Part C: Journal of Mechanical Engineering Science*, (pp. 395-401).

Peng, H. C., Long, F., & Ding, C. (2005). Feature selection based on mutual information: Criteria of max-dependency, max-relevance, and min-redundancy. *IEEE Transactions on Pattern Analysis and Machine Intelligence, 27*(8), 1226–1238. doi:10.1109/TPAMI.2005.159

Rahman, M., Wiratunga, N., Lothian, R., Chakraborti, S., & Harper, D. (2007). *Information gain feature selection for ordinal text classification using probability redistribution.* Paper presented at the In proceedings of the IJCAI07 workshop on texting mining and link analysis, Hyderabad IN.

Samuel, P. D., & Pines, D. J. (2005). A review of vibration-based techniques for helicopter transmission diagnostics. *Journal of Sound and Vibration, 282*(1-2), 475–508. doi:10.1016/j.jsv.2004.02.058

*The R Project for Statistical Computing*. (n.d.). Retrieved from http://www.r-project.org/

Vachtsevanos, G. L. F., Romer, M., Hess, A., & Wu, B. (2006). *Intelligent fault diagnosis and prognosis for engineering for engineering systems.* Hoboken, NJ: John Wiley Sons, Inc. doi:10.1002/9780470117842

Yu Lei, L. H. (2004). Efficient feature selection via analysis of relevance and redundancy. *Journal of Machine Learning Research, 5*, 1205–1224.

Zhang, B., Taimoor, K., Romano, P., & George, V. (2008). Blind deconvolution denoising for helicopter vibration signals. *IEEE/ASME Transactions on Mechatronics, 13*(5), 558–565. doi:10.1109/TMECH.2008.2002324

## KEY TERMS AND DEFINITIONS

**Diagnosis of Fault Levels:** A process of identifying the severity of a fault in a device (machine).

**Indicator:** A parameter that is sensitive enough to represent the health condition of a device (machine).

**Monotonic Trend:** A special relation between two variables; when one variable increases, the other variable either consistently increases and never decreases or consistently decreases and never increases.

**Monotonic Relevance:** A quantity measure of the monotonic trend between the two variables.

**Nominal Variable:** A nominal variable is the one that has two or more categories, but there is no intrinsic ordering to the elements. For example, gender is a nominal variable having two elements (male and female).

**Ordinal Information:** The information contained among different elements of an ordinal variable.

**Ordinal Variable:** An ordinal variable is the one whose elements are ordered in a meaningful sequence. For example, the grades for students, e.g. A+, A, A-, B+, …, F, is an ordinal variable. We know 'A+' is better than 'B+' but we cannot quantitatively say 'how much better'.

# Chapter 5
# Fault Detection and Isolation for Switching Systems using a Parameter–Free Method

**Assia Hakem**
*Lille 1 University, France*

**Komi Midzodzi Pekpe**
*Lille 1 University, France*

**Vincent Cocquempot**
*Lille 1 University, France*

## ABSTRACT

*This chapter addresses the problem of Fault Detection and Isolation (FDI) of switching systems where faulty behaviors are represented as faulty modes. The objectives of the online FDI are to identify accurately the current mode and to estimate the switching time. A data-based method is considered here to generate residuals for FDI. Conditions of two important properties, namely discernability between modes and switching detectability, are established. These conditions are different for the two properties. More specifically, it will be shown that a switching occurrence may be detected even if the two considered modes are not discernible. A vehicle rollover prevention example is provided for illustration purpose.*

## 1 INTRODUCTION

In this chapter we deal with a class of hybrid systems, namely switching systems. Switching systems consist of dynamical modes which are modeled in state space form, and a switching logic that governs the transitions between modes. Switching may be due to internal variables (in-

put, output, state variables, system faults), or external actions (human operators, environment conditions).

The dynamical modes may depict faulty behaviors. The objectives of online fault detection and isolation (FDI) are thus to recognize accurately the current mode and to estimate the switching time.

Indeed, if the system switches to a healthy mode, the system can continue to operate, but if the new mode is a failure mode, it has to be

DOI: 10.4018/978-1-4666-2095-7.ch005

recognized as soon as possible and then a corresponding action must be done to prevent the failure consequences on the system itself or its environment. FDI is an important task to ensure the system dependability in providing online pertinent informations on the system state not only to the human operators through a supervision interface, but also directly to the control system that may be re-configured adequately. Several methods related to FDI design methods have been proposed and many results are scattered in the literature. Model-based and model-free approaches are generally distinguished:

- In model-based methods, the principle is to generate mode-dedicated fault indicator called residual. This residual is calculated as a difference between the model and the process behaviors. A residual is close to zero in fault free case and differs from zero if a fault occurs. Several model-based FDI techniques (Patton, 1994; Simani et al., 2003) have been proposed in the literature: state observers have been used in (Palma, 2002), the parity-space approach has been proposed in (Chow and Willsky, 1984) and extended for switching system in (Cocquempot et al., 2004) and (Domlan et al., 2007), Kalman filters has been developed in (Frank and Ding, 1997).

- In model-free methods, the online available input and output data are analyzed to determine some behavior characteristics. Several techniques have been proposed in the literature as learning procedures in (Fussel and Balle, 1997), pattern recognition techniques in (Casimir et al., 2003) and Principal Component Analysis techniques in (Ding et al., 2010; Harkat et al., 2003). In some cases, the model structure (for instance, as in this chapter, linear structure) is known, but parameters are not known or are not easily identifiable.

This chapter presents a novel residual generation data-based method for FDI in linear switching systems where the parameters values of the linear models are not known. Previous publications have introduced the proposed method for different model structures (bilinear (Hakem et al., 2011) and linear (Pekpe, 2004). The evaluation form of the data-based residual is obtained by projecting an output matrix onto the kernel of an input Hankel matrix. The first contribution of that chapter is to extend the technique in the multiple mode case. Each mode is supposed to be linear and stable. It is shown how this residual can be used for switching detection, mode recognition and sensor FDI.

It is classically adopted that a necessary condition for mode recognition and switching detection, is that the modes must be discernable (or distinguishable). The discernability conditions between modes are actually the subject of intensive studies (Cocquempot et al., 2003; Domlan et al., 2007). The detectability of the switching is often likened to the discernability between modes. The second contribution of the chapter is to establish discernability and detectability conditions. Moreover, it will be proved that discernability and detectability have two different sets of conditions. As a consequence, even if two modes are not discernable, it will be possible to detect the mode switching.

The chapter is organized in 6 sections. Section 2 describes the proposed method objectives to be achieved and the needed assumptions. Section 3 depicts the proposed method for a simple case of switching systems. Section 4 is devoted to switching time estimation and current mode recognition. In section 5, a simulated example of a real world example is provided.

## 2 PROBLEM SETTING

Considering the following discrete-time switching system with $d$ modes:

$$\begin{cases} x_{k+1} = A_{\sigma_k} x_k + B_{\sigma_k} u_k \\ y_k = C_{\sigma_k} x_k + D_{\sigma_k} u_k + w_k \\ \sigma_k : \mathbb{N} \to \{1, 2, \ldots, d\} \end{cases} \qquad (1)$$

where $\sigma_k$ is the active mode index and $A_{\sigma_k} \in \mathbb{R}^{n_{\sigma_k} \times n_{\sigma_k}}$, $B_{\sigma_k} \in \mathbb{R}^{n_{\sigma_k} \times m}$, $C_{\sigma_k} \in \mathbb{R}^{\ell \times n_{\sigma_k}}$, $D_{\sigma_k} \in \mathbb{R}^{\ell \times m}$ are unknown system matrices, the vector $x_k \in \mathbb{R}^{n_{\sigma_k}}$ is the unknown system state. The vectors $u_k \in \mathbb{R}^m$ and $y_k \in \mathbb{R}^\ell$ are respectively the switching system inputs and outputs, where the system outputs are corrupted by a colored centered noise $w_k \in \mathbb{R}^\ell$. Notice that the order $n$ is subscripted by $\sigma_k$ which shows that modes may have different orders. $0_{a \times b}$ stands for the $a \times b$ zeros matrix.

## 2.1 Objectives

The aim consists in monitoring the switching system represented by (1) with unknown parameters, using only input-output data. The monitoring involves doing at each time instant the following tasks:

- Determine switching detection and active mode recognition conditions
- Estimate switching times $\tau$,
- Recognize active mode $\sigma_k$,
- Detect and isolate multiple sensor faults.

## 2.2 Assumptions

It is supposed that the following assumptions hold:

- Each matrix $A_{\sigma_k}$ is stable,
- The system stays a sufficiently long time in one mode. There is no Zeno phenomenon.

## 2.3 General Principle of the Data-Based Residual Generation Method

Let us first give a general principle of our residual generation method. Under stability conditions, it is possible to express the vector of outputs on a given time window as a function of the inputs and system parameters. The following expression is thus obtained.

$$Y = HU \qquad (2)$$

where $H$ depends only on system parameters, $U$ and $Y$ are matrices of inputs and outputs collected on a given time window. If the chosen time window is sufficiently large, we can then project Equation (2) on the right kernel $\Pi$ of $U$ ($U\Pi = 0$) and we can derive the residual:

$$\varepsilon = Y\Pi \qquad (3)$$

This residual equals zero in absence of disturbances and faults. When a fault occurs, $\varepsilon$ becomes different from zero and it can be used for FDI.

It is clear that no system parameter or state estimation is needed for residual computation since $\Pi$ depends only on inputs, which makes residual expression (3) independent on model parameters.

## 3 SENSOR FAULT DETECTION AND ISOLATION FOR ONE MODE

### 3.1 Data-Based Residual Method

A sensor fault detection and isolation method is provided here for one mode. This method will be extended in section 6 for switching systems. Since only one mode is active, the switching system can be represented without the subscript $\sigma_k$ and Equation (1) becomes:

$$\begin{cases} x_{k+1} = Ax_k + Bu_k \\ y_k = Cx_k + Du_k + f_k + w_k \end{cases} \qquad (4)$$

where $f_k \in \mathbb{R}^{\ell \times 1}$ represents sensor faults.

**Proposition 1:** The parameter-free residual for sensor fault detection and isolation is given by:

$$\bar{\varepsilon}_k = \bar{Y}_k \bar{\Pi}_k Z \in \mathbb{R}^\ell \qquad (5)$$

where $\bar{\Pi}_k \in \mathbb{R}^{L \times L}$ is the right kernel of the extended Hankel matrix $\bar{U}_k \in \mathbb{R}^{m(i+1) \times L}$ $\left( \bar{U}_k \bar{\Pi}_k = 0_{m(i+1) \times L} \right)$, this right kernel exists if the number of independent columns of $\bar{U}_k$ is greater than the number of independent rows (if all these columns and rows are independent this condition is $L > m(i+1) + \ell$). $\bar{U}_k$ and $\bar{Y}_k \in \mathbb{R}^{\ell \times L}$ are input and output matrices respectively. The vector $Z \in \mathbb{R}^\ell$ is a selective matrix. These matrices are defined as:

$$\bar{Y}_k = [y_{k-L+1} \cdots y_k] \qquad (6)$$

$$\bar{U}_k = \begin{bmatrix} \bar{u}_{k-L+1,i} & \bar{u}_{k-L+2,i} & \cdots & \bar{u}_{k,i} \end{bmatrix} \qquad (7)$$

with

$$\bar{u}_{k,i} = \begin{pmatrix} u_{k-i}^T & u_{k-i+1}^T & \cdots & u_k^T \end{pmatrix}^T \in \mathbb{R}^{m(i+1) \times 1},$$

$$Z = \begin{pmatrix} 0 & \cdots & 0 & 1 \end{pmatrix}^T$$

*Remark:* From Equation (5), it is clear that no model parameter is necessary to calculate the residual $\bar{\varepsilon}_k$, and only online input $(\bar{U}_k)$ and output $(\bar{Y}_k)$ measurements are used. Model param-

eters will be used in the following to show how Equation (5) may be obtained and to prove the sensitivity of $\bar{\varepsilon}_k$ to sensor faults.

By repeated substitution of Equation (4), the following relation can be obtained:

$$y_k = CA^i x_{k-i} + H_i \bar{u}_{k,i} + f_k + w_k \qquad (8)$$

where

$$H_i = \begin{bmatrix} CA^{i-1}B & \cdots & CB & D \end{bmatrix} \qquad (9)$$

is the Markov parameters matrix of order $i$.

Under stability hypothesis of system (4) $\|A^i\|$ tends to zeros (where $\|\cdot\|$ is a multiplicative norm), when $i$ tends to infinity:

$$\lim_{i \to \infty} \|A^i\| = 0 \qquad (10)$$

Therefore Equation (8) becomes:

$$y_k \cong H_i \bar{u}_{k,i} + f_k + w_k \qquad (11)$$

Concatenating Equation (11) on a window of size $L$ leads to:

$$\bar{Y}_k \cong H_i \bar{U}_k + F_k + W_k \qquad (12)$$

where $W_k$ and $F_k$ are constructed similarly as $\bar{Y}_k$.

Let us define $\bar{\Pi}_k$ the right kernel of the extended Hankel matrix $\bar{U}_k$. The proposed parameter-free residual for sensor fault detection and isolation in discrete-time linear switched systems is obtained by right multiplying Equation (12) by $\bar{\Pi}_k$ in order to eliminate the effect of inputs and consequently the matrix $H_i$ of system parameters.

The fault detectability condition is that $span(\bar{U}_k)$ is not included in $span(F_k)$. If this condition is not fulfilled, the fault cannot be detected since projecting on $\bar{\Pi}_k$ will eliminate the effect of faults on the residual.

Therefore, the computational form of the residual vector is given by the following relation:

$$\bar{\varepsilon}_k = \bar{Y}_k \bar{\Pi}_k Z \tag{13}$$

The evaluation form of the proposed residual is given by:

$$\bar{\varepsilon}_k \cong \underbrace{H_i \bar{U}_k \bar{\Pi}_k Z}_{=0} + F_k \bar{\Pi}_k Z + W_k \bar{\Pi}_k Z \tag{14}$$

$$\bar{\varepsilon}_k \cong F_k \bar{\Pi}_k Z + W_k \bar{\Pi}_k Z \tag{15}$$

The sensor fault vector appears in the evaluation form of the proposed residual (15), which makes $\bar{\varepsilon}_k$ suitable for sensor fault detection.

### 3.2 Sensitivity of Residual to Faults

Equation (15) shows that $\bar{\varepsilon}_k$ is a function of noise $w_k$, sensor fault $f_k$ and input $u_k$ collected in a time-window. The residual value in fault-free case and faulty case may be studied:

- **Fault-free case** $\Leftrightarrow F_k = 0_{\ell \times L}$: The residual (15) becomes: $\bar{\varepsilon}_k = \bar{\Pi}_k Z$, and its mathematical expectation is given by:

$$\begin{aligned}
E[\bar{\varepsilon}_k] &= E[W_k \bar{\Pi}_k Z] \\
&= E[W_k] E[\bar{\Pi}_k Z] \\
&= \underbrace{E[W_k]}_{=0} \bar{\Pi}_k Z = 0
\end{aligned} \tag{16}$$

since $w_k$ is a centered colored noise, $\bar{\Pi}_k Z$ is a deterministic signal independent of the noise $W_k$.

- **Faulty case** $\Leftrightarrow F_k \neq 0_{\ell \times L}$: The mathematical expectation given by

$$\begin{aligned}
E[\bar{\varepsilon}_k] &= E[F_k \bar{\Pi}_k Z] + E[W_k \bar{\Pi}_k Z] \\
&= F_k \bar{\Pi}_k Z + \underbrace{E[W_k]}_{=0} E[\bar{\Pi}_k Z]
\end{aligned} \tag{17}$$

becomes different from zero.

That proves that the residual mean changes when a sensor fault occurs.

### 3.3 Sensor Fault Isolation

The residual vector can be written as follows:

$$\bar{\varepsilon}_k = \begin{pmatrix} \bar{\varepsilon}_k(1) \\ \vdots \\ \bar{\varepsilon}_k(z) \\ \vdots \\ \bar{\varepsilon}_k(\ell) \end{pmatrix} = \begin{pmatrix} F_k(1,:)\bar{\Pi}_k Z + W_k(1,:)\bar{\Pi}_k Z \\ \vdots \\ F_k(z,:)\bar{\Pi}_k Z + W_k(z,:)\bar{\Pi}_k Z \\ \vdots \\ F_k(\ell,:)\bar{\Pi}_k Z + W_k(\ell,:)\bar{\Pi}_k Z \end{pmatrix}$$
$$(1 \leq z \leq \ell)$$

If only one single fault occurs in sensor $z$, it implies that $F_k(z,:)$ is not null, which makes only the corresponding residual component $\varepsilon_k(z)$ different from zero.

$$E[\bar{\varepsilon}_k] = \begin{pmatrix} 0 \\ \vdots \\ 0 \\ F_k(z,:)\bar{\Pi}_k Z \\ 0 \\ \vdots \\ 0 \end{pmatrix}$$

As a consequence the generated residual is structured with respect to the sensor faults, and faults can be easily isolated.

# 4 SWITCHING TIME ESTIMATION USING ONLY ONLINE INPUT-OUTPUT DATA

The switching time estimation is an online process, where the switching from the previous to the current mode is detected and its time of occurrence is estimated. The following proposition shows how the data-based proposed residual is used for switching time estimation.

**Proposition 2:** If $\exists \overline{\ell} \in \{1, 2, ..., \ell\}$, then the switching occurs at $\tau = k$. Where the time instant $k$ is the first instant that makes $E[\overline{\varepsilon}_k(\overline{\ell})] \neq 0$.

This proposition holds under detectability conditions which are detailed later in this chapter. Let s1 and s2 be the previous and the current modes respectively, with their corresponding online collected data $(U_{(s1),k-L+1:\tau}, Y_{(s1),k-L+1:\tau})$ and $(U_{(s2),\tau+1:k}, Y_{(s2),\tau+1:k})$ respectively.

Let construct the output vector:

$$Y_{k-L+1:k} = \left[Y_{(s1),k-L+1:\tau} \mid Y_{(s2),\tau+1:k}\right]$$

It yields Equation 18 where $H_{(s1),i}$ is defined as in (9), by putting $A_{s1}$, $B_{s1}$, $C_{s1}$, $D_{s1}$ instead of $A$, $B$, $C$, $D$ respectively (same for $H_{(s2),i}$),

where $U^* \in \mathbb{R}^{mi(i+1) \times L}$ is defined as the equations in Box 1.

The contribution of the initial state can be neglected, the reason is the following:

The state is multiplied by a term of general form $A_{s2}^j A_{s1}^{i-j}$ as shown in the Equation (18) where $0 \leq j \leq i$ so that the power sum is always $i$.

Using the sub-multiplicative norm theorem on the norm function $A_{s1}^j A_{s2}^{i-j} < A_{s1}^j A_{s2}^{i-j}$ and we have $A_{s1}^j A_{s2}^{i-j} < (max(A_{s1}, A_{s2}))^i$. It is supposed that $i$ is sufficiently large and $max(A_{s1}, A_{s2})^i$ is neglected, therefore $A_{s1}^j A_{s2}^{i-j}$ is neglected too.

In order to make $U_{(s1),k-L+1:k}$ appear, we add and subtract

$$H_{(s1),i} \left[0_{m(i+1) \times \tau - k + L} \mid U_{(s1,s2),\tau+1:\tau+i} \mid U_{(s2),\tau+i+1:k}\right]$$

in Equation (18), then another similar expression of the resulting input matrix $Y_{(\gamma,s1,s2),k-\frac{L}{2}+1:k}$ is given by Equation 19.

By projecting Equation (19) on $\Pi_{(s1,s2),k-L+1:k}$ which is the right kernel of

$$\left[U_{(s1),k-L+1:\tau} \mid U_{(s1,s2),\tau+1:\tau+i} \mid U_{(s2),\tau+i+1:k}\right],$$

and choosing its last column, the evaluation form of the residual proposed in Equation (5) is Equation 20.

*Equation 18.*

$$
\begin{aligned}
Y_{k-L+1:k} = &\left[C_{s1}A_{s1}^i x_{k-L-i+1} \mid \cdots \mid C_{s1}A_{s1}^i x_{\tau-i} \mid C_{s2}A_{s1}^i x_{\tau-i+1} \mid \right. \\
&C_{s2}A_{s2}A_{s1}^{i-1} x_{\tau-i+2} \mid \cdots \mid C_{s2}A_{s2}^{i-3}A_{s1}^3 x_{\tau-2} \mid C_{s2}A_{s2}^{i-2}A_{s1}^2 x_{\tau-1} \mid \\
&C_{s2}A_{s2}^{i-1}A_{s1} x_{\tau} \mid C_{s2}A_{s2}^i x_{\tau+1} \mid \cdots \mid C_{s2}A_{s2}^i x_{k-i-1} \mid \left. C_{s2}A_{s2}^i x_{k-i}\right] \\
&+ H_{(s1),i}\left[U_{(s1),k-L+1:\tau} \; 0_{m(i+1) \times k-\tau}\right] + \left[H_{(s1,s2),1} \cdots \mid H_{(s1,s2),i-1} \mid H_{(s1,s2),i}\right] U^* \\
&+ H_{(s2),i}\left[0_{m(i+1) \times L-k+\tau+i} \; U_{(s2),\tau+i+1:k}\right] + W_k
\end{aligned}
$$

*Box 1.*

$$U^* = \begin{bmatrix} 0 & | & U^\bullet & | & 0 \\ _{mi(i+1)\times\tau-k+\frac{L}{2}} & & & & _{mi(i+1)\times k-\tau-i+\frac{L}{2}} \end{bmatrix}$$

and

$$U^\bullet = \begin{bmatrix} \bar{u}_{\tau+1} & 0_{m(i+1)\times 1} & \cdots & \cdots & 0_{m(i+1)\times 1} \\ 0_{m(i+1)\times 1} & \bar{u}_{\tau+2} & \ddots & & \vdots \\ \vdots & \ddots & \ddots & \ddots & \vdots \\ \vdots & & \ddots & \ddots & 0_{m(i+1)\times 1} \\ 0_{m(i+1)\times 1} & \cdots & \cdots & 0_{m(i+1)\times 1} & \bar{u}_{\tau+i} \end{bmatrix}$$

$$H_{(s1,s2),1} = \begin{bmatrix} C_{s2}A_{s1}^{i-1}B_{s1} & | & C_{s2}A_{s1}^{i-2}B_{s1} & | & \cdots & | & C_{s2}B_{s1} & | & D_{s2} \end{bmatrix}$$

$$\vdots$$

$$H_{(s1,s2),i-1} = \begin{bmatrix} C_{s2}A_{s2}^{i-2}A_{s1}B_{s1} & | & C_{s2}A_{s2}^{i-2}B_{s1} & | & \cdots & | & C_{s2}B_{s2} & | & D_{s2} \end{bmatrix}$$

$$H_{(s1,s2),i} = \begin{bmatrix} C_{s2}A_{s2}^{i-1}B_{s1} & | & C_{s2}A_{s1}^{i-2}B_{s1} & | & \cdots & | & C_{s2}B_{s1} & | & D_{s2} \end{bmatrix}$$

If a switching occurs, in other words $s1 \neq s2$, then $E[\varepsilon_{(s1,s2),k}] \neq 0$, otherwise $(s1 = s2)$ this expectation is $E[\bar{\varepsilon}_{(s1,s2),k}] = 0$.

# 5 SWITCHING TIME ESTIMATION AND CURRENT MODE RECOGNITION USING ONLINE AND OFFLINE INPUT-OUTPUT DATA

The objective of this section is to estimate switching times and to recognize active mode using only collected data. In order to achieve this goal, we use

*Equation 19.*

$$Y_{(s1,s2),k-L+1:k} = W_k + H_{(s1),i}\begin{bmatrix} U_{(s1),k-L+1:\tau} & | & U_{(s1,s2),\tau+1:\tau+i} & | & U_{(s2),\tau+i+1:k} \end{bmatrix}$$
$$+ \begin{bmatrix} H_{(s1,s2),1} - H_{(s1),i} & | & \cdots & | & H_{(s1,s2),i-1} - H_{(s1),i} & | & H_{(s1,s2),i} - H_{(s1),i} \end{bmatrix} U^*$$
$$+ (H_{(s2),i} - H_{(s1),i})\begin{bmatrix} 0_{m(i+1)\times L-k+\tau+i} & | & U_{(s2),\tau+i+1:k} \end{bmatrix}$$

*Equation 20.*

$$\bar{\varepsilon}_{(s1,s2),k} = W_k \Pi_{(s1,s2),k-L+1:k} Z$$
$$+ \begin{bmatrix} H_{(s1,s2),1} - H_{(s1),i} & | & \cdots & | & H_{(s1,s2),i-1} - H_{(s1),i} & | & H_{(s1,s2),i} - H_{(s1),i} \end{bmatrix} U^* \Pi_{(s1,s2),k-L+1:k} Z$$
$$+ (H_{(s2),i} - H_{(s1),i})\begin{bmatrix} 0_{m(i+1)\times L-k+\tau+i} & | & U_{(s2),\tau+i+1:k} \end{bmatrix} \Pi_{(s1,s2),k-L+1:k} Z$$

*Algorithm 1. Input-output data matrices construction*

---

- "Labeled database $u_{(\gamma),\bullet}$ and $y_{(\gamma),\bullet}$ with known label $\gamma$":

An essential step for the proposed method is to collect a sufficient database ($\frac{L}{2}$ samples) from each mode offline, for example by constraining the system to operate in a healthy given mode (basically when the system is running for the first time). Therefore, the database $U_{(\gamma),1:\frac{L}{2}}$ and $Y_{(\gamma),1:\frac{L}{2}}$ is labeled by one element of the vector $\gamma = [1, 2, ..., d]$, these inputs-outputs database are organized in matrices as follows:

For input database: $U_{(\gamma),1:\frac{L}{2}} = \left[ \overline{u}_{(\gamma),i+1} \cdots \overline{u}_{(\gamma),p} \cdots \overline{u}_{(\gamma),i+\frac{L}{2}} \right] \in \mathbb{R}^{m(i+1)\times\frac{L}{2}}$

For output database: $Y_{(\gamma),1:\frac{L}{2}} = \left[ y_{(\gamma),1} \cdots y_{(\gamma),\frac{L}{2}} \right] \in \mathbb{R}^{\ell\times\frac{L}{2}}$.

Where $\overline{u}_{(\gamma),p} = \left( u^T_{(\gamma),p-i} \; u^T_{(\gamma),p-i+1} \; \cdots \; u^T_{(\gamma),p} \right)^T \in \mathbb{R}^{m(i+1)\times 1}$ with $p \in \left\{ 1, 2, ..., \frac{L}{2} \right\}$.

- "Labeled current mode data $u_{(\sigma_k),\bullet}$ and $y_{(\sigma_k),\bullet}$ with unknown label $\sigma_k$":

The next step is to collect data online from the current mode referred by $\sigma_k$, input-output current data are given as follows:

For current input: $U_{(\sigma_k),k-\frac{L}{2}+1:k} = \left[ \overline{u}_{(\sigma_k),k-\frac{L}{2}+1} \;\middle|\; \overline{u}_{(\sigma_k),k-\frac{L}{2}+2} \;\middle|\; \cdots \;\middle|\; \overline{u}_{(\sigma_k),k} \right] \in \mathbb{R}^{m(i+1)\times\frac{L}{2}}$

For current output: $Y_{(\sigma_k),k-\frac{L}{2}+1:k} = \left[ y_{(\sigma_k),k-\frac{L}{2}+1} \;\middle|\; y_{(\sigma_k),k-\frac{L}{2}+2} \;\middle|\; \cdots \;\middle|\; y_{(s_k),k} \right] \in \mathbb{R}^{\ell\times\frac{L}{2}}$.

---

offline labeled input-output measurements in addition to online data. As in the previous method model parameters $A_{\sigma_k}, B_{\sigma_k}, C_{\sigma_k}, D_{\sigma_k}$ are not used.

## 5.1 Data-Based Residual

In order to recognize the active mode, offline labeled data are used together with online data. These data are concatenated in the input-output matrices using Algorithms 1 and 2.

As a last step, the previous constructed data are used in Algorithm 2 to compute online the data-based residuals.

## 5.2 Mode Recognition

The mode recognition is an online process, where the current mode $\sigma_k$ is determined at each time

step. This operation is achieved without parameter estimation. The proposed residual is a data-based one, calculated using only input-output data collected online and concatenated with labeled database (with a known label $\gamma$). The following proposition shows how the data-based proposed residual is used for mode recognition.

**Proposition 3:** If all modes are discernable (see discernability condition for more details in the next subsection) and if there exists a mode $\gamma$ where $\gamma \in \left\{ 1, 2, ..., d \right\}$ whose residual is equal to zero ($E[\varepsilon_{(\gamma,\sigma_k),k}] \cong 0$), then the mode $\gamma$ is active ($\sigma_k = \gamma$).

If only one mode $\sigma_k$ is active without taking into account the switching occurrence, the evalu-

*Algorithm 2. Online computation of data-based residuals*

- $U_{(\gamma),1:\frac{L}{2}}$ and $U_{(\sigma_k),k-\frac{L}{2}+1:k}$ are offline and online inputs respectively.

- $Y_{(\gamma),1:\frac{L}{2}}$ and $Y_{(\sigma_k),k-\frac{L}{2}+1:k}$ are offline and online outputs respectively.

are concatenated together to form a new input and output matrices $U_{(\gamma,\sigma_k),k-\frac{L}{2}+1:k}$ and $Y_{(\gamma,\sigma_k),k-\frac{L}{2}+1:k}$ respectively defined as follows:

- $U_{(\gamma,\sigma_k),k-\frac{L}{2}+1:k} = [U_{(\gamma),1:\frac{L}{2}} \mid U_{(\sigma_k),k-\frac{L}{2}+1:k}] \in \mathbb{R}^{m(i+1)\times L}$ is the resulting input matrix.

- $Y_{(\gamma,\sigma_k),k-\frac{L}{2}+1:k} = [Y_{(\gamma),1:\frac{L}{2}} \mid Y_{(\sigma_k),k-\frac{L}{2}+1:k}] \in \mathbb{R}^{\ell\times L}$ is the resulting output matrix.

The resulting input and output matrices $U_{(\gamma,\sigma_k),k-\frac{L}{2}+1:k}$ and $Y_{(\gamma,\sigma_k),k-\frac{L}{2}+1:k}$ respectively are used for a data-based residual $\varepsilon_{(\gamma,\sigma_k),k}$ computation given by:

$$\varepsilon_{(\gamma,\sigma_k),k} = Y_{(\gamma,\sigma_k),k-\frac{L}{2}+1:k} \Pi_{(\gamma,\sigma_k),k-\frac{L}{2}+1:k} Z \in \mathbb{R}^{\ell} \quad (21)$$

Projection matrix $\Pi_{(\gamma,\sigma_k),k-\frac{L}{2}+1:k}$ defines the right kernel of resulting input matrix

$U_{(\gamma,\sigma_k),k-\frac{L}{2}+1:k}$, where $U_{(\gamma,\sigma_k),k-\frac{L}{2}+1:k} \Pi_{(\gamma,\sigma_k),k-\frac{L}{2}+1:k} = 0$.

This projection matrix may be computed as follows:

$$\Pi_{(\gamma,\sigma_k),k-\frac{L}{2}+1:k} = I_{L\times L} - U^T_{(\gamma,\sigma_k),k-\frac{L}{2}+1:k} (U_{(\gamma,\sigma_k),k-\frac{L}{2}+1:k} U^T_{(\gamma,\sigma_k),k-\frac{L}{2}+1:k})^{-1} U_{(\gamma,\sigma_k),k-\frac{L}{2}+1:k} \quad (22)$$

ation of an output matrix $Y_{k-L+1:k} = [y_{k-L+1} \cdots y_k] \in \mathbb{R}^{\ell\times L}$ is obtained here in terms of:

- The model parameters represented using the Markov parameter matrix of the mode $\sigma_k$ at the order $i$:
$H_{(\sigma_k),i} = [C_{\sigma_k} A_{\sigma_k}^{i-1} B_{\sigma_k} \mid \cdots \mid C_{\sigma_k} B_{\sigma_k} \mid D_{\sigma_k}] \in \mathbb{R}^{\ell\times m(i+1)}$

- Inputs $U_{k-L+1:k} = [\bar{u}_{k-L+1} \ \bar{u}_{k-L+2} \ \cdots \ \bar{u}_k] \in \mathbb{R}^{m(i+1)\times L}$

- And noise $W_k$ constructed similarly as $Y_{k-L+1:k}$.

By repeated substitutions of Equation (1) the following relation can be obtained:

$$y_k = C_{\sigma_k} A_{\sigma_k}^i x_{k-i} + H_{(\sigma_k),i} \bar{u}_k + w_k \quad (23)$$

Under stability condition: $A_{\sigma_k 2}^i \rightarrow 0, \forall \sigma_k$ being constant.

Therefore, if $i$ is chosen sufficiently large, it allows writing the Equation (23) as follows:

$$y_k \cong H_{(\sigma_k),i} \bar{u}_k + w_k \quad (24)$$

By concatenating the output $y_k$ defined by the Equation (24) on a window of size $L$ leads to the output matrix $Y_{k-L+1:k}$ defined as follows:

$$Y_{k-L+1:k} \cong H_{(\sigma_k),i} U_{k-L+1:k} + W_k \quad (25)$$

Let us replace the first $\frac{L}{2}$ vectors of $U_{k-L+1:k}$ and $Y_{k-L+1:k}$ by the corresponding mode database

$U_{(\gamma),1:\frac{L}{2}}$ and $Y_{(\gamma),1:\frac{L}{2}}$ respectively, so the general expression of $Y_{(\gamma,\sigma_k),k-\frac{L}{2}+1:k}$ defined in Equation (25) becomes:

$$Y_{(\gamma,\sigma_k),k-\frac{L}{2}+1:k} = H_{(\gamma),i}\begin{bmatrix} U_{(\gamma),1:\frac{L}{2}} & 0_{m(i+1)\times\frac{L}{2}} \end{bmatrix}$$
$$+H_{(\sigma_k),i}\begin{bmatrix} 0_{m(i+1)\times\frac{L}{2}} & U_{(\sigma_k),k-\frac{L}{2}+1:k} \end{bmatrix} + W_k \quad (26)$$

In order to make $U_{(\gamma,\sigma_k),k-\frac{L}{2}+1:k}$ appear, we add and subtract $H_{(\gamma),i}\begin{bmatrix} 0_{m(i+1)\times\frac{L}{2}} & | & U_{(\sigma_k),k-\frac{L}{2}+1:k} \end{bmatrix}$ from Equation (26), the output matrix $Y_{(\gamma,\sigma_k),k-\frac{L}{2}+1:k}$ defined in Equation (26) becomes:

$$Y_{(\gamma,\sigma_k),k-\frac{L}{2}+1:k} = H_{(\gamma),i}\begin{bmatrix} U_{(\gamma),1:\frac{L}{2}} & U_{(\sigma_k),k-\frac{L}{2}+1:k} \end{bmatrix}$$
$$+\left(H_{(\sigma_k),i} - H_{(\gamma),i}\right)\begin{bmatrix} 0_{m(i+1)\times\frac{L}{2}} & U_{(\sigma_k),k-\frac{L}{2}+1:k} \end{bmatrix} + W_k \quad (27)$$

By right multiplying Equation (27) by $\Pi_{(\gamma,\sigma_k),k-\frac{L}{2}+1:k}$, and choosing its last column, the evaluation form of the residual proposed in equation (21) is given by Equation 28.

From Equation (28), if $\exists\gamma \in \{1,2,...,d\}$, where the active mode is $\sigma_k = \gamma$, we have

$$H_{(\sigma_k),i} = H_{(\gamma),i},$$

then

$$E[\varepsilon_{(\gamma,\sigma_k),k}] = E[W_k\Pi_{(\gamma,\sigma_k),k-\frac{L}{2}+1:k} Z] \cong 0.$$

## 5.3 Mode Discernability

The following definition is a general one, which means that this definition holds whatever the used method.

*Definition 1:* Two modes $(a, b)$ are discernable, if for the same input, the output generated from $a$ and $b$ are different.

The following definition depends on the proposed method, and it is based on the previous general definition. The previous definition implies the following definition, since the proposed residual is calculated using input-output data only.

*Definition 2:* Two modes $(a, b)$ are discernable on an observation window $[k-L+1,k]$ if their corresponding parameter-free residuals mathematical expectation $E[\varepsilon_{a,k}]$ and $E[\varepsilon_{b,k}]$ are not simultaneously null.

This is a data-based definition for mode discernability condition.

A new mode discernability condition is given for switching systems with linear modes in the following proposition:

**Proposition 4:** Two modes $a$ and $b$ are discernable if and only if the following conditions hold simultaneously:

1. Inputs are persistently exciting,

*Equation 28.*

$$\varepsilon_{(\gamma,\sigma_k),k} = \left(H_{(\sigma_k),i} - H_{(\gamma),i}\right)\begin{bmatrix} 0_{m(i+1)\times\frac{L}{2}} & U_{(\sigma_k),k-\frac{L}{2}+1:k} \end{bmatrix}\Pi_{(\gamma,\sigma_k),k-\frac{L}{2}+1:k} Z + W_k\Pi_{(\gamma,\sigma_k),k-\frac{L}{2}+1:k} Z$$

2. Minimal realizations of modes $a$ and $b$ are different, that implies all the Markov parameters of both modes are not the same: there exists $j \in \mathbb{R}$ where $H_{(a),j}$ is different from $H_{(b),j}$.

Let us take two residuals $\varepsilon_{(s1,\sigma_k),k}$ and $\varepsilon_{(s2,\sigma_k),k}$ calculated using their corresponding database $\gamma = s1$ and $\gamma = s2$ respectively, residuals are given by Equations 29 and 30.

If the residuals of both modes s1 and s2 have their mathematical expectations simultaneously null ($E[\varepsilon_{(s1,\sigma_k),k}] = E[\varepsilon_{(s2,\sigma_k),k}] = 0$), then knowing that $\Pi_{(s1,\sigma_k),k-\frac{L}{2}+1:k} = \Pi_{(s2,\sigma_k),k-\frac{L}{2}+1:k}$, it can be deduced from Equations (29) and (30) the following involvement:

$$E[\varepsilon_{(s1,\sigma_k),k}] = E[\varepsilon_{(s2,\sigma_k),k}] = 0 \Rightarrow H_{(s1),i} = H_{(s2),i} \tag{31}$$

## 5.4 Switching Time Estimation

The switching time estimation is an online process, where the switching from the previous to the current mode is detected and its time of occurrence is estimated. The following proposition shows how the data-based proposed residual is used for switching time estimation.

**Proposition 5:** If the following two statements hold:

- $\exists \gamma \in \{1, 2, ..., d\}$, where the mode $q_{k-1} = \gamma$ is active in time window $[k - \frac{L}{2} - i, k - 1]$ and recognized at $k - 1$ instant,

- At $k$ instant, the current unknown mode $\sigma_k$ makes $E[\varepsilon_{(q_{k-1},\sigma_k),k}] \neq 0$, then the switching is estimated to be occured at $\tau = k$.

Notice that only one residual among d computed residuals detects the switching earlier than the others, which is the one calculated using database corresponding to the previous active mode $q_{k-1}$.

Let s1 and s2 be the previous and the current mode respectively, with their corresponding online collected data $(U_{(s1),k-\frac{L}{2}+1:\tau}, Y_{(s1),k-\frac{L}{2}+1:\tau})$ and $(U_{(s2),\tau+1:k}, Y_{(s2),\tau+1:k})$ respectively.

By following the same procedure that allowed us to find Equation (25), by taking into account the switching phenomenon and the concatenated database $(U_{(\gamma),1:\frac{L}{2}}, Y_{(\gamma),1:\frac{L}{2}})$, a more detailed resulting output is given by

$$Y_{k-L+1:k} = \left[ Y_{(\gamma),1:\frac{L}{2}} \mid Y_{(s1),k-\frac{L}{2}+1:\tau} \mid Y_{(s2),\tau+1:k} \right]$$

which leads to Equation 32.

*Equations 29 and 30*

$$\varepsilon_{(s1,\sigma_k),k} = W_k \Pi_{(s1,\sigma_k),k-\frac{L}{2}+1:k} Z + (H_{(\sigma_k),i} - H_{(s1),i}) \left[ 0_{m(i+1)\times\frac{L}{2}} \mid U_{(\sigma_k),k-\frac{L}{2}+1:k} \right] \Pi_{(s1,\sigma_k),k-\frac{L}{2}+1:k} Z \tag{29}$$

$$\varepsilon_{(s2,\sigma_k),k} = W_k \Pi_{(s2,\sigma_k),k-\frac{L}{2}+1:k} Z + (H_{(\sigma_k),i} - H_{(s2),i}) \left[ 0_{m(i+1)\times\frac{L}{2}} \mid U_{(\sigma_k),k-\frac{L}{2}+1:k} \right] \Pi_{(s2,\sigma_k),k-\frac{L}{2}+1:k} Z \tag{30}$$

*Equation 32*

$$
\begin{aligned}
Y_{k-L+1:k} &= H_{(\gamma),i} \left[ U_{(\gamma),1:\frac{L}{2}} \mid 0_{m(i+1)\times\frac{L}{2}} \right] \\
&+ \left[ 0_{\ell\times\frac{L}{2}} \mid C_{s1}A_{s1}^{i}x_{k-\frac{L}{2}-i+1} \mid \cdots \mid C_{s1}A_{s1}^{i}x_{\tau-i} \mid C_{s2}A_{s1}^{i}x_{\tau-i+1} \mid \right. \\
&\quad C_{s2}A_{s2}A_{s1}^{i-1}x_{\tau-i+2} \mid \cdots \mid C_{s2}A_{s2}^{i-3}A_{s1}^{3}x_{\tau-2} \mid C_{s2}A_{s2}^{i-2}A_{s1}^{2}x_{\tau-1} \mid \\
&\quad \left. C_{s2}A_{s2}^{i-1}A_{s1}x_{\tau} \mid C_{s2}A_{s2}^{i}x_{\tau+1} \mid \cdots \mid C_{s2}A_{s2}^{i}x_{k-i-1} \mid C_{s2}A_{s2}^{i}x_{k-i} \right] \\
&+ H_{(s1),i} \left[ 0_{m(i+1)\times\frac{L}{2}} \mid U_{(s1),k-\frac{L}{2}+1:\tau} \mid 0_{m(i+1)\times k-\tau} \right] \\
&+ \left[ H_{(s1,s2),1} \mid \cdots \mid H_{(s1,s2),i-1} \mid H_{(s1,s2),i} \right] U^{*} \\
&+ H_{(s2),i} \left[ 0_{m(i+1)\times L-k+\tau+i} \mid U_{(s2),\tau+i+1:k} \right] + W_{k}
\end{aligned}
\tag{32}
$$

The contribution of the initial state can be neglected, for the same reason detailed in section 4.

In order to make $U_{(\gamma,s1,s2),k-\frac{L}{2}+1:k}$ appear, we add and subtract

$$
H_{(\gamma),i} \left[ 0_{m(i+1)\times\frac{L}{2}} \mid U_{(s1),k-\frac{L}{2}+1:\tau} \mid U_{(s1,s2),\tau+1:\tau+i} \mid U_{(s2),\tau+i+1:k} \right]
$$

from Equation (32), so another similar expression of the resulting input matrix $Y_{(\gamma,s1,s2),k-\frac{L}{2}+1:k}$ is given by Equation 33.

By projecting equation (33) on $\Pi_{(\gamma,s1,s2),k-L+1:k}$, and choosing its last column, the evaluation form of the residual given by equation (21) is Equation 34.

If a switching occurs, in other words $s1 \neq s2$, let us study residual $\varepsilon_{(\gamma,s1),k}$ that corresponds to $\gamma = s1$. From the first property of the proposed residual, the expectation of this residual $\varepsilon_{(\gamma=s1,s1),k<\tau}$ is zero till $\tau - 1$, and when a switching occurs at $\tau$, this residual becomes Equation 35 thus $E[\varepsilon_{(\gamma,s1,s2),k}] \neq 0$ where the transition of the residual expectation from a zero to a non zero value, is explained by a switching occurrence.

*Equation 33*

$$
\begin{aligned}
Y_{(\gamma,s1,s2),k-L+1:k} &= W_{k} + H_{(\gamma),i} \left[ U_{(\gamma),1:\frac{L}{2}} \mid U_{(s1),k-\frac{L}{2}+1:\tau} \mid U_{(s1,s2),\tau+1:\tau+i} \mid U_{(s2),\tau+i+1:k} \right] \\
&+ (H_{(s1),i} - H_{(\gamma),i}) \left[ 0_{m(i+1)\times\frac{L}{2}} \mid U_{(s1),k-\frac{L}{2}+1:\tau} \mid 0_{m(i+1)\times k-\tau} \right] \\
&+ \left[ (H_{(s1,s2),1} - H_{(\gamma),i}) \mid \cdots \mid (H_{(s1,s2),i-1} - H_{(\gamma),i}) \mid (H_{(s1,s2),i} - H_{(\gamma),i}) \right] U^{*} \\
&+ (H_{(s2),i} - H_{(\gamma),i}) \left[ 0_{m(i+1)\times L-k+\tau+i} \mid U_{(s2),\tau+i+1:k} \right]
\end{aligned}
\tag{33}
$$

*Equation 34*

$$
\begin{aligned}
\varepsilon_{(\gamma,s1,s2),k} &= W_k \Pi_{(\gamma,s1,s2),k-\frac{L}{2}+1:k} Z \\
&+ (H_{(s1),i} - H_{(\gamma),i}) \left[ 0_{m(i+1)\times\frac{L}{2}} \mid U_{(s1),k-\frac{L}{2}+1:\tau} \mid 0_{m(i+1)\times k-\tau} \right] \Pi_{(\gamma,s1,s2),k-\frac{L}{2}+1:k} Z \\
&+ \left[ (H_{(s1,s2),1} - H_{(\gamma),i}) \mid \cdots \mid (H_{(s1,s2),i-1} - H_{(\gamma),i}) \mid (H_{(s1,s2),i} - H_{(\gamma),i}) \right] U^* \Pi_{(\gamma),s1,s2),k-\frac{L}{2}+1:k} Z \\
&+ (H_{(s2),i} - H_{(\gamma),i}) \left[ 0_{m(i+1)\times L-k+\tau+i} \mid U_{(s2),\tau+i+1:k} \right] \Pi_{(\gamma,s1,s2),k-\frac{L}{2}+1:k} Z
\end{aligned}
\tag{34}
$$

*Equation 35*

$$
\begin{aligned}
\varepsilon_{(\gamma,s1,s2),k} &= W_k \Pi_{(\gamma,s1,s2),k-\frac{L}{2}+1:k} Z + \\
&\left[ (H_{(s1,s2),1} - H_{(s1),i}) \mid \cdots \mid (H_{(s1,s2),i-1} - H_{(s1),i}) \mid (H_{(s1,s2),i} - H_{(s1),i}) \right] U^* \Pi_{(\gamma,s1,s2),k-\frac{L}{2}+1:k} Z \\
&+ (H_{(s2),i} - H_{(s1),i}) \left[ 0_{m(i+1)\times L-k+\tau+i} \mid U_{(s2),\tau+i+1:k} \right] \Pi_{(\gamma,s1,s2),k-\frac{L}{2}+1:k} Z
\end{aligned}
\tag{35}
$$

## 5.5 Switching Detectability

*Definition 3:* The switching between two non discernable modes is detected when their corresponding residuals is not zero, and the residuals of the other modes are not zero too (up to certain time delay).

This definition is a data-based switching detectability definition. In the following proposition we establish a new model-based detectability condition.

**Proposition 6**: The switching from a mode $a$ to a mode $b$ is detectable if the inputs are persistently excited and if one of the following conditions holds:

1. The two modes $a$ and $b$ are discernable.
2. If the two modes $a$ and $b$ are not discernable where $H_{a,\bullet} = H_{b,\bullet}$:

$$
\begin{aligned}
&(H_{(a,b),1} \neq H_{a,i}) \vee \cdots \vee (H_{(a,b),i-1} \neq H_{a,i}) \\
&\vee (H_{(a,b),i} \neq H_{a,i})
\end{aligned}.
$$

**Proof:** The two modes $a$ and $b$ are discernable: this case is studied in the switching estimation subsection where the switching is detected and the time of its occurrence is estimated.

The two modes $a$ and $b$ are not discernable: one can choose a database of label $\gamma = a$ or indifferently $\gamma = b$ to calculate data-based residual $\varepsilon_{(\gamma,a,b),k}$, because the mode $a$ and $b$ have the same Markov parameters $H_{a,\bullet} = H_{b,\bullet}$. As in Equation (35), the evaluation form of data-based residual $\varepsilon_{(\gamma,a,b),k}$ is given by Equation 36.

Since the modes $a$ and $b$ are not discernable, we have $H_{(b),i} - H_{(a),i} = 0$, then the data-based residual $\varepsilon_{(\gamma,a,b),k}$ given by the Equation (36) becomes Equation 37.

*Equation 36*

$$\begin{aligned} \varepsilon_{(\gamma,a,b),k} &= W_k \Pi_{(\gamma,a,b),k-\frac{L}{2}+1:k} Z + \\ &\left[ (H_{(a,b),1} - H_{(a),i}) \mid \cdots \mid (H_{(a,b),i-1} - H_{(a),i}) \mid (H_{(a,b),i} - H_{(a),i}) \right] U^* \Pi_{(\gamma,a,b),k-\frac{L}{2}+1:k} Z \\ &+ (H_{(b),i} - H_{(a),i}) \left[ 0_{m(i+1)\times L-k+\tau+i} \mid U_{(b),\tau+i+1:k} \right] \Pi_{(\gamma,a,b),k-\frac{L}{2}+1:k} Z \end{aligned} \tag{36}$$

*Equation 37*

$$\begin{aligned} \varepsilon_{(\gamma,a,b),k} &= W_k \Pi_{(\gamma,a,b),k-\frac{L}{2}+1:k} Z \\ &+ \left[ (H_{(a,b),1} - H_{(a),i}) \mid \cdots \mid (H_{(a,b),i-1} - H_{(a),i}) \mid (H_{(a,b),i} - H_{(a),i}) \right] U^* \Pi_{(\gamma,a,b),k-\frac{L}{2}+1:k} Z \end{aligned} \tag{37}$$

*Equation 38*

$$E[\varepsilon_{(\gamma,a,b),k}] = \left[ (H_{(a,b),1} - H_{(a),i}) \mid \cdots \mid (H_{(a,b),i-1} - H_{(a),i}) \mid (H_{(a,b),i} - H_{(a),i}) \right] U^* \Pi_{(\gamma,a,b),k-\frac{L}{2}+1:k} Z \tag{38}$$

The mathematical expectation of the obtained data-based residual $\varepsilon_{(\gamma,a,b),k}$ calculated using database $\gamma = a$ leads to Equation 38.

Where the mathematical expectation of an expression $\Pi_{(\gamma,a,b),k-\frac{L}{2}+1:k} Z$ multiplied by a centered Gaussian noise $W_k$ is zero. The switching is detected if the expectation of the data-based residual $E[\varepsilon_{(\gamma,a,b),k}]$ (38) is different from zero, in other words when at least one of the terms $(H_{(a,b),\bullet} - H_{(a),i})$ is different from zero, which may be expressed as:

$$E[\varepsilon_{(\gamma,a,b),k}] \neq 0 \Rightarrow (H_{(a,b),1} \neq H_{a,i}) \vee$$
$$\cdots \vee (H_{(a,b),i-1} \neq H_{a,i}) \vee (H_{(a,b),i} \neq H_{a,i}).$$

It is of prime importance to be able to detect the switching between two non discernable modes; this is the case when the fault does not change the healthy mode behavior (due to internal compensation, etc...) so that the healthy and the faulty modes have the same Markov parameters.

## 5.6 Example

In order to show the efficiency of the proposed approach, an academic example is proposed. The numerical model parameters are given in this chapter but they will not be used for residuals computation. Considering a switching system composed of three modes ($\sigma_k \in \{1, 2, 3\}$), where the third mode is not discernable from the second mode ($H_{(2),i} = H_{(3),i}$).

$$\begin{cases} x_{k+1} = A_{\sigma_k} x_k + B_{\sigma_k} u_k \\ y_k = C_{\sigma_k} x_k + D_{\sigma_k} u_k + w_k. \end{cases} \tag{39}$$

$$A_1 = \begin{pmatrix} -0.7 & 0 & 0 & 0 \\ 0 & 0.6 & 0 & 0 \\ 0 & 0 & 0.3 & 0 \\ 0 & 0 & 0 & 0.1 \end{pmatrix}, \ B_1 = \begin{pmatrix} 0.5 & 0.5 \\ 0.5 & 0.5 \\ 0.5 & 0.5 \\ 0.5 & 0.5 \end{pmatrix},$$

$$C_1 = \begin{pmatrix} 0.1 & 0 & 0 & 0 \\ 0 & 0.1 & 0 & 0 \\ 0 & 0 & 0.1 & 0 \\ 0 & 0 & 0 & 0.1 \end{pmatrix}, \ D_1 = \begin{pmatrix} 0.2 & 0.2 \\ 0.2 & 0.2 \\ 0.2 & 0.2 \\ 0.2 & 0.2 \end{pmatrix}$$

$$A_2 = \begin{pmatrix} -0.6 & 0 & 0 & 0 \\ 0 & 0.5 & 0 & 0 \\ 0 & 0 & 0.2 & 0 \\ 0 & 0 & 0 & 0.7 \end{pmatrix}, \ B_2 = \begin{pmatrix} 0.15 & 0.15 \\ 0.5 & 0.5 \\ 0.35 & 0.35 \\ 0.45 & 0.45 \end{pmatrix},$$

$$C_2 = \begin{pmatrix} 0.2 & 0 & 0 & 0 \\ 0 & 0.2 & 0 & 0 \\ 0 & 0 & 0.2 & 0 \\ 0 & 0 & 0 & 0.2 \end{pmatrix}, \ D_2 = \begin{pmatrix} 0.5 & 0.5 \\ 0.5 & 0.5 \\ 0.5 & 0.5 \\ 0.5 & 0.5 \end{pmatrix}$$

$$P = \begin{pmatrix} 1 & 3 & 2 & 1 \\ 5 & 7 & 2 & 1 \\ 3 & 0 & -2 & 5 \\ -2 & 4 & -9 & 1 \end{pmatrix},$$

$$A_3 = \mathrm{inv}(P)*A_2*P, \ B_3$$
$$= \mathrm{inv}(P)*B_2, \ C_3 = C_2*P, \ D_3 = D_2$$

The corresponding sequence for the simulated switching system is summarized in Table 1.

Where $T$ represents time intervals expressed in sample time unit.

Since the proposed data-based residual is computed in a time window, so $\varepsilon_{(\gamma,\sigma_k),k}$ cannot be

*Table 1.*

| T/Te | [0, 400] | [401, 600] | [601, 700] | [701, 900] | [901, 1000] |
|------|----------|------------|------------|------------|-------------|
| $\sigma_k$ | 1 | 3 | 2 | 1 | 2 |

calculated at the beginning. The real parameter-free residual starting from the black dashed lines is shown in Figure 1.

In Figure 1, the switching detection process of $\tau(1)$, $\tau(3)$ and $\tau(4)$ is successfully achieved. The switching at $\tau(2)$ between two no discernable modes 2 and 3 is detected also even if mode 2 and 3 have the same Markov parameters. This means that the condition $H_{(a),i} = H_{(b),i}$ is not sufficient for switching detection.

The three data-based residuals ($\varepsilon_{(\gamma(1),\sigma_k),k}(1)$, $\varepsilon_{(\gamma(2),\sigma_k),k}(1)$ and $\varepsilon_{(\gamma(3),\sigma_k),k}(1)$) are also used for first, second and third mode recognition respectively. The first mode is successfully recognized, unlike the second and the third modes since they are not discernable (same Markov parameters), which proves that discernability condition $H_{(a),i} = H_{(b),i}$ is needed to distinguish between two modes.

## 6 ALGORITHM

The mode recognition allows one for example to recognize the faulty mode, and its spread can be recovered. (Figure 2)

## 7 VEHICLE ROLLOVER PREVENTION EXAMPLE

The example of vehicle rollover prevention system is taken to illustrate the theoretical results. A robust switched controller is considered, to prevent instabilities due to abrupt changes in the center of gravity position. As a consequence changes in the vehicle parameters cause naturally a switching (Solmaz et al., 2007).

It is assumed that vertical changes in CG position is the only source of switching, which can be caused by vertical load shifts, passenger movements or cornering maneuver. The roll dy-

*Figure 1.* $\varepsilon_{(\gamma,\sigma_k),k}(1)$*, Switching detection and current mode recognition using data-based method*

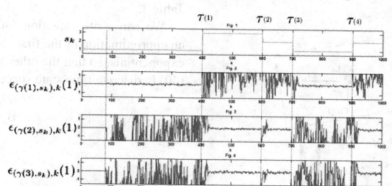

namics of a vehicle is significantly affected by the changes in the CG position.

The significant variation of the passenger, and/or load distribution is sometimes dangerous, which makes the consequence of the CG position no negligible (Solmaz et al., 2008).

The roll plane simplified model of a car is shown in Figure 3. The equation of motion for this model is given as (Solmaz et al., 2006):

$$J_{x_{eq},\gamma}\ddot{\overline{\phi}} + c\dot{\phi} + k\phi = \overline{m}\varpi_\gamma(a_y + g\phi) + u$$

(40)

*Figure 2. The general algorithm diagram*

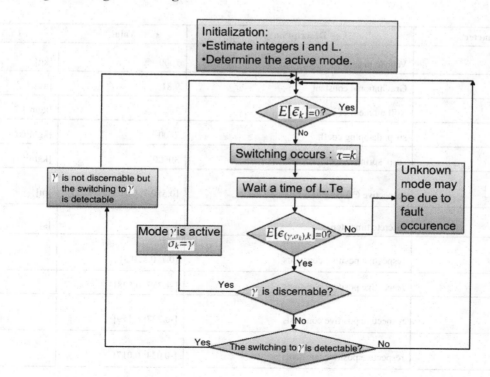

*Figure 3. Second order roll plane model*

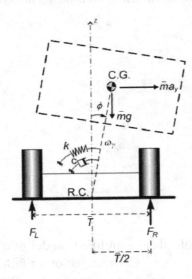

where the equivalent roll moment of inertia for each mode $\gamma$ is denoted by $J_{x_{eq},\gamma} = J_{xx} + m\varpi_\gamma^2$, $\phi$ is the roll angle and the active suspension actuators provides the roll torque input $u$.

Other parameters definitions are given in Table 2.

We can write Equation (40) differently by an approximation of the first order of the matrix exponentials, so that the other form of Equation (40) in discrete time state space is given by:

$$x_{k+1} = A_{d,\gamma} x_k + G_{d,\gamma} a_{y,k} + B_{d,\gamma} u_k \qquad (41)$$

where

$$A_{d,\gamma} = \begin{bmatrix} 1 & \Delta t \\ -\dfrac{\left(k - \bar{m}g\varpi_\gamma\right)\Delta t}{J_{x_{eq},\gamma}} & 1 - \dfrac{c\Delta t}{J_{x_{eq},\gamma}} \end{bmatrix},$$

$$G_{d,\gamma} = \begin{bmatrix} 0 & \dfrac{\bar{m}\varpi_\gamma\Delta t}{J_{x_{eq},\gamma}} \end{bmatrix}^T$$

*Table 2.*

| Parameter | Description | Value | Unit |
|---|---|---|---|
| $\bar{m}$ | Vehicle mass | 1000 | [kg] |
| g | Gravitational constant | 9.81 | [m/s²] |
| $J_{xx}$ | Roll moment of inertia | 300 | [kgm²] |
| c | susp. damping coeff. | 2500 | [kgm²/s] |
| k | susp. spring stifness | 30000 | [kgm²/s²] |
| $[\varpi_1; \varpi_2; \varpi_3]$ | respective CG heights | [0.5; 0.7; 0.9] | [m] |
| $\Delta t$ | discrete time step | 1 | [s] |
| $[\lambda_1; \lambda_2; \lambda_3]$ | respective positive constants | [0.7; 0.5; 0.3] | |
| $[v_1; v_2]$ | respective positive constants | [-1.759; 0.528] | |
| $[\eta_1; \eta_2]$ | respective positive constants | [-0.227; 0.159] | |
| $[\mu_1; \mu_2]$ | respective positive constants | [-0.034; 0.017] | |

*Figure 4. Lateral acceleration* $a_{y,k}$

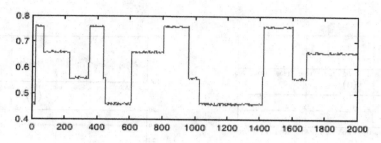

and

$$B_{d,\gamma} = \begin{bmatrix} 0 & \dfrac{\Delta t}{J_{x_{eq},\gamma}} \end{bmatrix}^{T}$$

The switched linear state feedback control structure proposed in (Solmaz et al., 2007) is given by:

$$u_{\gamma,k} = -K_{\gamma} x_{k} \qquad (42)$$

where $K_{\gamma} = \begin{bmatrix} \kappa_{\gamma,1} & \kappa_{\gamma,2} \end{bmatrix}$ with $\kappa_{\gamma,1},\ \kappa_{\gamma,2} \in \Re$, are fixed control gains corresponding to each CG height configuration. Then the closed loop system can be expressed as:

$$x_{k+1} = \tilde{A}_{d,\gamma} x_{k} + G_{d,\gamma} a_{y,k} \qquad (43)$$

where

$$\tilde{A}_{d,\gamma} = A_{d,\gamma} - B_{d,\gamma} K_{\gamma}$$

$$= \begin{bmatrix} 1 & \Delta t \\ -\dfrac{\left(k - \bar{m}g\varpi\gamma + \kappa_{\gamma,1}\right)\Delta t}{J_{x_{eq},\gamma}} & 1 - \dfrac{\left(c + \kappa_{\gamma,2}\right)\Delta t}{J_{x_{eq},\gamma}} \end{bmatrix}$$

$$\qquad (44)$$

Using the Lemma 3.1 of (Solmaz et al., 2007) and (Solmaz et al., 2008), the $K_{\gamma}$ elements are defined as follows:

$$\left. \begin{aligned} \kappa_{11} &= \frac{J_{x_{eq},1}\left(\lambda_{1}-1\right)\left(\lambda_{2}-1\right)}{\Delta t^{2}} - k + \bar{m}g\varpi_{1} \\ \kappa_{12} &= \frac{J_{x_{eq},1}\left(\lambda_{1}+\lambda_{2}-2\right)}{-\Delta t} - c \end{aligned} \right\} \text{ for } \gamma(c)=1$$

$$\left. \begin{aligned} \kappa_{21} &= \frac{J_{x_{eq},2}\left(\lambda_{3}-1\right)\left(\lambda_{2}-1\right)}{\Delta t^{2}} - k + \bar{m}g\varpi_{2} \\ \kappa_{22} &= \frac{J_{x_{eq},2}\left(\lambda_{3}+\lambda_{2}-2\right)}{-\Delta t} - c \end{aligned} \right\} \text{ for } \gamma(c)=2$$

$$\left. \begin{aligned} \kappa_{31} &= \frac{J_{x_{eq},3}\left(\lambda_{1}-1\right)\left(\lambda_{3}-1\right)}{\Delta t^{2}} - k + \bar{m}g\varpi_{3} \\ \kappa_{32} &= \frac{J_{x_{eq},3}\left(\lambda_{1}+\lambda_{3}-2\right)}{-\Delta t} - c \end{aligned} \right\} \text{ for } \gamma(c)=3$$

Where the number of switching occurrences is $r = 4$, $h \in \{1, 2, ..., 4\}$, switching instants are

*Figure 5. (a): Roll angle* $\phi_{k}$, *(b): Roll angular velocity* $\dot{\phi}_{k}$

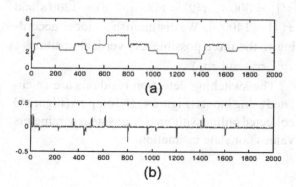

*Figure 6. (a):* $\sigma_k$*, (b):* $\bar{\varepsilon}_k(1)$*, (c):* $\bar{\varepsilon}_k(2)$*, (d):* $\bar{\varepsilon}_k(1) + \bar{\varepsilon}_k(2)$

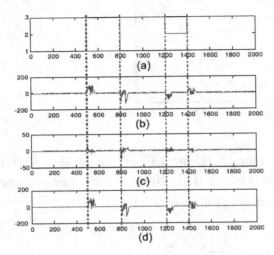

*Figure 8. (a):* $\varepsilon_{(2,\sigma_k),k}(1)$*, (b):* $\varepsilon_{(2,\sigma_k),k}(2)$*, (c):* $\sigma_k$

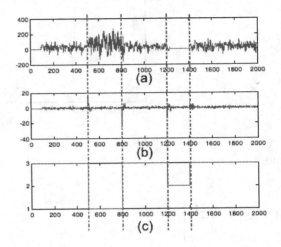

*Figure 7. (a):* $\varepsilon_{(1,\sigma_k),k}(1)$*, (b):* $\varepsilon_{(1,\sigma_k),k}(2)$*, (c):* $\sigma_k$

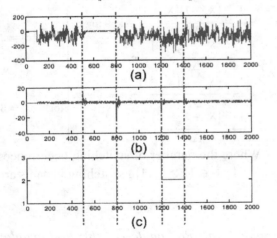

*Figure 9. (a):* $\varepsilon_{(3,\sigma_k),k}(1)$*, (b):* $\varepsilon_{(3,\sigma_k),k}(2)$*, (c):* $\sigma_k$

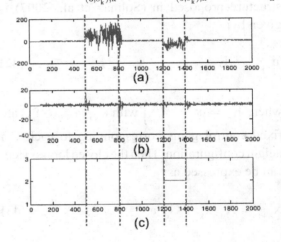

$\tau(1) = 500[s]$, $\tau(2) = 800[s]$, $\tau(3) = 1200[s]$ and $\tau(4) = 1400[s]$. We distinguish 3 modes according to the three possible CG vertical positions ( $\varpi_1$, $\varpi_2$ and $\varpi_3$ ).

The switching detection residuals are calculated using inputs (Figure 4) and outputs (Figure 5) collected online without using system parameters values nor state estimation.

From Figure 6, it is obvious that switching times are well estimated using the first and the second residual components of ( $\bar{\varepsilon}_k$ ).

From Figure 7, Figure 8 and Figure 9 which represent the behavior of the data-based residuals, one can notice that it is simple to recognize the modes and to estimate the switching time. The second component of each residual cannot be used for current mode recognition, because the second output is not sufficiently persistent

# 8 CONCLUSION

A data-based method for switching detection and mode recognition is proposed in this chapter. Both processes can be achieved successfully in both methods when the system switches between discernable modes. Discernability between modes requires a necessary condition related to the Markov parameters of the mode. Although when the system switches between non discernable modes, the switching is detected, which means that the non discernability condition does not imply the non switching detectability is not an sufficient condition for a non switching detectability. Finally the effectiveness of the proposed method is illustrated on the vehicle rollover prevention system.

# REFERENCES

Casimir, R., Boutleux, E., & Clerc, G. (2003). *Fault diagnosis in an induction motor by pattern recognition methods*. 4th IEEE International Symposium on Power Electronics and Drives, SDEMPED 24-26 August, Atlanta, Georgia, USA.

Chow, A. Y., & Willsky, A. (1984). Analyticaly redundancy and the design of robust failure detection systems. *IEEE Transactions on Automatic Control, 7*(29), 603–614. doi:10.1109/TAC.1984.1103593

Cocquempot, V., El Mezyani, T., & Staroswiecki, M. (2004). *Fault detection and isolation for hybrid systems using structured parity residuals*. IEEE/IFAC-ASCC'2004, 5th Asian Control Conference, Melbourne, Victoria, Australia, 20-23 June.

Cocquempot, V., Staroswiecki, M., & El Mezyani, T. (2003). Switching time estimation and fault detection for hybrid systems using structured parity residuals. In *Proceedings of the 15th IFAC Symposium on Fault Detection, Supervison and Safety of Technical Processes (SAFEPROCESS '03)*, (pp. 681-686). Washington, DC, USA.

Ding, S., Zhang, P., Ding, E., Naik, A., Deng, P., & Gui, W. (2010). On the application of PCA technique to fault diagnosis. *Tsinghua Science and Technology, 15*(2), 138–144. doi:10.1016/S1007-0214(10)70043-2

Domlan, E. A., Ragot, J., & Maquin, D. (2007). *Active mode estimation for switching systems*. 26th American Control Conference, New York City, NY, USA, ACC'2007, 9-13 July.

Frank, P. M., & Ding, X. (1997). Survey of robust residual generation and evaluation methods in observer-based fault detection systems. *Journal of Process Control, 7*(6), 403–424. doi:10.1016/S0959-1524(97)00016-4

Fussel, D., & Balle, P. (1997). *Combining neuro-fuzzy and machine learning for fault diagnosis of a DC motor*. American Control Conference, Albuquerque, NM, USA, 4-6 June.

Hakem, A., Pekpe, K. M., & Cocquempot, V. (2011). *Sensor fault diagnosis for bilinear systems using data-based residuals*. IEEE 50th International Conference on Decision and Control and European Control Conference (CDC-ECC), Orlando, FL, USA, December 12-15.

Harkat, M. F., Mourot, G., & Ragot, J. (2003). *Nonlinear PCA combining principal curves and RBF-networks for process monitoring*. 42th Conference of Decision and Control IEEE CDC, Lewis, FL, USA, 9-12 December.

Palma, L. B., Coito, F. J., & Silva, R. N. (2002). Adaptive observer based fault diagnosis approach applied to a thermal plant. *Proceedings of the 10th MED. Conference on Control and Automation*, Lisbon, Portugal, (pp. 115-123). 9-12 July.

Patton, R. (1994). Robust model-based fault diagnosis: The state of the art. *IFAC Symposium on Fault Detection, Supervision and Safety for Technical Processes SAFEPROCESS '94* (pp. 1-24). Helsinki, Finland, 13-15 June.

Pekpe, K. M., Mourot, G., & Ragot, J. (2004). *Subspace method for sensor fault detection and isolation-application to grinding circuit monitoring.* 11th IFAC Symposium on automation in Mining, Mineral and Metal processing, (MMM 2004), Nancy, France, 8-10 septembre.

Simani, S., Fantuzzi, C., & Patton, R. J. (2003). *Model-based fault diagnosis in dynamic systems using identification techniques.* New York, NY: Springer-Verlag, Inc.

Solmaz, S., Akar, M., & Shorten, R. (2006). On-line center of gravity estimation in automotive vehicles using multiple models and switching. In *Proceedings of the 9th International Conference on Control, Automation, Robotics and Vision ICARCV06,* Singapore, Dec. 5-8.

Solmaz, S., Akar, M., & Shorten, R. (2007). *Method for determining the center of gravity for an automotive vehicle.* Irish patent ref: (s2006/0162), European Patent Pending, February.

Solmaz, S., Akar, M., Shorten, R., & Kalkkuhl, J. (2008). Realtime multiple-model estimation of center of gravity position in automotive vehicles. *Vehicle System Dynamics Journal, 46*(9), 763–788. doi:10.1080/00423110701602670

## KEY TERMS AND DEFINITIONS

**Current Mode Recognition:** A process that specifies the active.

**Data-Based Methods:** Methods that use only input output data to achieve a special process, as fault detection and isolation process or switching detection and mode recognition process.

**Fault Detection:** A process that indicates whether a fault occurs or not.

**Fault Isolation:** A process that specifies the faulty component.

**Residual:** Fault signal indicator.

**Switching System:** Consists of dynamical modes which are modeled in state-space form, and a switching logic that governs the transitions between modes.

**Switching Time Estimation:** Is a process that estimates the time-instant when the system jumps from a mode to another mode.

# Section 4
# Data Driven Prognostics

# Chapter 6
# Data Driven Prognostics for Rotating Machinery

**Eric Bechhoefer**
*NRG Systems, USA*

## ABSTRACT

*A prognostic is an estimate of the remaining useful life of a monitored part. While diagnostics alone can support condition based maintenance practices, prognostics facilitates changes to logistics which can greatly reduce cost or increase readiness and availability. A successful prognostic requires four processes: 1) feature extraction of measured data to estimate damage; 2) a threshold for the feature, which, when exceeded, indicates that it is appropriate to perform maintenance; 3) given a future load profile, a model that can estimate the remaining useful life of the component based on the current damage state; and 4) an estimate of the confidence in the prognostic. This chapter outlines a process for data-driven prognostics by: describing appropriate condition indicators (CIs) for gear fault detection; threshold setting for those CIs through fusion into a component health indicator (HI); using a state space process to estimate the remaining useful life given the current component health; and a state estimate to quantify the confidence in the estimate of the remaining useful life.*

## INTRODUCTION

Condition based maintenance (CBM) systems have been shown to reduce costs by decreasing scheduled maintenance costs. However, CBM systems can be leveraged into far greater savings by developing a prognostics capability. The ability to estimate the remaining useful life (RUL) on a

component can greatly improve its availability and reduce logistics cost. Prognostics are the maturation of the CBM system.

Consider the effect on wind farm operations if a prognostic capability were available. Major maintenance events require a heavy lift crane. The availability of a crane can be limited and cost of rental is large. Once the decision is made to replace a gearbox, opportunistic maintenance can be performed on other low RUL/marginal

DOI: 10.4018/978-1-4666-2095-7.ch006

turbines. Alternatively, if the operator of a fleet of helicopter knows the RUL of his assets, he can deploy those aircraft that have the highest RUL and be assured that the aircraft will not need major maintenance while deployed.

The knowledge of a RUL allows the logistician to reduce inventory spares. It affects the man power needed for maintenance providers and facilitates more efficient operations. That said, there are currently few deployed prognostic health management (PHM) systems. While CBM is a maturing technology, PHM is relatively immature and difficult to implement.

The ability to estimate the RUL requires four pieces of information:

- An estimate of the current equipment health,
- A limit or threshold where it is appropriate to do maintenance,
- An estimate of the future equipment load, and
- A model to estimate the time from the current state to the limit/threshold based on projected load.

The current health of a system can be determined by CBM systems. Such systems measure features representing damage. For example, for a pump or generator, shaft order one acceleration is a measure of health. The limit for this vibration for some equipment can be found in standards such as ISO 10816 (2009). However, for gear, bearing, shaft, or equipment not covered by ISO standards, there are no formal or standardized limits. Future load for stationary equipment may be well known, but for helicopter or wind turbines, the load is a variable.

Damage models to predict future equipment health fall into two categories: physics of failure and data-driven. While physics of failure models are in their nature appealing, there is a cost associated with building the models, then validating and testing them. Further, the robustness of these models in application may not be satisfactory; how are material/manufacturing variance, unknown usage and maintenance accounted for in a real application? Data drive methods, while not capable of giving an absolute level of damage, can give a relative limit which may give acceptable performance.

Presented here is an end to end process for a data-driven prognostic for a vibration sensor. Descriptions of how CIs are generated for a gear run to failure test are given. The CIs are fused into a health indicator (HI) through a statistical process (to control the probability of false alarm). Given the current HI, the time until the HI reaches a predetermined value is calculated using Paris' Law, resulting in an estimation of the remaining useful life. Once the RUL is calculated, a bound can then be calculated and a confidence in the RUL is given. Finally, this process will be demonstrated on a spiral bevel gear.

## CONDITION INDICATORS FOR CONDITION MONITORING

### Feature Extraction to Improve Signal-to-Noise

Vibration signatures for machinery faults tend to be small relative to other vibration signatures. In the typical helicopter gearbox, the magnitude of gear mesh vibration is orders of magnitude larger than a fault feature. In this environment, it is difficult to find early stage faults with spectral analysis or root mean squares (RMS) of sensor voltage. This limits the availability to provide information useful for prognostics. Techniques to improve the signal-to-noise ratio are needed to remove tones associated with nominal components, while preserving the fault signatures.

Gear analysis is based on the use of time synchronous average (McFadden, 1987). Time synchronous averaging (TSA) is a signal processing technique that extracts periodic waveforms from noisy data. The TSA is well suited for gearbox analysis, where it allows the vibration signature of the gear under analysis to be distinguished from other gears and noise sources in the gearbox that are not synchronous with that gear. Additionally, the TSA compensates for variations in shaft speed. Without controlling for variation in shaft speed, there is spreading of spectral energy into an adjacent gear mesh bins. The TSA uses a keyphasor to account for the angular position of a shaft under analysis.

This phase information can be provided through a *n* per revolution tachometer signal (such as a Hall sensor or optical encoder, where the time at which the tachometer signal crosses from low to high is called the zero crossing) or though demodulation of gear mesh signatures (Comber & Gelman, 2007).

The model for vibration through a shaft in a gear box was given in McFadden as:

$$x(t) = \Sigma_{i=1:K} X_i (1 + a_i(t)) cos(2\pi i f_m(t) + \Phi_i) + b(t)$$

where:

$X_i$ is the amplitude of the *i*th mesh harmonic,
$f_m(t)$ is the average mesh frequency,
$a_i(t)$ is the amplitude modulation function of the *i*th mesh harmonic,
$\phi_i(t)$ is the phase modulation function of the *i*th mesh harmonic,
$\Phi_i$ is the initial phase of harmonic *i*, and
$b(t)$ is additive background noise.

The mesh frequency is a function of the shaft rotational speed: $f_m = Nf$, where $N$ is the number of teeth on the gear and $f$ is the shaft speed.

This vibration model assumes that $f$ is constant. In most systems, there is some wander in the shaft speed due to changes in load or feedback delay in the control system. This variation will result in smearing of amplitude energy in the frequency domain. The smearing effect, and non-synchronous

noise, is reduced by resampling the time domain signal into the angular domain: $m_x(\theta) = E[x(\theta)]$ $= m_x(\theta + \Theta)$. The variable $\Theta$ is the period of the cycle in which the gearbox operation is periodic, and E[] is the expectation (e.g. ensemble mean). This makes the assumption that $m_x(\theta)$ is stationary and ergodic. If this assumption is true, than non-synchronous noise is reduced by 1/sqrt(*rev*), where *rev* is the number of cycles measured for the TSA.

## TSA Techniques and Condition Indicators

The TSA is an example of angular resampling (McFadden, 1987, or Randall, 2011), where the number of data points in one shaft revolution ($r_n$) are interpolated into *m* number of data points, such that:

- For all shaft revolutions *n*, *m* is larger than *r*, and
- $m = 2^{ceiling\ (log2\ (r))}$ (typical for radix 2 Fast Fourier Transform).

Linear, bandwidth-limited linear interpolation, and spline techniques have been used (Bechhoefer & Kingsley, 2009). In this study, linear interpolation was used as it is considerable faster than spline or bandwidth-limited filtering, with no reduction in analysis performance of the TSA.

The TSA itself can be used for CIs. Typically, a CI is a statistic of a waveform (in the case the TSA). Common statistics are RMS, Peak-to-Peak, Crest Factor, Kurtosis and Skewness. For shaft, shaft order 1, 2, and 3 (first, second, and third shaft rate harmonic) can be used to determine shaft out of balance, bent shaft, and/or shaft coupling damage, respectively. Figure 1 outlines the process of generating the TSA and shaft CIs (example Matlab® code is provided in Bechhoefer & Kingsley, 2009).

*Figure 1. Generation of the TSA and selected CIs*

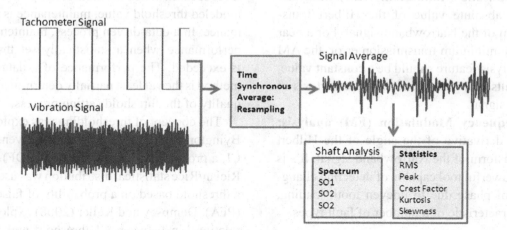

## Gear Fault Condition Indicators

There are at least six failure modes for gears (see ISO 10825, 2007): surface disturbances, scuffing, deformations, surface fatigue, fissures/cracks, and tooth breakage. Each failure mode can potentially generate a different fault signature. Additionally, relative to the energy associated with the gear mesh frequency and other noise sources, the fault signatures are typically small. A number of researchers have proposed analysis techniques to identify these different faults (Vecer et al, 2005, McFadden & Smith, 1985, Zakrajseky et al, 1993). Typically, these analyses are based on the operation of the TSA. Examples of analysis are:

- **Residual:** where shaft order 1, 2, and 3 frequencies, and the gear mesh harmonics of the TSA are removed. Faults such as a soft/broken tooth generate a 1 per rev impact in the TSA. In the frequency domain of the TSA, these impacts are expressed as multiple harmonic of the 1 per rev. The shaft order 1, 2 and 3 frequencies and gear mesh harmonics in the frequency domain, and then the inverse FFT is performed. This allows the impact signature to become prominent in the time domain. CIs are statistics

of this waveform (RMS, Peak-2-Peak, Crest Factor, Kurtosis).

- **Energy Operator:** a type of residual of the autocorrelation function. For a nominal gear, the predominant vibration is gear mesh. Surface disturbances, scuffing, etc, generate small higher frequency values which are not removed by autocorrelation. Formally, the EO is: $\mathbf{TSA}_{2:n-1}$ x $\mathbf{TSA}_{2:n-1}$ x $- \mathbf{TSA}_{1:n-2}$ x $\mathbf{TSA}_{3:n}$. The bold indicates a vector of TSA values. The CIs of the EO are the standard statistics of the EO vector.

- **Narrowband Analysis:** which operates the TSA by filtering out all frequencies except that of the gear mesh and with a given bandwidth. It is calculated by zeroing bins in of the frequency domain of the TSA, except the gear mesh. The bandwidth is typically 10% of the number of teeth on the gear under analysis. For example, a 23-tooth gear analysis would retain bins 21, 22, 23, 24, and 25, and there conjugate in frequency domain. Then the inverse FFT is taken, and statistics of waveform are taken on the time domain waveform. Narrowband analysis can capture sideband modulation of the gear mesh tone due to misalignment, or a cracked/broken tooth.

- **Amplitude Modulation (AM) analysis:** the absolute value of the Hilbert transform of the Narrowband signal. For a gear with minimum transmission error, the AM analysis feature should be a constant value. Faults will greatly increase the kurtosis of the signal.

- **Frequency Modulation (FM) analysis:** the derivative of the angle of the Hilbert transform of the Narrowband signal. It's is a powerful tool capable of detecting changes of phase due to uneven tooth loading, characteristic of a number of fault types.

For a more complete description of these analyses, see Vercer et al (2005) or McFadden & Smith (1985). Figure 2 is an example of the processing to generate the gear CIs for a spiral bevel gear with surface pitting and scuffing. This gear fault will be used throughout the chapter.

## THRESHOLD SETTING AND COMPONENT HEALTH

In a physics of failure prognostics method, modeling would estimate the CI generated for some level of fault. When the measured CI exceeds the modeled threshold value, maintenance is performance. In a data-driven process, maintenance is performance when a statistically set threshold is exceeded. The performance of a data-driven method is then at least partially determined by the quality of the threshold setting process.

The concept of thresholding was explored by Byington et al. (2003), where for a given, single CI, a probability density function (PDF) for the Rician/Rice statistical distribution was used to set a threshold based on a probability of false alarm (PFA). Dempsey and Keller (2008) explored the relationship between CI threshold and PFA to describe the receiver operating characteristics (ROC) of the CI for a given fault. Additionally, Dempsey used the ROC to evaluate the performance of the CI for a fault type. These methods support a data-driven approach for prognostics by formalizing a method for threshold setting.

Estimation of RUL given a threshold is complicated in that there are numerous failure modes for a gear. Further, no single CI has been identified that works with all fault modes. This suggests one of two possible architectures for a prognostics system:

*Figure 2. Process for generating gear CIs*

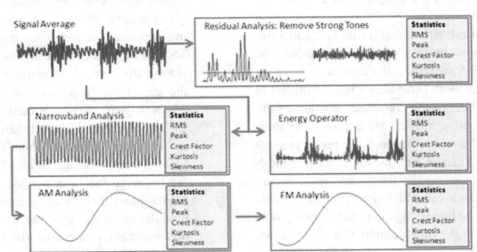

- Estimate the RUL for each Gear CI used, where the reported RUL is the minimum remaining useful life of each CI, or
- Fuse *n* number of CI into a gear health indicator (HI) and calculate the RUL based on the HI.

Computationally, the use of HIs is attractive. Health indicators (HI) provide decision-making tools for the end user on the status of system health. Health indicators consist of the integration of several condition indicators into one value that provides the health status of the component to the end user (Dempsey & Keller, 2008). Bechhoefer et al. (2007) highlighted a number of advantages of the HI over CIs, such as: controlling false alarm rate, improved detection, and simplification of user display. Bechhoefer, He, and Dempsey (2011) describe a threshold setting process for gear health, where the HI is a function of the CI distributions. They give a generalized process for threshold setting, where the HI is a function of distribution of CIs, regardless of the correlation between the CIs.

## Gear Health as a Function of Distributions

Prior to detailing the mathematical methods used to develop the HI, a nomenclature for component health is needed. To simplify presentation and knowledge creation for a user, a uniform meaning across all components in the monitored machine should be developed. Typically, the measured CI statistics (e.g. PDFs) are unique for each component type (due to different rates, materials, loads, etc.). This means that the critical values (thresholds) will be different for each monitored component. In the HI paradigm, we wish to derive an HI function that normalizes CIs. Such a function would be independent of the component and have a common reference value. Further, using guidance from GL Renewables (2007), the HI

function will be designed such that there are two alert levels, where:

- The HI ranges from 0 to 1, where the probability of exceeding an HI of 0.5 is the PFA,
- A warning alert is generated when the HI is greater than or equal to 0.75. Maintenance should be planned by estimating the RUL until the HI is 1.0.
- An alarm alert is generated when the HI is greater than or equal to 1.0. Continued operations could cause collateral damage.

Note that this nomenclature does not define a probability of failure for the component, or that the component fails when the HI is 1.0. Rather, it suggests a change in operator behavior to a proactive maintenance policy: perform maintenance prior to the generation of cascading faults. For example, by performing maintenance on a bearing prior to the bearing shedding extensive material, costly gearbox replacement can be avoided.

## Controlling for Correlation between CIs

All CIs have a probability distribution (PDF). Any operation on the CI to form a health index (HI) is then a function of distributions (Wackerly *et al.* 1996). Functions such as:

- The maximum of *n* CI (the order statistics),
- The sum of *n* CIs, or
- The norm of *n* CIs (energy)

are valid if and only if the distribution (e.g. CIs) are independent and identical (Wackerly et al. 1996). For Gaussian distribution, subtracting the mean and dividing by the standard deviation will give identical Z distributions. The issue of ensuring independence is much more difficult. In general, the correlation between CIs is non-zero. As an

*Table 1. Correlation coefficients for the six CIs used in the study*

| $\rho_{ij}$ | CI 1 | CI 2 | CI 3 | CI 4 | CI 5 | CI 6 |
|------|------|------|------|------|------|------|
| CI 1 | 1 | 0.84 | 0.79 | 0.66 | -0.47 | 0.74 |
| CI 2 | | 1 | 0.46 | 0.27 | -0.59 | 0.36 |
| CI 3 | | | 1 | 0.96 | -0.03 | 0.97 |
| CI 4 | | | | 1 | 0.11 | 0.98 |
| CI 5 | | | | | 1 | 0.05 |
| CI 6 | | | | | | 1 |

example, many of the correlation coefficients used in this study were near 1 (see Table 1).

This correlation between CIs implies that for a given function of distributions to have a threshold that operationally meets the design PFA, the CIs must be whitened (e.g. de-correlated). Fukinaga (1990) presents a whitening transform using the Eigenvector matrix multiplied by the square root for the Eigenvalues (diagonal matrix) of the covariance of the CIs: $A = \Lambda^{1/2} \Phi^T$, where $\Phi^T$ is the transpose of the eigenvalue matrix and and $\Lambda$ is the eigenvalue matrix. The transform is not orthonormal: the Euclidean distances are not preserved in the transform. While ideal for maximizing the distance (separation) between classes (such as in a Baysian classifier), the distribution of the original CI is not preserved. This property of the transform makes it inappropriate for threshold setting.

If the CIs represented a metric such as shaft order 1 magnitude (the magnitude of the acceleration of the first harmonic associated with the shaft RPM), then one can construct an HI which is the square of the normalized power (e.g. the square root of the sum of the acceleration magnitudes squared). This can be defined as normalized energy. Bechhoefer & Bernhard (2007) were able to whiten (e.g. remove correlation between) the CIs and establish a threshold for a given PFA.

A more general whitening solution can be found using Cholesky decomposition. The Cholesky decomposition of Hermitian, positive-definite

matrix results in $A = LL*$, where $L$ is a lower triangular, and $L*$ is its conjugate transpose. By definition, the inverse covariance is positive definite Hermitian. It then follows that if: $LL* = \Sigma^{-1}$, then $Y = L \times CI^T$. The vector $CI$ is the correlated CIs used for the HI calculation, and $Y$ is 1 to $n$ independent CI with unit variance (one CI representing the trivial case). The Cholesky decomposition, in effect, creates the square root of the inverse covariance. This in turn is analogous to dividing the CI by its standard deviation (the trivial case of one CI). In turn, $Y = L \times CI^T$ creates the necessary independent and identical distributions required to calculate the critical values for a function of distributions.

As an example of the importance of correlation on, consider a simple HI function: $HI = CI_1 + CI_2$. The CIs will be normally distributed with mean 0 and standard deviation of 1. The standard deviation of this HI is: $\sigma_{HI} = sqrt(\sigma^2_{CI_1} + \sigma^2_{CI_2} + 2 \rho_{CI_1,CI_2} \times \sigma_{HI} \times \sigma_{CI_1} \times \sigma_{CI_2})$, where $\rho_{CI_1,CI_2}$ is the correlation between $CI_1$ and $CI_2$. If one assumes $\rho_{CI_1,CI_2}$ is 0.0, then $\sigma_{HI} = 1.414$ (e.g. the sqrt(2)). For a PFA of $10^{-6}$, the threshold is then 6.722. Consider the case in which the observed correlation is closer to 1 (e.g. $\rho_{CI_1,CI_2}$ is 1.0): then the observed $\sigma_{HI} = 2$. For a threshold of 6.722, the operations PFA is $4 \times 10^{-4}$. This is 390 times greater than the designed PFA. This illustrates the effect of correlation on threshold setting.

## HI Based on Rayleigh PDFs

The CIs used for this example have Rayleigh-like PDFs (e.g. heavily tailed). Consequently, the HI function was designed using the Rayleigh distribution. The PDF for the Rayleigh distribution uses a single parameter, $\beta$, resulting in the mean ($\mu = \beta*(\pi/2)^{0.5}$) and variance ($\sigma^2 = (2 - \pi/2) * \beta^2$). The PDF of the Rayleigh is: $x/\beta^2 exp(x/2\beta^2)$. Note that when applying these equations to the whitening process, the value for $\beta$ for each CI will then be: $\sigma^2 = 1$, and $\beta = \sigma^2 / (2 - \pi/2)^{0.5} = 1.5264$

(for a more complete analysis, see Bechhoefer & Bernhard 2006).

A number of HI functions could be used, but experience has show (Bechhoefer, He, Dempsey, 2011) that the greatest signal to noise is achieve where the HI function is the norm of $n$ CIs. This represents the normalized energy of the CIs. If the CIs are independent and have identical distribution, it can be shown that the function defines a Nakagami PDF (Bechhoefer & Bernhard 2007). The statistics for the Nakagami are: $\eta = n$, and $\omega = 1/(2-\pi/2)*2*n$.

For this study, data was collected from experiments performed in the Spiral Bevel Gear Test facility at NASA Glenn. A description of the test rig and test procedure is given in Dempsey *et al*. (2002). Six CIs where used, so that: $\eta = 6$, and $\omega = 27.96$. For a PFA of $10^{-6}$, the threshold 10.882, with the HI function calculated as: $HI = .05/10.882 \times (\Sigma_{i=1:6} \mathbf{Y}_i^2)^{1/2}$.

The six CIs used for the HI calculation were: Residual RMS, Energy Operator RMS, FM0, NB KT, AM KT and FM RMS. These CIs were chosen because they exhibited good sensitivity to the fault. Residual Kurtosis and Energy Ratio also were good indicators, but were not chosen because:

- It has been the researcher's experience that these CIs become ineffective when used in complex gear boxes, and
- As the fault progresses, these CIs lose effectiveness. The residual kurtosis can in fact decrease, while the energy ratio will approach 1.

Covariance and mean values for the six CI were calculated by sampling healthy data from four gears prior to the fault propagating. This was done by randomly selecting 100 data points from each gear, and calculating the covariance and means over the resulting 400 data points. The selected CI's PDFs were not Gaussian, but exhibited a high degree of skewness. Because of this, the PDFs were "left shifted" by subtracting an offset such that the PDFs exhibited Rayleigh-like distributions. The estimated gear health is plotted in Figure 3, and the damage on the gear at the end of the test is seen in *Figure 4*.

The key issue with a data-driven prognostic is the appropriateness of the threshold. When the HI is 1.0, is the damage such that it is appropriate to do maintenance? From the example (Figure 4), an HI of 1 displays damage warranting maintenance. Because it is appropriate to performance

*Figure 3. Gear health and torque vs. time*

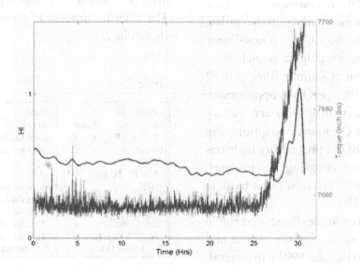

*Figure 4. Damage gear at HI of 1.5*

maintenance when the HI is 1.0 or greater, one can state that the RUL is the time from the current state until the estimated HI is 1.0.

## STATE SPACE MODELS FOR PROGNOSTICS

State-space representation of data provides a versatile and robust way to model systems. Starting with the definition of the states, and the basic principles underlying the characterization of phenomena under study, one can propagate the states as a data-driven stochastic process.

The choice of which type of state space model to use is driven by the nature of the system dynamics and noise source. If the phenomenology of the system has linear dynamics with Gaussian noise, a Kalman filter (KF) is used. If it is a non-linear process with Gaussian noise, a sigma-point Bayesian process (e.g. unscented Kalman filter - UKF) or extended Kalman filter (EKF) is appropriate. For non-linear dynamics with non-linear noise, we use a sequential Monte Carlo method employing sequential estimation of the probability distribution using "importance sampling" techniques. This method is generally referred to as particle filtering (PF) (Candy 2009).

A state space model estimates the state variable on the basis of measurement of the output and input control variables (Brogan 1991). In general,

a system plant can be defined by: $\mathbf{x} = \mathbf{Ax} + \mathbf{Bu}$, and $\mathbf{y} = \mathbf{Hx}$, where $\mathbf{x}$ is the state variable, $\mathbf{x}$ is the rate of change of the state variable, and $\mathbf{y}$ is the output of the system.

An observer is a subsystem used to reconstruct the state space of the plant. The model of the observer is the same as that of the plant, except that one adds an additional term which includes the estimated error to account for inaccuracies in the $\mathbf{A}$ and $\mathbf{B}$ matrixes. This means that any hidden state (such as RUL) can be reconstructed if we can model the plant (e.g. failure propagation) successfully.

The observer is defined as: $E[\mathbf{x}] = E[\mathbf{Ax}] + \mathbf{Bu} + \mathbf{K}(\mathbf{y} - E[\mathbf{Hx}])$, where $E[\mathbf{x}]$ is the estimate state derivative, and $E[\mathbf{Hx}]$ is the expectation of the system output. The matrix $\mathbf{K}$ is called the Kalman gain matrix (linear, Gaussian case). It is a weighting matrix that maps the differences between the measured output $\mathbf{y}$ and the estimated output $E[\mathbf{Hx}]$. A KF is used to optimally set the Kalman Gain matrix.

A KF is a recursive algorithm that optimally filters the measured state based on *a priori* information such as the measurement noise, the unknown behavior of the state, and relationship between the input and output states (e.g. the plant), and the time between measurements. Computationally, it is attractive because it can be designed with no matrix inversion and it is a one-step, iterative process. The filtering process is given as the predictions shown in Table 2.

*Table 2.*

| *Prediction* | where: |
|---|---|
| $\mathbf{X}_{t\mid t-1} = \mathbf{AX}_{t-1\mid t-1}$ State | $t\mid t-1$ is the condition statement (e.g. given the information at t-1) |
| $\mathbf{P}_{t\mid t-1} = \mathbf{A}\,\mathbf{P}_{t-1\mid t-1}\mathbf{A}^T + \mathbf{Q}$ Covariance | $\mathbf{X}$ is the state information ($x$, $dx/dt$, $dx^2/dt^2$) |
| *Gain* | $\mathbf{A}$ is the state transition matrix |
| $\mathbf{K} = \mathbf{P}_{t\mid t-1}\mathbf{H}^T\,[\mathbf{HP}_{t\mid t-1}\mathbf{H}^T + \mathbf{R}]^{-1}$ | $\mathbf{Y}$ is the measured data |
| *Update* | $\mathbf{K}$ is the Kalman Gain |
| $\mathbf{P}_{t\mid t} = (\mathbf{I} - \mathbf{KC})\,\mathbf{P}_{t\mid t-1}$ State Covariance | $\mathbf{P}$ is the state covariance matrix |
| | $\mathbf{Q}$ is the process noise model |
| $\mathbf{X}_{t\mid t} = \mathbf{X}_{t\mid t-1} + \mathbf{K}(\mathbf{Y} - \mathbf{H}\,\mathbf{X}_{t\mid t-1})$ State Update | $\mathbf{H}$ is the measurement matrix |
| | $\mathbf{R}$ is the measurement variance |

For nonlinear systems with Gaussian noise (UKF or EKF), the state prediction is a function of $\mathbf{X}_{t|t-1}$ and the state transition matrix $\mathbf{A}$, and $\mathbf{H}$ is the derivative of the state with respect to the measurement.

For non-linear, non-Gaussian noise problems, particle filters (PF) are attractive. PF is based on representing the filtering distribution as a set of particles. The particles are generated using sequential importance re-sampling (a Monte Carlo technique), where a proposed distribution is used to approximate a posterior distribution by appropriate weighting. In this example, the state update is nonlinear and the measurement noise is Gaussian. As such, an extended Kalman filter was used.

## System Dynamics for Estimating the RUL

The state space model can be constructed as a parallel system to the plant (e.g. the system under study). This requires an appropriate model to simulate the system dynamics. In general, failure modes propagating in mechanical systems are difficult to model at a level of fidelity that would generate any meaningful results (e.g. Health and RUL based on physics of failure). One needs a generalized, data-driven process that can model the plant adequately enough to generate RUL with small error.

Since 1953, a number of fault growth theories have been proposed, such as: net area stress theories, accumulated strain hypothesis, dislocation theories, and others (Frost et al 1999). Through substitution of variables, most of these theories can be generalized by the Paris' Law: $da/dN = D(\Delta K)^n$. Paris' Law governs the rate of crack growth in a homogenous material, where:

- $da/dN$ is the rate of change of the half-crack length,
- $D$ is a material constant of the crack growth equation,

- $\Delta K$ is the range of strain K during a fatigue cycle,
- $n$ is the exponent of the crack growth equation.

The range of strain, $\Delta K$ is given as: $\Delta K = 2\sigma\alpha(\pi a)^{1/2}$, where

- $\sigma$ is gross strain,
- $\alpha$ is a geometric correction factor, and
- $a$ is the half-crack length.

These variables are specific to a given material and test article. In practice, the variables are unknown. This requires some simplifying assumptions to be made to facilitate analysis. For many materials, the crack growth exponent is 2 (Frost et al. 1999). The geometric correction factor $\alpha$, is set to 1 (a constant which will accounted for in the calculation of $D$), which allows Paris' law to be reduced to: $da/dN = D(4\sigma^2\pi a)$.

Taking the inverse $da/dN$ gives the rate of change in cycles per change in crack length, or: $dN/da = 1/[D(4\sigma^2\pi a)]$. Integrating over crack length give the number of cycles (for near synchronous systems, RUL is N x rpm): $N = 1/[D(4\sigma^2\pi a)]$ $(\ln(a_f) - \ln(a_o))$, where the current measured crack is $a_o$ and the final crack length $a_f$. Since the crack length is unknown, the current state, HI, will be used as a surrogate for $a_o$ while $a_f$ will be 1.0 (the RUL is the time from the current HI state until the HI is 1.0). $N$ is the RUL times some constant (RPM for example). The material crack constant, $D$, can be estimated as: $D = da/dN/(4\sigma^2\pi a)$. Gross strain cannot generally be measured, thus, an appropriate surrogate value (e.g. torque, or yaw misalignment) will be used.

The use of Paris's law for the calculation of RUL was given by (Bechhoefer & Bernhard 2008) and (Orchard et al. 2007), but lacked a measure of confidence (e.g. how good was the prognostics). Confidence is an important requirement for a PHM system (Vachtsevanos et al., 2006).

This example implementation was mechanized using two Kalman filters: one filters the measured HI, and calculated *dHI/dt* (rate of crack growth) and one estimates the unknown parameter *D*. The unknown *D* can be found via an extended Kalman filter, where an estimate of HI (e.g. *a*) and strain σ is derived from the current estimate of *D*. The state prediction of σ and *a* are:

$$\hat{a} = \frac{da/dN}{\hat{D}\left(4\sigma^2\pi a\right)}$$

and

$$\hat{\sigma} = \sqrt{\frac{da/dN}{\hat{D}\left(4\pi a\right)}}$$

where $\hat{D}$ is the state estimator for D. The measurement matrix *H*, which is the Jacobian is:

$$H = \begin{bmatrix} \dfrac{-da/dN}{4D^2\pi a}\sqrt{\dfrac{da/dN}{\hat{D}\left(4\pi a\right)}} & 0 \\ \dfrac{-da/dN}{4D^2\pi\sigma^2} & 0 \end{bmatrix}$$

## A Prognostic and Confidence in the Prognostics

In practice, a prognostic or PHM capability would be used to schedule maintenance or assist in assets management and logistic support. The asset owner/operator will make decisions which effect the operational availability and future revenues based on the PHM system. They will need an intuitive, simple display that conveys information on: current health, RUL, and confidence in the RUL prediction.

Model confidence is essential in any RUL prediction (Vachtsevanos et al., 2006). For any RUL calculation, given 1 hour of nominal usage, the RUL should decrease by 1 (e.g. *dN/dt* is approximately -1: one hour of life is consumed for each hour of operation). Further, a measure of model drift or convergence is the second derivative *d²N/dt²*: a value close to zero indicates convergence. When these conditions are met, the model used for calculation of the RUL is consistent, and is indicative of a good estimate of the RUL of the component.

One can use visual cues for the prognostics based on model convergence. Visual cues, such as color, can indicate the confidence in the RUL:

- **Low Confidence:** Yellow, abs(*dN/dt*-1) > 3 and abs(*d²N/dt²*) > 0.5
- **Medium Confidence:** Blue abs*(dN/dt*-1) > 2 and abs(*d²N/dt²*) > 0.5
- **High Confidence:** Green, abs(*dN/dt*-1) < 2 and abs(*d²N/dt²*) < 0.5

A key requirement of the prognostic model is the ability to predict what the health of the component will be some time in the future. For a given state space mode, the RUL or any predicted health is an expectation based on the current state and future usage (e.g. damage or strain). Paris' Law is driven by delta strain: changes in strain will affect the RUL. Future health is then based on the mean strain and a bound on that strain to give a range on the RUL (one benefit in a PF model is a direct distribution of the RUL). This strain information could be based on forecast weather or usage for a wind turbine or type of mission for a helicopter. The health at any time in the future is then: $a_f = \exp(ND(4\sigma^2\pi) + \ln(a_o))$.

## Test Article and a Prognostic Example

Data used for this example was provided by the Spiral Bevel Gear Test facility at NASA. A description of the test rig and test procedure is given in Dempsey *et al*. (2002). The tests consisted of running the gears under load through a "back-to-

*Figure 5. Initial, low-confidence prognostics*

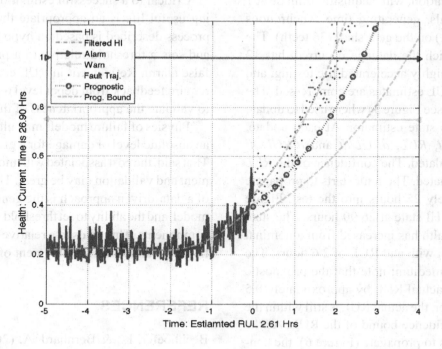

*Figure 6. High-confidence prognostics*

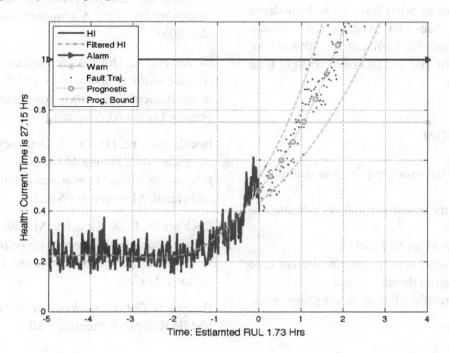

back" configuration, with acquisitions made at 1 minute intervals, generating time synchronous averages (TSA) on the gear shaft (36 teeth). The pinion, on which the damage occurred, has 12 teeth. This is highly accelerated life testing, and as such, the RUL estimates are compressed. The calculated HI (see Figure 3) where used to update sequential, the state estimator. At each update, the *HI, dHI/dt, RUL, dRUL/dt* and *d²RUL/dt²* and were calculated. The confidence of the *RUL* was then evaluated. The fault starts to propagate at approximately 25 hours into the test. Figure 5 displays the HI state at 26.90 hours. The state estimate of health has increased from a nominal value of.2 to.4, with and RUL of 2.6 hours. The confidence is medium: note that the prognostic is lagging the actual RUL by approximately 0.5 hours. However, the actual RUL is still within the estimated confidence bound of the RUL. As the fault continues to propagate (Figure 6), the confidence in the prognostics has improved and the estimate RUL is concurrent with the actual RUL.

In practice, it is anticipated that the time period of the RUL will be thousands of hours for equipment such as wind turbines and hundreds of hours for devices such as helicopter transmissions (see Bechhoefer, Bernhard, He, 2008, where a prognostics of 100 to 150 hours of flight time was observed).

## CONCLUSION

Data-driven prognostics require four conditions:

- The ability to extract a feature related to damage,
- A process to set thresholds,
- A fault model to propagate the current state to the desired threshold, and
- A measure of confidence in the prognostics.

Critical to a successful estimation of remaining useful life is an appropriate threshold. The process described is based on hypothesis testing and sets a threshold relative to a probability of false alarm. Refinement in RUL estimation will require feedback from depot level repair services to validate the appropriateness of the threshold.

Physics of failure models may ultimately give an absolute level of damage for a given CI value. That said, the cost associated with model development and validation may be great. The advantage of a data-driven approach is the generality of the model, and the ability to set threshold with nominal components. This leads to a relative low application cost and a faster deployment of systems.

## REFERENCES

Bechhoefer, E., & Bernhard, A. (2007). *A generalized process for optimal threshold setting in HUMS*. IEEE Aerospace Conference, Big Sky.

Bechhoefer, E., Bernhard, A., & He, D. (2008). *Use of Paris law for prediction of component remaining life*. IEEE Aerospace Conference, Big Sky. 2008

Bechhoefer, E., Duke, A., & Mayhew, E. (2007). *A case for health indicators vs. condition indicators in mechanical diagnostics*. American Helicopter Society Forum 63, Virginia Beach.

Bechhoefer, E., He, D., & Dempsey, P. (2011). *Gear threshold setting based on a probability of false alarm*. Annual Conference of the Prognostics and Health Management Society.

Bechhoefer, E., & Kingsley, M. (2009). *A review of time synchronous average algorithms*. Annual Conference of the Prognostics and Health Management Society

Brogan, W. (1991). *Modern control theory*. Upper Saddle River, NJ: Prentice Hall.

Byington, C., Safa-Bakhsh, R., Watson, M., & Kalgren, P. (2003). *Metrics evaluation and tool development for health and usage monitoring system technology*. HUMS 2003 Conference, DSTO-GD-0348.

Candy, J. (2009). *Bayesian signal processing: Classical, modern, and particle filtering methods*. Hoboken, NJ: John Wiley & Sons.

Combet, L., & Gelman, L. (2007). An automated methodology for performing time synchronous averaging of a gearbox signal without seed sensor. *Mechanical Systems and Signal Processing, 21*(6), 2590–2606. doi:10.1016/j.ymssp.2006.12.006

Dempsey, P., & Afjeh, A. (2002). *Integrating oil debris and vibration gear damage detection technologies using fuzzy logic*. NASA Technical Memorandum 2002-211126.

Dempsy, P., & Keller, J. (2008). *Signal detection theory applied to helicopter transmissions diagnostics thresholds*. NASA Technical Memorandum 2008-215262.

Frost, N., March, K., & Pook, L. (1999). *Metal fatigue* (pp. 228–244). Mineola, NY: Dover Publications.

Fukunaga, K. (1990). *Introduction to statistical pattern recognition* (p. 75). London, UK: Academic Press.

ISO 10816-3:2009. (2009). *Mechanical vibration – Evaluation of machine vibration by measurement on non-rotating parts*.

ISO 10825. (2007) *Gears -- Wear and damage to gear teeth – Terminology*.

McFadden, P. (1987). A revised model for the extraction of periodic waveforms by time domain averaging. *Mechanical Systems and Signal Processing, 1*(1), 83–95. doi:10.1016/0888-3270(87)90085-9

McFadden, P., & Smith, J. (1985). A signal processing technique for detecting local defects in a gear from a signal average of the vibration. *Proceedings - Institution of Mechanical Engineers, 199*(4).

Orchard, M., & Vachtsevanos, G. (2007). A particle filtering approach for on-line failure prognosis in a planetary carrier plate. *International Journal of Fuzzy Logic and Intelligent Systems, 7*(4), 221–227. doi:10.5391/IJFIS.2007.7.4.221

Randal, R. B. (2011). *Vibration-based condition monitoring*. West Sussex, UK: John Wiley & Sons. doi:10.1002/9780470977668

Renewables, G. L. (2007). *Guidelines for the certification of condition monitoring systems for wind turbines*. Retrieved from http://www.gl-group.com/en/certification/renewables/CertificationGuidelines.php

Vachtsevanos, G., Lewis, F. L., Roemer, M., Hess, A., & Wu, A. (2006). *Intelligent fault diagnosis and prognosis for engineering systems* (1st ed.). Hoboken, NJ: John Wiley & Sons, Inc. doi:10.1002/9780470117842

Vecer, P., Kreidl, M., & Smid, R. (2005). Condition indicators for gearbox condition monitoring systems. *Acta Polytechnica, 45*(6).

Wackerly, D., Mendenhall, W., & Scheaffer, R. (1996). *Mathematical statistics with applications*. Belmont, CA: Buxbury Press.

Zakrajsek, J., Townsend, D., & Decker, H. (1993). *An analysis of gear fault detection method as applied to pitting fatigue failure damage*. NASA Technical Memorandum 105950.

# Section 5
# Degradation Modeling

# Chapter 7
# Identifying Suitable Degradation Parameters for Individual–Based Prognostics

**Jamie Coble**
*Pacific Northwest National Laboratory, USA*

**J. Wesley Hines**
*The University of Tennessee, USA*

## ABSTRACT

*The ultimate goal of most prognostic systems is accurate prediction of the remaining useful life of individual systems or components based on their use and performance. Traditionally, individual-based prognostic methods form a measure of degradation which is used to make lifetime estimates. Degradation measures may include sensed measurements, such as temperature or vibration level, or inferred measurements, such as model residuals or physics-based model predictions. Often, it is beneficial to combine several measures of degradation into a single prognostic parameter. Parameter features such as trendability, monotonicity, and prognosability can be used to compare candidate prognostic parameters to determine which is most useful for individual-based prognosis. By quantifying these features for a given parameter, the metrics can be used with any traditional optimization technique to identify an appropriate parameter. This parameter may be used with a parametric extrapolation model to make prognostic estimates for an individual unit. The proposed methods are illustrated with an application to simulated turbofan engine data.*

## INTRODUCTION

Unforeseen equipment failure is costly, both in terms of equipment repair costs and lost revenue. Discovery of unanticipated pressure vessel head

degradation at the Davis-Besse nuclear plant led to a 25-month outage and estimated repair costs exceeding $600 million (Union of Concerned Scientists, 2009). In September, 2008, a turbine generator malfunction at the D.C. Cook nuclear plant resulted in a fire which led to eventual manual plant shutdown. Turbine repairs totaled $332 mil-

DOI: 10.4018/978-1-4666-2095-7.ch007

lion, in addition to lost revenue during the one-year outage (World Nuclear News, 2008). Enterprise server downtime can be even more costly, resulting in a possible loss of $6.4 million *per hour* for brokerage operations and $2.6 million *per hour* for credit card authorization services (Feng, 2003). Traditional maintenance strategies fall into one of four categories: corrective, preventive, proactive, and condition-based. Corrective maintenance occurs only when equipment fails or malfunctions, such as replacing car tires only when they will no longer hold air. While this maintenance strategy avoids any unnecessary maintenance by only repairing components or systems which have already failed, it also reduces equipment uptime, resulting in lost revenue. Alternatively, preventive, or periodic, maintenance occurs on a time-based schedule, such as replacing car tires after 30,000 miles of use. It is completed every cycle, regardless of need, and is intended to occur frequently enough to preclude any failures from occurring in service. Clearly, this maintenance strategy can result in a significant amount of unnecessary maintenance but reduces the occurrence of failure. Often, preventive maintenance regimes are less costly than corrective due to the added cost of failure during operation and associated damage to other components; a tire blow out while driving may require towing or replacement of damaged wheels. A third maintenance strategy, proactive maintenance, attempts to remove failure modes from a system by performing ongoing maintenance to reduce the probability of the fault or redesigning the system to remove the fault. Maintaining proper alignment in a car to reduce the probability of uneven tire wear is a form of proactive maintenance intended to increase the lifetime of the tires. However, sometimes this is not feasible due to design or cost limitations. Condition-based maintenance can provide a more elegant and cost-effective maintenance strategy. Condition-based maintenance involves monitoring equipment health and performing maintenance actions on an as needed basis. Condition-based maintenance

can be facilitated by a health monitoring system. Health monitoring systems commonly employ several modules that perform specific functions, including but not limited to: system monitoring, fault detection, fault diagnostics, prognostics, and operations and maintenance planning. System monitoring and fault detection modules are used to determine if a component or system is operating in a nominal and expected way. If a fault or anomaly is detected by the monitoring system, the diagnostic system determines the type and, in some cases, the severity of the fault. The prognostics module uses all the available information—including sensed system measurements, monitoring system residuals, and fault detection and diagnostic results—to estimate the remaining useful life (RUL) of the system or component along with associated confidence bounds. With this information in hand, a system could be shut down for maintenance or system operation may be adjusted to mitigate the effects of failure or to slow the progression of failure, thereby extending the equipment life until it is more convenient to perform maintenance.

The ultimate goal of most prognostic systems is accurate prediction of the RUL of individual systems or components based on their specific use and performance, termed individual-based prognostics. As equipment degrades, measured parameters of the system tend to change; these sensed measurements, or appropriate transformations thereof, may be used to characterize the system degradation. Traditionally, individual-based prognostic methods use some measure or prediction of degradation to make RUL estimates. Degradation measures may include sensed measurements, such as temperature or vibration level, or inferred measurements, such as model residuals or physics-based model predictions. Often, it is beneficial to combine or fuse several measures of degradation into a single prognostic parameter to provide a more robust prognostic model. Selection of an appropriate prognostic parameter is key for making useful individual-based RUL estimates, but methods and guidelines to aid in this selection have not been fully developed. Typically, identification of a prognostic parameter is left to expert analysis, visual inspection of available data, and physics-based knowledge of the degradation mechanisms. This approach is tedious and costly, and scales with the number of available data sources and possible fault modes. By formulating this problem in a mathematical framework, an optimal or near-optimal prognostic parameter can be automatically identified from all the available data sources.

This chapter presents a set of metrics that characterize the suitability of a prognostic parameter. Parameter features such as monotonicity, prognosability, and trendability can be used to compare candidate prognostic parameters to determine which is most useful for individual-based prognosis. Monotonicity characterizes the underlying positive or negative trend of the parameter, which addresses the common assumption that physical systems do not self-heal. Prognosability gives a measure of the variance in the critical failure value of a degradation parameter for a population of systems or components, which improves confidence in the estimate of failure. Finally, trendability indicates the degree to which the developed degradation parameters of a population of systems have the same underlying shape and can be described by the same functional form. These three intuitive metrics can be formalized to give a quantitative measure of prognostic parameter suitableness. The combination of the three measures, the suitability, can then be used as a fitness function to optimize the development of a prognostic parameter.

Once identified, an appropriate prognostic parameter may be used with a parametric extrapolation model, such as the general path model (GPM), to make RUL estimates for specific systems or components. This algorithm is derived in the framework of linear regression models. A dynamic Bayesian updating methodology is introduced to incorporate prior information in the GPM regression parameters, thereby capitalizing on all available information. Incorporating prior knowledge into the regression is particularly useful when only a few observations of the degradation parameter are available or the available observations are contaminated with high noise levels. The proposed parameter optimization and RUL estimation methods will be illustrated with an application to the simulated turbofan engine data proved in the 2008 Prognostics and Health Management Conference Prognostics Challenge and additional data from the same simulation available at the NASA Prognostics Data Repository (Saxena and Goebel, 2008).

## BACKGROUND

Conventional reliability analysis uses only failure time data to estimate a time to failure distribution for a population of components resulting in average life estimates. As equipment becomes more reliable, fewer failure times may be available, even with accelerated life testing. Although failure time data becomes more sporadic as equipment reliability rises, often other measures are available which may contain some information about equipment degradation. Lu and Meeker (1993) developed the GPM to assess equipment reliability using these degradation measures, or appropriate functions thereof. This method was originally proposed to move reliability analysis from failure time to failure mechanism analysis. It has since been extended to prognostic applications (Upadhyaya et al., 1994; Chinnam, 1999; Engel et al., 2000; Brown et al., 2007). The GPM assumes that there is some underlying parametric model which describes component degradation. The model may be derived from physical principles or directly constructed from available historical degradation data. Typically, this degradation model accounts for both population (fixed) effects and individual (random) effects. Most commonly, the fitted model is extrapolated to some known failure threshold to estimate the RUL of a particular component. By combining the historically determined degradation model with the appropriate degradation measures from the current component or system, the GPM can be used to make truly individual-based prognostic estimates.

Appropriate degradation parameters do not have to be a directly measured parameter. These parameters could be obtained through a function of several measured variables that provide a quantitative prediction of degradation. It could also be an empirical model prediction of a key system degradation that cannot be measured. For example, pump impeller thickness may be an appropriate degradation parameter but there may not be an unobtrusive method to directly measure it. However, there may be related measurable variables that can be used to predict the impeller thickness, such as cavitation induced vibration, pressure, and flow rate. In this case, the degradation measure is not a directly measurable parameter, but a function of several measurable parameters. Several measures of key degradation sources may be combined to provide a robust and accurate prognostic parameter for estimating RUL.

Commonly, identification of prognostic parameters is left to expert analysis and engineering judgment. If any first-principle, or physics of failure, information is available about the failure mechanism, this knowledge can also be used to inform degradation parameter construction. In the absence of any specific engineering knowledge of the failure mechanism, however, parameters are typically identified through visual inspection of the available data. Both of these methods are time consuming, tedious, and expensive; and the effort necessary for identifying parameters compounds with additional data sources, fault modes, failure mechanisms, and confounding factors such as discrete operating conditions. Because the effort needed to identify appropriate prognostic parameters from data can quickly make the problem intractable for a manual approach, an automated

approach which results in an optimal, or near-optimal, degradation parameter is very attractive.

## METHODOLOGY

As suggested by the "No Free Lunch" Theorem, no one prognostic algorithm is ideal for every situation (Koppen, 2004). A variety of models have been developed for application to specific situations or specific classes of systems. The efficacy of these algorithms for a new process depends on the type and quality of data available, the assumptions inherent in the algorithm, and the assumptions which can validly be made about the system. This discussion focuses on the general path model, an algorithm which attempts to characterize the lifetime of a specific component based on measures of degradation collected or inferred from the system.

### The General Path Model

The general path model (GPM), also called degradation modeling, was first proposed by Lu and Meeker (1993) to move reliability analysis methods from failure-time analysis to failure-process analysis. Traditional methods of reliability estimation use failure times recorded during normal use or accelerated life testing to estimate a time of failure (TOF) distribution for a population of identical components. In contrast, the original development of the GPM used degradation models to estimate the failure times of censored data so that they could be combined with actual failure data to form a TOF distribution.

GPM analysis begins with some assumption of an underlying functional form of the degradation path for a specific fault mode. The degradation, $y_{ij}$, of the $i^{th}$ unit at time $t_j$ is given by:

$$y_{ij} = \eta(t_j, \phi, \theta_i) + \varepsilon_{ij},$$

where $\varphi$ is a vector of fixed (population) effects, $\theta_i$ is a vector of random (individual) effects for the $i^{th}$ component, and $\varepsilon_{ij} \sim N(0, \sigma^2_e)$ is the standard measurement error term. Application of the GPM methodology involves several assumptions. First, the degradation data must be describable by a function, $\eta$; this function may be derived from physics-of-failure models or from the degradation data itself. In order to fit this model, the second assumption is that historical degradation data from a population of identical components or systems is available. This data should be collected under similar use (or accelerated degradation test) conditions and should reasonably span the range of individual variations between components. Because GPM uses degradation measures instead of failure times, it is also not necessary that all historical units are run to failure; censored data contains information useful to GPM forecasting. The final assumption of the GPM model is that there exists some defined critical level of degradation, $D$, beyond which a component no longer meets its design specifications, i.e. the component has failed. Therefore, some components should be run to failure in order to quantify this degradation level. Alternatively, engineering judgment may be used if the nature of the degradation parameter is explicitly known.

The GPM reliability methodology has a natural extension to estimation of remaining useful life of an individual component or system; the degradation path model, $y_i$, can be extrapolated to the failure threshold, $D$, to estimate the individual component's time of failure. This type of degradation extrapolation was proposed early on by Upadhyaya, et al. (1994). In that work, the authors used both neural networks and nonlinear regression models to predict the RUL of a small induction motor. The prognostic methodology used for the current research is described below.

Exemplar degradation paths are used to fit the assumed parametric model and determine the critical failure threshold. These parameter estimates are used to evaluate the random-effects distribu-

tions, to determine the mean population random effects, the mean time to failure (MTTF) and their associated standard deviations, and to estimate the noise variance in the degradation paths. The MTTF distribution can be used to estimate the time of failure for any component which has not yet been degraded.

As data is collected during component use, the degradation model is fit for the individual component. This specific model can be used to project a time of failure for the component. The standard deviation of the parameters can be estimated through traditional linear regression techniques. Well-established methods for inverse prediction can be used to project a prediction interval about the estimated failure time (Tamhane and Dunlop, 1999).

The methodology described considers only the data collected on the current unit to fit the degradation model. However, prior information is available from the historic degradation paths used for initial model fitting, including the mean degradation path and associated distributions. This data can provide valuable knowledge for fitting the degradation model of an individual component, particularly when only a few data points have been collected or the collected data suffers from excessive noise. Bayesian updating methods have been developed to incorporate this additional information into the fitted model.

## Incorporating Prior Information through Bayesian Updating

Bayesian updating methods can be used to include prior information in regression problems. This discussion will focus on application to linear regression models. However, as discussed above, the GPM methodology can be applied to nonlinear regression problems as well as other parametric modeling techniques such as neural networks. Other Bayesian updating methods could be applied to these types of models, but such application is beyond the scope of this chapter. For a complete discussion of Bayesian statistics including other Bayesian update methods, the interested reader is referred to (Lindley and Smith, 1972), (Gelman et al., 2004), and (Carlin and Lewis, 2000). In addition, work by Robinson and Crowder (2000) focuses on Bayesian methods for nonlinear regression GPM.

A linear regression model is given by:

$$Y = bX,$$

where $Y$ is the degradation measure, $X$ is the matrix of predictors, and $b$ is the vector of regression coefficients which relates $X$ and $Y$. It is important to note that the linear regression model is not necessarily a linear model. The data matrix $X$ typically is a measure of time or duty cycles and can be populated with any function thereof, including higher order terms, interaction terms, and functions such as *sin(x)* or $e^x$ resulting in a nonlinear model. The model parameters are estimated from:

$$b = \left( X^T \Sigma_y^{-1} X \right)^{-1} X^T \Sigma_y^{-1} Y,$$

where $\Sigma_y$ is the variance-covariance noise matrix for the response observations. It is convenient to assume that the noise in the degradation measurements is constant and uncorrelated. Some a priori knowledge of the noise variance is available from the exemplar degradation paths. If this assumption is not valid for a particular problem, then other methods of estimating the noise variance must be used. The assumption of uncorrelated noise allows the variance-covariance matrix to be a diagonal matrix consisting of noise variance estimates and a priori knowledge variance estimates. If this assumption is not valid, including covariance terms is trivial; again these terms can be estimated from historical degradation paths.

If prior information is available for a specific model parameter, i.e. $\beta_j \sim N(\beta_{jo}, \sigma^2_{\beta jo})$ based on the population estimates, then the matrix $X$ should be appended with an additional row with value one at the $j^{th}$ position and zero elsewhere, and the $Y$ matrix should be appended with the *a priori* value of the $j^{th}$ parameter, $\beta_{jo}$:

$$X^* = \begin{bmatrix} X \\ 0 \cdots 0 \ 1 \ 0 \cdots 0 \end{bmatrix},$$
$$Y^* = \begin{bmatrix} Y \\ \beta_{j0} \end{bmatrix}.$$

Finally, the variance-covariance matrix is augmented with a final row and column of zeros, with the variance of the prior information in the diagonal element.

$$\Sigma_y^* = \begin{bmatrix} \sigma_y^2 & 0 & \cdots & 0 \\ 0 & \ddots & 0 & \vdots \\ 0 & \cdots & \sigma_y^2 & 0 \\ 0 & \cdots & 0 & \sigma_{\beta_{j_o}}^2 \end{bmatrix}.$$

If knowledge is available about multiple regression parameters, the matrices should be appended multiple times with one additional row for each parameter.

After a priori knowledge is used to obtain a posterior estimate of degradation parameters, this estimate becomes the new prior distribution for the next estimation of degradation parameters. The variance of this new knowledge is estimated as:

$$\frac{1}{\sigma^2_{post\beta_j}} = \frac{n}{\sigma_y^2} + \frac{1}{\sigma^2_{prior\beta_j}},$$

where $\sigma^2_{post\beta j}$ is the posterior estimate of the variance in $\beta_j$, $\sigma^2_{prior\beta j}$ is the prior estimate of the variance in $\beta_j$, and $n$ is the number of observations used to fit the current model.

The importance of including prior information in model estimation, particularly when few observations are available, will be highlighted in the example application presented later.

## Selecting an Optimal Prognostic Parameter

Identification of an appropriate prognostic parameter is key for applying a GPM prognostic model to a system. An ideal prognostic parameter has at least three identified qualities: monotonicity, prognosability, and trendability (Coble and Hines, 2009). Monotonicity characterizes the underlying positive or negative trend of the parameter. This is an important feature of a prognostic parameter because it is generally assumed that systems do not undergo self-healing, which would be indicated by a non-monotonic parameter. However, this assumption is not valid for some components such as batteries, which may experience some degree of self-repair during short periods of nonuse. The monotonic trend is generally considered valid when considering an entire system, even if individual components or sub-systems may experience some self-repair. Prognosability gives a measure of the variance in the critical failure value of a population of systems. Ideally, failure should occur at a crisp, well-defined degradation level. A wide spread in critical failure values can make it difficult to accurately extrapolate a prognostic parameter to failure. Finally, trendability indicates the degree to which the parameters of a population of systems have the same underlying shape and can be described by the same functional form. These three intuitive metrics can be formalized to give a quantitative measure of prognostic parameter suitability. Ideally, these metrics would each range from zero to one, one indicating a very high score on that metric and zero indicating that the parameter is not suitable according to the particular metric.

Monotonicity can be easily related to the slope of the prognostic parameter through time. It is a straightforward measure given by:

*Monotonicity*

$$= mean \left( \left| \frac{\# \, pos \, d/dx}{n-1} - \frac{\# \, neg \, d/dx}{n-1} \right| \right),$$

where $\# \, pos \, d/dx$ is the number of positive derivatives, $\# \, neg \, d/dx$ is the number of negative derivatives, and $n$ is the number of observations in a particular unit history. The monotonicity of a population of parameters is given by the average difference of the fraction of positive and negative derivatives for each path. Parameter slopes can be determined through a variety of methods; it is important to choose a method that is robust to noise in the parameters.

Prognosability is calculated as the deviation of the final failure values for each path divided by the mean range of the path. This is exponentially weighted to give the desired zero to one scale:

*Prognosability*

$$= \exp \left( - \frac{std \left( failure values \right)}{mean \left( \left| ranges \right| \right)} \right),$$

where *std(failurevalues)* is the standard deviation of the population of parameter values at failure and *mean(|ranges|)* is the mean of the ranges of each individual parameter, calculated as the difference between the parameter value at beginning of life and the value at failure. This measure encourages well-clustered failure values, i.e. small standard deviation of failure values, and large parameter ranges.

Finally, characterizing the trendability of a population of parameters poses significant difficulty compared to the other two metrics. A candidate parameter is trendable if the same underlying parametric functional form can model each parameter in the population. Trendability is characterized across a population by resampling prognostic parameters with respect to the fraction of total lifetime. This results in each prognostic parameter containing exactly 100 observations, with each observation corresponding to 1% of lifetime. Resampling the prognostic parameters to 100 observations gives a balance between the robustness of the suitability metric and the complexity of the calculation. The linear correlation is calculated across the population of prognostic parameters, and the trendability is given by the smallest absolute correlation across all pairs of exemplar prognostic parameters, $i$ and $j$:

$$Trendability = \min\left(\left|corrcoef_{ij}\right|\right).$$

Automated methods for identifying prognostic parameters are possible with a formalized set of metrics to characterize their suitability. By defining a fitness function as a weighted sum of the three metrics:

$$fitness = w_m monotonicity + w_p prognosability + w_t trendability,$$

a set of prognostic parameters can be compared to determine the most suitable one. Here, the constants $w_m$, $w_p$, and $w_t$ control how important each metric is in the optimization. For most applications, these constants can each be identically one to give equal weight to each parameter feature. However, depending on the application, different weights may be appropriate. For instance, as discussed previously, monotonicity may not be an appropriate prognostic parameter feature for some applications, such as for battery health monitoring. In that case, monotonicity may be excluded from the parameter fitness calculation by giving it a weight of zero. Prescription of appropriate weights is left to engineering judgment based on the equipment under surveillance; there is no practical limitation on the value of the weights. The fitness function can be used with traditional optimization techniques such as gradient descent,

genetic algorithms, and machine learning methods to identify useful prognostic parameters by optimizing a combination of information sources. In this work, the fitness function is used to optimize the weights in a weighted average of prognostic parameter inputs. Discussion of optimization methods is beyond the scope of the current chapter. The interested reader is referred to (Horst and Hoang, 1996) and (Schneider, 2006) for information on deterministic and stochastic optimization methods, respectively.

## Combined Monitoring and Prognostics System

Figure 1 gives a combined monitoring, fault detection, and prognostics system similar to the one used in this research. The monitoring system employs an Auto-Associative Kernel Regression (AAKR) model for monitoring and the Sequential Probability Ratio Test (SPRT) for fault detection. AAKR is a non-parametric, data-driven model that can be thought of as an error-correction routine; the model uses historical data to estimate the "correct" value of new observations which may be faulty due to system degradation, sensor faults, data acquisition problems, etc. The outputs of an AAKR model are the error-corrected estimates of the inputs. The interested reader is referred to (Hines et al., 2008) for details on the AAKR modeling method. Monitoring system residuals are the difference between the AAKR "corrected" values and the measured values. SPRT is a statistical test that looks at a sequence of residuals to determine if the time series of data is more likely from a nominal distribution or a pre-specified faulted distribution (Wald, 1945). When a fault is detected, a diagnostic tool is used to identify and, in some cases, isolate the cause of the fault. Systems will likely degrade in different ways depending on the type of fault, and different prognostic models will be applicable to each fault mode. Then, information from the system, such as original data, the monitoring system residuals, and the results of the

*Figure 1. Combined monitoring and prognostic system*

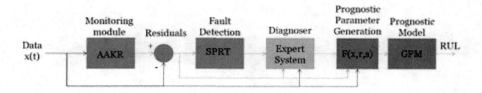

fault detection and isolation routines, can be used to develop the prognostic parameter and prognostic model. Monitoring system residuals are natural candidates for prognostic parameters because they inherently characterize the deviation between the current system operation and normal conditions.

## APPLICATION AND RESULTS

### Data Set Description

The Prognostics and Health Management Society (PHM) Challenge data set consists of 218 cases of multivariate data that track from nominal operation through fault onset to system failure. Data were provided which modeled the damage propagation of aircraft gas turbine engines using the Commercial Modular Aero-Propulsion System Simulation (C-MAPSS). This engine simulator allows faults to be injected in any of the five rotating components and gives output responses for 58 sensed engine variables. The PHM Challenge data set included 21 of these 58 output variables as well as three operating condition indicators. Each simulated engine was given some initial level of wear which would be considered within normal limits, and faults were initiated at some random time during the simulation. Fault propagation was assumed to evolve in an exponential way based on common fault propagation models and the results seen in practice. Engine health was determined as the minimum health margin of the rotating equipment, where the health margin was a function of efficiency and flow for that particular

component; when this health indicator reached zero, the simulated engine was considered failed. The interested reader is referred to (Saxena et al., 2008) for a more complete description of the data simulation.

The data have three operational variables – altitude, Mach number, and throttle resolver angle (TRA) – and 21 sensor measurements. Initial data analysis resulted in the identification of six distinct operational settings; based on this result, the operating condition indicators were collapsed into one indicator which fully defined in which of the six modes the engine was operating. In addition, ten sensed variables were identified whose residuals changed in a meaningful way through time and were well correlated to each other. In this way, the 24 sensor data set was reduced to 11 variables, with original variable numbers: 1 (the operating condition indicator), 5, 6, 7, 12, 14, 17, 18, 20, 23, and 24. Because the nature of the prognostics challenge was to develop a prognostic algorithm with no knowledge of the system under test or the variables available, no effort is made to physically relate these eleven sensors. However, they are considered suitable for auto-associative modeling due to the strong inter-correlations within the group.

The GPM method uses degradation information, either directly measured or inferred, to estimate the system RUL. Initial analysis of the raw data does not reveal any trendable degradation parameter. That is, no sensed measurement has an identifiable trend toward failure. Figure 2 is a plot of the eleven variables that were determined to statistically change with time, grouped only by

*Figure 2. Eleven variables chosen for system monitoring*

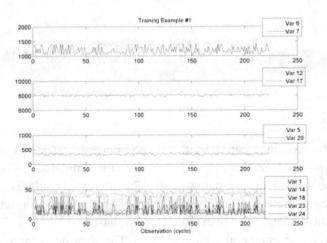

range of values for improved visual analysis. As the figure shows, there is no obvious trend toward failure in the raw data. These variables were used to develop a monitoring and prognostics system, similar to that described above. For this application, all faults are assumed to result from the same fault mode; therefore, no diagnostic engine is used.

## Finding a Prognostic Parameter through Expert Analysis

Visual inspection (VI) of the residuals suggests that an appropriate parameter might be a weighted average of the residuals for variables 6, 7, 14, 18, and 20. Table 1 gives the parameter suitability metrics for each of the residuals. The five residuals with total suitability (given by the sum of the three suitability metrics) greater than 1.5 were combined to create the final parameter, which has greater suitability metrics than any one constituent residual. The five residuals are weighted by the inverse of their average range and summed to give the prognostic parameter identified through visual inspection; the resulting parameters for all 218 training cases are shown in Figure 3. By scaling the residuals, the relative importance of each contributor to the prognostic parameter is equal. Because no engineering judgment can be

made concerning the physical characteristics of the system for this application, weighting each input equally is a reasonable approach. The inputs can be scaled with respect to several measures, including the standard deviation, the input range, the correlation to RUL, etc. By combining several similar residuals, the spread in the failure value is reduced. This is sometimes referred to

*Table 1. Prognostic parameter suitability metrics for individual residuals*

| Variable | Monoton-icity | Prognos-ability | Trend-ability | Suit-ability |
|---|---|---|---|---|
| 3 | 0.435 | 0.370 | 0.000 | 0.805 |
| 5 | 0.537 | 0.613 | 0.018 | 1.168 |
| 6 | 0.604 | 0.727 | 0.291 | 1.622 |
| 7 | 0.749 | 0.818 | 0.726 | 2.293 |
| 12 | 0.654 | 0.314 | 0.001 | 0.968 |
| 14 | 0.782 | 0.851 | 0.814 | 2.447 |
| 17 | 0.639 | 0.282 | 0.000 | 0.921 |
| 18 | 0.703 | 0.731 | 0.649 | 2.083 |
| 20 | 0.625 | 0.737 | 0.476 | 1.838 |
| 23 | 0.447 | 0.497 | 0.000 | 0.945 |
| 24 | 0.457 | 0.510 | 0.000 | 0.967 |
| VI Pa-rameter | 0.846 | 0.891 | 0.903 | 2.640 |

*Figure 3. Prognostic parameter identified by visual inspection*

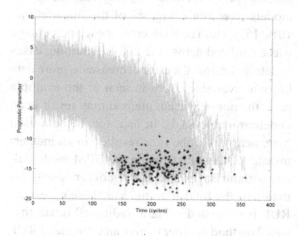

as parameter bagging and is a common variance reduction technique. Identification of this parameter involved several weeks of expert analysis of the available data.

A second order polynomial model can be used to model the degradation parameter. While an exponential model may be more physically appropriate, and was certainly found to be after the data simulation method was made public, the quadratic model is more robust to noise and better describes the data fit for the chosen prognostic parameter. For the methodology proposed, the model must be linear in parameters; however, simple exponential models, such as $y=exp(ax+b)$ parameterized as $ln(y) = ax +b$, cannot be used with negative y-values, because the natural logarithm of a negative number is undefined in the real number system. This adds unnecessary complexity to the modeling method. Quadratic equations, on the other hand, are naturally linear in parameters and can be used without significant concern for the effects of noise on the model fit. Shifting the prognostic parameter to the positive quadrant eliminates the problem of taking the logarithm of negative values; however, the quadratic fit results in a lower fitting error than the exponential fit, with mean squared errors of 1.53 and 2.33 respectively. Because of its robustness

to noise and reduced modeling error, the quadratic fit is chosen for this research.

Figure 4 gives an example of a polynomial fit of the prognostic parameter with the time the model crosses the critical failure threshold indicated. The threshold of -13.9 was chosen as the upper 95% level of the distribution of failure values for the known failed cases. This gives an estimated system reliability of 95%, which is a conservative estimate of failure time and reduces the possibility of overestimating RUL and experiencing an in-service failure. The time between the last sample and when the critical failure threshold is crossed gives the estimate of RUL, as indicated by the shaded area. For this case, the estimated RUL is exactly correct, with an estimated remaining life of 36 cycles.

This prognostic parameter was used to develop a GPM prognostic model with Bayesian updating. The model was developed using the prognostic parameter resulting from 218 training cases which ran from beginning of life to failure, and was tested with 218 test cases which ran from beginning of life to some point after a failure-inducing fault occurred but before actual failure. The predictive results for each of these test cases are shown in Figure 5. The GPM model developed with this prognostic parameter resulted in a mean

*Figure 4. Prognostic parameter trending and RUL estimation*

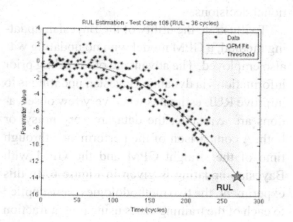

*Figure 5. Visual inspection RUL estimate results with Bayesian updating*

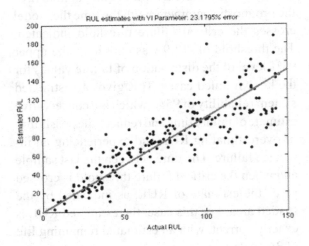

absolute percent error (MAPE) of 23.2%. While this result is larger than would be accepted in practice, it is important to note that in approximately half of the 218 test cases the RUL is predicted to within 10% accuracy. The large MAPE is due to poor performance on predictions made early in life which are far from the failure time. This is an expected and acceptable feature of most prognostic models; larger errors early in life can be tolerated if estimates become more accurate as the equipment approaches failure. The models converge to the true RUL as more data is collected; therefore, the predictions become more accurate when they will be used to make operational decisions.

To illustrate the utility of the Bayesian updating method, a GPM model with no updating was also employed. The advantage of including prior information via dynamic Bayesian updating is to improve RUL estimates when very few observations are available, the data are very noisy, or both. A comparison of the performance through time of the straight GPM and the GPM with Bayesian updating is given in Figure 6. In this experiment, the two methodologies were applied to each of the training cases using only a fraction

of the full lifetime. The models were applied to subsets of each lifetime in 5% increments, i.e. the models were run using 5% of the full lifetime, 10%, 15%, etc. The RUL error at each percentage was calculated across the 218 full training cases to determine how the error decreases as more data become available. As was seen in the example case, the non-Bayesian method may result in an undeterminable RUL. In fact, for the data used here, nearly half the runs resulted in an indeterminate RUL estimate using the GPM methodology without Bayesian updating for runs using less than half the total lifetime. For these cases, the RUL is estimated using a traditional reliability-based method in order to give an estimate of RUL prediction error. The mean residual life is found at each time using a Weibull fit of the failure times and the current lifetime. The Weibull distribution is commonly used for reliability analysis because it is flexible enough to model decreasing, constant, and increasing failure rates (Abernethy, 1996). Mean residual life (MRL) is found by:

$$MRL(t) = \frac{1}{R(t)} \int_{t}^{\infty} R(s)ds$$

where $R(t)$ is the reliability function at time $t$. In practice, the prognostic method would likely fall

*Figure 6. GPM results with and without Bayesian updating*

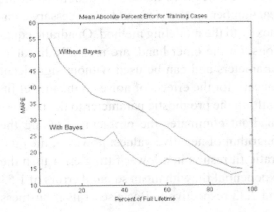

back to a more rudimentary method such as this if the individual-based model did not produce a reasonable answer. The GPM/Type I model which does not include prior information gives an average error of approximately 55% when only 5% of the full lifetime is available and relies on the Type I method for approximately half of the cases. Conversely, the GPM/Bayes method gives approximately 25% error and is able to predict an RUL for every case. As Figure 6 shows, the average error of both methods decreases as more data becomes available and eventually converges to approximately equal error values when the available data overpowers the prior information.

## Identifying an Optimal Prognostic Parameter with Genetic Algorithms

A genetic algorithm (GA) routine was used to optimize the weighting coefficients in a weighted sum of the eleven monitoring system residuals. GA optimization is a stochastic search technique which does not rely on gradient information for optimization. Given enough time, stochastic methods are guaranteed to find the global optimal solution; given a set amount of time, they are able to find near-optimal solutions (Schneider, 2006). GAs evaluate a population of potential solutions at each iteration, or generation, keeping the best solutions for the next generation and filling the rest of the population through biologically-inspired reproduction methods such as crossover and mutation. A full discussion of GA is available in (Haupt and Haupt, 2004).

For this application, the fitness function was given by the sum of the three suitability metrics, with $w_m$, $w_p$, and $w_t$ all equal to one:

*fitness = monotonicity + prognosability + trendability.*

This gives equal weight to each of the three parameter suitability measures. These weights were chosen based on the results of preliminary studies, which suggested that all three suitability metrics were equally important for parameter selection. While the visual inspection parameter described earlier involved several weeks of expert analysis, the GA optimization involved only a few minutes of supervised time to set up and approximately 1.68 hours of unsupervised computer runtime. While the time needed for the GA optimization to run will scale with the number of possible inputs, it involves mainly computer runtime and is only a fraction of the time needed for parameter identification through expert opinion. A method for alleviating this computational burden by pre-screening the possible inputs and only including those which are expected to contribute is outlined in (Coble, 2010).

The parameter suitability metrics for the VI and GA parameters are nearly equivalent, with monotonicity of 0.846 and 0.941, prognosability of 0.891 and 0.895, and trendability of 0.903 and 0.931, respectively. The suitability of each parameter is calculated as the sum of the three suitability metrics, as it was determined in the fitness function, giving an overall suitability of 2.640 for the VI parameter and 2.766 for the GA parameter. The fitness of the GA-optimized parameter is practically equivalent to that of the parameter identified via visual inspection. GPM models developed with the two parameters performed similarly, with MAPE of 23.2% and 20.9% for the VI and GA parameters, respectively.

These GA-optimized prognostic parameters were also used to develop GPM prognostic models. Figure 7 gives the results for the prognostic model developed with the GA parameter. Overall model performance is comparable to that seen with the VI parameter described earlier, and it is nonsensical to claim one parameter as best based on the model performance. However, this application has shown that the proposed automated parameter identification method can identify prognostic parameters comparable to those identified by visual inspection and expert analysis in a fraction of the time.

*Figure 7. RUL estimates for GA-Optimized Parameter with Bayesian updating*

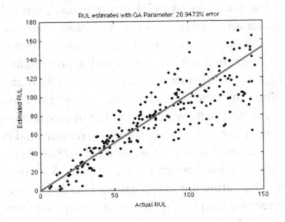

## CONCLUSION

Estimation of remaining useful life is a burgeoning field in the reliability and maintenance community. Prognostics is a key component in a full health monitoring system, which typically includes system monitoring, fault detection and diagnostics, failure prognostics, and operations planning. While system monitoring, fault detection, and diagnostics are well-established fields, prognostics is still in its infancy. Despite prognostics' relative youth as a research area, much attention has been given in the last decade to the development of algorithms for predicting remaining useful life (RUL). Individual-based prognostic algorithms typically use measures of system health to estimate the lifetime of a specific component operating in its specific environment. The most common individual-based prognostic algorithm, and the one employed in this research, is the general path model (GPM), which tracks a measure of system health estimated with a prognostic parameter. The prognostic parameter may be a direct measurement of system health, such as tire tread depth, or it may be inferred from other measurements made on the system, such as tire pressure or operating temperature. A parametric model is fit to the prognostic parameter and extrapolated to a pre-

defined critical failure level. The RUL is estimated as the difference between the current time and the time at which the extrapolated model crosses the failure threshold. This research proposed a modification to the traditional GPM method which uses Bayesian updating methods to incorporate prior information about the expected prognostic parameter path. Incorporating prior information allows the GPM to be applied to systems with very few observations or whose observations are contaminated by noise.

The performance of GPM prognostics depends primarily on the identification of an appropriate prognostic parameter. Ideally, a prognostic parameter should have three key qualities: monotonicity, prognosability, and trendability. Monotonicity is a measure of the general positive or negative trend of the prognostic parameter. Damage is generally assumed to be cumulative and irreversible. Because the prognostic parameter can be considered an indicator of system health, if the parameter were not monotonic, it would indicate some measure of self-healing is taking place. This assumption is not valid for some specific components, such as batteries which do experience some increased capacity after a period of rest. However, it is usually considered valid when dealing with an entire system. Prognosability characterizes how well clustered the failure values are for a population of systems. Because a critical failure threshold must be identified for the GPM model, failure values should be well clustered in order to give a crisp failure value and reduce RUL uncertainty. Finally, trendability measures how well each parameter for a population of systems can be described by the same underlying function. The GPM depends on fitting a parametric model to the prognostic parameter and extrapolating it to failure, so it is key that the same parametric function be applicable to the entire population of systems with the same type of failure mechanism.

By formalizing these metrics so that the suitability of a candidate prognostic parameter can be quantified, it is straightforward to apply conventional optimization methods to identify an optimal, or near-optimal, prognostic parameter from many possible data sources including sensed data, monitoring system residuals, environmental and operational conditions, fault detection and diagnostic results, etc.

The PHM challenge data application utilized the results of a condition monitoring and fault detection system to characterize the degradation in a given system. Prognostic parameters were generated from a subset of the monitoring system residuals; monitoring system residuals are natural components of a prognostic parameter because they inherently measure the deviation of a system from normal operation. This example application suggests that the proposed prognostic parameter selection method performs comparably to expert analysis in a fraction of the time.

## ACRONYMS AND ABBREVIATIONS

**AAKR:** Auto-Associative Kernel Regression
**C-MAPSS:** Commercial Modular Aero-Propulsion System Simulation
**GA:** Genetic Algorithm
**GPM:** General Path Model
**MAPE:** Mean Absolute Percent Error
**MRL:** Mean Residual Life
**MTTF:** Mean Time to Failure
**PHM:** Prognostics and Health Management Society
**RUL:** Remaining Useful Life
**TOF:** Time of Failure
**TRA:** Throttle Resolver Angle
**VI:** Visual Inspection

## REFERENCES

Abernethy, R. B. (1996). *The new Weibull handbook* (2nd ed.). North Palm Beach, FL: Abernethy.

Brown, D. W., Kalgren, P. W., Byington, C. S., & Roemer, M. J. (2007). Electronic prognostics – A case study using global positioning system (GPS). *Microelectronics and Reliability, 47,* 1874–1881. doi:10.1016/j.microrel.2007.02.020

Carlin, B. P., & Louis, T. A. (2000). *Bayes and empirical Bayes methods for data analysis* (2nd ed.). Boca Raton, LA: Chapman and Hall/CRC.

Chinnam, R. B. (1999). On-line reliability estimation of individual components, using degradation signals. *IEEE Transactions on Reliability, 48*(4), 403–412. doi:10.1109/24.814523

Coble, J. (2010). *Merging data sources to predict remaining useful life – An automated method to identify prognostic parameters* (Doctoral dissertation). Retrieved from http://trace.tennessee.edu/utk_graddiss/683

Coble, J., & Hines, J. W. (2009, April). *Fusing data sources for optimal prognostic parameter selection.* Paper presented at the Sixth American Nuclear Society International Topical Meeting on Nuclear Plant Instrumentation, Control, and Human-Machine Interface Technologies NPIC&HMIT 2009, Knoxville, TN.

Engel, S., Gilmartin, B., Bongort, K., & Hess, A. (2000). Prognostics, the real issues involved with predicting life remaining. *Proceedings of the 2000 IEEE Aerospace Conference,* (pp. 457-469).

Feng, W. C. (2003). Making a case for efficient supercomputing. *Queue,* Oct, 54 – 64.

Gelman, A. JCarlin, J., Stern, H., & Rubin, D. (2004). *Bayesian data analysis,* 2nd ed. Boca Raton, LA: Chapman and Hall/CRC.

Haupt, R. L., & Haupt, S. E. (2004). *Practical genetic algorithms.* Hoboken, NJ: John Wiley & Sons Ltd.

Hines, J. W., Garvey, D., Seibert, R., & Usynin, A. (2008). Technical review of on-line monitoring techniques for performance assessment: *Vol. 2. Theoretical issues (NUREG/CR-6895)*. Rockville, MD: Nuclear Regulatory Commission.

Horst, R., & Hoang, T. (1996). *Global optimization: Deterministic approaches*. New York, NY: Springer.

Lindely, D. V., & Smith, A. F. (1972). Bayes estimates for linear models. *Journal of the Royal Statistical Society. Series B. Methodological, 34*(1), 1–41.

Lu, C. J., & Meeker, W. (1993). Using degradation measures to estimate a time-to-failure distribution. *Technometrics, 35*(2), 161–174.

Robinson, M. E., & Crowder, M. T. (2000). Bayesian methods for a growth-curve degradation model with repeated measures. *Lifetime Data Analysis, 6*, 357–374. doi:10.1023/A:1026509432144

Saxena, A., & Goebel, K. (2008). *C-MAPSS data set*. NASA Ames Prognostics Data Repository. Retrieved from http://ti.arc.nasa.gov/project/prognostic-data-repository

Saxena, A., Goebel, K., Simon, D., & Eklund, N. (2008, October). *Damage propagation modeling for aircraft engine run-to-failure simulation*. Paper presented at International Conference on Prognostics and Health Management, Denver, CO.

Schneider, J. J. (2006). *Stochastic optimization*. Berlin, Germany: Springer-Verlag.

Tamhane, A. C., & Dunlop, D. D. (1999). *Statistics and data analysis: From elementary to intermediate*. Upper Saddle River, NJ: Prentice Hall.

Union of Concerned Scientists. (n.d.). *Davis-Besse outage report*. Retrieved from http://www.ucsusa.org/assets/documents/nuclear_power/davis-besse-ii.pdf

Upadhyaya, B. R., Naghedolfeizi, M., & Raychaudhuri, B. (1994). Residual life estimation of plant components. *P/PM Technology*, June, 22 – 29.

Wald, A. (1945). Sequential tests of statistical hypotheses. *Annals of Mathematical Statistics, 16*(2), 117–186. doi:10.1214/aoms/1177731118

World Nuclear News. (2008). *Cook 1 restart September at the earliest*. Retrieved from http://www.world-nuclear-news.org/C-Cook_1_restart_September_at_the_earliest-0212088.html

# Chapter 8
# Modeling Multi-State Equipment Degradation with Non-Homogeneous Continuous-Time Hidden Semi-Markov Process

**Ramin Moghaddass**
*University of Alberta, Canada*

**Ming J Zuo**
*University of Alberta, Canada*

**Xiaomin Zhao**
*University of Alberta, Canada*

## ABSTRACT

*The multi-state reliability analysis has received great attention recently in the domain of reliability and maintenance, specifically for mechanical equipment operating under stress, load, and fatigue conditions. The overall performance of this type of mechanical equipment deteriorates over time, which may result in multi-state health conditions. This deterioration can be represented by a continuous-time degradation process with multiple discrete states. In reality, due to technical problems, directly observing the actual health condition of the equipment may not be possible. In such cases, condition monitoring information may be useful to estimate the actual health condition of the equipment. In this chapter, the authors describe the application of a general stochastic process to multi-state equipment modeling. Also, an unsupervised learning method is presented to estimate the parameters of this stochastic model from condition monitoring data.*

DOI: 10.4018/978-1-4666-2095-7.ch008

## INTRODUCTION

In the domain of reliability and maintenance, significant attention has been paid in recent years to equipment with discrete multi-state health conditions. *Multi-state models*, particularly for mechanical equipment under stress or load conditions that deteriorates or degrades over time, can properly model the transitions among the intermediate health states, ranging from perfect functioning to complete failure. These intermediate health states can either be directly observable (*completely observable states*) or partially observable (*partially observable states*). Currently, due to the complexity of the degradation process and other technical problems, most mechanical equipment is only partially observable through *condition monitoring* (CM) techniques (Moghaddass & Zuo, 2011b), such as vibration analysis and signal processing. Therefore, modeling multi-state equipment with partially observable states involves two types of processes; namely, the degradation process and the observation process. The degradation process deals with the behavior of transitions between health states, while the observation process deals with the stochastic relationship between the degradation states and condition monitoring information. Details for different types of degradation and observation processes can be found in (Moghaddass & Zuo, 2011b).

The primary step before using a *condition-based diagnostic and prognostic* method for multi-state equipment with unobservable states is to determine the type of stochastic models to be used for degradation and observation process modeling. The subsequent step is to estimate the characteristic parameters of these models from real-time condition monitoring data. These parameters are needed to characterize the degradation and observation processes. Dealing with the observation process is more straightforward than evaluating the degradation process. With regard to the *observation process*, the relationship between the condition monitoring information and the actual health state is investigated. A common approach to study the observation process for a piece of mechanical equipment is to artificially create damage that corresponds to the particular discrete health states, and then extract condition monitoring data for each health condition. *Machine learning* techniques and statistical data-driven methods have been developed to explore this relationship in the literature (Worden & Manson, 2007; Zhang et al., 2011). However, there is less work involved in the degradation process modeling especially for multi-state equipment (Wu et al., 2010; Petrovic et al., 2011).

In this chapter, we focus on degradation process modeling for multi-state equipment with unobservable states. From both the theoretical and practical points of view, we first consider the characteristics of the continuous-time stochastic models that are available for multi-state equipment modeling in the literature, and then present generalized results for *multi-state degradation modeling* using the non-homogenous continuous-time hidden semi-Markov process (NHCTHSMP). We will show that NHCTHSMP is capable of overcoming some of the limitations of reported methods, while covering much of the previously research, since it makes more general assumptions for the equipment being studied. Finally, we present an estimation method to find the unknown parameters of an NHCTHSMP that is applicable to the modeling of multi-state equipment with unobservable states. In this chapter, we aim to present a general method that can model the stochastic characteristics of the degradation and observation processes of a piece of multi-state equipment.

With regard to the observation process, based on the approach taken in other studies (Moghaddass & Zuo, 2011a; Moghaddass & Zuo, 2011b), we assumed that there is a stochastic relationship between the unobservable states (hidden states) and the condition monitoring (CM) indicator that can be represented by an observation probability matrix. A more elaborated description of this matrix will be presented in the third part of this chapter. The remainder of this chapter is organized as follows: In the second part, we review the previous work on degradation modeling for multi-state equipment, and provide complementary definitions and explanations for several stochastic models that are applicable to multi-state equipment modeling. Thereafter, we use the basics of the non-homogenous Markov renewal theory to formulate several types of degradation processes. In the fourth part of this chapter, the modeling of multi-state equipment with unobservable states through NHCTHSMP is presented. A parameter estimation method is then described. A simple, simulation-based numerical experiment is discussed. Finally, we explore the direction of future research and end this chapter by conclusion.

## BACKGROUND

The degradation process between two discrete unobservable states in a piece of multi-state equipment conforms to either a Markovian or a non-Markovian structure (Moghaddass & Zuo, 2011b). In *continuous-time Markovian degradation*, the transition between two states depends only on the two states involved in that transition, and is conditionally independent of the history of the process. If all transitions out of state *i* are Markovian, the sojourn time distribution at state *i* follows an exponential distribution. In this case, the degradation process follows a continuous-time Markov process (CTMP). On the other hand, in *continuous-time non-Markovian degradation*, the transition between two states depends on factors in the history of the process, such as the age of the equipment and/or the sojourn time at the current state. An example of non-Markovian degradation is a *semi-Markov degradation*, in which the sojourn time at each state follows an arbitrary distribution (non-exponential). It should be noted that, if at least one transition out of state *i* is non-

Markovian, the sojourn time distribution at this state no longer follows an exponential distribution, and the process is no longer CTMP. In addition, if all transitions are semi-Markovian, then the degradation process follows a continuous-time semi-Markov process (CTSMP).

The continuous-time semi-Markov process with discrete states occurs in one of three forms: a homogenous continuous-time semi-Markov process (HCTSMP), an explicit-duration or variable duration continuous-time semi-Markov process (ED-CTSMP), or a nonhomogeneous continuous-time semi-Markov process (NHCTSMP). In this section, we briefly review the differences between these three forms. Finally, we present how NHCTSMP, as a powerful and flexible modeling tool, can cover the other three degradation models (CTMP, HCTSMP, and ED-CTSMP).

In multi-state equipment with CTMP structure as reported in (Moghaddass & Zuo, 2011b; Lin & Makis, 2004; Kim et al., 2010), the equipment continuously degrades from any current state *i* to a degraded state *j* with a fixed time-homogeneous transition rate of $\lambda_{i,j}$. The transition between the two states is realized according to the event that occurs first in a competition among all transitions out of the current state. Based on this assumption, the sojourn time distribution at any state *i* follows an exponential distribution with transition rate $\sum_{j \in E} \lambda_{i,j}$. Therefore, the characteristic parameters for the degradation process are the constant transition rates $\lambda_{i,j}, (i,j) \in E$.

In multi-state equipment with HCTSMP structure as reported in (Moghaddass & Zuo, 2011a; Lisnianski & Levitin, 2003), the equipment continuously degrades from its current state *i* to a degraded state *j* with a time nonhomogeneous transition rate of $\lambda_{i,j}(t)$, where *t* is the sojourn time at state *i*. Here, the time to each transition follows a non-exponential distribution. Based on this assumption, the sojourn time distribution at state *i* follows an arbitrary distribution (non-exponential), with a transition rate equal to $\sum_{j \in E} \lambda_{i,j}(t)$. Therefore, the characteristic parameters for the degradation process are all those elements that characterize the transition rates $\lambda_{i,j}(t)$, $(i, j) \in E$. Based on this structure, the transitions depend on the sojourn time at the current state. It is clear that CTMP is a special case of HCTSMP, where its transition rates are constant (independent of $t$).

In multi-state equipment with ED-CTSMP structure as reported in (Moghaddass & Zuo, 2011b; Hongzhi et al., 2011), the equipment stays at its current state $i$ following an arbitrary continuous-time distribution, and then transits to a degraded health state $j$ with a fixed probability of $p_{i,j}$. Many examples of this structure with discrete-time domain for the sojourn time (ED-DTSMP) are reported in (Dong et al., 2006; Dong & He, 2007). The main difference between HCTSMP and ED-CTSMP is that in ED-CTSMP, the sojourn time (holding time) distribution at each state is explicitly defined, while in HCTSMP, the distribution of the sojourn time at each state is not directly known (or explicitly defined); however, it can be indirectly calculated considering all transitions out of each state. This will be discussed further in this chapter. The characteristic parameters for the degradation process in ED-CTSMP are the constant transition probabilities $(p_{i,j}, (i, j) \in E)$ and all elements that characterize the sojourn time distribution at each state $(F_i(t), i \in E)$. It should be noted that the sojourn time distribution at state $i$ can be represented by its corresponding transition rate $\lambda_i(t)$.

Finally, in multi-state equipment with NHCTSMP degradation structure as reported in (Moghaddass & Zuo, 2011b), the equipment continuously degrades from its current state $i$ to a degraded state $j$ with a time nonhomogeneous transition rate of $\lambda_{i,j}(s, t)$, where $s$ is the process time (age of the equipment) that state $i$ is reached, $t$ is the sojourn time at state $i$, and $t+s$ is the process time (or the total age of the equipment). Based on this process, the conditional sojourn time distribution at state $i$, given that state $i$ has been reached at time $s$, follows an arbitrary distribution with a nonhomogeneous transition rate equal to $\sum_{j \in E} \lambda_{i,j}(s, t)$, which may depend on both $s$ and $t$.

The main difference between NHCTSMP and HCTSMP is that in NHCTSMP, transitions between two states can depend on the total age of the equipment (process time) and/or the time spent at the current state (sojourn time). It is clear that CTMP, ED-CTSMP, and HCTSMP are special cases of NHCTSMP. Here, the characteristic parameters for the degradation process in NHCTSMP are all elements that characterize the transition rate function between two states $(\lambda_{i,j}(s, t), (i, j) \in E)$. A detailed description of this type of transition rate will be provided later in this chapter.

Although CTMP and ED-CTSMP (or ED-DTSMP) have been widely used in the literature for modeling the degradation process of multi-state equipment with unobservable states, they are subject to several practical limitations (Moghaddass & Zuo, 2011b). For CTMP, the constant transition rates and the exponential sojourn time distribution limit its application with regard to practical cases. Using this process to model the degradation process of multi-state equipment has the consequence that degradation between states and aging of the equipment are ignored. Also for ED-CTSMP (or ED-DTSMP), the fixed probability of transitions, regardless of the age of the equipment or the time spent at the current state, limits the use of this model in practical cases (Moghaddass & Zuo, 2011b; Peng & Dong, 2011; Peng & Dong, 2010; Dong & Peng, 2011). In addition, it is not always easy to find an explicit sojourn time distribution for each state (Moghaddass & Zuo, 2011b). For this type of process, the probability of staying at each state is not affected by the history of the process before reaching the current state. To overcome the aforementioned limitations, the fact that the transitions between states should be affected by the age and deterioration over time needs to be taken into consideration. Although both HCTSMP and NHCTSMP consider the time spent at the current state (transitions depend on the sojourn time at each state), there is surprisingly very limited reported work using these processes for degradation modeling. It should be noted that NHCTSMP covers more cases on the basis that transitions depend on the total age of the equipment and/or the time spent at each state. This shows that NHCTSMP not only covers CTMP, ED-CTSMP, and HCTSMP, but it also complies with more practical assumptions. That is the main reason for our focus on NHCTSMP in this chapter.

In this chapter, we do not review discrete degradation models, such as discrete-time Markov chain (Morcous et al., 2003; De Rango et al., 2011), explicit-duration discrete-time semi-Markov process (Dong & He, 2007), and discrete-time aging Markovian degradation (Peng & Dong, 2010; Chen & Wu, 2007). Details about the limitations of these methods can be found in (Moghaddass & Zuo, 2011b). In this chapter, we consider a general case, in which any transition in the form of CTMP, ED-CTSMP, HCTSMP, and NHCTSMP, can exist for a single piece of equipment. We first use the basics of the NHCTSMP to model the degradation process of multi-state equipment. We adopt a general approach to defining the transition rate functions between two states, and show how transition rate functions can be employed to model NHCTSMP. We also report some of our previous findings in (Moghaddass & Zuo, 2011a; Moghaddass & Zuo, 2011b), while providing general details for employing NHCTHSMP for multi-state degradation modeling. Finally, we report the unsupervised estimation method in (Moghaddass & Zuo, 2011b) to find the unknown parameters of a NHCTHSMP using real-time *condition monitoring* data.

## MULTI-STATE DEGRADATION MODELING WITH MARKOV RENEWAL PROCESS

Regardless of the type of the degradation process, equipment with multi-state condition goes through a sequence of health states from perfect functioning to complete failure. If we consider a change in the health state as a transition, the process $(X_n, T_n)$ can be considered a Markov renewal process, where $X_n$ and $T_n$ are respectively the starting state and the time at the $n$th transition. In *homogeneous Markov renewal process* (HMRP), the interarrival times between two states are i.i.d. random variables with an arbitrary distribution while in a *nonhomogeneous Markov renewal process* (NHMRP), the interarrival times between two states are independent random variables with an arbitrary distribution, not necessarily following identical distributions. The classical Markov renewal theory described in (Janssen et al., 2002; López Droguett et al., 2008; Janssen & Manca, 2001; D'Amico et al., 2011; Moura et al., 2009) has a powerful and flexible mathematical structure which can efficiently be employed for multi-state degradation modeling. We will use the basics of the Markov renewal theory to define all four degradation models described in this chapter and present some general results on the application of nonhomogeneous Markov renewal processes in multi-state degradation modeling. Most of the notation adopted here follows that of (Blasi et al., 2004; Moghaddass & Zuo, 2011b). The list of the notation used in this chapter is as follows:

$N$ : The number of states of the equipment

$E = \{1, ... N\}$ : The finite state space

$X_n$ : The state of the equipment at the $n$th transition $(X_n : \Omega \rightarrow E)$

$T_n$ : The time (age of the equipment) at the $n$th transition $(T_n : \Omega \rightarrow [0, \infty])$

$Z_t$ : The state of the equipment at time $t$ (degradation process at time $t$)

$Y_t$ : The output of the observation process (condition monitoring indicator value) at time $t$

$\theta$ : The set of characteristic parameters

$B$ : Observation probability matrix

$M$ : The number of possible distinct observation values (symbols) for the CM indicator

$V = \{v_1, ..., v_M\}$ : The condition monitoring indicator space

$b_i(j)$ : The probability of getting CM indicator $v_j$ while the equipment is in state $i$; an element of the observation probability matrix $B$

$\Gamma$ : Characteristic parameters corresponding to the degradation process

$\xi_{i,j}$ : The type of the transition between state $i$ and $j$

$\lambda_{i,j}(s,t)$ : Degradation transition rate between state $i$ and state $j$ at time $(s+t)$, given the time at which the equipment entered state $i$ $(s)$

$T_{i,j}^s$ : Time to transition from state $i$ to state $j$, given the time at which the equipment entered state $i$ $(s)$

$f_{i,j}(s,t)$ : Probability density function of $T_{i,j}^s$

$F_{i,j}(s,t)$ : Cumulative density function of $T_{i,j}^s$

$R_{i,j}(s,t)$ : Survival function of $T_{i,j}^s$ $\left( R_{i,j}(s,t) = 1 - F_{i,j}(s,t) \right)$

$FS_i$ : The set of states to which a degradation transition may occur from state $i$

$BS_i$ : The set of states from which a degradation transition may occur to state $i$

$K$ : The number of historical sequences of observations

$l_k$ : The number of observation points for the $k$th sequence of observations

$t_1, ..., t_{l_k}$ : The observation time points for the $k$th sequence of observations

$\Delta$ : Constant time between two observation points

$O^{(k)}$ : $k$th sequence of observations

$Q^{(k)}$ : $k$th sequence of states

$O_i^{(k)}$ : The CM indicator value of the $k$th sequence of observations at time $t_i$

$q_i^{(k)}$ : The state of the equipment at time $t_i$ for sequence $O^{(k)}$

$n_{Q^{(k)}}$ : The number of transitions within $O^{(k)}$

$Q_n$ : The sequence of states between the $(n-1)^{\text{th}}$ and $(n)^{\text{th}}$ transitions for an arbitrary sequence $Q$

## Preliminaries of NHCTSMP

Homogenous and nonhomogeneous Markov renewal processes are distinguished based on certain factors. The most common way to describe these processes is through the kernel function $(Q)$. Using the kernel function as the fundamental describer of the process, other important measures can be calculated. Based on the Markov renewal theory, the kernel function for the homogeneous Markov renewal process and the nonhomogeneous Markov renewal process are different. The $(X, T)$ process is called a homogenous Markov renewal process (HMRP), if its kernel has the property seen in Equation 1 (Blasi et al., 2004) and the $(X, T)$ process is called a nonhomogeneous Markov renewal process (NHMRP), if its kernel function has the property shown in Equation 2 (Blasi et al., 2004).

The process $Z_t = (Z_t, t \in \Re_0^+)$ follows a homogeneous continuous-time semi-Markov process for HMRP and a nonhomogeneous continuous-time semi-Markov process for NHMRP. Note that $Z_t = X_{N_t}$, where $N_t = \sup\{n : T_n \leq t\}$. The main difference between HMRP and NHMRP is that in NHMRP, the kernel depends on both the time of the last transition $(s)$ and the process time $(t)$ while in HMRP, the kernel depends only on the time of the last transition $(s)$. Now based on kernel $Q = [Q_{i,j}]$ associated with NHMRP, we define five important measures for multi-state equipment which can be expressed in terms of the kernel function. These measures will be used

*Equations 1 and 2*

$$Q_{i,j}(t) = \Pr\left(X_{n+1} = j, T_{n+1} - T_n \leq t \mid X_n = i, T_n = s, (X_c, T_c), 0 \leq c < n\right)$$
$$= \Pr\left(X_{n+1} = j, T_{n+1} - T_n \leq t \mid X_n = i\right) \tag{1}$$

$$Q_{i,j}(s,t) = \Pr\left(X_{n+1} = j, T_{n+1} \leq t \mid X_n = i, T_n = s, (X_c, T_c), 0 \leq c < n\right)$$
$$= \Pr\left(X_{n+1} = j, T_{n+1} \leq t \mid X_n = i, T_n = s\right) \tag{2}$$

later to characterize the four types of multi-state degradation processes described in the previous section.

The first measure is the embedded *transition probability matrix* $(P = [p_{i,j}])$ which provides the one step transition probabilities of the nonhomogeneous embedded Markov chain as:

$$p_{i,j}(s) = \Pr\left(X_{n+1} = j \mid X_n = i, T_n = s\right)$$
$$= \lim_{t \to \infty} Q_{i,j}(s,t), \qquad (i,j) \in E \tag{3}$$

Now, given that state $i$ is reached at time $s$, the probability of the state subsequently occupied can be calculated from Equation (3). In other words, $p_{i,j}$ shows the one-step probability of transition between state $i$ and state $j$. The second measure is the sojourn time at state $i$, given that state $i$ is

reached at time $s$, which is a random variable with the following cumulative distribution function (CDF):

$$H_i(s,t) = P\left(T_{n+1} - T_n \leq t \mid X_n = i, T_n = s\right)$$
$$= \sum_{j \in E} Q_{i,j}(s, t+s), \quad i \in E \tag{4}$$

Now it is possible to define the sojourn time cumulative distribution at state $i$, given that state $i$ is reached at time $s$, in terms of the process time $t$ (total age of the equipment) as:

$$Q_i(s,t) = \Pr\left(T_{n+1} \leq t \mid X_n = i, T_n = s\right),$$
$$= \sum_{j \in E} Q_{i,j}(s,t), \quad i \in E \tag{5}$$

*Table 1. The differences between the four types of transition*

| Transition Type | $p_{i,j}(s)$ | $H_i(s,t)$ | $G_{i,j}(s,t)$ | $H_i(s,t)$ and $G_{i,j}(s,t)$ |
|---|---|---|---|---|
| $\{\xi_{i,k} = 1 \mid \forall k \in FS_i\}$ | Constant | Exponential | Exponential | Equivalent |
| $\{\xi_{i,k} = 2 \mid \forall k \in FS_i\}$ | Constant | Non-Exponential (Explicitly defined) | Non-Exponential (Explicitly defined) | Equivalent |
| $\{\xi_{i,k} = 3 \mid \forall k \in FS_i\}$ | Constant | Non-Exponential | Non-Exponential | Different |
| $\{\xi_{i,k} = 4 \mid \forall k \in FS_i\}$ | Dependent on $s$ | Non-Exponential (Dependent on $s$) | Non-Exponential (Dependent on $s$) | Different |

which gives the probability that the process will leave state $i$ before process time $t$, given that state $i$ is reached at time $s$ . The relationship between Equation (4) and Equation (5) is as $H_i(s,t) = Q_i(s, s + t)$ or $Q_i(s,t) = H_i(s, t - s)$. The fourth measure is the conditional sojourn time distribution given that the state subsequently occupied is state $j$ and state $i$ is reached at time $s$. This random variable has the following CDF:

$$G_{i,j}(s,t)$$
$$= \Pr\left(T_{n+1} - T_n \le t \mid X_n = i, X_{n+1} = j, T_n = s\right)$$
$$= \begin{cases} \dfrac{Q_{i,j}(s, s+t)}{p_{i,j}(s)} & \text{if } p_{i,j}(s) \ne 0 \\ \\ 1 & \text{if } p_{i,j}(s) = 0 \end{cases}.$$
$$i, j \in E$$
$$(6)$$

Now it is also possible to define the conditional sojourn time distribution at state $i$ in terms of the process time $t$, given that the state that the equipment enters next is $j$ and state $i$ is reached at time $s$ as:

$$W_{i,j}(s,t)$$
$$= \Pr\left(T_{n+1} \le t \mid X_n = i, X_{n+1} = j, T_n = s\right)$$
$$= \begin{cases} \dfrac{Q_{i,j}(s,t)}{p_{i,j}(s)} & \text{if } p_{i,j}(s) \ne 0 \\ \\ 1 & \text{if } p_{i,j}(s) = 0 \end{cases}$$
$$i, j \in E$$
$$(7)$$

which gives the probability that the process will leave state $i$ before process time $t$, given that the next state is $j$ and state $i$ is reached at time $s$. The relationship between Equation (6) and Equation (7) is as $G_i(s,t) = W_i(s, s+t)$ or $W_i(s,t) = G_i(s, t - s)$. We will use the above measures as well as the kernel function to differentiate four types of the degradation processes described in the previous section of this chapter.

## Transition Types

The use of transition rate functions in modeling transitions between states in a semi-Markov process can be found in (Moura et al., 2009, Becker, 2000). In this section, we use a general definition for the transition rate which can define the four types of the degradation models introduced earlier in this chapter. Then we will introduce the relationship between the transition rates, the kernel function, and other important measures described in the previous section. As mentioned earlier, with respect to the dependency to the deterioration and the age of the equipment, transitions between two states (in continuous domain) can be in any of the four types of CTMP, HCTSMP, ED-CTSMP, and NHCTSMP. Our purpose is to generalize the definition for transition rate to model the multi-state equipment in the sense that all four types of degradation transition may be present for a single piece of equipment.

For the stochastic process associated with the transition between state $i$ and state $j$, given that the equipment is in state $i$ at time $u$, the probability that it transits to state $j$ in an infinitesimal time interval $(u, u+du)$ is the transition rate of the process at time $u$ denoted by $\lambda_{i,j}(u)$. This can be expressed as follows:

$$\lambda_{i,j}(u).du = \Pr\left\{ Z_{u+du} = j \mid Z_u = i \right\} \to \lambda_{i,j}(u)$$
$$= \lim_{du \to 0} \frac{\Pr\left\{ Z_{u+du} = j \mid Z_u = i \right\}}{du}.$$
$$(8)$$

To cover all four types of transitions, the following general definition can be used as the transition rate function between state $i$ and state $j$ as Equation 9 where $\lambda_{i,j}(s,u)$ is the transition rate function between state $i$ and $j$, given that the equipment reached state $i$ at time $s$. By the general definition given in Equation (9), transition between two state may depend on the two states involved in transition, the time spent at the current state (sojourn time at the current state), and/or the total age of the equipment. Similar definition for transition rates was used in (Becker, 2000). We will show how this general definition for transition rate has the flexibility to cover all four types of transitions described earlier.

*Type I - CTMP:* As discussed earlier, this type of transition is represented by a constant transition rates $\lambda_{i,j}, (i, j) \in E$. Therefore:

$$\lambda_{i,j}(s, u) = \lim_{du \to 0} \frac{\int_u^{u+du} f_{i,j}(\tau)d\tau}{du \times \int_u^{\infty} f_{i,j}(\tau)d\tau} \to \lambda_{i,j}(s, u)$$

$$= \frac{f_{i,j}(u)}{R_{i,j}(u)} = \lambda_{i,j}, \quad (i, j) \in E, (s, u) \in [0, \infty]$$

(10)

Equation (10) verifies that transitions are independent of the sojourn time at the current state and the process time (the total age of the equipment).

*Type II - ED-CTSMP:* As discussed earlier, this type of transition can be represented by a time-dependent transition rate $\lambda_i(t), (i) \in E$ and one-step transition probabilities $p_{i,j}, (i, j) \in E$ as:

$$\lambda_{i,j}(s, u) = \lim_{du \to 0} p_{i,j} \frac{\int_u^{u+du} f_i(\tau)d\tau}{du \times \int_u^{\infty} f_i(\tau)d\tau} \to \lambda_{i,j}(s, u)$$

$$= p_{i,j} \frac{f_i(u)}{R_i(u)} = p_{i,j}\lambda_i(u), \quad (i, j) \in E, (s, u) \in [0, \infty]$$

(11)

Equation (11) verifies that transitions depend on the two state involved in transition and the time spent at the current state.

*Type III - HCTSMP:* As discussed earlier, this type of transition can be represented by a time-dependent transition rate $\lambda_{i,j}(u), (i, j) \in E$:

$$\lambda_{i,j}(s, u) = \lim_{du \to 0} \frac{\int_u^{u+du} f_{i,j}(\tau)d\tau}{du \times \int_u^{\infty} f_{i,j}(\tau)d\tau} \to \lambda_{i,j}(s, u)$$

$$= \frac{f_{i,j}(u)}{R_{i,j}(u)} = \lambda_{i,j}(u), \quad (i, j) \in E, (s, t) \in [0, \infty]$$

(12)

Equation (12) verifies that transitions are independent of the time that the equipment enters state *i*. This type of transition deals with those that depend on the two states involved in transitions and the time spent on the last state or the sojourn time at the last state. This means that transitions are not affected by the total age of the equipment (process time). In other words, the degradation process from state *i* to state *j* occurs only when the equipment reaches state *i* first. An example of this type of transition is given in (Moghaddass & Zuo, 2011a; Vaurio, 1997).

*Type IV - NHCTSMP:* As discussed earlier, this type of transition can be represented by a time-dependent transition rate $\lambda_{i,j}(s, u)$. Therefore, we have:

*Equation 9*

$$\lambda_{i,j}(s, u)\, du = \Pr\left\{ u \le T_{n+1} - T_n \le u + du \cap X_{n+1} = j \mid u \le T_{n+1} - T_n \cap X_n = i \cap T_n = s \right\} \quad (9)$$

$$\lambda_{i,j}(s,u) = \lim_{du \to 0} \frac{\int\limits_{u}^{u+du} f_{i,j}(s,\tau)d\tau}{du \times \int\limits_{u}^{\infty} f_{i,j}(s,\tau)d\tau} \to \lambda_{i,j}(s,u)$$

$$= \frac{f_{i,j}(s,u)}{R_{i,j}(s,u)}, \quad (i,j) \in E, (s,t) \in [0,\infty]$$

(13)

This type of transition deals with those transitions that depend on the two states involved in transitions, the total age of the equipment, and/or the sojourn time at the current state. Here, if the transition rates depend only on the process time (total age of the equipment) which is $s+u$, then $\lambda_{i,j}(s,u) = \dfrac{f_{i,j}(s+u)}{R_{i,j}(s+u)}$, where $f_{i,j}(u)$ is the probability density function of time to transition from state $i$ to state $j$. With this type of transition, the probability of transition to a degraded state increases as equipment ages. An example of this transition is a catastrophic failure which becomes more likely as the equipment ages (aging Markovian process). Another example of this type of transition is the proportional hazard model used in (Ghasemi et al., 2011), where transition from any state to the failure state is a function of states involved in that transition and the total age of the equipment. Now, it can be verified that the general definition given for transition rate in Equation (9) has the flexibility to describe different types of transitions from Type 1 to Type 4. This general form of the transition rate function will be used in this chapter as the fundamental describer of degradation transition between states.

## Modeling NHCTSMP with Transition Rate Functions

As discussed, transition rate functions can be used to represent the transition between two states. In this section, we present the mathematical expressions for the important measures of a NHCTSMP in terms of transition rates. Here we provide the results for the general case of NHCTSMP as well as the special cases of CTMP, ED-CTSMP, and HCTSMP. The kernel function for NHCTSMP is in the form of Equation (2) and is a function of $s$ and $t$ as Equation 14.

The expression of kernel in terms of transition rates for NHCTSMP is:

$$\begin{aligned}
Q_{i,j}(s,t) &= \Pr\left(X_{n+1} = j, T_{n+1} \leq t \mid X_n = i, T_n = s\right) \\
&= \Pr\left\{(T_{i,j}^s \leq (t-s)) \bigcap_{z \neq j} (T_{i,j}^s < T_{i,z}^s)\right\} \\
&= \int\limits_{0}^{t-s} f_{i,j}(s,u) \prod_{z \neq j} R_{i,z}(s,u) \; du \\
&= \int\limits_{0}^{t-s} \lambda_{i,j}(s,u) \prod_{z} R_{i,z}(s,u) \; du \\
&= \int\limits_{0}^{t-s} \lambda_{i,j}(s,u) \exp(-\int\limits_{0}^{u} \sum_{z} \lambda_{i,z}(s,x)\,dx)du,
\end{aligned}$$

$$(i,j) \in E$$

(15)

*Equation 14*

$$Q(s,t) = \begin{bmatrix} Q_{1,2}(s,t) & Q_{1,3}(s,t) & .. & Q_{1,N-1}(s,t) & Q_{1,N}(s,t) \\ - & Q_{2,3}(s,t) & .. & Q_{2,N-1}(s,t) & Q_{2,N}(s,t) \\ - & - & .. & Q_{3,N-1}(s,t) & Q_{3,N}(s,t) \\ \vdots & \vdots & .. & \vdots & \vdots \\ & & .. & - & Q_{N-1,N}(s,t) \end{bmatrix}$$

(14)

which can be simplified to the cases shown in Equation 16 where $\xi_{i,k}$ is the type of transition from state $i$ to state $k$. The expressions for the embedded transition probabilities, sojourn time, and conditional sojourn time distributions are given in Equations (17)-(19).

Note that $Q_i(s,t)$ and $W_{i,j}(s,t)$ can be calculated in a way similar to Equations (18)-(19), respectively. Once the transition rate functions between any two states are given, one can calculate the important measures of a NHCTSMP from Equations (15)-(19). These important measures provide information on the stochastical properties of the degradation process associated with the equipment.

## Which Transition Type to Use?

It is interesting to note that in most studies on multi-state equipment reported in the literature, researchers considered a same type of transition for all transitions between states. Then based on their assumptions associated with the type of the transition, the parameters of the corresponding degradation model are estimated using historical data. However, in reality transitions between states can have different structures. That is why the general definition for transition rate can deal with this limitation. The unique properties of each type of transition such as the dependency to the process time or sojourn time can be used to determine which transition type should be used to model a transition between two states. However, such information is not always available or sometimes

*Equations 16 and 17*

$$
Q_{i,j}(s,t) = \begin{cases}
\dfrac{\lambda_{i,j}}{\sum\limits_{z} \lambda_{i,z}}\left(1 - \exp(-\sum\limits_{z} \lambda_{i,z}(t-s))\right) & \{\xi_{i,k}=1 \mid \forall k \in FS_i\} \\[3ex]
p_{i,j} \times \left(1 - \exp(-\int\limits_{0}^{t-s} \lambda_i(x)dx)\right) & \{\xi_{i,k}=2 \mid \forall k \in FS_i\} \\[3ex]
\int\limits_{0}^{t-s} \lambda_{i,j}(u)\exp(-\int\limits_{0}^{u}\sum\limits_{z}\lambda_{i,z}(x)\,dx)du & \{\xi_{i,k}=3 \mid \forall k \in FS_i\} \\[3ex]
\int\limits_{0}^{t-s} \lambda_{i,j}(s,u)\exp(-\int\limits_{0}^{u}\sum\limits_{z}\lambda_{i,z}(s,x)\,dx)du & \{\xi_{i,k}=4 \mid \forall k \in FS_i\}
\end{cases}
\tag{16}
$$

$$
p_{i,j}(s) = \begin{cases}
\dfrac{\lambda_{i,j}}{\sum\limits_{z} \lambda_{i,z}} & \{\xi_{i,k}=1 \mid \forall k \in FS_i\} \\[3ex]
p_{i,j} & \{\xi_{i,k}=2 \mid \forall k \in FS_i\} \\[3ex]
\int\limits_{0}^{\infty} \lambda_{i,j}(u)\exp(-\int\limits_{0}^{u}\sum\limits_{z}\lambda_{i,z}(x)\,dx)du & \{\xi_{i,k}=3 \mid \forall k \in FS_i\} \\[3ex]
\int\limits_{0}^{\infty} \lambda_{i,j}(s,u)\exp(-\int\limits_{0}^{u}\sum\limits_{z}\lambda_{i,z}(s,x)\,dx)du & \{\xi_{i,k}=4 \mid \forall k \in FS_i\}
\end{cases}
\tag{17}
$$

*Equations 18 and 19*

$$H_i(s,t) = \begin{cases} 1 - \exp(-\sum_z \lambda_{i,z} t) & \{\xi_{i,k} = 1 \mid \forall k \in FS_i\} \\[2mm] 1 - \exp(-\int_0^t \lambda_i(x)dx) & \{\xi_{i,k} = 2 \mid \forall k \in FS_i\} \\[2mm] \int_0^t \sum_z \lambda_{i,z}(u) \exp(-\int_0^u \sum_z \lambda_{i,z}(x)\,dx)du = 1 - \exp(-\int_0^t \sum_z \lambda_{i,z}(x)\,dx) & \{\xi_{i,k} = 3 \mid \forall k \in FS_i\} \\[2mm] \int_0^t \sum_z \lambda_{i,z}(s,u) \exp(-\int_0^u \sum_z \lambda_{i,z}(s,x)\,dx)du = 1 - \exp(-\int_0^t \sum_z \lambda_{i,z}(s,x)\,dx) & \{\xi_{i,k} = 4 \mid \forall k \in FS_i\} \end{cases} \tag{18}$$

$$G_{i,j}(s,t) = \begin{cases} 1 - \exp(-\sum_z \lambda_{i,z} t) & \{\xi_{i,k} = 1 \mid \forall k \in FS_i\} \\[2mm] 1 - \exp(-\int_0^t \lambda_i(x)dx) & \{\xi_{i,k} = 2 \mid \forall k \in FS_i\} \\[2mm] \dfrac{\int_0^t \sum_z \lambda_{i,j}(u) \exp(-\int_0^u \sum_z \lambda_{i,z}(x)\,dx)du}{\int_0^\infty \lambda_{i,j}(u) \exp(-\int_0^u \sum_z \lambda_{i,z}(x)\,dx)du} & \{\xi_{i,k} = 3 \mid \forall k \in FS_i\} \\[4mm] \dfrac{\int_0^t \lambda_{i,j}(s,u) \exp(-\int_0^u \sum_z \lambda_{i,z}(s,x)\,dx)du}{\int_0^\infty \lambda_{i,j}(s,u) \exp(-\int_0^u \sum_z \lambda_{i,z}(s,x)\,dx)du} & \{\xi_{i,k} = 4 \mid \forall k \in FS_i\} \end{cases} \tag{19}$$

very complicated to obtain. In Table 1, based on the formula given in Equations (15)-(19), we present the main differences between the 4 types of transitions. This table can be considered as a guideline to determine when to use each transition type, if sufficient information is available.

In order to be able to use the information provided in Table 1, decision makers need to find the empirical distributions of $p_{i,j}$, $H_i$ and $G_{i,j}$ from training data and then compare their results with those in Table 1. We point out here that the above course of action is complicated and requires huge amount of supervised training data. When no information is available to determine the transition type, one can assume a certain transition

type for each transition and after the parameters of the degradation model are estimated with a parameter estimation method, it can be checked whether or not the corresponding properties given in Table 1 are fulfilled. An efficient and easy-to-implement approach to determine the types of transitions between the health states need to be taken into account in future research work.

## Multi-State Equipment Modeling Using NHCTHSMP

In the previous sections of this chapter, we focused on the degradation process modeling. As mentioned earlier, the equipment under study has multiple health states, which are not directly observable. These states are partially observable through condition monitoring. The following definition shows how NHCTSMP can be used for modeling both the degradation process and the observation process, when the health states are not directly observable.

*Definition.* Let $V = \{v_1, ..., v_M\}$ be the observation space with $M$ possible values and let us also define the random variable $U_n : \Omega \rightarrow V$, where $U_n$ is the output of the observation process at the $n$th transition. The relationship between the degradation process and the observation process can be defined as:

$$\Pr(U_n = v_j \mid (U_c, X_c), 0 \leq c \leq n, \ X_n) \\ = \Pr(U_n = v_j \mid X_n = i) = b_i(j)$$

$$(20)$$

Let us define $Y_t(Y_t \in V)$ as the output of the observation process at time $t$, and $Y_t = U_{N_t}$ where $N_t = \sup\{n : T_n \leq t\}$. Given that $Z_t$ follows a nonhomogeneous continuous-time semi-Markov process, the $(Z, Y)$ process is a non-homogeneous continuous-time hidden semi-Markov process (NHCTHSMP). This NHCTHSMP will be used in the remaining sections of this chapter as the basic tool for multi-state equipment modeling which includes both degradation and observation processes. The main assumptions for the multi-state equipment with unobservable states and the descriptions for the parameters of the associated NHCTHSMM will be presented in the next three subsections. The assumptions are adopted from (Moghaddass & Zuo, 2011b).

## Assumptions

1. The equipment has $N$ levels of degradation states ranging from perfect functioning (level 1) to complete failure (level $N$). At each health state, the equipment has a certain level of operational performance and physical properties.

2. The health state of the equipment is not directly observable except state $N$ which is the failure state and is directly observable through the observation process (self-announcing).

3. Over time, the equipment degrades from its current state to one of its degraded states due to the degradation process. Transitions are left-to-right only, which means that the equipment can not transit to a healthier state over time, that is, no repair or maintenance is considered. Therefore, at each state, the equipment is subject to multiple competing deterioration processes.

4. Degradation transitions between two states depend on the states involved in the transitions, the time spent at the current state, and/ or the total age of the equipment. Therefore, each degradation transition can follow an arbitrary distribution. The degradation transitions are represented by transition rate functions using the general formula given in Equation (9). NHCTSMP structure is used to formulate the degradation process.

5. Only a single indicator is used for health condition monitoring. This indicator is extracted at particular observation points. A stochastic relationship exists between the health state of the equipment and this condition monitoring indicator. The condition monitoring indicator can get any of the $M$ possible observation values. Therefore, the relationship between the actual health states and the observation values is shown by an observation probability matrix. The elements of this observation probability matrix represent the probability of observing each discrete indicator value under different health states. The size of this matrix is $N \times M$.

6. The data required in order to estimate the parameters of a NHCTHSMP associated with the multi-state equipment are multiple

temporal sequences of observations which are not labeled with the actual health states.

7. The time interval between two consecutive observation points is constant and denoted by $\Delta$.

8. The equipment is not repairable and is replaced after failure. Therefore, it is as good as new after replacement.

## Parameters of the NHCTHSMP

To model the multi-state equipment with NHCTHSMP under the assumptions given in the previous section, certain parameters are used which can fully characterize both the degradation process and the observation process. This set of parameters which are adopted from (Moghaddass & Zuo, 2011b) are described below:

1. The number of discrete degradation states is $N$.

2. The degradation transition between state $i$ and $j$ is represented by a transition rate function $\lambda_{i,j}(s,t)$, where $t$ is the sojourn time at state $i$ and $s$ is time point when the equipment entered state $i$. Here, $(s+t)$ is the process time or the total age of the equipment. The number of coefficients required to characterize each transition rate function varies with respect to the type of the degradation process associated with that transition. The set of all transition rate functions in the model as well as their types is represented by $\Gamma$.

3. The number of possible distinct observation values (symbols) for the CM indicator is $M$. These observation symbols are denoted by $V = \left\{ v_1, v_2, ..., v_M \right\}$.

4. The observation distribution is shown by an observation probability matrix $B$. The element at the $i$th row and the $j$th column of this matrix is $b_i(j) = \Pr\left(Y_t = v_j \mid Z_t = i\right)$ for $t>0$. The size of this matrix is $N \times M$.

The compact notation $\theta = (N, \Gamma, V, B)$ is used to indicate the characteristic parameters of the NHCTHSMP model. In other words, the described multi-state equipment with unobservable degradation states is fully characterized if $\theta$ is known.

## ESTIMATION OF NHCTHSMP PARAMETERS FOR MULTI-STATE EQUIPMENT

As discussed earlier in this chapter, to use condition monitoring for equipment diagnostic and prognostic, the structure and parameters of the associated degradation and observation processes need to be determined. Parameter estimation (learning problem) refers to the estimation of the parameters of the model using historical observations. As discussed earlier, the described multi-state equipment has a parameters set denoted by $\theta = (N, \Gamma, V, B)$. Depending on the available information regarding the degradation and the observation processes associated with the equipment, some of these parameters may be unknown. In this chapter, we assume that the number of degradation states ($N$) and the possible values of the condition monitoring indicator ($V$) are known. Therefore, the remaining set of unknown parameters that need to be estimated are $(\Gamma, B)$.

The first group of unknown parameters to be estimated ($\Gamma$) deals with parameters representing the transition distributions between states. In other words, these parameters characterize the degradation process. Since it is assumed that the time to transition between states could follow an arbitrary continuous-time distribution, a distinct transition rate function is needed for each possible transition. Also, the type of each transition should be clearly defined. Depending on the distribution of transitions between states, the number of unknown parameters for each transition can vary. For example, if a Weibull distribution is used to represent a transition, 2 parameters (shape and scale) need to be estimated for this particular transition.

The second group of parameters to be estimated ($B$) represents the stochastic relationship between the health state of the equipment and the observation process. In this chapter, this relationship is represented by a matrix called the observation probability matrix. The entries of this matrix are the unknown parameters of the model. This matrix has $N$ (number of states) rows and $M$ columns. It should be noted that it is assumed that failure state (state $N$) is directly observable through the observation process. Without loss of generality, we assign the observation value $v_M$ to the failure state. Therefore, all elements in the last row and last column of this observation probability matrix are zero except the entry in the last row and the last column which is equal to 1. This means that the probability of observing $v_M$ in the case of failure is 1 and the probability of observing any other observation values $(v_1, ..., v_{M-1})$ in the case of failure is zero.

Depending on the type of the available historical information, estimation methods can be classified to unsupervised estimation methods and supervised estimation methods. Unsupervised estimation methods are used when condition monitoring information are not labeled with the actual health state. In this chapter, we summarize the unsupervised estimation method given in (Moghaddass & Zuo, 2011b) to estimate the unknown parameters of the degradation process and observation process for multi-state equipment. We also present how the kernel function can be employed throughout the estimation procedure.

## Unsupervised Estimation Method

Let us assume that there are $K$ sequences of observations available to be used for estimation. In the corresponding *maximum likelihood* optimization

problem, the product of the probability of these $K$ observation sequences (likelihood function) is to be maximized as:

$$Max \quad L = \prod_{k=1}^{K} \Pr(O^{(k)} \mid \theta) \overset{L'=\log(L)}{\Rightarrow} Max \quad L'$$
$$= \log\left(\prod_{k=1}^{K} \Pr(O^{(k)} \mid \theta)\right) = \sum_{k=1}^{K} \log\left(\Pr(O^{(k)} \mid \theta)\right),$$

(21)

which yields to

$$\theta^* = \arg\max_{\theta} \left( \sum_{k=1}^{K} \log\left( \Pr(O^{(k)} \mid \theta) \right) \right).$$

In order to optimize Equation (21) with respect to $\theta$, the Baums' auxiliary function (Rabiner, 1989) is employed which can be iteratively used to find the optimum value of the log-likelihood function. The Baum's auxiliary function for multiple sequences of observations can be expressed as:

$$\omega(\theta_{old}, \theta) =$$
$$\sum_{k=1}^{K} \sum_{Q^{(k)}} \log\left(\Pr(O^{(k)}, Q^{(k)} \mid \theta)\right) \times \Pr(Q^{(k)} \mid O^{(k)}, \theta_{old}),$$

(22)

where $Q^{(k)}$ is an arbitrary sequence of states with a same length as $O^{(k)}$. It has been shown that maximizing $\omega(\theta_{old}, \theta)$ leads to increasing the likelihood function as

$$\left(\theta_{new} = \arg\max_{\theta} \left[\omega(\theta_{old}, \theta)\right]\right)$$
$$\Rightarrow \Pr(O \mid \theta_{new}) \geq \Pr(O \mid \theta_{old})$$

(Rabiner, 1989). By considering an initial estimate for $\theta$, Equation (22) can be iteratively optimized until a predefined stopping criterion is fulfilled. Now the relationship between $\theta$ and

*Equation 23*

$$
\omega(\theta_{old}, \theta) =
$$
$$
\sum_{k=1}^{K} \left( \sum_{Q^{(k)}} \log\left(\Pr(Q^{(k)} \mid \theta)\right) \times \Pr(Q^{(k)} \mid O^{(k)}, \theta_{old}) + \sum_{Q^{(k)}} \left( \sum_{t=1}^{l_k} \log\left(b_{q_t^{(k)}}(O_t^{(k)})\right) \right) \times \Pr(Q^{(k)} \mid O^{(k)}, \theta_{old}) \right) \tag{23}
$$

Equation (22) should be clearly defined. We simplify Equation (22) to directly express it in terms of $\theta$. Since

$$
\Pr(O^{(k)}, Q^{(k)} \mid \theta) = \Pr(Q^{(k)} \mid \theta) \times \Pr(O^{(k)} \mid Q^{(k)}, \theta),
$$

$$
\Pr(O^{(k)} \mid Q^{(k)}, \theta) = \prod_{t=1}^{l_k} b_{q_t^{(k)}}(O_t^{(k)}),
$$

and

$$
\Pr(Q^{(k)} \mid O^{(k)}, \theta) = \frac{\Pr(Q^{(k)}, O^{(k)} \mid \theta)}{\Pr(O^{(k)} \mid \theta)},
$$

we have Equation 23.

Equation (23) can be divided to two parts as $\omega(\theta_{old}, \theta) = \omega_1(\theta_{old}, \theta) + \omega_2(\theta_{old}, \theta)$ in a way that the first term depends only $\Gamma$ and the second term depends only $B$ from $\theta = (\Gamma, B)$. This will finally enables us to independently estimate these two sets of unknown parameters. To simplify Equation (23), we first show how to calculate $\Pr(Q^{(k)} \mid \theta)$ which is the probability of a random sequence $Q^{(k)}$. Now let $Q_n$ be an arbitrary sequence of states between the *(n-1)*th and the *n*th transitions in $Q^{(k)}$. We can characterize $Q_n$ based on four elements which are the starting time point $(T_{n-1})$, the ending time point $(T_n)$, the state at the starting point $(X_{n-1})$, and the state at the ending point $(X_n)$. Thus, the following relationship holds true between $Q^{(k)}$ and $Q_n$ as:

$$
\Pr(Q^{(k)} \mid \theta) = \prod_{n=1}^{n_{Q^{(k)}}} \Pr(X_n, T_n \mid X_{n-1}, T_{n-1}, \theta)
$$
$$
= \prod_{n=1}^{n_{Q^{(k)}}} \Pr(Q_n \mid Q_{n-1}, \theta)
$$
$$
\tag{24}
$$

Now, we can express $\omega_1(\theta_{old}, \theta)$ as Equation 25.

Now, based on the assumption that at most one transition can occur in an observation interval $\Delta$, we can discretize Equation (25) and rewrite it in an equivalent form (from optimization perspective) as follows:

$$
\omega_{1,1}(\theta_{old}, \theta) = \sum_{k=1}^{K} \sum_{i=1}^{N} \sum_{j \in FS_i} \sum_{t=0}^{l_k} \sum_{d=0}^{l_k - t} \Pr(O^{(k)} \mid \theta_{old})^{-1},
$$
$$
\times \log\left(\varepsilon_t^1(i, j, d)\right) \times \varepsilon_t^2(i, j, d, O^{(k)}) \tag{26}
$$

where we get Equation (27).

In order to simplify Equation (26), we consider all transitions out of each state as a group and express each state separately. This results in *N-1* independent sub-equations as Equation (28) where $\omega_{1,1}(\theta_{old}, \theta) = \sum_{r=1}^{N-1} \omega_{1,1}^r(\theta_{old}, \theta)$. We point out here that each $\omega_{1,1}^r(\theta_{old}, \theta)$ contains only transitions out of state $r$. It should be noted all elements in Equation (28) depends only on the elements of the degradation process ($\Gamma$) and is not affected by the observation process. The relationship between

*Equation 25*

$$\omega_1(\theta_{old}, \theta) = \sum_{k=1}^{K} \left( \sum_{Q^{(k)}} \log\left(\Pr(Q^{(k)} \mid \theta)\right) \times \Pr(Q^{(k)} \mid O^{(k)}, \theta_{old}) \right)$$

$$\xrightarrow{\Pr(Q^{(k)}|\theta) = \prod_{n=1}^{n_Q(k)} \Pr(Q_n|Q_{n-1},\theta)} \sum_{k=1}^{K} \sum_{Q^{(k)}} \left( \sum_{n=1}^{n_Q(k)} \left(\log(\Pr(Q_n \mid Q_{n-1}, \theta))\right) \times \prod_{n=1}^{n_Q(k)} \Pr(Q_n \mid Q_{n-1}, O^{(k)}, \theta_{old}) \right)$$

$$= \sum_{k=1}^{K} \sum_{Q^{(k)}} \left( \sum_{n=1}^{n_Q(k)} \left[ \log(\Pr(Q_n \mid Q_{n-1}, \theta)) \times \prod_{m=1}^{n_Q(k)} \Pr(Q_m \mid Q_{m-1}, O^{(k)}, \theta_{old}) \right] \right)$$

$$= \sum_{k=1}^{K} \sum_{Q^{(k)}} \sum_{n=1}^{n_Q(k)} \left( \log\left(\Pr\left(Q_n \mid Q_{n-1}, \theta\right)\right) \times \Pr(Q_n \mid Q_{n-1}, O^{(k)}, \theta_{old}) \times \prod_{\substack{m=1, \\ m \neq n}}^{n_Q(k)} \Pr(Q_m \mid Q_{m-1}, O^{(k)}, \theta_{old}) \right) \tag{25}$$

$$\xrightarrow{\sum_{Q^{(k)}} \prod_{\substack{m=1, \\ m \neq n}}^{n_Q(k)} \Pr(Q_m|Q_{m-1},O^{(k)},\theta_{old})=1} \sum_{k=1}^{K} \sum_{Q_n} \Pr(O^{(k)} \mid \theta_{old})^{-1} \times \left( \log\left(\Pr\left(Q_n \mid \theta\right)\right) \times \Pr(Q_n, O^{(k)}, \theta_{old}) \right)$$

$$= \sum_{k=1}^{K} \sum_{i=1}^{N} \sum_{j=1}^{N} \int_{0}^{t_{l_k}} \int_{0}^{t_{l_k}-t} \Pr(O^{(k)} \mid \theta_{old})^{-1} \times \log\left(\Pr\left(X_n = j, T_n = t+u \mid X_{n-1} = i, T_{n-1} = t, \theta\right)\right)$$

$$\times \Pr\left(X_n = j, T_n = t+u, X_{n-1} = i, T_{n-1} = t, O^{(k)}, \theta_{old}\right) du\, dt$$

the elements of Equation (28) and $\Gamma$ is described later in this section. The second term of Equation (23) which involves only the elements of $B$ is Equation (29) where $\sum_{j=1}^{M} b_i(j) = 1$, $i \in E$. Equation (29) can be maximized by adding the Lagrange multiplier $\vartheta(\sum_{j=1}^{M} b_i(j) - 1)$, $i \in E$, and setting the associated derivative of $\omega_2(\theta_{old}, \theta) + \vartheta(\sum_{j=1}^{M} b_i(j) - 1)$, $i \in E$ with respect to each $b_i(w)$ and $\vartheta$ equal to zero. Thus, we have Equation (30) where $\delta_{O_t^{(k)}, w}$ is equal to 1 when the

$t$th observation value of $O^{(k)}$ is equal to $v_w$, and 0 otherwise. Now, at each step of the stepwise optimization problem, all entries of the observation probability matrix $B$ can be directly estimated using Equation (30).

To be able to optimize Equation (28) and use Equation (30) to find $\theta_{new}$, the concept of *Expectation-maximization* (EM) (Dempster et al., 1977) is used. The summary of all steps for the unsupervised estimation procedure in order to find the unknown parameters of a NHCTHSMP associated with the multi-state equipment is illustrated as follows:

*Equation 27*

$$\begin{cases} \varepsilon_t^1(i, j, d) = \Pr(X_n = j, (t+d-1)\Delta \leq T_n \leq (t+d)\Delta \mid X_{n-1} = i, T_{n-1} = t\Delta, \theta) \\ \varepsilon_t^2(i, j, d, O^{(k)}) = \Pr(X_n = j, (t+d-1)\Delta \leq T_n \leq (t+d)\Delta, X_{n-1} = i, T_{n-1} = t\Delta, O^{(k)}, \theta_{old}) \end{cases} \tag{27}$$

*Equations 28, 29, and 30*

$$\omega_{1,1}^{r}(\theta_{old},\theta) = \sum_{k=1}^{K} \sum_{j\in FSr} \sum_{t=0}^{l_k} \sum_{d=0}^{l_{k-t}} \Pr(O^{(k)}\mid\theta_{old})^{-1} \times \log(\varepsilon_t^1(r,j,d)) \times \varepsilon_t^2(r,j,d,O^{(k)}) \quad r\in(1,...,N-1) \tag{28}$$

$$\omega_2(\theta_{old},\theta) = \sum_{k=1}^{K}\left[ \Pr(O^{(k)}\mid\theta_{old})^{-1} \times \left( \sum_{j=1}^{N}\sum_{t=1}^{l_k} \Pr(q_t^{(k)}=j,O^{(k)}\mid\theta_{old}) \times \log\left( b_{q_t^{(k)}}(O_t^{(k)}) \right) \right) \right] \tag{29}$$

$$\left[ \frac{\partial}{\partial b_i(w)} \left( \sum_{k=1}^{K}\left[ \Pr(O^{(k)}\mid\theta_{old})^{-1} \times \left( \sum_{j=1}^{N}\sum_{t=1}^{l_k} \Pr(q_t^{(k)}=j,O^{(k)}\mid\theta_{old}) \times \log\left( b_{q_t^{(k)}}(O_t^{(k)}) \right) \right) \right] + \vartheta\left( \sum_{j=1}^{M} b_i(j)-1 \right) \right) = 0 \right.$$

$$\left. \frac{\partial}{\partial\vartheta} \left( \sum_{k=1}^{K}\left[ \Pr(O^{(k)}\mid\theta_{old})^{-1} \times \left( \sum_{j=1}^{N}\sum_{t=1}^{l_k} \Pr(q_t^{(k)}=j,O^{(k)}\mid\theta_{old}) \times \log\left( b_{q_t^{(k)}}(O_t^{(k)}) \right) \right) \right] + \vartheta\left( \sum_{j=1}^{M} b_i(j)-1 \right) \right) = 0 \right] \Rightarrow \tag{30}$$

$$b_i(w) = \frac{\sum_{k=1}^{K}\left[ \Pr(O^{(k)}\mid\theta_{old})^{-1} \times \sum_{t=1}^{l_k} \Pr(q_t=i,O^{(k)}\mid\theta_{old})\delta_{O_t^{(k)},w} \right]}{\sum_{k=1}^{K}\left[ \Pr(O^{(k)}\mid\theta_{old})^{-1} \times \sum_{t=1}^{l_k} \Pr(q_t=i,O^{(k)}\mid\theta_{old}) \right]} =$$

$$\frac{\sum_{k=1}^{K}\left[ \Pr(O^{(k)}\mid\theta_{old})^{-1} \times \sum_{t=1}^{l_k} \gamma_t(i,O^{(k)})\delta_{O_t^{(k)},w} \right]}{\sum_{k=1}^{K}\left[ \Pr(O^{(k)}\mid\theta_{old})^{-1} \times \sum_{t=1}^{l_k} \gamma_t(i,O^{(k)}) \right]} \qquad w\in\{1,...,M\}, i\in E$$

**Step 1:** Set an initial estimates for $\Gamma$ and $B$ and let $\theta_{old} = (\Gamma_{initial}, B_{initial})$.

**Step 2:** Use re-estimation formula given in Equation (30) and update $B_{new}$.

**Step 3:** Optimize all $N-1$ equations in the form of Equation (28) to find $\Gamma_{new}$ (parameters of transition distributions). Update $\theta$ as $\theta_{new} = (\Gamma_{new}, B_{new})$.

**Step 4:** Find the average log-likelihood function using Equation (21) as $\frac{L'}{K}$. If $\frac{\left(L'(new)-L'(old)\right)}{K} < \varepsilon$, terminate the algorithm and output $\theta^* = \theta_{new}$, otherwise set $\theta_{old} = \theta_{new}$, and go back to step 2.

It is expected that with iteration, the log-likelihood function will increase. To perform the above steps, the expression of all elements used in Equations (21), (28) and (30) should be defined in terms of $\theta$. For notational convenience, we show how to calculate each of these elements for a single observation sequence $O$ with $l$ observation points and constant observation interval $\Delta$. However the results can be extended to multiple sequences of observations and variable observation intervals. We report the results given in (Moghaddass & Zuo, 2011b) to calculate these elements and also present the relationships between these measures and the kernel function.

The first measure used in Equation (21), (26), and (28) is $\Pr(O\mid\theta)$, which is the probability of observing the sequence of observations $O = O_1 O_2 ....O_l$. This measure can be calculated by modifying the *forward-backward procedure* given in (Rabiner, 1989). The forward variables are

$$\alpha_t(i) = \Pr(O_1 O_2 ....O_t, q_t = i\mid\theta),$$

the joint probability of being at state $i$ at the $t$th observation point and observing sequence of $O_1 O_2 ....O_t$, and

$$u_t(i) = \Pr(O_1 O_2...O_t, q_1...q_{t-1} \neq i, q_t = i \mid \theta),$$

the joint probability of reaching state $i$ for the first time at the $t$th observation point and observing sequence of $O_1 O_2 ....O_t$. All $\alpha_t(i)$ s and $u_t(i)$ s can be calculated iteratively using two steps which are Initialization and Induction. In the Initialization step, the initial values of specific forward variables are determined as the starting point. Later, in the Induction step, other forward variables are iteratively calculated based on the previously calculated forward variables. It should be noted that at both of these steps, we discretize continuous-time to discrete-time points that correspond to the observation time points. We note that all following steps are based on the assumption that the equipment is at state 1 (perfectly functioning) at time zero. The detailed equations of these two steps are given below:

Initialization step:

$$\begin{cases} u_0(1) = 1 \\ u_t(1) = 0 & 1 \le t \le l \\ u_0(i) = 0 & 2 \le i \le N \\ u_1(i) = G_{1-i}(0,1,0) \times b_i(O_1) & i \in \{ FS_1 \} \\ \alpha_1(i) = G_{1-i}(0,1,0) \times b_i(O_1) & i \in \{ FS_1, 1 \} \end{cases}$$

$$(31)$$

Induction step:

$$u_t(i) =$$
$$\sum_{j \in BS_i} \sum_{z=0}^{t-1} \left( u_z(j) \times G_{j-i}(z,t,0) \times \prod_{w=z+1}^{t-1} b_j(O_w) \right) \times b_i(O_t),$$
$$2 \le i \le N, 2 \le t \le l$$

$$(32)$$

$$\alpha_t(i) = \sum_{z=0}^{t} \left( u_z(i) \times G_{i-i}(z,t,0) \times \prod_{w=z+1}^{t} b_i(O_w) \right).$$
$$1 \le i \le N, 2 \le t \le l$$

$$(33)$$

Starting from the Initialization step, $u_t(i)$ and $\alpha_t(i)$ can be iteratively calculated for $1 \le i \le N$ and $1 \le t \le l$. After the last forward variables $\alpha_l(1),...., \alpha_l(N)$ are calculated from Equations (31)-(33), we can use the following Termination step to find the probability of observation $O$.

Termination step:

$$P(O \mid \theta) = \sum_{i=1}^{N} \alpha_l(i). .$$

$$(34)$$

Now, consider a backward variable

$$\beta_t^d(i) =$$
$$\Pr(O_{t+1}....O_l \mid q_1...q_{t-d} \neq i, q_{t-d+1}...q_t = i, \theta).$$

This backward variable is the conditional probability of future observations $(O_{t+1}....O_l)$, given that the equipment reached state $i$, $(d-1)\Delta$ unit earlier than the current time $(t)$ and is still at state $i$. Similar to the forward variables, we start from an Initialization step and proceed with the Induction step to find all possible backward variables $\beta_t^d(i)$ s. Unlike the forward variables, the Induction step for the backward variables is initiated from the last observation time ($t=l$) and ends at $t=1$.

Initialization step:

$$\begin{cases} \beta_l^{l+1}(1) = 1 \\ \beta_l^d(i) = 1 & 1 \le i \le N, 1 \le d \le l \end{cases}.$$

$$(35)$$

Induction step:

$$\beta_t^d(i) = G_{i-i}(t, t+1, d-1) \times \beta_{t+1}^{d+1}(i) \times b_i(O_{t+1}) +$$

$$\sum_{j \in FS_i} G_{i-j}(t, t+1, d-1) \times \beta_{t+1}^1(j) \times b_j(O_{t+1})$$

$$1 \leq i \leq N, 1 \leq t \leq l-1, 1 \leq d \leq t$$

$$(36)$$

Now that the first backward variables $\beta_1^1(1)$, ..., $\beta_1^1(N)$ are calculated from Equation (36), we can use the Termination step to find the probability of observation $O$. It should be noted that since both Equation (34) and Equation (36) provide a same result, they both can be used to calculate $\Pr(O \mid \theta)$.

Termination step:

$$\Pr(O \mid \theta) = G_{1-1}(0, 1, 0) \times \beta_1^2(1) \times b_1(O_1)$$

$$+ \sum_{j \in FS_1} G_{1-j}(0, 1, 0) \times \beta_1^1(j) \times b_j(O_1)$$

$$(37)$$

The $G_{i-j}(t, t+d, d_0)$, which is an element used in Equations (31)-(36) is the conditional probability of transition from state $i$ (at $t$th observation point) to state $j$ (at $(t+d)$th observation point), given that the equipment reached state $i$ at $((t-d_0)$th) observation point and stayed there for $d_0 \Delta$ units of time. $G$ can be expressed in terms of the kernel function for $1 \leq j \neq i \leq N$ as Equation 38 and for $1 \leq j = i \leq N$ as Equation 39.

Also $G$ can be expressed in terms of transition functions for $1 \leq j \neq i \leq N$ as Equation 40 and for $1 \leq j = i \leq N$ as Equation 41.

It can be verified from Equations (38)-(41) that $G_{i-j}(t, t+d, d_0)$ depends only on the transition rate functions between state $i$ and all $j \in FS_i$.

The next step is to calculate the two measure used in Equation (28) which are $\varepsilon_t^1(i, j, d)$ and $\varepsilon_t^2(i, j, d, O)$. $\varepsilon_t^1(i, j, d)$ is the probability of reaching state $j$ in the time interval $[(t+d-1)\Delta, (t+d)\Delta]$, given that the equipment reached state $i$ at the $t$th observation point. $\varepsilon_t^2(i, j, d, O)$ is joint probability of reaching state $i$ at time $t$, reaching state $j$ in the time interval $[(t+d-1)\Delta, (t+d)\Delta]$, and observing sequence $O$. After $\alpha$, $u$, and $\beta$ are calculated from Equations (31)-(36), the two measures used in Equation (28) can be calculated as follows:

$$\varepsilon_t^1(i, j, d) = G_{i-j}(t, t+d, 0)$$
$$= Q_{i,j}(t\Delta, (t+d)\Delta) - Q_{i,j}(t\Delta, (t+d-1)\Delta),$$
$$(i, j) \in E, 1 \leq t \leq l, 1 \leq d \leq l-t$$

$$(42)$$

$$\varepsilon_t^2(i, j, d, O) = u_t(i) \times G_{i-j}(t, t+d, 0) \times \beta_{t+d}^1(j)$$
$$\times \prod_{w=t+1}^{t+d-1} b_i(O_w) \times b_j(O_{t+d}) \quad (i, j) \in E,$$
$$1 \leq t \leq l, 1 \leq d \leq l-t,$$

$$(43)$$

The last measure that needs to be calculated is $\gamma_t(i)$ used in Equation (30). This measure is the joint probability of being at state $i$ at time $t$ and observing the full sequence of observations $O_1 O_2 \ldots O_l$. This measure can be calculated as:

$$\gamma_t(i, O) = \Pr(q_t = i, O \mid \theta)$$

$$= \sum_{v=0}^t u_v(i) \times G_{i-i}(v, t, 0) \times \beta_t^{t-v+1}(i)$$

$$\times \prod_{w=v+1}^t b_i(O_w) \quad i \in E, 1 \leq t \leq l,$$

$$(44)$$

where $\sum_{i=1}^N \gamma_t(i) = \Pr(O \mid \theta)$. Now that all elements of the optimization problem are approximated, the iterative algorithm described in this chapter can be used to find the unknown parameters of the NHCTHSMP associated with the multi-state equipment.

*Equations 38 and 39*

$$G_{i-j}(t, t+d, d_0) = \Pr(X_{n+1} = j, (t+d-1)\Delta \leq T_{n+1} \leq (t+d)\Delta \mid X_n = i, T_n =$$

$$(t-d_0)\Delta, t\Delta \leq T_{n+1}) = \frac{\Pr(X_{n+1} = j, (t+d-1)\Delta \leq T_{n+1} \leq (t+d)\Delta \mid X_n = i, T_n = (t-d_0)\Delta)}{\Pr(t\Delta \leq T_{n+1} \mid X_n = i, T_n = (t-d_0)\Delta)} =$$

$$\frac{Q_{i,j}((t-d_0)\Delta, (t+d)\Delta) - Q_{i,j}((t-d_0)\Delta, (t+d-1)\Delta)}{1 - Q_i((t-d_0)\Delta, t\Delta)} \qquad (38)$$

$$0 \leq t \leq l, 0 \leq d \leq l-t, t-l \leq d_0 \leq t$$

---

$$G_{i-j}(t, t+d, d_0) = \Pr((t+d)\Delta \leq T_{n+1} \mid X_n = i, T_n = (t-d_0)\Delta, t\Delta \leq T_{n+1}) =$$

$$\frac{\Pr((t+d)\Delta \leq T_{n+1} \mid X_n = i, T_n = (t-d_0)\Delta)}{\Pr(t\Delta \leq T_{n+1} \mid X_n = i, T_n = (t-d_0)\Delta)} = \frac{1 - Q_i((t-d_0)\Delta, (t+d)\Delta)}{1 - Q_i((t-d_0)\Delta, t\Delta)} \qquad (39)$$

$$0 \leq t \leq l, 0 \leq d \leq l-t, t-l \leq d_0 \leq t$$

## NUMERICAL EXAMPLE

In this section, a numerical example similar to the one used in (Moghaddass & Zuo, 2011b) is used to show how NHCTHSMP can model the degradation process and the observation process of a piece of multi-state equipment. Also, the correctness and the efficiency (CPU time) of the proposed estimation procedure are demonstrated.

*Example Description:* A piece of mechanical equipment is operating under four levels of health conditions, which are normal condition, slight damage, medium damage, and failure. The equipment may pass through all degradation states before failure. However, due to the existence of random shocks, it can also directly fail from any state. All states are unobservable except the failure state which is completely observable. The transition diagram of this equipment is shown in Figure 1.

Since the health condition of the equipment is not directly observable, a single condition monitoring indicator is monitored periodically for this equipment. The objective is to find the parameters of the NHCTHSMP associated with this equip-

ment. Monte Carlo Simulation is first employed to generate random sequences of observations. Then, the generated random observation sequences are used as the input for the estimation procedure. The results of the estimation procedure will then be compared with those which were originally used in Monte Carlo simulation to generate the observation sequences. It is assumed that originally the observed condition monitoring indicator has 8 possible discrete outcomes ($M$=8). Therefore, the observation probability matrix ($B$) has 4 rows corresponding to the 4 states and 8 columns corresponding to the 8 possible discrete outcomes. For transition distributions, we used the Weibull distribution as it is the most commonly used distribution to represent degradation (Boutros & Liang, 2011). It is assumed that the transition from state 2 to state 4 depends on the total age of the equipment ($\xi_{2,4} = 4$), whereas all other transitions depend on the sojourn time of the last state ($\xi_{1,2} = \xi_{1,4} = \xi_{2,3} = \xi_{3,4} = 3$). Figures 2 and 3 present how transition rate functions vary with respect to each transition. The expressions for these transition rate functions are as follows:

*Equations 40 and 41*

$$G_{i-j}(t, t+d, d_0)$$

$$= \left( \int_{(d+d_0-1)\Delta}^{(d+d_0)\Delta} (f_{i,j}((t-d_0)\Delta, \tau) \times \prod_{z \neq j} \int_\tau^\infty f_{i,z}((t-d_0)\Delta, x)\, dx\,)d\tau \right) \times \left( \prod_z \int_{d_0\Delta}^\infty f_{i,z}((t-d_0)\Delta, \tau)\, d\tau \right)^{-1}$$

$$= \left( \int_{(d+d_0-1)\Delta}^{(d+d_0)\Delta} (f_{i,j}((t-d_0)\Delta, \tau) \times \prod_{z \neq j} R_{i,z}((t-d_0)\Delta, \tau))\, d\tau \right) \times \left( \prod_z R_{i,z}((t-d_0)\Delta, d_0\Delta) \right)^{-1}$$

(40)

$$= \left( \int_{(d+d_0-1)\Delta}^{(d+d_0)\Delta} \lambda_{i,j}((t-d_0)\Delta, \tau) \times e^{-\int_0^\tau \lambda_{i,j}((t-d_0)\Delta, x)\, dx} \times \prod_{z \neq j} e^{-\int_0^\tau \lambda_{i,z}((t-d_0)\Delta, x)\, dx}\, d\tau \right) \times \left( \prod_z e^{-\int_0^{d_0\Delta} \lambda_{i,z}((t-d_0)\Delta, x)\, dx} \right)^{-1}$$

$$= \left( \int_{(d+d_0-1)\Delta}^{(d+d_0)\Delta} \lambda_{i,j}((t-d_0)\Delta, \tau) \times e^{-\int_0^\tau \sum_z \lambda_{i,z}((t-d_0)\Delta, x)\, dx}\, d\tau \right) \times \left( \prod_z e^{-\int_0^{d_0\Delta} \lambda_{i,z}((t-d_0)\Delta, x)\, dx} \right)^{-1}$$

---

$$G_{i-j}(t, t+d, d_0)$$

$$= \left( \int_{(d+d_0)\Delta}^\infty \prod_z f_{i,j}((t-d_0)\Delta, \tau)\, d\tau \right) \times \left( \prod_z \int_{d_0\Delta}^\infty f_{i,z}((t-d_0)\Delta, \tau)\, d\tau \right)^{-1}$$

$$= \left( \prod_z R_{i,z}((t-d_0)\Delta, (d+d_0)\Delta) \right) \times \left( \prod_z R_{i,z}((t-d_0)\Delta, d_0\Delta) \right)^{-1}$$

(41)

$$= \left( \prod_z e^{-\int_0^{(d+d_0)\Delta} \lambda_{i,z}((t-d_0)\Delta, x)\, dx} \right) \times \left( \prod_z e^{-\int_0^{d_0\Delta} \lambda_{i,z}((t-d_0)\Delta, x)\, dx} \right)^{-1}$$

$$= \left( e^{-\int_0^{(d+d_0)\Delta} \sum_z \lambda_{i,z}((t-d_0)\Delta, x)\, dx} \right) \times \left( e^{-\int_0^{d_0\Delta} \sum_z \lambda_{i,z}((t-d_0)\Delta, x)\, dx} \right)^{-1}$$

$$\lambda_{i,j}(s,t) =$$

$$\begin{cases} \dfrac{\beta_{i,j}}{\alpha_{i,j}} \left( \dfrac{t+s}{\alpha_{i,j}} \right)^{\beta_{i,j}-1} & (i,j) = (2,4) \\[2mm] \dfrac{\beta_{i,j}}{\alpha_{i,j}} \left( \dfrac{t}{\alpha_{i,j}} \right)^{\beta_{i,j}-1} & i \in E, j \in FS_i, (i,j) \neq (2,4) \end{cases}$$

(45)

Therefore, the set of parameters of the numerical examples that need to be estimated are as the equations in Box 1.

Here, the elements in the $i$th row and $j$th column of the scale and shape parameter matrices are the parameter of degradation transition from state $i$ to $j$. The total number of parameters associated with the degradation process for the example is five for scale parameters and five for shape parameters corresponding to the five possible degradation transitions. The total number of parameters associated with the observation process for the example is 32 (4×8) corresponding to four states and eight possible observation values.

*Figure 1. Multi-state equipment with 4 discrete health states*

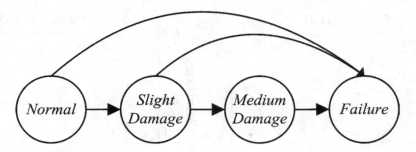

However, since the failure state is assumed to be directly observable with associated observation value 8, the entries of the last row and the last column of this observation probability matrix are known. Therefore, the total number of unknown parameters associated with the observation process for this example is 21 (3×7). We use Matlab 2008 on a stand-alone PC with CPU 2.3 GHz and 16 GB of RAM for our experiment. The *fminsearch* function in Matlab is used throughout the estimation procedure.

*Initial Estimates:* For the transition distributions, we used the initial value of 40 for all scale parameters and the initial value of 1 for all shape parameters. Also for the observation probability matrix, we used the initial value of $(\frac{1}{M-1})$ for all entries except those in the last row and the last column, which are related to the observable failure state.

*Error of Estimation*: To be able to evaluate the correctness of the estimation method and also to see the improvement in estimation procedure (with respect to log-likelihood improvement) after each step of the proposed estimation method, we used a well-known measure called *Root-Mean-Square Error* (RMSE), which measures the difference between the estimated and the actual values. RMSE is calculated as

$$RMSE(\theta, \hat{\theta}) = \sqrt{\sum_{i=1}^{n} \frac{(\theta_i - \hat{\theta}_i)^2}{n}},$$

where $\theta_i$ and $\hat{\theta}_i$ are respectively the actual and the estimated values of the $i$th parameter and $n$ is the total number of parameters. Because the ranges of the scale parameters, shape parameters, and entries of the observation probability matrix are different with each other, we separately cal-

*Box 1.*

$$\Gamma : \begin{cases} \alpha(1,2) = 24, \ \alpha(1,4) = 29, \alpha(2,3) = 20, \alpha(2,4) = 30, \alpha(3,4) = 8 \\ \beta(1,2) = 6, \quad \beta(1,4) = 9, \quad \beta(2,3) = 8, \quad \beta(2,4) = 5, \quad \beta(3,4) = 4 \end{cases}$$

$$B = [b_i(j)] = \begin{bmatrix} 0.45 & 0.35 & 0.15 & 0.05 & 0 & 0 & 0 & 0 \\ 0.05 & 0.1 & 0.3 & 0.35 & 0.1 & 0.05 & 0.05 & 0 \\ 0 & 0 & 0.05 & 0.05 & 0.2 & 0.25 & 0.45 & 0 \\ 0 & 0 & 0 & 0 & 0 & 0 & 0 & 1 \end{bmatrix}$$

*Figure 2. Transition rate functions of Type 3*

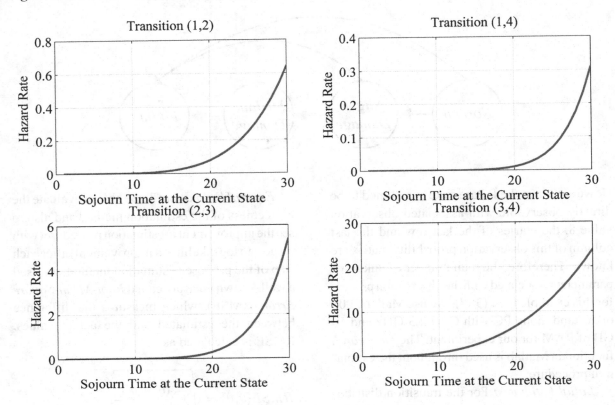

culate RMSE for each of these parameters and finally calculate the average RMSE considering all these three types of parameters.

*Random Sequence Generation:* The multi-state equipment with degradation states under the continuous-time hidden semi-Markov assumptions is simulated by the *Monte Carlo simulation* method. The state of the equipment, the sojourn time at each state, and the condition monitoring indicator values are extracted from simulation. The output of the simulation consists in multiple sequences of observations with unknown states which are generated according to the successive visited states. Since throughout the chapter, transition rate functions were considered as the fundamental describer of the degradation process for single equipment, here we introduce a technique to generate random sequences based on a

NHCTHSMP in terms of transition rate functions. A random sequence involves both the degradation process (successive sequence of visited states and the time of each transition) and the observation process (successive observed condition monitoring indicator). In order to do this, the triplet $(X_n, T_n, U_n)$ is generated for $1 \leq n \leq n_f$, where $n_f = \{n \mid X_n = N\}$. Our method is based on the *inverse transform technique* [Yura & Hanson, 2011] which is defined as follows:

**Definition:** Let $a$ be a uniform random variable in the range [0, 1]. If $b = F^{-1}(a)$, then $b$ is a random variable with CDF $F_B(b) = F$.

Now, it is possible to define the cumulative hazard function between state $i$ and $j$ based on the corresponding cumulative transition rate function as:

*Figure 3. Transition rate functions for transition (2,4)*

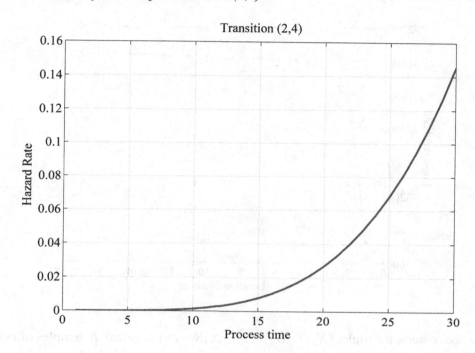

Now, based on the inverse transform technique, we can generate random number $t_{i,j}^s$ for the time to transition from state $i$ to state $j$, given that state $i$ is reached at time $s$. Let $a$ be a uniform random variable in the range $[0, 1]$, then a random number can be generated as:

$$1 - F_{i,j}(s, t) = \exp\left(-\Lambda_{i,j}(s, t)\right) \to \Lambda_{i,j}(s, t)$$
$$= -\log(1 - a) \to t_{i,j}^s = \Lambda_{i,j}^{-1}(s, -\log(1 - a)). \tag{47}$$

The following steps summarized all steps to generate a random sequence $(X_n, T_n)$:

**Step 1:** Let $X_0 = 1$, $T_0 = 0$, and $c = 0$ and move on to Step 2.

**Step 2:** Let $i = X_c$. For all $j \in FS_i$, generate random number from Equation (47) given that $s = T_c$.

$$\Lambda_{i,j}(s, t) = \int_o^t \lambda_{i,j}(s, u)\, du. \tag{46}$$

**Step 3:** Let $c = c + 1$ and find the state and the time of the next transition as $X_c = \arg\min_{j \in FS_i} t_{i,j}^s$ and $T_c = T_{c-1} + \min_j \left(t_{i,j}^s\right)$. This step is based on the fact that the next transition is realized according the event that occurs first in a competition among all possible transitions out of state $i$.

**Step 4:** If $X_n = N$, terminate the algorithm and output $(X_n, T_n)$, otherwise move back to Step 2.

The following steps shows how to generate $U_n$, which is the corresponding value of the observation process associated with $(X_n, T_n)$.

Let $a$ be a random variable in the range $[0,1]$. The process $U_n$ is equal to $v_m$, if

$$\sum_{z=1}^{m-1} b_{X_n}(z) < a \le \sum_{z=1}^{m} b_{X_n}(z), \qquad 1 \le m \le M.$$

*Figure 4. Log-likelihood improvement*

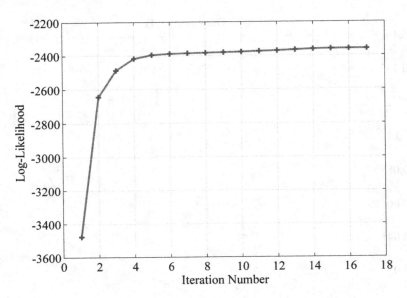

Using the above steps, the triplet $(X_n, T_n, U_n)$ can be generated as a random sequence. Now, based on the predetermined observation interval $\Delta$, we can generate $K$ samples of observations as the input to evaluate the parameter estimation method.

*Figure 5. RMSE of the estimated parameters*

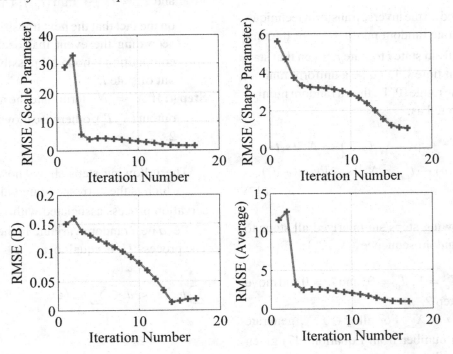

*Discussions of the Results*: The result of the estimation procedure with respect to the likelihood improvement and the error of estimation is presented in Figures 4-5. As shown in Figure 4, the value of the log-likelihood function is increased with the iteration of the algorithm. However, the rate of increase is lower in higher iterations. Also, as can be seen in Figure 5, the estimated values of each set of parameters get closer to their actual values after each iteration. The final RMSE value for scale parameters, shape parameters, and elements of the observation probability matrix are respectively 1.62, 1.06, and 0.02, whereas the final average RMSE for all three sets of parameters is 0.90. From computational point of view, the estimation procedure took 4.23 hours which is reasonable considering the fact that in general, the unsupervised estimation problem is an offline procedure.

## FUTURE RESEARCH DIRECTIONS

The ultimate objective of condition monitoring for multi-state equipment is to develop appropriate diagnostic and prognostic methods and tools which can efficiently facilitate decision makers to make appropriate and timely maintenance decisions. In our future research work, we aim to develop diagnostic and prognostic methods for multi-state equipment under the NHCTHSMP structure. Also, developing the diagnostic and prognostic methods when multiple failure modes exist is to be investigated in our future research work.

## CONCLUSION

In this book chapter, we present a general stochastic model using nonhomogeneous continuous-time hidden semi-Markov process (NHCTHSMP) to model the degradation process and the observation process of a piece of multi-state equipment with unobservable states. The detailed mathematical structure for the NHCTHSMP associated with the multi-state equipment was described. Important measures of a NHCTHSMP based on the associated kernel function and the transition rate function were illustrated. Finally an estimation method was presented which can be used to estimate the unknown parameters of a NHCTHSMP using condition monitoring information. A simple numerical example was provided to describe the application of NHCTHSMP in modeling the degradation process and the observation process of multi-state equipment with unobservable states.

## ACKNOWLEDGMENT

This work is financially supported by the Natural Science and Engineering Research Council of Canada (NSERC).

## REFERENCES

Becker, G., Camarinopoulos, L., & Zioutas, G. (2000). A semi-Markovian model allowing for inhomogenities with respect to process time. *Reliability Engineering & System Safety*, *70*(1), 41–48. doi:10.1016/S0951-8320(00)00044-2

Blasi, A., Janssen, J., & Manca, R. (2004). Numerical treatment of homogeneous and non-homogeneous semi-Markov reliability models. *Communications in Statistics Theory and Methods*, *33*(3), 697–714. doi:10.1081/STA-120028692

Boutros, T., & Liang, M. (2011). Detection and diagnosis of bearing and cutting tool faults using hidden Markov models. *Mechanical Systems and Signal Processing*, *25*(6), 2102–2124. doi:10.1016/j.ymssp.2011.01.013

Chen, A., & Wu, G. S. (2007). Real-time health prognosis and dynamic preventive maintenance policy for equipment under aging Markovian deterioration. *International Journal of Production Research*, *45*(15), 3351–3379. doi:10.1080/00207540600677617

D'Amico, G., Janssen, J., & Manca, R. (2011). Duration dependent semi-Markov models. *Applied Mathematical Sciences, 5*(42), 2097–2108.

De Rango, F., Veltri, F., & Marano, S. (2011). Channel modeling approach based on the concept of degradation level Discrete-Time Markov chain: UWB system case study. *IEEE Transactions on Wireless Communications, 10*(4), 1098–1107. doi:10.1109/TWC.2011.012411.091590

Dempster, A. P., Laird, N. M., & Rubin, D. B. (1977). Maximum likelihood from incomplete data via the EM algorithm. *Journal of the Royal Statistical Society. Series B. Methodological, 39*(1), 1–38.

Dong, M., & He, D. (2007). Hidden semi-Markov model-based methodology for multi-sensor equipment health diagnosis and prognosis. *European Journal of Operational Research, 178*(3), 858–878. doi:10.1016/j.ejor.2006.01.041

Dong, M., He, D., Banerjee, P., & Keller, J. (2006). Equipment health diagnosis and prognosis using hidden semi-Markov models. *International Journal of Advanced Manufacturing Technology, 30*(7-8), 738–749. doi:10.1007/s00170-005-0111-0

Dong, M., & Peng, Y. (2011). Equipment PHM using non-stationary segmental hidden semi-Markov model. *Robotics and Computer-integrated Manufacturing, 27*(3), 581–590. doi:10.1016/j.rcim.2010.10.005

Ghasemi, A., Yacout, S., & Ouali, M.-S. (2010). Parameter estimation methods for condition-based maintenance with indirect observations. *IEEE Transactions on Reliability, 59*(2), 426–439. doi:10.1109/TR.2010.2048736

Hongzhi, T., Jianmin, Zh., Xisheng, J., Yunxian, J., Xinghui, Zh., & Liying, C. (2011). Experimental study on gearbox prognosis using total life vibration analysis. *Prognostics and System Health Management Conference* (PHM-Shenzhen), (pp. 1-6).

Janssen, J., Blasi, A., & Manca, R. (2002). Discrete time homogeneous and non-homogeneous semi-Markov reliability models. *Proceedings of the Third International Conference on Mathematical Methods in Reliability,* Trondheim, Norway.

Janssen, J., & Manca, R. (2001). Numerical solution of non homogeneous semi Markov processes in transient case. *Methodology and Computing in Applied Probability, 3*(3), 271–293. doi:10.1023/A:1013719007075

Kim, M. J., Makis, V., & Jiang, R. (2010). Parameter estimation in a condition-based maintenance model. *Statistics & Probability Letters, 80*(21-22), 1633–1639. doi:10.1016/j.spl.2010.07.002

Lin, D., & Makis, V. (2004). On-line parameter estimation for a failure-prone system subject to condition monitoring. *Journal of Applied Probability, 41*(1), 211–220. doi:10.1239/jap/1077134679

Lisnianski, A., & Levitin, G. (2003). *Multi-state system reliability: Assessment, optimization and applications*. Singapore: World Scientific.

López Droguett, E., das Chagas Moura, M., Magno Jacinto, C., & Feliciano Silva Jr., M. (2008). A semi-Markov model with Bayesian belief network based human error probability for availability assessment of down hole optical monitoring systems. *Simulation Modelling Practice and Theory, 16*(10), 1713–1727. doi:10.1016/j.simpat.2008.08.011

Moghaddass, R., & Zuo, M. J. (2011a). A parameter for a multi-state deteriorating system with incomplete information. *Proceedings of the 7th International Conference on Mathematical Methods in Reliability: Theory, Methods, and Applications (MMR 2011)* (pp. 34-32), Beijing, China.

Moghaddass, R., & Zuo, M. J. (2011b). A parameter estimation method for condition monitored equipment under multi-state deterioration. Reliability Engineering and System Safety. 106, 94-103. doi:10.1016/j.ress.2012.05.004.

Morcous, G., Lounis, Z., & Mirza, M. S. (2003). Identification of environmental categories for Markovian deterioration models of bridge decks. *Journal of Bridge Engineering, 8*(6), 353–361. doi:10.1061/(ASCE)1084-0702(2003)8:6(353)

Moura, M. C., & Droguett, E. L. (2009). Mathematical formulation and numerical treatment based on transition frequency densities and quadrature methods for non-homogeneous semi-Markov processes. *Reliability Engineering & System Safety, 94*(2), 342–349. doi:10.1016/j.ress.2008.03.032

Peng, Y., & Dong, M. (2010). A prognosis method using age-dependent hidden semi-Markov model for equipment health prediction. *Mechanical Systems and Signal Processing, 25*(1), 237–252. doi:10.1016/j.ymssp.2010.04.002

Peng, Y., & Dong, M. (2011). A hybrid approach of HMM and grey model for age-dependent health prediction of engineering assets. *Expert Systems with Applications, 38*(10), 12946–12953. doi:10.1016/j.eswa.2011.04.091

Petrovic, G., Marinkovic, Z., & Marinkovic, D. (2011). Optimal preventive maintenance model of complex degraded systems: A real life case study. *Journal of Scientific and Industrial Research, 70*(6), 412–420.

Rabiner, L. R. (1989). A tutorial on hidden Markov models and selected applications in speech recognition. *Proceedings of the IEEE, 77*(2), 257–286. doi:10.1109/5.18626

Vaurio, J. K. (1997). Reliability characteristics of components and systems with tolerable repair times. *Reliability Engineering & System Safety, 56*(1), 43–52. doi:10.1016/S0951-8320(96)00133-0

Worden, K., & Manson, G. (2007). Damage identification using multivariate statistics: Kernel discriminant analysis. *Inverse Problems in Engineering, 8*(1), 25–46. doi:10.1080/174159700088027717

Wu, J., Ng, T., Xie, M., & Huang, H. Z. (2010). Analysis of maintenance policies for finite life-cycle multi-state systems . *Computers & Industrial Engineering, 59*(4), 638–646. doi:10.1016/j.cie.2010.07.013

Yura, H. T., & Hanson, S. G. (2011). Digital simulation of an arbitrary stationary stochastic process by spectral representation. *Journal of the Optical Society of America. A, Optics, Image Science, and Vision, 28*(4), 675–685. doi:10.1364/JOSAA.28.000675

Zhang, D., Li, W., Xiong, X., & Liao, R. (2011). Evaluating condition index and its probability distribution using monitored data of circuit breaker. *Electric Power Components and Systems, 39*(10), 965–978. doi:10.1080/15325008.2011.552091

## ADDITIONAL READING

Archer, G. E. B., & Titterington, D. M. (2002). Parameter estimation for hidden Markov chains. *Journal of Statistical Planning and Inference, 108*(1-2), 365–390. doi:10.1016/S0378-3758(02)00318-X

Barbu, V. S., & Limnios, N. (2008). *Semi-Markov chains and hidden semi-Markov models toward applications: Their use in reliability and DNA analysis*. New York, NY: Birkhauser.

Bijak, K. (2008). Genetic algorithms as an alternative method of parameter estimation and finding most likely sequences of states of Hidden Markov chains for HMMs and hybrid HMM/ANN models. *Fundamental Informaticae, 86*(1-2), 1–17.

Chauhan, S., Wang, P., Lim, C., & Anantharaman, V. (2008). A computer-aided MFCC-based HMM system for automatic auscultation. *Computers in Biology and Medicine, 38*(2), 221–233. doi:10.1016/j.compbiomed.2007.10.006

Chen, J., & Jiang, Y. C. (2011). Development of hidden semi-Markov models for diagnosis of multiphase batch operation. *Chemical Engineering Science, 66*(6), 1087–1099. doi:10.1016/j.ces.2010.12.009

Daming, L., & Makis, V. (2004). Filters and parameter estimation for a partially observable system subject to random failure with continuous-range observations. *Advances in Applied Probability, 36*(4), 1212–1230. doi:10.1239/aap/1103662964

Dong, M., & He, D. (2007). A segmental hidden semi-Markov model (HSMM)-based diagnostics and prognostics framework and methodology. *Mechanical Systems and Signal Processing, 21*(5), 2248–2266. doi:10.1016/j.ymssp.2006.10.001

Giordana, N., & Pieczynski, W. (1997). Estimation of generalized multi-sensor hidden Markov chains and unsupervised image segmentation. *IEEE Transactions on Pattern Analysis and Machine Intelligence, 19*(5), 465–475. doi:10.1109/34.589206

Guédon, Y. (1999). Computational methods for discrete hidden semi-Markov chains. *Applied Stochastic Models in Business and Industry, 15*(3), 195–224. doi:10.1002/(SICI)1526-4025(199907/09)15:3<195::AID-ASMB376>3.0.CO;2-F

Guédon, Y. (2003). Estimating hidden semi-Markov chains from discrete sequences. *Journal of Computational and Graphical Statistics, 12*(3), 604–639. doi:10.1198/1061860032030

Guédon, Y. (2005). Hidden hybrid Markov/semi-Markov chains. *Computational Statistics & Data Analysis, 49*(3), 663–688. doi:10.1016/j.csda.2004.05.033

Huang, X. D., & Jack, M. A. (1989). Semi-continuous hidden Markov models for speech signals. *Computer Speech & Language, 3*(3), 239–251. doi:10.1016/0885-2308(89)90020-X

Ibe, O. C., & Wein, A. S. (1992). Availability of systems with partially observable failures. *IEEE Transactions on Reliability, 41*(1), 92–96. doi:10.1109/24.126678

Lam, C. T., & Yeh, R. H. (1994). Optimal replacement policies for multistate deteriorating systems. [NRL]. *Naval Research Logistics, 41*(3), 303–315. doi:10.1002/1520-6750(199404)41:3<303::AID-NAV3220410302>3.0.CO;2-2

Levinson, S. E. (1986). Continuously variable duration hidden Markov models for automatic speech recognition. *Computer Speech & Language, 1*(1), 29–45. doi:10.1016/S0885-2308(86)80009-2

Liang, Y., Liu, X., Lou, Y., & Shan, B. (2011). An improved noise-robust voice activity detector based on hidden semi-Markov models. *Pattern Recognition Letters, 32*(7), 1044–1053. doi:10.1016/j.patrec.2011.02.015

Lin, D., & Makis, V. (2003). Recursive filters for a partially observable system subject to random failure. *Advances in Applied Probability, 35*(1), 207–227. doi:10.1239/aap/1046366106

Liporace, A. (1982). Maximum likelihood estimation for multivariate observations of Markov sources. *IEEE Transactions on Information Theory, 28*(5), 729–734. doi:10.1109/TIT.1982.1056544

Yu, S., & Kobayashi, H. (2003). A hidden semi Markov model with missing data ad multiple observation sequences for mobility tracking. *Signal Processing, 83*(2), 235–250. doi:10.1016/S0165-1684(02)00378-X

Yu, S. Z. (2010). Hidden semi-Markov models. *Artificial Intelligence, 174*(2), 215–243. doi:10.1016/j.artint.2009.11.011

## KEY TERMS AND DEFINITIONS

**Condition Monitoring:** A process of monitoring condition parameters of a device (machinery) for the purpose of maintenance.

**Multi-State Degradation Process:** A degradation process with multiple discrete health states (conditions).

**Observation Process:** A process that deals with the stochastic relationship between condition monitoring data and health status of a device (degradation process).

**Partially Observable Health States:** Health states of a device that are not directly observable (identifiable).

**Semi-Markov Kernel:** A fundamental describer of a semi-Markov process representing the probability of transitions between states within a certain interval.

**Transition Rate:** The instantaneous rate of state transition between two states.

**Unsupervised Parameter Estimation:** A type of parameter estimation method from unlabeled training data.

# Chapter 9
# Stochastic Fatigue of a Mechanical System Using Random Transformation Technique

**Seifedine Kadry**
*American University of the Middle East, Kuwait*

## ABSTRACT

*In this chapter, a new technique is proposed to find the probability density function (pdf) of a stress for a stochastic mechanical system. This technique is based on the combination of the Probabilistic Transformation Method (PTM) and the Finite Element Method (FEM) to obtain the pdf of the response. The PTM has the advantage of evaluating the probability density function pdf of a function with random variable, by multiplying the joint density of the arguments by the Jacobien of the opposite function. Thus, the "exact" pdf can be obtained by using the probabilistic transformation method (PTM) coupled with the deterministic finite elements method (FEM). In the method of the probabilistic transformation, the pdf of the response can be obtained analytically when the pdf of the input random variables is known. An industrial application on a plate perforated with random entries was analyzed followed by a validation of the technique using the simulation of Monte Carlo.*

## INTRODUCTION

Fatigue crack growth is one of the most important factors in the design of the steel structures. Numerous experiment and researches have been performed for the prediction of fatigue crack growth (Kanninen et al. 1985). In the past, fatigue analysis was largely the domain of the development engineer, who used measurements taken from prototype components to predict the fatigue behavior. This gave rise to the traditional "Build it, Test It, Fix It" approach to fatigue design illus-

DOI: 10.4018/978-1-4666-2095-7.ch009

trated in Figure 1. This approach is known to be very costly as an iterative design cycle is centered on the construction of real prototype components. This inhibits the ability to develop new concepts and reduces confidence in the final product due to a low statistical sample of tests. It is also common to find early products released with 'known' defects or product release dates being delayed whilst durability issues were addressed. A more desirable approach is to conduct more testing based on computer simulations. Computational analysis can be performed relatively quickly and much earlier in the design cycle.

Confidence in the product is therefore improved because more usage scenarios can be simulated. It is not recommended, however, that these simulations completely replace prototype testing. It will always remain desirable to have prototype signoff tests to validate the analysis performed and improve our future modeling techniques. However, the number of prototype stages, and hence the total development time, can be reduced.

*Figure 1. The build it, test it, fix it method of design*

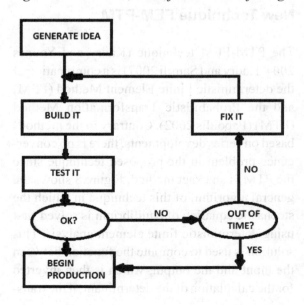

## Stochastic Fatigue

The fatigue process of mechanical components under service loading is stochastic in nature. Life prediction and reliability evaluation is still a challenging problem despite extensive progress made in the past decades. A comprehensive review of early developments can be found in (Yao et al. 1986). Compared to fatigue under constant amplitude loading, the fatigue modeling under variable amplitude loading becomes more complex both from deterministic and probabilistic points of view. An accurate deterministic damage accumulation rule is required first, since the frequently used linear Palmgren-Miner's rule may not be sufficient to describe the physics (Fatemi and Yang 1998). Second, an appropriate uncertainty modeling technique is required to include the stochasticity in both material properties and external loadings, which should accurately represent the randomness of the input variables and their covariance structures. In addition to the above difficulties, such a model should also be computationally and experimentally inexpensive. The last characteristic is the main reason for the popularity of simpler models despite their inadequacies.

Fatigue crack growth under random and variable amplitude loading has been analyzed by many different authors (Sunder and Prakash 1996, Schijve et al. 2004). Some of them pay special attention to the simulation of representative load histories of defined random loading processes (Van Dijk 1975, Chang 1981, and Schütz, 1989). Others analyze the effect of overloads and their distribution throughout the load history of the crack growth life (Wheatley et al. 1999, Lang and Marci 1999, Pommier and Freitas 2002).

## Probabilistic Transformation Method (PTM)

The Probabilistic transformation Method is based on the following theorem (Papoulis 2002):

Suppose that $X$ is a continuous random variable with *pdf* $f_X(x)$ and $A \subset \Re$ is the one–dimensional space where $f_X(x) > 0$, is differentiable and monotonic. Consider the random variable $Y = u(X)$, where $y = u(x)$ defines a one-to-one transformation that maps the set $A$ onto a set $B \subset \Re$ so that the equation $y = u(x)$ can be uniquely solved for $x$ in terms of $y$, say $x = u^{-1}(y)$. Then, the *pdf* of $Y$ is (Figure 2):

$$f_Y(y) = f_X\left[u^{-1}(y)\right]|J|, \qquad y \in B \qquad (1)$$

where, $J = \dfrac{dx}{dy} = \dfrac{du^{-1}(y)}{dy}$ is the transformation Jacobean, which must be continuous for all points $y \in B$.

We notice in the previous theorem that the function *u(.)* should be monotonic. For the general case of non-monotonic functions, the following theorem allows us to consider a piecewise monotonic transformation.

*Figure 2. Transformation method*

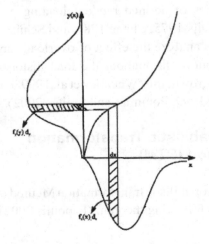

## Finite Element Method (FEM)

The finite element method (FEM) is used for finding approximate solutions of partial differential equations (PDE) as well as of integral equations such as the heat transport equation. The solution approach is based either on eliminating the differential equation completely (steady state problems), or rendering the PDE into an approximating system of ordinary differential equations, which are then solved using standard techniques such as finite differences, Runge-Kutta, etc.

In solving partial differential equations, the primary challenge is to create an equation that approximates the equation to be studied, but is numerically stable, meaning that errors in the input data and intermediate calculations do not accumulate and cause the resulting output to be meaningless. There are many ways of doing this, all with advantages and disadvantages. The Finite Element Method is a good choice for solving partial differential equations over complex domains (like mechanical system), when the domain changes (as during a solid state reaction with a moving boundary), when the desired precision varies over the entire domain, or when the solution lacks smoothness.

## New Technique FEM-PTM

The PTM-FEM technique (Kadry and Younes 2004, Kadry and Samili 2007) is a combination of the deterministic Finite Element Method (FEM) and the Probabilistic Transformation Method (PTM) (Papoulis 2002). Contrary to the methods based on series developments, there is no convergence problem in the proposed technique since the PTM is an exact method. Figure 3 shows the general algorithm of this technique in which the stochastic equation of equilibrium is solved first using deterministic finite element analysis. This solution is used to compute the function between the input and the output, which is then inverted for the calculation of the determinant of the trans-

*Figure 3. General algorithm of PTM-FEM*

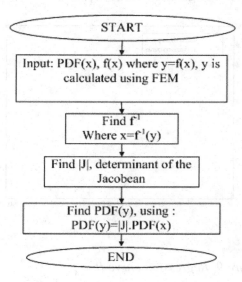

formation Jacobean. Finally, the response *pdf* at any point in the domain can be deduced by using the probabilistic transformation method. This is simply defined by multiplying the input *pdf* by the Jacobean of the inverse mechanical function. For small number of random variables, this approach has the advantage of giving a closed-form of the density function of the response, which is very helpful for reliability analysis of mechanical systems.

The advantage of the PTM-FEM technique in the context of static system is clear: it gives the *pdf*, which is the most complete characteristic in probabilistic analysis, of the response in a closed-form expression, contrary to other numerical methods like perturbation and spectral methods which give only first and second moments of the response under some conditions.

## Application: Perforated Plate under Tension

The PTM-FEM approach is applied to a thin perforated plate (Figure 4) fixed at one end and under tension on the other end (Figure 5). The plate, of thickness $t = 1$ mm, has for half-width $B = 100$

mm and for hole radius $R = 30$ mm. The applied tensile force $T = 10$ kN is uniformly distributed over the plate edge. The material is isotropic with Young's modulus $E = 210$ GPa, Poisson's ratio $v = 0.3$ and yield stress $f_Y = 200$ MPa. In this application, the hole radius $R$ is considered as uncertain variable.

The software Matlab has been used to the discretization of this structure with the finite element mesh shown in the Figure 6.

Figure 7 gives a 3D representation of the plate displacement. It is to be noted that the lower edge

*Figure 4. Perforated plate*

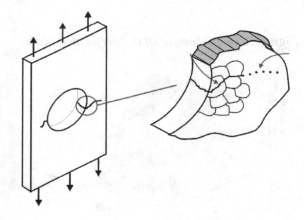

*Figure 5. Perforated plate*

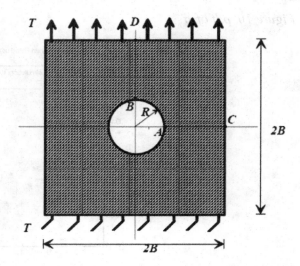

*Figure 6. Discretization of the perforated plate*

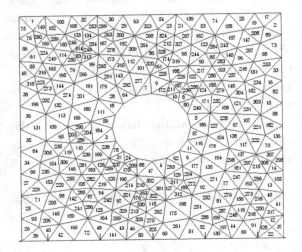

*Figure 8. Regression curve $\sigma_A$-R*

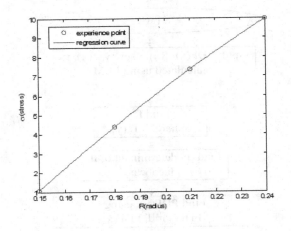

*Figure 9. Regression curve R-$\sigma_A$*

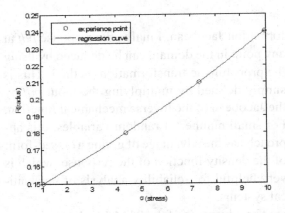

*Figure 7. Displacement of the perforated plate*

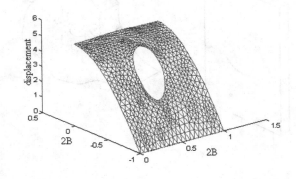

*Figure 10. pdf of $\sigma_A$*

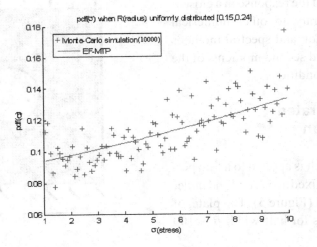

has been fixed while the tensile force is applied on the upper edge.

For this plate, the maximum stress $(\sigma_A)$ is localized at the hole perimeter on the axis of symmetry: i.e. point $A$ in Figure 5. In order to apply the PTM method, we have built a regression curve (Figure 8) relating the tensile stress to the radius of the hole (random variable of the system). This curve leads to the following relationship:

$$\sigma_A = 81.36 - 1839.70\,R$$
$$+12335.05\,R^2 - 24259.87\,R^3$$

For the computation of the transformation Jacobean, the inverse function is required. The third degree polynomial is also applied for regression (Figure 9), leading to:

$$R = 0.141878 + 8.094 \times 10^{-3}\,\sigma_A$$
$$-1.1988 \times 10^{-5}\,\sigma_A^2 + 1.5984 \times 10^{-5}\,\sigma_A^3$$

Using the PTM-FEM, the *pdf* of the maximum tensile stress when the radius uniformly distributed in the range [0.15, 0.24] is (Figure 10):

$$f_{\sigma_A}(\sigma_A) = |J|\,f_R(R)$$
$$= \left(0.008094 - 0.000024\,\sigma_A + 0.000048\,\sigma_A^2\right)\frac{100}{9}$$

## CONCLUSION

In this chapter a new technique is developed to find the "exact" probability density function of the stress for a perforated plate with random parameter. Our technique is verified with 10000 Monte-Carlo simulations. In future work, the extension of our technique for multi-dimension variables should be analyzed and applied to a more complex structure.

## REFERENCES

Chang, J. B. (1981). Round-robin crack growth prediction on center cracked tension specimen under random spectrum loading. In Chang, J. B., & Hudson, C. M. (Eds.), *Methods and models for predicting fatigue crack growth under random loading, ASTM STP 748* (pp. 4–40). doi:10.1520/STP28332S

Fatemi, A., & Yang, L. (1998). Cumulative fatigue damage and life prediction theories: A survey of the state of the art for homogeneous materials. *International Journal of Fatigue, 20*, 9–34. doi:10.1016/S0142-1123(97)00081-9

Kadry, S., & Samili, K. (2007). One dimensional transformation method in reliability analysis. *Journal of Basic and Applied Sciences, 3*(2).

Kadry, S., & Younes, R. (2004). Etude probabiliste d'un systeme mecanique a parametres incertains par une technique basee sur la methode de transformation. *Proceeding of CanCam*, Canada.

Kanninen, M. F., & Popelar, C. H. (1985). *Advanced fracture mechanics*. New York, NY: Oxford University Press.

Lang, M., & Marci, G. (1999). Influence of single and multiple overloads on fatigue crack propagation. *Fatigue of Engineering Materials and Structures, 22*, 257–271. doi:10.1046/j.1460-2695.1999.00165.x

Papoulis, A. (2002). *Probability, random variables and stochastic processes* (4th ed.). Boston, MA: McGraw-Hill.

Pommier, S., & Freitas, M. D. (2002). Effect on fatigue crack growth of interactions between overloads. *Fatigue of Engineering Materials and Structures, 25*, 709–722. doi:10.1046/j.1460-2695.2002.00531.x

Schijve, J., Skorupa, M., Skorupa, A., Machniewicz, T., & Gruszczynsky, P. (2004). Fatigue crack growth in aluminium alloy D16 under constant and variable amplitude loading. *International Journal of Fatigue, 26*(1), 1–15. doi:10.1016/S0142-1123(03)00067-7

Schütz, W. (1989). Standardized stress-time histories—An overview. In Potter, J. M., & Watanabe, R. T. (Eds.), *Development of fatigue loading spectra, ASTM STP 1006* (pp. 3–16). doi:10.1520/STP10346S

Sunder, R., & Prakash, R. V. (1996). A study of fatigue crack growth in lugs under spectrum loading. In Mitchell, M. R., & Landgraf, R. W. (Eds.), *Advances in fatigue lifetime predictive techniques* (*Vol. 3*, p. 1292). ASTM STP. doi:10.1520/STP16141S

Van Dijk, G. M. (1975). *Introduction to a fighter aircraft loading standard for fatigue evaluation.* Amsterdam, The Netherlands: National Aerospace Lab.

Wheatley, G., Wu, X. Z., & Estrin, Y. (1999). Effects of a single tensile overload on fatigue crack growth in a 316L steel. *Fatigue of Engineering Materials and Structures, 22*, 1041–1051. doi:10.1046/j.1460-2695.1999.00225.x

Yao, J. T. P., Kozin, F., Wen, Y. K., Yang, J. N., Schueller, G. I., & Ditlevsen, O. (1986). Stochastic fatigue, fracture and damage analysis. *Structural Safety, 3*, 231–267. doi:10.1016/0167-4730(86)90005-6

# Chapter 10
# Degradation Based Condition Classification and Prediction in Rotating Machinery Prognostics

**Chao Liu**
*Tsinghua University, P. R. China*

**Dongxiang Jiang**
*Tsinghua University, P. R. China*

## ABSTRACT

*A transition stage exists during the equipment degradation, which is between the normal condition and the failure condition. The transition stage presents small changes and may not cause significant function loss. However, the transition stage contains the degradation information of the equipment, which is beneficial for the condition classification and prediction in prognostics. The degradation based condition classification and prediction of rotating machinery are studied in this chapter. The normal, abnormal, and failure conditions are defined through anomaly determination of the transition stage. The condition classification methods are analyzed with the degradation conditions. Then the probability of failure occurrence is discussed in the transition stage. Finally, considering the degradation processes in rotating machinery, the condition classification and prediction are carried out with the field data.*

## INTRODUCTION

Due to the increasing requirements of the equipment reliability in reality, prognostics has attracted more attention in recent researches. Different from diagnostics, prognostics concentrates more on the possibility of fault occurrence and the remaining

useful life of the equipment (Byington, Roemer, and Galie, 2002). The prognostic methods are usually based on the assumption that failures are caused by the component's aging and degradation. And the parameters indicating the degradation are available with the failure onset extraction of the continuously monitored data in mechanical system

DOI: 10.4018/978-1-4666-2095-7.ch010

applications (Spieler, Staudacher, and Fiola, 2008; Zaita, Buley, and Karlsons, 1998).

Condition classification and prediction are two important aspects in prognostics, where the equipment's conditions include normal, abnormal and faulty states (Jiang, and Liu, 2011). Usually, the normal state is defined with the condition when the equipment is firstly installed or after maintenance. The failure condition is determined when the equipment is greatly affected by the component failure such as efficiency reduction, vibration increment and so on. The abnormal state is defined when the equipment is operating in some kind of abnormality but the failure has not been observed. The abnormal state is a transition stage in life cycle of the equipment, and it's beneficial for the prognostics as the degradation information is included in the transition stage.

This chapter analyzes the degradation based condition classification and prediction approaches in prognostics. The remaining sections are organized as follows. The basic concepts of condition classification and prediction in prognostics are firstly discussed. Then the anomaly is defined to detect the degradation of the equipment. Finally, the condition classification and prediction are discussed based on the degradation processes.

## BACKGROUND

Model-driven and data-driven methods are two important approaches taken by prognostics (Heng, Zhang, Tan, and Mathew, 2009; Byington, and Stoelting, 2004; Goebe, Saha, and& Saxena, 2008). Model-driven prognostics is established by the mathematical model of the physical component or statistical model of the certain failure mode. Consequently, model-driven prognostics presents higher accuracy but with a specific application range. Data-driven prognostics is implemented by analyzing the monitoring data as well as the history data (Schwabacher, and Goebel, 2007). Extensive adaptability characteristic makes the

data-driven prognostics with lower accuracy as fault mechanism is not considered. Other prognostics methods are also studied in many cases where evolutionary prognostics is a promising approach to predict the equipment's fault onset and indicate the possibility of the failure (Roemer, Byington, Kacprzynski, and Vachtsevanos, 2006), and it is data-based.

The conditions in condition monitoring are usually classified into normal, and various faulty types. Yang (2005) used condition classification to study the healthy and faulty states in a small reciprocating compressor. One normal condition and four faulty conditions of the roller bearing were classified in (Jack, and Nandi, 2002). In fault diagnostics, 14 faulty types of the turbo pump were classified by Yuan, and Chu (2006), while normal states and abnormal states were not considered. Kinds of condition classification methods are studied including the linear and nonlinear classifiers. S. J. Sixon consider the five common classifiers: Euclidean distance to centroids, linear discriminant analysis, quadratic discriminant analysis, learning vector quantization and support vector machines (Dixon, and Brereton, 2009), with the results that the accuracy of the classifiers depends on the structure of the data set.

In machine operation processes, once the machine departs from the normal state, there exists a transition state between the normal state and the failure condition. Usually the intermittent state has little effect on the equipment, and may not cause significant functional loss. However, the intermittent states contain the information of equipment degradation. And estimating the abnormal state between the normal and failure conditions is essential in prognostics.

The prediction of the failure time and the probability of the failure occurrence are usually considered in prognostics. Time series forecasting is widely applied in predicting the failure time. Based on the past observations of the same variable, the forecasting model is established to describe the underlying relationship of the data, and then the extrapolation is applied to predict the data in the following observations. Kinds of time series forecasting methods are proposed, and five categories are generalized in Gooijer, and Hyndman, 2006): exponential smoothing, state space and Kalman filter, nonlinear models, ARCH/GARCH models, and neural network. Confidence estimate is an important step in the time series forecasting.

Time series forecasting can acquire high accuracy in the short time forecasting, usually with one-step prediction. However, in equipment condition prediction, a longer time is necessary to predict the remaining life and the failure time. The time series forecasting has difficulty to predict the equipment condition in a longer period. The uncertainty grows quickly as the forecasting time increases, which limits the application of time series forecasting in equipment condition prediction.

Another aspect in estimating the equipment condition in prognostics is the probability of the failure occurrence. It is different from the confidence estimate in the time series forecasting. The probability of the failure occurrence is the estimate of the current state. The probability estimate of the failure occurrence can illustrate the healthy degree or abnormal degree of the equipment. Then the prediction can be acquired with the abnormal degree to a certain failure. This chapter tries to establish the anomaly variable in the feature space and propose the probability estimate approaches of the failure occurrence.

## ANOMALY DEFINITION AND DEGRADATION DETECTION

In equipment degradation process, the component will undergo a transition stage before the failure occurs. The transition stage between the normal condition and failure condition is the basis of evolutionary prognosis. Detecting the transition stage can understand the status of the equipment

and may predict the equipment condition in the future. Anomaly is applied to distinguish the transition stages.

To define the anomaly, the following attributes are necessary:

- It can reflect the transition stage in the degradation process;
- It can distinguish the abnormal conditions from the normal and failure conditions;
- It can demonstrate different levels in equipment degradation process.

If the anomaly is defined in one dimensional space, the situation is simple and shown in Figure 1. The parameter of the equipment signifies the machine condition. At first, the value of the parameter is in normal condition where the equipment is running well. As the time goes on, the parameter goes into the abnormal condition at $t1$, and then the parameter goes into the faulty condition at $t2$.

Usually, the fault onset cannot be signified by a single parameter. And some faults which have similar degradation mechanisms are difficult to be distinguished by one parameter. Therefore, the feature space is applied with the extracted parameters. The parameters can be the measured variables, the processed measured variables and so on. To determine the parameters used in feature space, the feature extraction approaches are applied.

To establish the standard of the equipment's condition, the anomaly is defined with the combination of multiple feature parameters:

$$D_{i,j} = \sum_{j=1}^{n} \lambda_j (P_{i,j} - \bar{P}_j)^2 , \qquad (1)$$

where, $\lambda_j$ is the coefficient according to the impact on fault occurrences of the $jth$ parameter, $P_{i,j}$ is the measured value at count $i$ of the $jth$ parameter, $\bar{P}_j$ is the typical value of the $jth$ normal state.

*Figure 1. Typical degradation process*

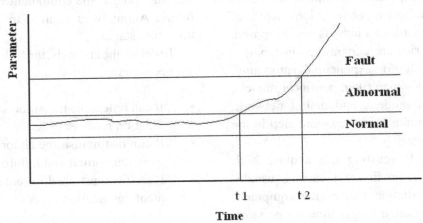

Here, the normal state $\bar{P}_j$ is defined as the initially stable state of the equipment when it is firstly installed or after maintenance.

For the faulty state of each degradation processes, $F_j$ is acquired to express the fault level. Then, the rule of abnormal state is defined:

$$\alpha_1 \left| F_j - \bar{P}_j \right| \leq \left| D_{i,j} - \bar{P}_j \right| \leq \alpha_2 \left| F_j - \bar{P}_j \right|,$$
(2)

where, $\alpha_1$ is the lower limit of the abnormal state, $\alpha_2$ is the upper limit of the abnormal state.

## DEGRADATION BASED CONDITION CLASSIFICATION

Considering the situation in degradation, the abnormal condition is adjacent to the normal condition and the failure condition. Therefore, the classifiers with nonlinear boundaries are preferred. Learning Vector Quantization and Support Vector Machine are discussed in this section.

## Learning Vector Quantization (LVQ)

The LVQ is widely used for pattern recognition in many areas which has great adaptability for complex boundaries. The LVQ is a class of learning algorithms for nearest prototype classification (Kohonen, Hynninen, Kangas, Laaksonen, and Torkkola, 1995; Kohonen, 2001) introduced by Kohonen (1986), and it has a good adaptability for complex bounds. An LVQ neural network contains an input layer, a competitive layer, and an output layer. The input layer contains the same number of nodes as the dimension of the input feature vector, the competitive layer contains the Kohonen neurons, and the output layer is linearly related to the competitive layer. In the training process of the LVQ neural network, the Euclidean distance from the input layer to the Kohonen layer is calculated as (Liu, Zuo, Zeng, Vroman, and Rabensolo, 2010):

$$d_i = \parallel w_i - x \parallel = \left( \sum_{j=1}^{N} \left( w_{i,j} - x_j \right)^2 \right)^{\frac{1}{2}},$$
(3)

where $x$ is the input vector, and $w$ is the weight vector of the competitive layer. The nearest node is determined to be the winner, and its weight vector is adjusted according to whether the winning node is in the class of the target output vector.

If the winner is the correct class, then

$$w_{i+1} = w_i + \alpha(x - w_i),  \tag{4}$$

where $\alpha$ is the learning parameter.

If the winner is not the correct class, then

$$w_{i+1} = w_i - \gamma(x - w_i),  \tag{5}$$

where is $\gamma$ the learning parameter.

## Support Vector Machine (SVM)

The Support Vector Machine was firstly proposed by Vapnik based on the idea that mapping the input vectors into a high dimensional feature space through nonlinear mapping. In the high-dimensional space, the classification with hyperplane can be found and the computational complexity will not increase significantly. Detailed description of SVM can be found in (Vapnik, 1999).

SVM can be used for two-class pattern recognition and multi-class pattern recognition. For multi-class pattern recognition, two kinds of approaches are proposed. One is multi-output SVM algorithm (Weston, and Watkins, 1999). The other is the combination of multiple two-class classifiers for multi-class pattern recognition. The multi-output SVM algorithm implements the classification using one SVM classifier. The classifier takes all classes into account and constructs the multi-class model. Only one classifier in used in multi-output SVM algorithm, the classification function is therefore very complex which may be difficult to solve with large computational complexity. Problems also exist in the training speed and the classification accuracy. The combination of multiple two-class classifiers is more and more applied in multi-class pattern recognition including one-against-others (Vapnik, 1998), one-against-one Kressel, 1999)

and decision directed acyclic graph (DDAG) (Platt, Cfisfianini, and Shawe, 2000). Compared with multi-output SVM algorithm, combination of multiple two-class classifiers can acquire high training speed and better classification accuracy.

As the shortcomings of the multi-output SVM algorithm, combination of multiple two-class classifiers is applied for multi-class classification in this work.

## DEGRADATION BASED CONDITION PREDICTION

Based on the anomaly definition and classification, the abnormal condition can be determined. Furthermore, evaluating the abnormal state can get more information about the equipment condition. One example is that, the abnormal state near the normal state is different from the abnormal state near the failure condition, and the two situations play different impacts on the maintenance strategy. The abnormal state near the normal state is close to the healthy state and can last a relatively long time before failure. While the abnormal state near the failure condition is critical to the equipment, and the maintenance needs to be carried out in a short time. Therefore, distinguishing different situations in the abnormal state is a promising method to realize the condition prediction in prognostics.

## 1. Principles of the Probability Estimate with Abnormal Condition Classification

Figure 2 shows three degradation conditions in two dimensional feature space. Three failure conditions exist in the equipment. The degradation condition 1 lies between the normal state and the failure condition II, which means that the equipment is deteriorating from the normal state to the failure condition II. And the probability of failure occurrence in condition 1 can be calculated using the Euclidean distance to the failure condition II.

*Figure 2. Probability estimate in degradation*

## 2. Probability Estimate of the Abnormal Condition Using SVM

The probability estimate is not outputted in standard SVM. Vapnik suggests a method for mapping the outputs of SVM to probabilities by decomposing the feature space into a direction orthogonal to the separating hyperplane, and all of the other dimensions of the feature space. The direction orthogonal to the separating hyperplane is parameterized by $t$, while all of the other directions are parameterized by a vector $\mathbf{u}$. In full generality, the posterior probability depends on both $t$ and $\mathbf{u}$: $P(y = 1 \mid t, \mathbf{u})$. The probability is fitted with a sum of cosine terms:

$$P(y = 1 \mid t, \mathbf{u}) = a_0(\mathbf{u}) + \sum_{n=1}^{N} a_n(\mathbf{u}) \cos(nt)$$

(6)

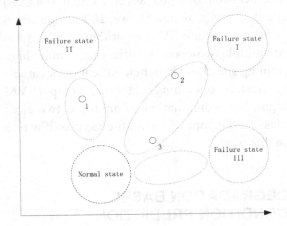

*Figure 2. Probability estimate in degradation*

The similar situation is performed in degradation condition 2. While in degradation condition 3, the distance to the failure condition III is closer than that to failure condition I. If the previous assumption is applied, it will be concluded that the probability of the failure condition III is higher. But the actual situation is that the degradation condition 3 is the abnormal condition between the normal condition to failure condition I. Therefore, the anomaly determination and the condition classification are necessary to get the correct degradation condition and the probability estimate.

Using the distance to the failure conditions can distinguish the degradation condition, and acquire the probability of the failure occurrence. But errors may exist in the situations like degradation condition 3 in Figure 2. If the anomaly is defined and the condition classifier is established previously, adding the condition classification results into the probability estimate can improve the prediction accuracy.

Another advantage using the condition classification results based on anomaly definition is that, the probability can be generalized and the probabilities among abnormal states can be generalized.

The coefficients of the cosine expansion will minimize a regularized function (Vapnik, 1998). It's a linear equation for the $a_n$ that depends on the value of $\mathbf{u}$ for the current input.

J. C. Platt discussed the above method, figured out the limitations and proposed another method to estimate the probability, which was firstly proposed by Hastie, Tibshiran, and Tibshirani (1990). And the probability estimate is improved in SVM using a more flexible version of the Gaussion fit and Bayes' rule (Wu, Lin, and Weng, 2004).

Given training samples,

$$x_i = R^n, i = 1, 2, \dots, m,$$

labeled by $y_i = \{+1, -1\}$, the class probability $\Pr(y = 1 \mid x)$ is approximated by a sigmoid function (Wu, Lin, and Weng, 2004):

$$\Pr(y = 1 \mid x) \approx P_{A,B}(f) \equiv \frac{1}{1 + e^{Af+B}},$$

(7)

where, $f = f(x)$ is the decision values, A and B are the parameters to be optimized.

The best parameter setting $Z^* = (A^*, B^*)$ is determined by solving the regularized maximum likehood problem (with $N_+$ of the $y_i$'s positive, and $N_-$ negative):

$$\min_{z=(A,B)} F(z) = -\sum_{i=1}^{m} \big(t_i \log(p_i) + (1-t_i)\log(1-p_i)\big),$$

for

$$p_i = P_{A,B}(f_i),$$

and

$$t_i = \begin{cases} \dfrac{N_+ + 1}{N_+ + 2} & if\ y_i = +1 \\[2mm] \dfrac{1}{N_- + 2} & if\ y_i = -1 \end{cases}, i = 1, 2, \ldots, m \tag{8}$$

The probability estimates of test samples are calculated to implement the degradation condition prediction in the following section.

## CASE OF CONDITION CLASSIFICATION AND PREDICTION IN ROTATING MACHINERY

The data measured in field in some type of engine is analyzed in this section, the degradation features of the different faults are analyzed in (Jiang, and

*Figure 3. Features of three degradation processes*

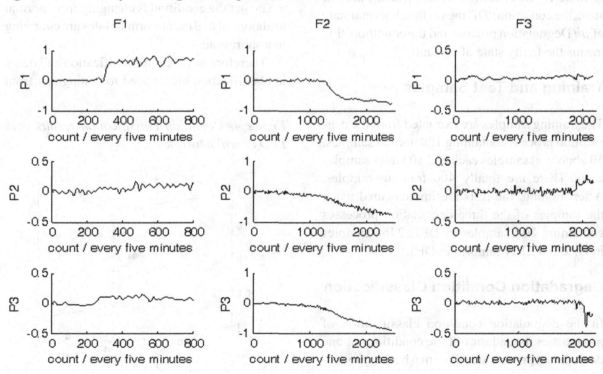

195

*Table 1. Degradation conditions to be classified*

| Condition | DP1 | DP2 | DP3 |
|---|---|---|---|
| 1 | Nor | Nor | Nor |
| 2 | D1 | - | - |
| 3 | - | D2 | - |
| 4 | - | - | D3 |
| 5 | F1 | - | - |
| 6 | - | F2 | - |
| 7 | - | - | F3 |

Liu, 2011). Figure 3 shows the extracted features. Here, three parameters are selected to form the feature space and they are numbered P1, P2 and P3. Three faults are analyzed which are numbered F1, F2 and F3 and their degradation processes are numbered DP1, DP2 and DP3.

Three conditions are defined in each degradation process: normal, abnormal and faulty. In the three degradation processes, the normal states are consistent, while the abnormal and faulty states are different. Therefore seven states are formed in Table 1. The condition "Nor" mean the normal state, the condition "Di" means the abnormal state of *ith* Degradation process and the condition "Fi" means the faulty state of *ith* fault.

## Training and Test Samples

The training samples are extracted from the degradation process containing 100 normal samples, 50 abnormal samples each and 50 faulty samples each. There are totally 400 training samples. After training, the tests are implemented using the samples of the three degradation processes containing 800 samples in DP1, 2460 samples in DP2 and 2164 samples in DP3.

## Degradation Condition Classification

In the degradation condition classification of prognostics, boundaries of the conditions in one degradation process are adjacent where the bound-

ary of the normal condition is adjacent to the boundary of the abnormal condition and the boundary of the faulty condition is adjacent to the boundary of the abnormal condition. The abnormal rule is defined that, when $\left| D_{i,j} - \bar{P}_j \right| \leq \alpha_1 \left| F_j - \bar{P}_j \right|$, the condition is to be classified into the normal region; when $\left| D_{i,j} - \bar{P}_j \right| \geq \alpha_2 \left| F_j - \bar{P}_j \right|$, the condition is to be classified to fault region. The abnormal region is between the two conditions. The lower limit and the upper limit are selected as: $\alpha_1 = 0.2$, $\alpha_2 = 0.8$.

Although the abnormal rule is presented, it still needs detailed analysis to distinguish the boundaries of the adjacent areas in specific degradation processes. As the complex boundaries of the adjacent areas and the boundaries between the normal and abnormal are difficult to distinguish accurately. Furthermore, if the condition close to the boundaries of the normal states is classified into the abnormal state, there's no severe consequence because the states close to the boundaries mean that the condition is changing from normal to abnormal and the abnormal states are emerging in a short time.

Therefore advance in classification and delay in classification are defined to distinguish from

*Figure 4. Condition classification results with LVQ neural network*

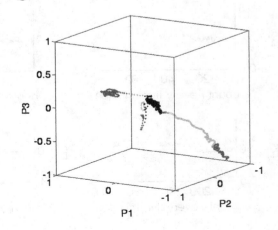

*Table 2. Classification results with LVQ neural network*

| Degradation Process | Number of test samples | Advance in classification | | Delay in classification | | Error classification | |
|---|---|---|---|---|---|---|---|
| | | Num. (counts) | Ratio (%) | Num. (counts) | Ratio (%) | Num. (counts) | Ratio (%) |
| DP1 | 800 | 13 | 1.62 | 0 | 0 | 0 | 0 |
| DP2 | 2460 | 10 | 0.41 | 77 | 3.13 | 3 | 0.12 |
| DP3 | 2164 | 37 | 1.71 | 14 | 0.65 | 32 | 1.48 |

*Figure 5. Frame of the one-against-others SVMs*

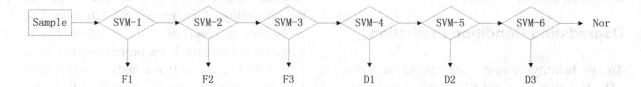

the error classification. Both of the two defined variables are in the state of change from normal to abnormal and from abnormal to faulty. In the adjacent areas of classification boundaries, if the sample belonging to the previous state is classified to the later one, it is determined as advance in classification. Otherwise it will be determined as delay in classification.

LVQ neural network and SVM are applied to classify the degradation conditions. The principles of the two classifiers are discussed in the previous section.

Figure 4 shows the classification results with LVQ neural network, where the normal state is in black, the abnormal state is in green and the faulty state is in blue. Table 2 shows the classification errors.

Based on the one-against-others SVMs in Figure 5, the degradation condition classification is implemented using six SVMs. Then the classification effects are evaluated using the test samples listed in the previous section. Figure 6 shows the classification results. Table 3 shows the classification errors.

From the classification errors in Table 2 and Table 3, good accuracy is acquired using the LVQ

neural network and SVMs. The maximal classification error of LVQ neural network is 3.13% in DP2. The maximal classification error of SVMs is 2.22% in DP3.

Considering the two classifiers, SVM performs better in the degradation condition classification. It means that the boundaries classified by SVMs are more accurate. Another probable reason is that

*Figure 6. Degradation condition classification results with SVMs*

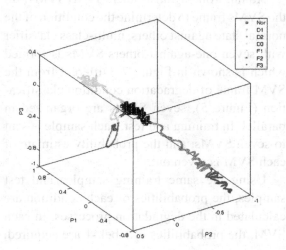

*Table 3. Classification results with SVMs*

| Degradation Process | Number of test samples | Advance in classification | | Delay in classification | | Error classification | |
|---|---|---|---|---|---|---|---|
| | | Num. (counts) | Ratio (%) | Num. (counts) | Ratio (%) | Num. (counts) | Ratio (%) |
| DP1 | 800 | 16 | 2.00 | 0 | 0 | 0 | 0 |
| DP2 | 2460 | 12 | 0.49 | 16 | 0.65 | 0 | 0 |
| DP3 | 2164 | 48 | 2.22 | 9 | 0.42 | 9 | 0.42 |

SVM is more efficient in classification problems with small training samples.

## Degradation Condition Prediction

The probability estimates are carried out using LIBSVM (Wu, Lin, and Weng, 2004), the probabilities of different degradation conditions are used for degradation condition prediction.

As each degradation process goes through the variations from normal to abnormal, or from abnormal to fault, which is adaptable for equipment with significant degradation mechanism. That's to say, if the equipment is encountering significant degradation from the normal operation process, the possibility of emergence of some certain abnormal state will increase. It's similar for the variation of state from the abnormal condition to the failure condition. The probability estimate of emergence in a state forms the basis of the condition prediction.

Adding a one-against-others SVM (SVM-7) to the SVMs frame to determine the condition of the normal state against others, multi-class classifier with seven one-against-others SVMs is formed which is shown in Figure 7. Different from the SVM frame of degradation condition classification (Figure 5), seven SVMs are organized in parallel. In training and test, each sample is sent to seven SVMs, and the probability estimate of each SVM is given out.

Using the same training samples and test samples, the probabilities of each condition are calculated in the degradation processes. In each SVM, the probabilities of label -1 are acquired.

Figure 8 shows the probability estimates of Degradation Process 2 (DP2). The value illustrates the possibility of emergence of each state. The probability of condition "Nor" is close to 1 at counts 800-1300, while the probability of condition "D2" is close to 0 at counts 800-1000 and increases after the count 1000. Based on this character, the condition prediction can be given out that the equipment is encountering degradation

*Figure 7. SVMs for degradation condition prediction*

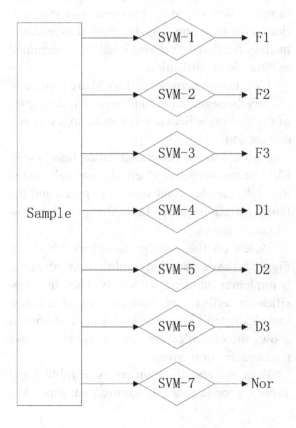

towards abnormal condition "D2". Similarly, when the equipment is running at count 1500-1800, the probability of condition "F2" is increasing which means the possibility of emergence of "F2" increases. The prediction of the "F2" emergence can be concluded based on this character. Using this method, once the condition monitoring data is acquired in field, the condition prediction can be given out using the SVMs classifier. And the possibility of the failure occurrence is acquired.

For DP2, the occurrence of the abnormal state "D2" is predicted at count 1303 and the occurrence of the faulty state "F2" is predicted at count 1900. The criterion used is that $p(y = i \mid \Lambda) \geq 0.5$, where $\Lambda$ is the domain of the seven conditions listed in Table 1.

## FUTURE RESEARCH DIRECTIONS

As the increasing demand of the reliability of rotating machinery, the research in the prognostics has attracted more and more research interests. The research direction in the future may lie in the following aspects:

### 1. The Probability Estimate of Failure Occurrence

The confidence estimate has been carried out with the time series forecasting results. The confidence estimate illustrates the reliability of the forecasting results and can be referenced in the maintenance strategy. The proposed method with probability estimate of the failure occurrence gives out the evaluation of the degradation condition. Combining the confidence estimate and the probability is the promising direction in prognostics.

### 2. The Dimensionality Reduction in Complicated Equipment

With the increase of the complexity in the equipment, more features are necessary to distinguish the degradation processes. The dimensions of the feature space are therefore large in complicated equipment with multiple failure modes. The basic principle is that, the higher dimensional feature space can get higher accuracy in condition classification. However, this kind of classifier is difficult to understand if higher dimensional space is applied, especially for the users in field. Therefore, dimensionality reduction is necessary in the complicated equipment. And preserving the structure of the continuous degradation process at the same time is a challenge in the dimensionality reduction of prognostics.

## CONCLUSION

The abnormal condition exists in the transition stage from the normal condition to the failure condition during the equipment degradation process,

*Figure 8. Probability estimates of degradation process*

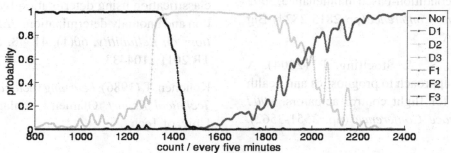

which contains the degradation information and can be applied in prognostics. With the defined anomaly parameter, the condition classification in degradation processes of the equipment is analyzed in this chapter.

Using the condition classification results in degradation processes, the probability estimate of failure occurrence is discussed. Different from the confidence estimate of time series forecasting, probability estimate of the abnormal condition can indicate the healthy degree or failed degree of the equipment as well as the probable failure mode.

Based on the proposed methods, the condition classification and prediction are carried out using the field data of rotating machinery. LVQ neural network and SVM are applied in condition classification. Both of the classifiers achieve good accuracy, while SVM performs better. Using the probability outputs in SVM, the condition prediction is realized in degradation process, and the probability of failure occurrence is acquired, which also indicates the failure mode.

## ACKNOWLEDGMENT

This work is supported by the National Natural Science Foundation of China (No. 51174273) and (No. 60979014).

## REFERENCES

Byington, C. S., Roemer, M. J., & Galie, T. (2002). Prognostic enhancements to diagnostic systems for improved condition-based maintenance. *2002 IEEE Aerospace Conference* (pp. 2815-2824). Big Sky, MT, USA.

Byington, C. S., & Stoelting, P. (2004). A model-based approach to prognostics and health management for flight control actuators. *2004 IEEE Aerospace Conference* (pp. 3551-3562). Big Sky, MT.

Dixon, S. J., & Brereton, R. G. (2009). Comparison of performance of five common classifiers represented as boundary methods: Euclidean distance to centroids, linear discriminant analysis, quadratic discriminant analysis, learning vector quantization and support vector machines, as dependent on data structure. *Chemometrics and Intelligent Laboratory Systems*, *95*, 1–17. doi:10.1016/j.chemolab.2008.07.010

Goebe, K. L., Saha, B., & Saxena, A. (2008). A comparison of three data-driven techniques for prognostics. *62nd Meeting of the Society for Machinery Failure Prevention Technology (MFPT)* (pp. 119-131). VA, USA.

Gooijer, J. G. D., & Hyndman, R. J. (2006). 25 years of time series forecasting. *International Journal of Forecasting*, *22*, 443–473. doi:10.1016/j.ijforecast.2006.01.001

Hastie, T., & Tibshirani, R. (1990). *Classification by pairwise coupling, Technical report*. Stanford University and University of Toronto.

Heng, A., Zhang, S., Tan, A. C. C., & Mathew, J. (2009). Rotating machinery prognostics: state of the art, challenges and opportunities. *Mechanical Systems and Signal Processing*, *23*, 724–739. doi:10.1016/j.ymssp.2008.06.009

Jack, L. B., & Nandi, A. K. (2002). Fault detection using support vector machines and artificial neural network, augmented by genetic algorithms. *Mechanical Systems and Signal Processing*, *16*, 373–390. doi:10.1006/mssp.2001.1454

Jiang, D. X., & Liu, C. (2011). Machine condition classification using deterioration feature extraction and anomaly determination. *IEEE Transactions on Reliability*, *60*(1), 41–48. doi:10.1109/TR.2011.2104433

Kohonen, T. (1986). *Learning vector quantization. Technical Report*. Otaniemi, Finland: Helsinki Univ. of Tech.

Kohonen, T. (2001). *Self organization maps.* New York, NY: Springer-Verlag.

Kohonen, T., Hynninen, J., Kangas, J., Laaksonen, J., & Torkkola, K. (1995). *LVQ Pak: The learning vector quantization program package.* Helsinki, Finland: Helsinki Univ. of Tech.

Kressel, U. (1999). *Pairwise classification and support vector machines.* Cambridge, MA: MIT Press.

Liu, J., Zuo, B., Zeng, X., Vroman, P., & Rabensolo, B. (2010). Nonwoven uniformity identification using wavelet texture analysis and LVQ neural network. *Expert Systems with Applications, 37,* 2241–2246. doi:10.1016/j.eswa.2009.07.049

Platt, J. C., Cfisfianini, N., & Shawe, T. J. (2000). *Large margin DAGs for multi-class classification.* Cambridge, MA: MIT Press.

Roemer, M. J., Byington, C. S., Kacprzynski, G. J., & Vachtsevanos, G. (2006). An overview of selected prognostic technologies with application to engine health management. *51st ASME Turbo Expo* (pp. 707-715). Barcelona, Spain.

Schwabacher, M., & Goebel, K. (2007). A survey of artificial intelligence for prognostics. *Working Notes of 2007 AAAI Fall Symposium: AI for Prognostics* (pp. 107-114). Arlington, VA, USA.

Spieler, S., Staudacher, S., & Fiola, R. (2008). Probabilistic engine performance scatter and deterioration modeling. *Journal of Engineering for Gas Turbines and Power, 130,* 1–9. doi:10.1115/1.2800351

Vapnik, V. N. (1998). *Statistical learning theory.* New York, NY: John Wiley & Sons.

Vapnik, V. N. (1999). An overview of statistical learning theory. *IEEE Transactions on Neural Networks, 10*(5), 988–999. doi:10.1109/72.788640

Weston, J., & Watkins, C. (1999). *Multi-class support vector machines.* Paper presented at the ESANN99, Brussels, Belgium.

Wu, T. F., Lin, C. J., & Weng, R. C. (2004). Probability estimates for multi-class classification by pairwise coupling. *Journal of Machine Learning Research, 5,* 975–1005.

Yang, B. S. (2005). Condition classification of small reciprocating compressor for refrigerators using artificial neural networks and support vector machines. *Mechanical Systems and Signal Processing, 19,* 371–390. doi:10.1016/j.ymssp.2004.06.002

Yuan, S., & Chu, F. L. (2006). Support vector machines-based fault diagnosis for turbo-pump rotor. *Mechanical Systems and Signal Processing, 20,* 939–952. doi:10.1016/j.ymssp.2005.09.006

Zaita, V., Buley, G., & Karlsons, G. (1998). Performance deterioration modeling in aircraft gas turbine engines. *Journal of Engineering for Gas Turbines and Power, 120,* 344–349. doi:10.1115/1.2818128

## ADDITIONAL READING

Agarwal, K., Shivpuri, R., & Zhu, Y. (2011). Process knowledge based multi-class support vector classification (PK-MSVM) approach for surface defects in hot rolling. *Expert Systems with Applications, 38*(6), 7251–7262. doi:10.1016/j.eswa.2010.12.026

Carr, M. J., & Wang, W. (2010). Modeling failure modes for residual life prediction using stochastic filtering theory. *IEEE Transactions on Reliability, 59,* 346–355. doi:10.1109/TR.2010.2044607

Chang, C. C., & Lin, C. J. (2010) *LIBSVM: A library for support vector machines.* Department of Computer Science, National Taiwan University. Retrieved from http://www.csie.ntu.edu.tw/~cjlin/libsvm

Davison, C. R., & Birk, A. M. (2001). Development of fault diagnosis and failure prediction techniques for small gas turbine engines. *2001 International Gas Turbine & Aeroengine Congress & Exhibition* (pp. 1-8). New Orleans, LA, USA.

Davison, R., & Birk, A. M. (2000). Steady state performance simulation of auxiliary power unit with faults for component diagnosis. *2000 International Gas Turbine & Aeroengine Congress & Exhibition* (pp. 1-9). Munich, Germany.

Dietterich, T. G. (1998). Approximate statistical tests for comparing supervised classification learning algorithms. *Neural Computation, 10,* 1895–1923. doi:10.1162/089976698300017197

Elman, J. (1990). Finding structure in time. *Cognitive Science, 14,* 179–211. doi:10.1207/s15516709cog1402_1

Francois, A., & Patrick, F. (1995). Improving the readability of time-frequency and time-scale representations by the reassignment method. *IEEE Transactions on Signal Processing, 43,* 1068–1089. doi:10.1109/78.382394

Gao, X. Z., & Ovaska, S. J. (2002). Genetic algorithm training of elman neural network in motor fault detection. *Neural Computing & Applications, 11,* 37–39. doi:10.1007/s005210200014

Jardine, K. S., Lin, D. M., & Banjevic, D. (2006). A review on machinery diagnostics and prognostics implementing condition-based maintenance. *Mechanical Systems and Signal Processing, 20,* 1483–1510. doi:10.1016/j.ymssp.2005.09.012

Kothamasu, R., Huang, S. H., & Verduin, W. H. (2006). William H. System health monitoring and prognostics - A review of current paradigms and practices. *International Journal of Advanced Manufacturing Technology, 28,* 1012–1017. doi:10.1007/s00170-004-2131-6

Li, R., Meng, G., Gao, N., & Xie, H. (2007). Combined use of partial least-squares regression and neural network for residual life estimation of large generator stator insulation. *Journal of Measurement Science and Technology, 18,* 2074–2075. doi:10.1088/0957-0233/18/7/038

Mahanty, R. N., & Dutta Guppta, P. B. (2004)... *IEE Proceedings. Generation, Transmission and Distribution, 151,* 201–204. doi:10.1049/ip-gtd:20040098

Martin, K. F. (1994). A review by discussion of condition monitoring and fault-diagnosis in machine-tools. *International Journal of Machine Tools & Manufacture, 34,* 527–551. doi:10.1016/0890-6955(94)90083-3

Rabelo Baccarini, L. M., Rocha e Silva, V. V., & de Menezes, B. R. (2011). SVM practical industrial application for mechanical faults diagnostic. *Expert Systems with Applications, 38*(6), 6980–6984. doi:10.1016/j.eswa.2010.12.017

Sikorska, J. Z., Hodkiewicz, M., & Ma, L. (2011). Prognostic modelling options for remaining useful life estimation by industry. *Mechanical Systems and Signal Processing, 25*(5), 1803–1836. doi:10.1016/j.ymssp.2010.11.018

Song, Y. H., Xuan, Q. Y., & Johns, A. T. (1997). Protection scheme for E H V transmission systems with thyristor controlled series compensation using radial basis function neural networks. *Electric Machines and Power Systems, 25,* 553–565. doi:10.1080/07313569708955759

Spieler, S., Staudacher, S., Fiola, R., Sahm, R., & Weißschuh, M. (2008). Probabilistic engine performance scatter and deterioration modeling. *Journal of Engineering for Gas Turbines and Power, 130,* 1–9. doi:10.1115/1.2800351

Taylor, J. W. (2003). Exponential smoothing with a damped multiplicative trend. *International Journal of Forecasting, 19*(4), 715–725. doi:10.1016/S0169-2070(03)00003-7

Tenenbaum, J., de Silva, V., & Langford, J. (2000). A global geometric framework for nonlinear dimensionality reduction. *Science, 290,* 2319–2323. doi:10.1126/science.290.5500.2319

Tse, P. W., & Atherton, D. P. (1999). Prediction of machine deterioration using vibration based fault trends and recurrent neural networks. *Journal of Vibration and Acoustics, 121,* 355–361. doi:10.1115/1.2893988

Wang, W. (2007). A prognosis model for wear prediction based on oil-based monitoring. *The Journal of the Operational Research Society, 57,* 887–893. doi:10.1057/palgrave.jors.2602185

Widodo, A., & Yang, B. S. (2007). Support vector machine in machine condition monitoring and fault diagnosis. *Mechanical Systems and Signal Processing, 21,* 2560–2574. doi:10.1016/j.ymssp.2006.12.007

# Section 6
# Diagnostics

# Chapter 11
# A Temporal Probabilistic Approach for Continuous Tool Condition Monitoring

**Omid Geramifard**
*National University of Singapore, Singapore*

**Jian-Xin Xu**
*National University of Singapore, Singapore*

**Junhong Zhou**
*Singapore Institute of Manufacturing Technology, Singapore*

## ABSTRACT

*In this chapter, a temporal probabilistic approach based on hidden semi-Markov model is proposed for continuous (real-valued) tool condition monitoring in machinery systems. As an illustrative example, tool wear prediction in CNC-milling machine is conducted using the proposed approach. Results indicate that the additional flexibility provided in the new approach compared to the existing hidden Markov model-based approach improves the performance. 482 features are extracted from 7 signals (three force signals, three vibration signals and acoustic emission) that are acquired for each experiment. After the feature extraction phase, Fisher's discriminant ratio is applied to find the most discriminant features to construct the prediction model. The prediction results are provided for three different cases, i.e. cross-validation, diagnostics, and prognostics. The possibility of incorporating an asymmetric loss function in the proposed approach in order to reflect and consider the cost differences between an under- and over-estimation in tool condition monitoring is also explored and the simulation results are provided.*

## INTRODUCTION

Tool Condition monitoring (TCM) is a challenging task in industrial environments. TCM reduces the downtime of machinery for maintenance purposes (Jun-Hong, Chee Khiang, Lewis, & Zhao-Wei, 2009). Consequently, TCM reduces the mainte-

nance cost while improving the performance of the machine. Furthermore, TCM increases the quality of the product.

The idea of Continuous TCM is to regularly assess the health status of the tool based on a continuous metric. In other words, instead of setting some thresholds and differentiating dis-

DOI: 10.4018/978-1-4666-2095-7.ch011

tinct health states as various (ordinal) classes, we would like to monitor the health status of the tool using a continuous measure. This task allows us to have a smoother decision maker system for the condition based maintenance. It also enables us to incorporate different quality thresholds for different applications using the same condition based maintenance system e.g. to satisfy and guarantee different qualities in various products.

Temporal probabilistic models can be identified as a group of probabilistic graphical models which can be unrolled over time. They take into account the dependencies among the states as well as observations over the time, and use the conditional independencies to make inferences (Neapolitan, 2003; Russell & Norvig, 2009). The essential motivations for applying these models to the tool condition monitoring are as follows. Firstly, the tool condition prediction based on the non-intrusively sensed information from the machines, is a task which is inherently uncertain and probabilistic (Rao, 1996). Secondly, there is temporal information lying in the sequentially sensed data which can be captured using temporal models. Last but not least, the output of these models is a probability distribution over states that can be directly used either to do the decision making or to find the expected value of the tool condition.

One of the simplest temporal probabilistic models, commonly used for discrete TCM, is called hidden Markov Model (HMM) (Atlas, Ostendorf, & Bernard, 2000; Zhu, Hong, & Wong, 2008). In (Geramifard, Xu, Zhou, & Li, 2010), a single HMM-based approach is used to do the continuous health assessment in a CNC-milling machine. However, one of the deficiencies of using the HMM is its fixed duration distribution (Geometric distribution). The duration distribution indicates the probability of staying in one state for different possible durations. Having a fixed duration distribution may lead to unsatisfactory prediction results in cases that the assumption

of having a geometric duration distribution does not hold. In lots of real applications, the duration distribution is not geometric. Hence, to improve the prediction performance in TCM, a more complex temporal probabilistic model, namely, hidden semi-Markov model (HSMM) can be used to dissolve the aforementioned fixed duration distribution problem in the HMM-based approach.

In HSMM, the idea is to use temporal information in a more effective way than in HMM by keeping track of the duration of staying at each state (Yu, 2010). Both HMMs and HSMMs are previously used in (Atlas, et al., 2000; Dong & He, 2007; Zhu, et al., 2008) to recognize different fault types and states (discrete TCM). However, in the way these models are used, they do not provide any relation between the actual physical states and the hidden states in them. Contrary to the existing approaches based on HMM or HSMM, the HSMM-based approach introduced in this chapter provides an explicit relationship between the actual physical states and its hidden state values. Using the HSMM, the aforementioned deficiency of HMM-based approach is also remedied.

Another issue that we would like to address in this chapter is how to incorporate an asymmetric loss function into our approach, to consider the dramatic cost differences between an over- and an under- estimation of the tool condition. To this end, we can take the advantage of having flexible duration distributions in the HSMM and try to modify the overall skewness of the duration distributions based on a given asymmetric loss function and training dataset.

This chapter is organized as follows. In section II, HMM and its graphical concept used in tool condition monitoring are introduced. In section III, a single HSMM-based approach for continuous health assessment is proposed. Then a computationally efficient version of forward-backward algorithm for inference in the implemented HSMM as well as state estimation variables is given in section IV. Diagnosis and Prognosis procedures using the computationally efficient forward-backward algorithm are discussed in section V. Section VI provides information about the extracted and selected features from the acquired signals from the conducted experiments. Afterwards, simulation results based on the experimental data are provided and compared in section VIII. The possibility of incorporating an asymmetric loss function in the proposed approach to reflect and consider the cost differences between an under- and over- estimation in TCM is explored and provided in section IX along with its simulation results. Finally, the chapter is concluded in section X.

## Hidden Markov Model

Hidden Markov model is one of the simplest dynamic Bayesian network. This model has only one discrete hidden state variable, and a set of discrete or continuous observation nodes (Darwiche, 2009). Figure 1 depicts the transition graph of the HMM and its graphical model being used in TCM, respectively.

Both HMMs and HSMMs are used in (Atlas, et al., 2000; Dong & He, 2007; Zhu, et al., 2008)

to recognize different fault types and states. However, the way these models are used, they do not provide any relation between the actual physical states and the hidden states in them. Each HMM or HSMM is used to recognize only one health state or fault. Therefore, even though each of the HMMs or HSMMs corresponds to one specific health state or fault, there is no explicit connection between the actual physical status of the tool and the hidden state values within those models.

In (Geramifard, et al., 2010), an HMM-based approach is proposed which provides the aforementioned relationship between the physical health states and the hidden state values for tool wear prediction purpose. In this approach, as depicted in Figure 1, there are only two possible transitions (excluding the last state which is modeled as an attraction point) from each health state $(H_i)$, either staying in the same wearing state $(H_i)$ with probability of $p_i$ or going to the next wearing state $(H_{i+1})$ with probability of $1-p_i$. Thus, the probability of staying in $H_i$ for exact $d$ time steps in the implemented HMM can be calculated as

*Figure 1. a) Transition graph of the HMM used in TCM. It depicts the concept of using HMM in TCM. b) Graphical model of the HMM. $S_t$ and $O_t$ are the hidden state and observations (measured features) at time t, respectively (Geramifard, et al., 2010).*

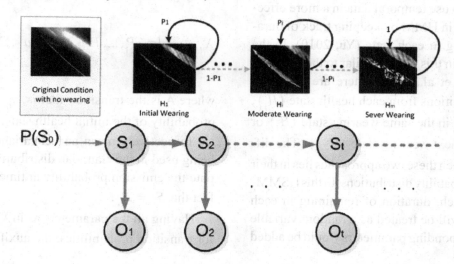

$$P(\text{staying in } H_i \text{ for exact } d \text{ time steps}) = p_i^d (1 - p_i),$$

$$(1)$$

in which $p_i$ is the probability of self-transition at $i$th state $H_i$. As it can be understood from (1), this probability is a *geometric* distribution. However, it is known that lots of processes in nature are not abiding geometric distribution (Sansom & Thomson, 2001). Therefore, in order to add more flexibility to the probability distribution function, duration factor is added to HMM and transition probabilities are redefined based on these durations in the succeeding section.

In the next section, an approach based on HSMM is proposed to tackle problem of continuous tool condition monitoring. The proposed approach provides an explicit relationship between the actual physical states and the hidden state values while having a flexible duration distribution in contrast to the case of HMM-based approaches. As an illustrative example, tool wearing prediction of the cutter in a CNC-milling machine is conducted.

## HIDDEN SEMI-MARKOV MODEL-BASED APPROACH

The idea behind hidden semi-Markov model (HSMM) is to use temporal data in a more effective way than in HMM by keeping track of duration of staying in each state (Yu, 2010). In the HSMM used in this work, similar to the HMM in (Geramifard, et al., 2010), there are only two possible transitions from each health state $(H_i)$, either staying in the same wearing state $(H_i)$ or going to the next wearing state $(H_{i+1})$. The difference between these two approaches lies in their transition probability distributions. In this HSMM-based approach, duration of remaining in each health state will be treated as a random variable and its corresponding parameters would be added

to the parameter set needed to be determined for the model in the training (parameter estimation) procedure. Furthermore, for simplicity, the duration variables are assumed to have normal distributions.

Similar to (Geramifard, et al., 2010), firstly the measured tool wearing of the cutters in the training set are uniformly discretized into $m$ ordinal classes (clusters), $\{H_1, ..., H_m\}$. These ordinal classes corresponds to different health stages of the cutters from initial wearing conditions to severely worn out. Moreover, the labels of these classes are real-valued and correspond to the mean wearing value in each class. These real-valued labels are later used as the values that hidden states of the both HSMM and HMM can take at each time step.

As defined in (Geramifard, et al., 2010), the parameter set to be determined for HMM-based approach is as follows,

$$A = \begin{bmatrix} p_1 & 1-p_1 & 0 & 0 & \cdots & 0 \\ 0 & p_2 & 1-p_2 & 0 & \cdots & 0 \\ \cdot & & \cdot & & & \\ \cdot & & & \cdot & & \cdot \\ \cdot & & & & \cdot & \cdot \\ 0 & 0 & \cdots & 0 & p_{m-1} & 1-p_{m-1} \\ 0 & 0 & 0 & \cdots & 0 & 1 \end{bmatrix},$$

$$\lambda_{HMM} = \{\pi_0, P_1, ..., P_{m-1}, \mu_1, ..., \mu_m, \Sigma_1, ..., \Sigma_m\}$$

where A is the transition matrix, $\pi_0$ is the prior probability of the initial health state, $\mu_i$ and $\Sigma_i$ are respectively mean and covariance matrices being used in the Gaussian distributions to compute the emission probability at time $t$ given the fact that $S_t = H_i$.

Having all the parameters as in $\lambda_{HMM}$, except for transition probabilities, the auxiliary parameters that are needed to be defined to formulate HSMM are as follows,

- $d_i$ is defined as a random variable having a normal distribution which corresponds to duration distribution in the $i$th health state (hidden state). Consequently, two more parameters would be added to the parameter set for each health state, $\mu_{d_i}$ and $\sigma_{d_i}$ as the mean and standard deviation of the duration variable $d_i$ for each health state.

- For implementation purposes, $d_{max}$ is also needed to be defined as the overall maximum possible duration.

As mentioned, in each time step, the implemented HSMM has only two possible options of either remaining at the same health state as the previous time step or proceeding to the next health state. The probability of remaining in the same state would be generated using the duration distribution, and the transition probability would only be used at a time that the duration of staying at one health state is over and the health state is going to change. That is why this model is called hidden semi-Markov model, because it only uses the Markov transition when the duration of staying in one state is over (Yu, 2010). Therefore, the transition matrix $A$ would be different from the transition matrix in the HMM-based approach. Out of two possible transitions at each time step, the

only valid transition after the duration of staying in one state has passed is the option of going to the next state. This assumption corresponds to each row of the transition matrix, $A$, having only one non-zero element, which indicates the transition to the next wearing stage. Figure 2 schematizes the HSMM transition graph.

Based on the assumptions that we have made, and using the duration distributions to find the probability of staying at each health state, the parameter set and the transition matrix used in (Geramifard, et al., 2010) may be converted as follows,

$$A = \begin{bmatrix} 0 & 1 & 0 & 0 & \dots & 0 \\ 0 & 0 & 1 & 0 & \dots & 0 \\ \cdot & & \cdot & & & \cdot \\ \cdot & & & \cdot & & \cdot \\ \cdot & & & & \cdot & \cdot \\ 0 & 0 & \dots & 0 & 0 & 1 \\ 0 & 0 & 0 & \dots & 0 & 1 \end{bmatrix},$$

$$\lambda_{HSMM} = \{\underbrace{\pi_0, \mu_1, \dots, \mu_m, \Sigma_1, \dots, \Sigma_m}_{\text{HMM Parameters}},$$
$$\underbrace{\mu_{d_1}, \dots, \mu_{d_m}, \sigma_{d_1}, \dots, \sigma_{d_m}, d_{max}}_{\text{Duration Distribution Parameters}}\},$$

where $A$ is the transition matrix and $\lambda_{HSMM}$ is the parameter set of the implemented HSMM, where

*Figure 2. Schematic transition graph of the implemented HSMM*

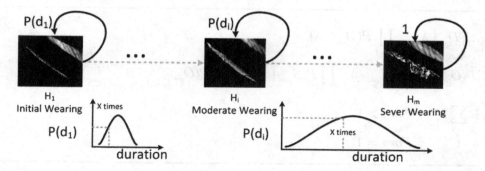

209

$\pi_0$ is the prior probability of the initial health state. $\mu_i$ and $\Sigma_i$ are respectively mean and covariance matrices being used in the Gaussian distributions to compute the emission probability at time $t$ given the fact that $S_t = H_i$. $\mu_{d_i}$ and $\sigma_{d_i}$ are the parameters of the Gaussian distribution representing the duration variable at the $i$th health state denoted as $d_i$. All the required parameters can be identified using *maximum log likelihood* approach on the complete dataset that is provided as the training set.

Assuming that all experiments are independently distributed, joint probability of observing training sequences given the parameters of the HSMM can be written as Equation 2.

Where $D_{seq}$ is the *seq*th experiment, $N$ is the number of experiments, $t_i^{seq}$ is the starting time step of the $i$th health stage in each specific sequence, $m$ is the number of health states, $k_j^{seq}$ is the observed duration of the $j$th health state in each specific sequence, $O_{t_i^{seq}:t_{i+1}^{seq}-1}$ is the observations from the time step entering to the $i$th state till transition to next state in the *seq*th experimental data and $S_{t_i^{seq}:t_{i+1}^{seq}-1}$ indicates the state sequence during that time. Assuming conditional independence between the observations in one sequence given their corresponding states and the duration variables following Gaussian distributions, (2) can be written as Equation 3.

Furthermore, assuming the emission probability is a *multivariate Gaussian* variable, (3) can be rewritten as Equation 4 where $\chi$ is the dimension of observation (input) space. Finally based on (2) and (4), the log likelihood of the joint probability of $P(D_{1:N} \mid \lambda)$ can be written as Equation 5.

Now, by setting the partial derivatives of the log likelihood to zero, the parameters of the HSMM can be found as follows,

*Table 2. Shares of extracted features from each signal in the set of selected features. $F_x$, $F_y$ and $F_z$ are force signals in different directions. $V_x$, $V_y$ and $V_z$ are vibration signals in different directions and $AE$ is the acoustic emission signal.*

| Signal | $F_x$ | $F_y$ | $F_z$ | $V_x$ | $V_y$ | $V_z$ | $AE$ | Total |
|---|---|---|---|---|---|---|---|---|
| Statistical Features | 6 | 2 | 6 | - | - | - | - | 14 |
| Wavelet Features | 11 | 5 | 5 | 0 | 3 | 0 | 0 | 24 |
| Total Share | 17 | 7 | 11 | 0 | 3 | 0 | 0 | 38 |

*Equation 2*

$$
\begin{aligned}
P(D_1, D_2, \ldots, D_N \mid \lambda) &= \prod_{seq=1}^{N} P(D_{seq} \mid \lambda) \\
&= \pi_0 P(d_1) P(O_{t_1^{seq}:t_2^{seq}-1} \mid S_{t_1^{seq}:t_2^{seq}-1}) \times \prod_{i=2}^{m} P(S_{t_i^{seq}} \mid S_{t_i^{seq}}) P(d_i) P(O_{t_i^{seq}:t_{i+1}^{seq}-1} \mid S_{t_i^{seq}:t_{i+1}^{seq}-1}) \\
t_i^{seq} &= \begin{cases} 1 + \sum_{j=1}^{i-1} k_j^{seq} & For\ i > 1 \\ 1 & For\ i = 1 \end{cases}
\end{aligned}
$$

$$(2)$$

*Table 3. Shares of different wavelet decomposition levels of each signal in the set of selected features*

| Signal | Level | | | | |
|---|---|---|---|---|---|
| | $L_1$ | $L_2$ | $L_3$ | $L_4$ | $L_5$ |
| $F_x$ | 0 | 1 | 2 | 3 | 5 |
| $F_y$ | 1 | 1 | 1 | 1 | 1 |
| $F_z$ | 1 | 1 | 1 | 1 | 1 |
| $V_x$ | 0 | 0 | 0 | 0 | 0 |
| $V_y$ | 0 | 0 | 0 | 1 | 2 |
| $V_z$ | 0 | 0 | 0 | 0 | 0 |
| $AE$ | 0 | 0 | 0 | 0 | 0 |

$$\frac{\partial \bar{L}}{\partial \sigma_{d_i}} = 0 \rightarrow \sum_{seq=1}^{N} \left( \frac{-1}{\sigma_{d_i}} + \frac{(k_i^{seq} - \mu_{d_i})^2}{\sigma_{d_i}^3} \right)$$

$$= 0 \Rightarrow \sigma_{d_i} = \sqrt{\frac{\sum_{seq=1}^{N} (k_i^{seq} - \mu_{d_i})^2}{N}},$$

$$(7)$$

$$\frac{\partial \bar{L}}{\partial \mu_i} = 0 \rightarrow \sum_{seq=1}^{N} \sum_{t=t_i^{seq}}^{t_{i+1}^{seq}-1} \frac{1}{2} (\Sigma_i^{-1} + \Sigma_i^{-T})(O_t^{seq} - \mu_i)(-1)$$

$$= 0 \rightarrow \sum_{seq=1}^{N} \sum_{t=t_i^{seq}}^{t_{i+1}^{seq}-1} \Sigma_i^{-1}(O_t^{seq} - \mu_i)$$

$$= 0 \rightarrow \sum_{seq=1}^{N} \sum_{t=t_i^{seq}}^{t_{i+1}^{seq}-1} O_t^{seq} - \mu_i \sum_{seq=1}^{N} k_i^{seq}$$

$$= 0 \Rightarrow \mu_i = \frac{\sum_{seq=1}^{N} \sum_{t=t_i^{seq}}^{t_{i+1}^{seq}-1} O_t^{seq}}{\sum_{seq=1}^{N} k_i^{seq}}$$

$$(8)$$

$$\frac{\partial \bar{L}}{\partial \mu_{d_i}} = 0 \rightarrow \sum_{seq=1}^{N} \frac{(k_i^{seq} - \mu_{d_i})}{\sigma_{d_i}^2},$$

$$= 0 \Rightarrow \mu_{d_i} = \frac{\sum_{seq=1}^{N} k_i^{seq}}{N}$$

$$(6)$$

*Equation 3*

$$P(D_{seq} \mid \lambda) = \pi_0 \frac{1}{\sqrt{2\pi\sigma_{d_1}^2}} e^{\left(-\frac{(k_1^{seq} - \mu_{d_1})^2}{2\sigma_{d_1}^2}\right)} \times \prod_{j=t_1^{seq}}^{t_2^{seq}-1} P(O_j^{seq} \mid S_j^{seq})$$

$$\times \prod_{i=2}^{m} P(S_{t_i^{seq}}^{seq} \mid S_{t_i^{seq}-1}^{seq}) \frac{1}{\sqrt{2\pi\sigma_{d_i}^2}} e^{\left(-\frac{(k_i^{seq} - \mu_{d_i})^2}{2\sigma_{d_i}^2}\right)} \times \prod_{t=t_i^{seq}}^{t_{i+1}^{seq}-1} P(O_t^{seq} \mid S_t^{seq})$$

$$t_i^{seq} = \begin{cases} 1 + \sum_{j=1}^{i-1} k_j^{seq} & For\ i > 1 \\ 1 & For\ i = 1 \end{cases}.$$

$$(3)$$

*Table 4. Prediction error rate in case II using HMM and HSMM-based approaches in terms of MSE*

| Approach | Mean Square Error | | | |
|---|---|---|---|---|
| | *09BX3* | *18SC3* | *33PN6* | **Total** |
| **HSMM-based Approach** | **314.2978** | 136.0284 | **263.9663** | **238.0975±91.9070** |
| **HMM-based Approach** | 341.2336 | **135.3554** | 297.5691 | 258.0527±108.4787 |

*Table 5. Prognosis error rate for HMM and HSMM-based approaches in terms of average MSE with different prediction horizons on the testing data*

| Approach | Prediction Horizon (Time Steps Ahead) | | | | | | | | |
|---|---|---|---|---|---|---|---|---|---|
| | **1** | **2** | **3** | **4** | **5** | **6** | **7** | **8** | **9** |
| HSMM-based | 311.374 | 314.077 | 317.910 | 322.409 | 327.579 | 333.124 | 339.546 | 346.508 | 353.997 |
| HMM-based | 422.307 | 491.458 | 602.735 | 680.120 | 732.523 | 771.166 | 802.260 | 829.000 | 852.779 |

$$\frac{\partial \overline{L}}{\partial \Sigma_i} = 0 \rightarrow -\sum_{seq=1}^{N} [\frac{k_i^{seq}}{2} \frac{1}{|\Sigma_i|} |\Sigma_i| (\Sigma_i^{-1})^T$$

$$-\sum_{t=t_i^{seq}}^{t_{i+1}^{seq}-1} \frac{1}{2} \Sigma_i^{-1} (O_t^{seq} - \mu_i)(O_t^{seq} - \mu_i)^T \Sigma_i^{-1}]$$

$$-\sum_{seq=1}^{N} \frac{k_i^{seq}}{2} \Sigma_i^{-1}$$

$$+\Sigma_i^{-1} [\frac{1}{2} \sum_{seq=1}^{N} \sum_{t=t_i^{seq}}^{t_{i+1}^{seq}-1} (O_t^{seq} - \mu_i)(O_t^{seq} - \mu_i)^T] \Sigma_i^{-1}.$$

$$= 0 \rightarrow \sum_{seq=1}^{N} \frac{k_i^{seq}}{2}$$

$$= [\frac{1}{2} \sum_{seq=1}^{N} \sum_{t=t_i^{seq}}^{t_{i+1}^{seq}-1} (O_t^{seq} - \mu_i)(O_t^{seq} - \mu_i)^T] \Sigma_i^{-1}$$

$$\Rightarrow \Sigma_i = \frac{\sum_{seq=1}^{N} \sum_{t=t_i^{seq}}^{t_{i+1}^{seq}-1} (O_t^{seq} - \mu_i)(O_t^{seq} - \mu_i)^T}{\sum_{seq=1}^{N} k_i^{seq}}$$

(9)

In this implementation of HSMM, the initial probability $\pi_0$ and $d_{max}$ are also defined as follows,

$$\pi_0(i) = \frac{1}{m},$$

$$d_{max} = \max_i \{\mu_{d_i} + 2\sigma_{d_i}\}.$$

$d_{max}$ is used as the upper bound limit of duration variables and $\pi_0$ is defined as uniform distribution in the case that no prior information about this distribution is available. In the implementation of the HSMM, an $m \times d_{max}$ probability distribution matrix is computed and stored based on the probability distributions parameterized by the mean and variance of the duration within each health state. The $i$th row, $j$th column element of the probability distribution matrix indicates the probability of staying in the $i$th health state for $j$th time steps computed by the relevant normal distribution. Each row of the probability distribution matrix is later normalized by the summation of its elements. Thus, the duration of staying in each health state is regarded as a bounded positive integer that may be modeled by a truncated normal distribution.

After estimating the parameters of the HSMM based on the training set by the given formulas in (6)-(9). We would like to estimate the health states for a newly given data (test data). This task can be done using inference algorithms. The inference in HSMM can be done by means of *Forward-Backward* algorithm. An Efficient Forward-Backward

*Equations 4 and 5*

$$P(D_{seq} \mid \lambda) = \pi_0 \frac{1}{\sqrt{2\pi\sigma_{d_1}^2}} \exp(-\frac{(k_1^{seq} - \mu_{d_1})^2}{2\sigma_{d_1}^2}) \prod_{j=t_1^{seq}}^{t_2^{seq}-1} \frac{1}{(2\pi)^{\chi/2} \mid \Sigma_1 \mid^{1/2}} e^{(-\frac{1}{2}(O_j^{seq} - \mu_1)^T \Sigma_1^{-1}(O_j^{seq} - \mu_1))}$$

$$\times \prod_{i=2}^{m} \frac{1}{\sqrt{2\pi\sigma_{d_i}^2}} e^{(-\frac{(k_i^{seq} - \mu_{d_i})^2}{2\sigma_{d_i}^2})} \prod_{t=t_i^{seq}}^{t_{i+1}^{seq}-1} \frac{1}{(2\pi)^{\chi/2} \mid \Sigma_i \mid^{1/2}} e^{(-\frac{1}{2}(O_t^{seq} - \mu_i)^T \Sigma_i^{-1}(O_t^{seq} - \mu_i))} \tag{4}$$

$$\overline{L} = \sum_{seq=1}^{N} L_{seq} = \sum_{seq=1}^{N} \ln(P(D_{seq} \mid \lambda))$$

$$L_{seq} = \ln(P(D_{seq} \mid \lambda)) = \ln(\pi_0) - \sum_{i=1}^{m} [\frac{1}{2}\ln(2\pi) + \ln(\sigma_{d_i}) + \frac{(k_i^{seq} - \mu_{d_i})^2}{2\sigma_{d_i}^2}]$$

$$- \sum_{i=1}^{m} [\frac{k_i^{seq}\chi}{2}\ln(2\pi) + \frac{k_i}{2}\ln(\mid \Sigma_i \mid) + \sum_{t=t_i^{seq}}^{t_{i+1}^{seq}-1} \frac{1}{2}(O_t^{seq} - \mu_i)^T \Sigma_i^{-1}(O_t^{seq} - \mu_i)]. \tag{5}$$

algorithm for HSMM (explicit-duration hidden Markov model) is introduced in (Yu & Kobayashi, January 2003). Here, a similar Forward-Backward algorithm is used to do inference in the implemented HSMM for TCM.

## COMPUTATIONALLY SIMPLIFIED FORWARD-BACKWARD ALGORITHM

Forward-Backward algorithm is a recursive algorithm that can be used in Markov models to answer the inference problems in them. A computationally simplified implementation of this algorithm for tool condition monitoring purpose is introduced in this section. This algorithm uses two auxiliary variables called forward and backward variables to compute the required probabilities, recursively.

### Forward-Backward Variables

Forward variable is defined as the joint probability of being at $i$th state at time $t$ and remaining in that state for the next $k$ steps while observing the inputs from time step 1 to $t$. According to (Yu &

Kobayashi, January 2003), forward variable can be formulated as follows,

$$\alpha_t(i,k) \equiv P(O_{1:t}, (s_t, \tau_t) = (H_i, k)), \tag{10}$$

where $O_{1:t} = \{O_1, O_2, ..., O_t\}$ are the (input) observations from time step 1 to $t$, $(s_t, \tau_t)$ is the pair of hidden state of the model at time step $t$ and its remaining duration $(\tau_t)$ at that state from time step $t$ onwards. $H_i$ is the $i$th health state value that state variable in the model can take, $H_i \in \{H_1, H_2, ..., H_m\}$. From law of total probability, (10) can be computed as follows,

$$\alpha_t(i,k) = \alpha_{t-1}(i, k+1)P(O_t \mid s_t = H_i)$$
$$+ \sum_{j=1, j\neq i}^{m} \alpha_{t-1}(j, 1)a_{ij}P(O_t \mid s_t = H_i)P((S_t, \tau_t)$$
$$= (H_i, k)), \tag{11}$$

where $m$ is the number of health state values, $a_{ij}$ is the transition probability from the $i$th health state $(H_i)$ to $j$th health state $(H_j)$ and $d_i$ is the duration of remaining in $i$th health state $(H_i)$. The initial condition for the recursive equation in (11) is

$$\alpha_1(i,k) = \pi_0(i)P(O_1 \mid s_1 = H_i)P((S_1, \tau_1) = (H_i, k)).$$

Based on the assumptions that are made about graduality of the wearing process and consequently the possible transitions in the implemented model, (11) can be simplified as the equation in Box 1.

The second auxiliary variable to be defined is called *backward* variable. According to (Yu & Kobayashi, January 2003), the backward variable is defined as

$$\beta_t(i,k) \equiv P(O_{t+1:T} \mid (s_t, \tau_t) = (H_i, k)),$$

where $(s_t, \tau_t) = (H_i, k)$ indicates that the health status is supposed to remain in this state for the next $k$ time steps and then will transit to another

state $j, j \neq i$. Therefore, $\beta_t(i,k)$ can be written as Equation 12.

Since (12) is a backward recursive formula, its initial condition has been defined as

$$\beta_T(i,k) = 1, \; k \geq 1.$$

Based on the assumptions that are made on the graduality of the wearing process and consequently the model's specific form of transition matrix in our implemented HSMM, the *backward* variable can be simplified further and be rewritten as the equation in Box 2.

## State Estimation

In order to estimate the state values of the state variables at each time step based on the observations using the HSMM model, three auxiliary variables are further defined. Similar to auxiliary variables defined in (Yu & Kobayashi, January 2003), these variables would help to simplify the

*Table 1. List of statistical extracted features from each force signal. X prefix in the No. indicates the force signal channel that the features are extracted from. X can be replaced by Y and Z to indicate the features extracted from the other two force signal channels (Jie, Hong, Rahman, & Wong, 2002).*

| No. | Feature | No. | Feature |
|---|---|---|---|
| $X_1$ | Residual Error | $X_9$ | Sum of the Squares of Residual Errors |
| $X_2$ | First Order Differencing | $X_{10}$ | Peak Rate of Cutting Forces |
| $X_3$ | Second Order Differencing | $X_{11}$ | Total Harmonic Power |
| $X_4$ | Maximum Force Level | $X_{12}$ | Average Force |
| $X_5$ | Total Amplitude of Cutting Force | $X_{13}$ | Variable Force |
| $X_6$ | Combined Incremental Force changes | $X_{14}$ | Standard Deviation |
| $X_7$ | Amplitude Ratio | $X_{15}$ | Skewness |
| $X_8$ | Standard Deviation of the Force Components in Tool Breakage Zone | $X_{16}$ | Kurtosis |

*Box 1.*

$$\alpha_t(i,k) = \begin{cases} \alpha_{t-1}(i,k+1)P(O_t \mid s_t = H_i) + \alpha_{t-1}(i-1,1)P(O_t \mid s_t = H_i)P((S_t, \tau_t) = (H_i, k)) & For\, i > 1 \\ \alpha_{t-1}(i,k+1)P(O_t \mid s_t = H_i) & For\, i = 1 \end{cases}.$$

*Equation 12*

$$\beta_t(i,k) = \begin{cases} P(O_{t+1} \mid s_{t+1} = H_i)\beta_{t+1}(i, k-1) & For\, k > 1 \\ \sum_{j=1, j\neq i}^{m} a_{ij} P(O_{t+1} \mid s_{t+1} = H_j) \times \sum_{d\geq 1} P((S_{t+1}, \tau_{t+1}) = (H_j, d))\beta_{t+1}(j, d) & For\, k = 1 \end{cases}$$

(12)

*Box 2.*

$$\beta_t(i,k) = \begin{cases} P(O_{t+1} \mid s_{t+1} = H_i)\beta_{t+1}(i, k-1) & For\, k > 1 \\ P(O_{t+1} \mid s_{t+1} = H_{i+1}) \times \sum_{d=1}^{D_{max}} P((S_{t+1}, \tau_{t+1}) = (H_{i+1}, d))\beta_{t+1}(i+1, d) & For\, k = 1 \end{cases}$$

state estimation computation for either current time (diagnosis) or future (prognosis). The three variables are defined and further simplified based on the assumptions that are made in the TCM problem statement.

The first variable to be defined is $\zeta_t(i)$, which is the joint probability of observing $O_{1:T}$ and transition from $i$th health state $(H_i)$ to its next health state at time $t$. $\zeta_t(i)$ can be written as

$$\zeta_t(i) \equiv P(O_{1:T}, S_{t-1} = H_i, S_t = H_{i+1}). \quad (13)$$

The joint probability in (13) can be calculated as follows,

$$\zeta_t(i) = \alpha_{t-1}(i,1)(1-p_i)P(O_t \mid S_t = H_{i+1}) \times \sum_{k\geq 1} P(d_{i+1} = k)\beta_t(i+1, k),$$

where $P(d_{i+1} = k)$ is the probability of staying at $H_{i+1}$ for the exact $k$ time steps.

The second auxiliary variable that can be used for state estimation of $S_t$ based on the whole observation sequence $O_{1:T}$ is $\gamma_t(i)$. $\gamma_t(i)$ is the joint probability of $O_{1:T}$ and $S_t = H_i$, which can be written as follows,

$$\gamma_t(i) \equiv P(O_{1:T}, S_t = H_i).$$

Moreover, based on law of total probability, $P(O_{1:T}, S_t = H_i, S_{t+1} = H_i)$ can be written as

$$\begin{aligned} P\left(O_{1:T}, S_t = H_i, S_{t+1} = H_i\right) \\ = P\left(O_{1:T}, S_t = H_i\right) - P\left(O_{1:T}, S_t = H_i, S_{t+1} = H_{i+1}\right) \\ = P\left(O_{1:T}, S_{t+1} = H_i\right) - P\left(O_{1:T}, S_t = H_{i-1}, S_{t+1} = H_i\right) \end{aligned}$$

(14)

Consequently, based on (14), a *backward* recursive formula can be derived for $\gamma_t(i)$ as follows,

$$\gamma_t(i) = \gamma_{t+1}(i) + \zeta_{t+1}(i) - \zeta_{t+1}(i-1),$$

and its initial condition can be calculated as follows,

$$\gamma_T(i) = \sum_{k=1}^{D_{max}} \alpha_T(i, k). \quad (15)$$

The third auxiliary variable is $\xi_t(i,k)$. In order to find the probability of being at one state in future, the model must be unrolled on the time horizon while there are no more observations. Therefore, similarly to the forward variable defined in (11), $\xi_t(i,k)$ is defined as a form of forward variable for $t'$ ($t' > T$) that only considers the observations up to time step $T$. It can be written in a recursive form as follows,

$$\xi_{t'}(i,k) \equiv P(O_{1:T},(s_{t'},\tau_{t'}) = (H_i,k)), t' > T$$
$$\xi_{t'}(i,k) = \xi_{t'-1}(i,k+1)n$$
$$+ \xi_{t'-1}(i-1,1)P(d_i = k \mid S_{t'} = H_i), t' > T$$
with initial condition : $\xi_T(i,k) = \alpha_T(i,k)$

$$(16)$$

## DIAGNOSTICS AND PROGNOSTICS

Diagnosis is the task of predicting the health state at time $T$ given all the observations from time step 1 to $T$. In the realm of Bayesian Networks, this task is called *filtering* or *monitoring* and it can be written in a probabilistic manner as follows,

$$P(S_T = H_i \mid O_{1:T},\lambda) = \frac{P(O_{1:T},S_T \mid \lambda)}{P(O_{1:T} \mid \lambda)}$$
$$\rightarrow P(S_T = H_i \mid O_{1:T},\lambda) = \frac{\gamma_T(i)}{P(O_{1:T} \mid \lambda)}.$$
$$(17)$$

Probability of being at each health state at time $T$ can be calculated using (15) and (17). It is worth mentioning that the denominator in (17) is not required to be calculated. It is a normalizing factor that can be calculated after finding $\gamma_T(i)$ for all the health states as follows,

$$\sum_{i=1}^{m} P(S_T = H_i \mid O_{1:T},\lambda)$$
$$= 1 \rightarrow \frac{1}{P(O_{1:T} \mid \lambda)}\sum_{i=1}^{m} \gamma_T(i) \qquad (18)$$
$$= 1 \Rightarrow P(O_{1:T} \mid \lambda) = \sum_{i=1}^{m} \gamma_T(i).$$

Consequently, based on (15) and (18), (17) can be rewritten as

$$P(S_T = H_i \mid O_{1:T},\lambda) = \frac{\gamma_T(i)}{\displaystyle\sum_{i=1}^{m} \gamma_T(i)}. \qquad (19)$$

Finally, the continuous output of the model, $C_T$ that corresponds to the expected amount of tool wear at the current time step $T$, can be calculated based on (19) as follows,

$$C_T = \sum_{i=1}^{m} P(S_T = H_i \mid O_{1:T}) \times H_i. \qquad (20)$$

Prognosis is the task of predicting the future health state at time $t'$ ($t' > T$) while the observation data is only available up to the current time $T$. Similar to the diagnostics case, prognostics can be written in a probabilistic manner as

$$P(S_{t'} = H_i \mid O_{1:T},\lambda) = \frac{P(O_{1:T},S_{t'} = H_i \mid \lambda)}{P(O_{1:T} \mid \lambda)}$$
$$= \frac{\displaystyle\sum_{k=1}^{d_{max}} P(O_{1:T},(S_{t'},\tau_{t'}) = (H_i,k))}{P(O_{1:T})}.$$
$$(21)$$

Using $\xi_t(i,k)$ defined in (16), (21) can be rewritten and calculated as

$$P(S_{t'} = H_i \mid O_{1:T}, \lambda) = \frac{\sum_{k=1}^{d_{max}} \xi_{t'}(i, k)}{P(O_{1:T})}$$

$$= \frac{\sum_{k=1}^{d_{max}} \xi_{t'}(i, k)}{\sum_{i=1}^{m} \sum_{k=1}^{d_{max}} \xi_{t'}(i, k)} \tag{22}$$

Similar to (18), $P(O_{1:T})$ is a normalizing factor that is replaced by $\sum_{i=1}^{m} \sum_{k=1}^{d_{max}} \xi_{t'}(i, k)$.

After computing $P(S_{t'} = H_i \mid O_{1:T}, \lambda)$, the continuous output $C_{t'}(t' > T)$ of the model, which corresponds to the expected amount of tool wear at time step $t'$ in future, can be calculated as follows,

$$C_{t'} = \sum_{i=1}^{m} P(S_{t'} = H_i \mid O_{1:T}) \times H_i. \tag{23}$$

## DATASET AND FEATURES

The dataset is obtained through real-time sensing on a milling machine which consists of 7 signals (three force signals, three vibration signals and Acoustic Emission). The experimental data comprises cutting process of 6 cutters which are *07BX1, 09BX3, 18SC3, 31PN4, 33PN6* and *34PT1*. The cutters are different with one another by their cutter geometry and coating but they are all *6mm* Alignment-Tool carbide ball-nose end with three flutes.

In all the cutting processes, *Inconel 718*, which is used in Jet engines, is used as the work-piece material. During the cutting process, the upper face of the material is cut with horizontal lines from the top edge to the bottom edge. After 320 cutting times, the cutter will start again from the top edge of the material for another 320 cuts. The cutting face is 112.5*mm* wide and 78*mm* high.

Therefore one cutter will travel for $1125 \times 320$ = 36000*mm* = 36*m*. The cuts are 0.25*mm* deep and the duration for one cut is around 4 seconds.

The running condition for all the experiments is set up as follows,

- Spindle speed of 10360 rounds per minute (rpm) - which is considered as high speed,
- Feed rate of 1.555*m/min*.

In this section, the extracted features (i.e. statistical features and wavelet features) from the acquired signals are introduced. Then, Fisher's discriminant ratio (FDR) is applied to find the most discriminant features out of 482 extracted features (48 statistical features and 434 average energy of wavelet coefficients). Constructing the prediction models using only the selected features prevents unnecessary complexity in the models and consequently improves the prediction results. In addition, requirement to extract only the selected features from the newly conducted experiments greatly reduces the feature extraction computation costs for online applications.

## Statistical Features

16 statistical features are extracted from force signals in each direction (X, Y and Z) which makes the total of 48 statistical features for each experiment. A list of these features is shown in Table 1.

## Wavelet Features

Signals that are captured using the sensors mounted on the milling machine have non-stationary characteristic, hence it would be a wise decision to use wavelet or multi-resolution approaches to extract features from these signals. In this work, Daubechies wavelet 8 (D8) is applied to three force signals and discrete Meyer wavelet is used for three vibrations and AE signals, all wavelets with 5 decomposition levels (Mallat, 2008). Thus, 62 ($2^1 + 2^2 + 2^3 + 2^4 + 2^5 = 62$) coefficient vectors are acquired for each signal, which sum up to 434 for all the 7 signals. The average energy of these coefficient vectors are used as the 434 extracted wavelet features. According to (Zhu, et al., 2008), the average energy can be written as

$$E_j = \frac{1}{N_j} \sum_{c=1}^{N_j} [d_{j,c}^n(t)]^2 ,$$

where $j$ is the scale, $d_{j,c}^n(t)$ is the wavelet packet coefficients of the signal, and $N_j$ is the number of coefficients at each scale and $t$ is the discrete time.

## Feature Selection

The idea of feature selection in a classification domain is to find a proper subset of features that explicitly discriminate the classes based on the training set. In order to do so those features must have similar values for the samples in one class and distinct values for the samples from two different classes. One of the ways that this idea has been formulated is called Fisher's Discriminant Ratio (FDR) (Duda, Hart, & Stork, 2001). FDR is a metric that shows how discriminative a feature is. It is a ratio of scatter between ($SB$) and the scatter within ($SW$). A modified version of FDR (Zhu, et al., 2008) is as follows,

$$FDR(f_i) = \frac{SB^{f_i}}{SW^{f_i}} = \frac{\sum_{k=1}^{K} \sum_{j=1}^{K} (\mu_k^{f_i} - \mu_j^{f_i})^2}{\sum_{k=1}^{K} SW_k^{f_i}} ,$$

(24)

where $K$ is the number of classes, $f_i$ is the $i$th element (feature) in the observation vector $O = [f_1 f_2 \cdots f_\chi]^T$, $\mu_k^{f_i}$ is the mean value of $f_i$ in the $k$th class and $SW_k^{f_i}$ is the scatter within the $k$th class measured for $f_i$. In order to use the FDR

criterion for feature selection in the continuous TCM, the output space must be discretized into several clusters, and samples must be assigned to those clusters (classes) based on their original corresponding outputs. Two approaches can be taken for this task, *hard-discretization* and *soft-discretization (Ebert-Uphoff, 2009)*. Hard-discretization means that after discretization each sample will be only assigned to one of the clusters (classes) and it cannot have multiple memberships in different classes. However, in the soft-discretization, samples can be members of different clusters (classes) with various degrees of membership. In this work, soft-discretization is implemented using Gaussian mixture model (GMM). GMM tries to model the given data which in this case is the output set (tool wear) by a mixture of Gaussian functions. The parameters of these Gaussian functions will be estimated by *expectation-maximization* technique. To read more details on GMM see (Bilmes, 1997; Bouman, Shapiro, Cook, Atkins, & Cheng, 1998; Duda, et al., 2001).

Similar to any clustering problem, one of the main issues while using GMM is to find a proper number of Gaussian functions that fits the data. Therefore, various numbers of Gaussian functions are explored from 1 to 10 and then the model with minimum Bayesian Information Criterion (BIC) is chosen as the appropriate GMM for clustering the output set. BIC is a measure to assess goodness of fit in an estimated statistical model taking into account the complexity of the model using a penalizing term to penalize the complex models that may overfit the data (Box, Jenkins, & Reinsel, 1994). Exploration results are depicted in Figure 3. Here, the data collected from three of the cutters which are *07BX1*, *31PN4* and *34PT1* are chosen to be the training set. Hence the feature selection process is performed only on the data from these three cutters.

From Figure 3, it can be understood that the minimum of BIC is achieved by the model with 3 Gaussian functions. Thus, the number of Gaussian functions to be used is set to 3. After applying

*Figure 3. Bayesian information criterion for GMMs with various number of mixtures*

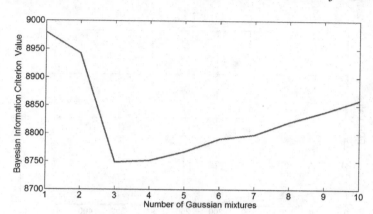

the GMM to the output set, the posterior probability can be computed for every sample in the training set. $p(m_k \mid y_t^j)$ is the posterior probability indicating the actual tool wear at time $t$ in the $j$th experiment, $y_t^j$, is generated by the $k$th Gaussian function, $m_k$. In order to apply the soft-discretization technique, all of the 3 posterior probabilities available for each sample, are used as its membership function in a fuzzy manner.

In soft-discretization approach after finding the posterior probabilities, $\mu_k^{f_i}$ and $SW_k^{f_i}$ can be calculated in a weighted form as follows,

$$\mu_k^{f_i} = \frac{\text{weighted sum of } i\text{th feature in } k\text{th cluster}}{\text{effective cardinality of } i\text{th cluster}}$$

$$= \frac{\sum\limits_{j=1}^{N}\sum\limits_{t=1}^{T_j} p(m_k \mid y_t^j) f_i^j(t)}{\sum\limits_{j=1}^{N}\sum\limits_{t=1}^{T_j} p(m_k \mid y_t^j)},$$

$$SW_k^{f_i} = \frac{\sum\limits_{j=1}^{N}\sum\limits_{t=1}^{T_j} [p(m_k \mid y_t^j) f_i^j(t) - \mu_k^{f_i})]^2}{\sum\limits_{j=1}^{N}\sum\limits_{t=1}^{T_j} p(m_k \mid y_t^j)},$$

where $f_i^j(t)$ is the value of the $i$th feature at time $t$ in the $j$th experimental sequence. The calculated means and scatter within measures can be used in (24) to find FDR for each feature. Since the FDR value shows how discriminant each feature is, the corresponding FDR value of each feature can be used to rank all the features. Figure 4 shows the FDR values computed using the given approach. The FDR values are sorted in a descending manner. As it can be understood from Figure 4, 38 seem to be an appropriate number of features to choose out of 482, since there is a knee at that point in the curve. Thus, 38 features with the highest FDR value are adopted to be used in the prediction model.

Table 2 shows shares of extracted features from each signal in the set of selected features and Table 3 lists shares of various wavelet levels of each signal in the set of selected features.

As it can be understood from both Table 2 and Table 3, features extracted from the force signals are majorly selected as the most discriminant features. On the other hand, none of the features extracted from acoustic emission are selected which suggests that this signal may not be useful for diagnostics and prognostics in milling machines comparing to the force signal.

*Figure 4. Corresponding FDR values of the features after sorting all the features in a descending manner*

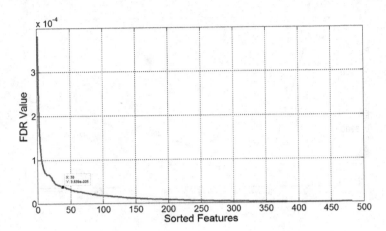

## DIAGNOSTICS AND PROGNOSTICS RESULTS

In this section, performance of the HMM and HSMM-based approaches are compared in diagnostics and prognostics of the cutter's wearing metric in a milling machine. Their performance is compared in three cases. Case I, can be regarded as a cross-validation phase conducted to identify an appropriate number of hidden state values for both HMM and HSMM-based approaches. In Case II, which is the diagnosis task, performance of HMM and HSMM are compared based on *mean square error* (MSE). Case III compares the ability of HMM and HSMM-based approaches in prognostics. A short description of the three cases is as follows,

- **Case 1:** Cross-validation, this task can be done in two modes, i.e. *leave one flute out* and *leave one experiment out.*
- **Case 2:** Testing diagnosis ability, testing the models for diagnostics on the experiments that are excluded from the training set.
- **Case 3:** Testing prognosis ability, using the models for prognostics purposes by unrolling the model over the time horizon (this

task can be done with different prediction horizons).

It is noteworthy that in all three cases, the collected data from the three cutters *07BX1, 31PN4* and *34PT1* which was used for feature selection process is regarded as the training set.

### Case 1: Cross-Validation

In this case, two modes can be considered for cross-validation i.e. *leave one flute out* and *leave one experiment out.* Here, since the training set is limited to only 3 experiments (*07BX1, 31PN4* and *34PT1*), *leave one flute out* is conducted. Both models are trained on two flutes out of three flutes, in each experiment of the training set and then tested on the excluded flute. In the cross-validation phase, various values for hyper-parameter of the models i.e. number of hidden state values (in HMM and HSMM) are explored. Figure 5 depicts the cross-validation error of the excluded flute in terms of mean square error using the HMM and HSMM-based approaches with different number of possible hidden state values.

From Figure 5, it can be understood that 14 may be selected as an appropriate number of health states in both models on this dataset. Hence, the

*Figure 5. Cross-validation error rate in both HMM and HSMM-based approaches with various number of hidden state values (health states)*

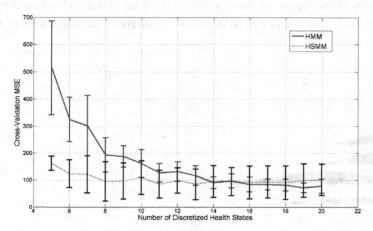

number of hidden state values (health states) of both models is set to 14 in the succeeding cases. In addition, Figure 5 shows that HSMM-based approach has generally outperformed HMM-based approach in terms of average error rate in cross-validation case with different number of hidden state values.

## Case 2: Diagnostics

This case is to compare diagnosis accuracy of the two approaches on the testing set (the data collected from the three cutters *09BX3*, *18SC3* and *33PN6*), which are excluded from the training set data collected from cutters *07BX1*, *31PN4* and *34PT1*. In this simulated experiment, at each time step (cut), the data from the beginning of the experiment up to the simulated current time step is given to both models in order to predict the tool wear of the three flutes at that time step. Number of health states is set to 14 in both models as suggested from the cross-validation results in case I. In this work, since the desired accuracy in the quality of the ultimate work-piece depends on the maximum tool wear of the three flutes, the maximum wearing value of the three flutes at each time step (cut) is regarded as the desired outcome

at each time step. Table 4 shows the diagnosis error of the three approaches on the testing set.

All the parameters computed during the training phase of HSMM-based approach are depicted in Figure 6. For visualization simplicity the covariance matrices are assumed to be diagonal in this figure.

## Case 3: Prognostics

In this case, both models are unrolled over the future time steps for prognostics. Different prediction horizons are studied and the corresponding prediction accuracies are provided in Table V.

From Table 5, it can be seen that the prediction error rate gets larger as the prediction horizon increases. It is also noteworthy, that the prediction error rate of the HSMM-based approach with prediction horizon of 9 is less than the acquired average prognosis error rate from the HMM-based approach with prediction horizon of 1. This fact indicates that HSMM-based approach is more efficient in capturing the temporal information, which results into smaller error rates in both diagnostics and prognostics.

It is noteworthy that the error indicator (MSE) used in this section indicates predictions deviation

*Figure 6. Computed parameters of the HSMM-based approach in case II. The Diagrams on the left depict all the duration distributions of various health states with truncated Gaussian distributions. On the right, the diagrams show the mean and variance of all the observed selected features within each health state. The covariance matrices are assumed to be diagonal for simplicity in visualization.*

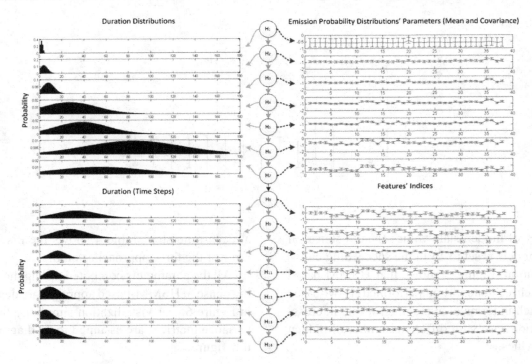

from the exact actual wearing values. However, it does not recognize the difference between an under- and over-estimation and can be regarded as a symmetric loss function (Hastie, Tibshirani, & Friedman, 2009). In the next section, an asymmetric loss function will be given and the performance of the approaches is compared based on their ability in minimizing the total loss.

## Asymmetric Loss Function

All the error rates given in the previous section are based on square error that can be regarded as a symmetric loss function, not distinguishing between the over- and under-estimations. But in reality in most of tool condition monitoring systems, it is preferred to be pessimistic rather than optimistic. For example, in the case of cutter condition monitoring in a CNC-milling machine, it is preferred

to change the cutter sooner rather than using the cutter longer and having a defected work-piece as a result. Thus, it is better to over-estimate the tool wear of the cutter rather than under-estimate. This preference can be formulated using an asymmetric loss function. In this section, an arbitrary asymmetric function is given as follows,

$$\text{err } (10^{-3}\text{mm})$$
$$=\text{estimated tool wear-actual tool wear,}$$
$$loss(err) = \begin{cases} e^{(-err/8)} - 1 & err < 0 \\ e^{(err/10)} - 1 & err \geq 0 \end{cases}$$

$$(25)$$

Now, the performance of the two approaches can be compared based on the given asymmetric loss function. One of the main advantages of the proposed HSMM-based approach, which is its duration flexibility, can be exploited in this case. Since the loss function is asymmetric and the HSMM duration distribution is flexible it can be made asymmetric to incorporate the asymmetricity of the loss function in the predictive model in order to improve the performance. In the example of CNC-milling machine given here, the values of estimated and actual tool wear are recorded and used in scale of $10^{-3}mm$.

In the previous section, as we had a symmetric square error loss function, we assumed for simplicity that duration distributions may be modeled with Gaussian distributions. However, when we want to use an asymmetric loss function, consequently we would like different sides of the peak in the Gaussian functions to be also asymmetric corresponding to the difference indicated in the loss function.

Therefore, an asymmetric version of Gaussian function (Genton, M. G., 2004) which has a parameter called asymmetry to control the amount of asymmetricity in the function is adopted in this section. The newly defined duration distribution $P(d_i)$ is,

$$P(d_i = x) = \begin{cases} \dfrac{1}{\sqrt{2\pi}\sigma_i} \exp(-\dfrac{q^2}{2\sigma_{L_i}^2}) & q < 0 \\ \dfrac{1}{\sqrt{2\pi}\sigma_i} \exp(-\dfrac{q^2}{2\sigma_{R_i}^2}) & q \geq 0 \end{cases},$$

$$q = x - \overline{d}_i \ , \sigma_{L_i} = \sigma_i(1 + A) \ , \sigma_{R_i} = \sigma_i(1 - A)$$

(26)

where $A$ is the asymmetry factor, $\overline{d}_i$ is the duration at the peak in the $i$th duration distribution function, $\sigma_i$ is the average root mean square width, $\sigma_{L_i}$ and $\sigma_{R_i}$ are left and right width (deviation), respectively.

In this chapter, since the right side of the Gaussian distribution corresponds to the longer probable durations which may cause under-estimation, the asymmetry parameter should favor the left side

of the peak. Thus the asymmetry factor may vary between (0, 1). As indicated in (26), once the asymmetry parameter is set, then the right and left width can be computed based on asymmetry and the average width.

Without loss of generality, the duration samples from the training set can be sorted based on their values in an ascending order. Then, the log-likelihood of the $i$th duration distribution, $L_{d_i}$, can be written as

$$L_{d_i} = -\frac{n_{d_i}}{2} log(2\pi) - n_{d_i} log(\sigma_i)$$
$$- \sum_{j=1}^{n_{L_i}} \frac{(x_j - \overline{d}_i)^2}{2\sigma_{L_i}^2} - \sum_{j=n_{L_i}+1}^{n_{d_i}} \frac{(x_j - \overline{d}_i)^2}{2\sigma_{R_i}^2} \quad (27)$$

where $x_j$ is the $j$th duration sample, $n_{d_i}$ is the total number of duration samples for the $i$th health state and

$$n_{L_i} = n_{d_i} \times \text{General left side Percentage (GLSP)}$$

is the number of samples assumed to be generated from the left side of the peak in the $i$th duration distribution. GLSP is an auxiliary hyper-parameter of the model that can be found through cross-validation and it indicates the general percentage of the samples in the duration distributions generated from the left side of the peaks in the duration distributions. Figure 7 depicts the effects of asymmetry and GLSP factors on the estimated asymmetric Gaussian distributions.

Assuming to find proper values for $A$ and GLSP as the hyper-parameters of the duration distributions based on cross-validation phase. $\overline{d}_i$ and $\sigma_i$ may be estimated using maximum likelihood method. Substituting $\sigma_{L_i}$ and $\sigma_{R_i}$ in (27) with their parametric form in (26), (27) can be rewritten as

*Figure 7. a) Effect of asymmetry factor on the estimated asymmetric Gaussian distribution while GLSP factor is arbitrarily set to 30%. b) Effect of GLSP factor on the estimated asymmetric Gaussian distribution while asymmetry factor is arbitrarily set to 0.3.*

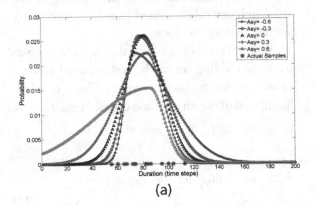

(a)

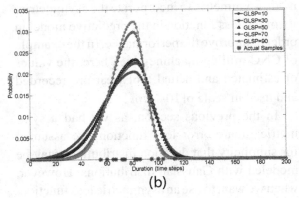

(b)

$$L_{d_i} = -\frac{n_{d_i}}{2} \log(2\pi) - n_{d_i} \log(\sigma_i)$$

$$- \frac{\sum_{j=1}^{n_{L_i}} (x_j - \overline{d}_i)^2}{2\sigma_i^2 (1+A)^2} - \frac{\sum_{j=n_{L_i}+1}^{n_{d_i}} (x_j - \overline{d}_i)^2}{2\sigma_i^2 (1-A)^2}$$

(28)

By setting the derivatives of (28) to zero, $\overline{d}_i$ and $\sigma_i$ can be estimated as follows,

$$\frac{\partial L_{d_i}}{\partial \overline{d}_i} = 0 \rightarrow \frac{\sum_{j=1}^{n_{L_i}} (x_j - \overline{d}_i)}{\sigma_i^2 (1+A)^2} + \frac{\sum_{j=n_{L_i}+1}^{n_{d_i}} (x_j - \overline{d}_i)}{\sigma_i^2 (1-A)^2} = 0$$

$$\rightarrow (1-A) \sum_{j=1}^{n_{L_i}} (x_j - \overline{d}_i) + (1+A) \sum_{j=n_{L_i}+1}^{n_{d_i}} (x_j - \overline{d}_i) = 0$$

$$\rightarrow \sum_{j=1}^{n_{d_i}} x_j - n_{d_i} \overline{d}_i + A(n_{d_i} - 2n_{L_i})\overline{d}_i + A[\sum_{j=n_{L_i}+1}^{n_{d_i}} x_j - \sum_{j=1}^{n_{L_i}} x_j]$$

$$= 0 \Rightarrow \overline{d}_i = \frac{S + A(S - 2S_L)}{n_{d_i} + A(n_{d_i} - 2n_{L_i})}, S = \sum_{j=1}^{n_{d_i}} x_j, S_L = \sum_{j=1}^{n_{L_i}} x_j,$$

(29)

$$\frac{\partial L_{d_i}}{\partial \sigma_i} = 0 \rightarrow -\frac{n_{d_i}}{\sigma_i} + \frac{\sum_{j=1}^{n_{L_i}} (x_j - \overline{d}_i)^2}{\sigma_i^3 (1+A)^2}$$

$$+ \frac{\sum_{j=n_{L_i}+1}^{n_{d_i}} (x_j - \overline{d}_i)^2}{\sigma_i^3 (1-A)^2} = 0 \rightarrow -\frac{n_{d_i}}{\sigma_i}$$

$$+ \frac{1}{\sigma_i^3}[\frac{\sum_{j=1}^{n_{L_i}} (x_j - \overline{d}_i)^2}{(1+A)^2} + \frac{\sum_{j=n_{L_i}+1}^{n_{d_i}} (x_j - \overline{d}_i)^2}{(1-A)^2}] = 0$$

$$\Rightarrow \sigma_i = \sqrt{\frac{1}{n_{d_i}}[\frac{\sum_{j=1}^{n_{L_i}} (x_j - \overline{d}_i)^2}{(1+A)^2} + \frac{\sum_{j=n_{L_i}+1}^{n_{d_i}} (x_j - \overline{d}_i)^2}{(1-A)^2}]}$$

(30)

Now, diagnostics accuracy of the two approaches is re-examined based on the asymmetric loss function given in (25) to show the applicability of the HSMM-based approach while having an asymmetric loss function. But, before proceeding to the diagnostics part, the appropriate values for the two newly added hyper-parameters must be adopted through cross-validation.

## Case 1: Asymmetric Cross-Validation

In this section, the number of hidden state values is set to 14 so that the results may be comparable with the models used in the previous sections. The hyper-parameters that are considered to be explored in this case are the asymmetry factor *A* and GLSP. Figure 8 depicts the corresponding median of the total loss on the cross-validated sets for every value taken by *A* and GLSP.

As can be seen in Figure 8, A= 0.7 and GLSP= 90% leads to the least median for the cross-validation total loss. Therefore these values are adopted for the hyper-parameters of the HSMM-based approach in diagnostics.

## Case 2: Asymmetric Diagnostics

Here, diagnostics accuracy of the HMM and HSMM-based approaches are compared based on the given asymmetric loss function in (23). Similar to diagnostics case in the previous section, the testing set is the data collected from the three cutters *09BX3*, *18SC3* and *33PN6*. All the models are trained based on the data collected from cutters *07BX1*, *31PN4* and *34PT1*.

In this simulated experiment, at each time step (cut), the data from the beginning of the experiment up to the current time step is given to both models in order to predict the tool wear of the three flutes at that time step. Number of health states is set to 14 in all models as suggested in the previous section. In this case, similar to the

*Figure 8. The median of cross-validated total loss for every value taken by A and GLSP*

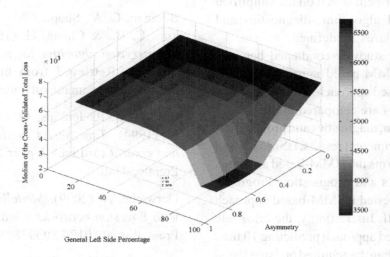

*Table 6. Prediction error rate in case II using HMM and HSMM-based approaches in terms of total loss for the given loss function in (25)*

| Approach | Total loss | | | |
|---|---|---|---|---|
| | *09BX3* | *18SC3* | *33PN6* | **Average Total Loss** |
| **Asymmetric HSMM-based** | **2147** | **705** | **1672** | **1508 ± 734** |
| **Symmetric HSMM-based** | 2719 | 843 | 1794 | 1785 ± 938 |
| **HMM-based** | 4826 | 883 | 2289 | 2666 ± 1998 |

previous section, the maximum wearing value of all 3 flutes at each time step (cut) within each experiment is regarded as the desired outcome. As an indicator to show how much the asymmetric function adopted for the duration distributions in the HSMM-based approach has improved the performance, the results of HSMM with symmetric Gaussian function is also included along with the HMM-based approach. Table 6 shows the diagnosis accuracies of the approaches based on the total loss on the testing set.

## CONCLUSION

In this chapter, a hidden semi-Markov model based approach was introduced for continuous diagnosis and prognosis. Also, a computationally efficient version of forward-backward algorithm for application of HSMM in continuous health condition monitoring is described. Based on the simplified forward-backward algorithm, diagnostics and prognostics procedures are defined.

A comparative study is conducted between the suggested HSMM-based approach and the existing HMM-based approach. Performances of the two approaches are compared in three cases i.e. cross-validation, diagnostics and prognostics. Based on the experimental results, HSMM-based approach outperforms the HMM-based approach in both diagnostics and prognostics. Prognosis ability of the suggested HSMM-based approach is tested in case III. Interestingly, the error rate of the HSMM-based approach predicting 10 time steps ahead is less than the acquired one step ahead average prognosis error rate from the HMM-based approach which indicates how powerful HSMM is compared to HMM in capturing the underlying temporal information.

In the last section, applicability of the asymmetric Gaussian functions as the duration distributions in the proposed HSMM-based approach is explored. Using the asymmetric Gaussian function, the asymmetric loss functions can be incorporated into the approach in order to recognize the preference between the under and over estimations. Experimental results also indicate that using asymmetric Gaussian functions as the duration distribution functions improve the performance substantially while using an asymmetric loss function.

## REFERENCES

Atlas, L., Ostendorf, M., & Bernard, G. D. (2000). *Hidden Markov models for monitoring machining tool-wear.* Paper presented at the IEEE International Conference on Acoustics, Speech, and Signal Processing.

Bilmes, J. (1997). *A gentle tutorial on the EM algorithm and its application to parameter estimation for Gaussian mixture and hidden Markov models.* Berkeley: University of California.

Bouman, C. A., Shapiro, M., Cook, G. W., Atkins, C. B., & Cheng, H. (1998). *Cluster: An unsupervised algorithm for modeling Gaussian mixtures.* Retrieved from http://dynamo.ecn.purdue.edu/~bouman/software/cluster

Box, G. E. P., Jenkins, G. M., & Reinsel, G. C. (1994). *Time series analysis: Forecasting and control* (3rd ed.). Upper Saddle River, NJ: Prentice-Hall.

Darwiche, A. (2009). *Modeling and reasoning with Bayesian networks.* Cambridge University Press. doi:10.1017/CBO9780511811357

Dong, M., & He, D. (2007). A segmental hidden semi-Markov model(HSMM)-based diagnostics and prognostics framework and methodology. *Mechanical Systems and Signal Processing, 21,* 2248–2266. doi:10.1016/j.ymssp.2006.10.001

Duda, R. O., Hart, P. E., & Stork, D. G. (Eds.). (2001). *Pattern classification* (2nd ed.). Wiley-Interscience.

Ebert-Uphoff, I. (2009). *A probability-based approach to soft discretization for Bayesian networks*. Georgia Institute of Technology.

Geramifard, O., Xu, J. X., Zhou, J. H., & Li, X. (2010, 7-10 Dec. 2010). *Continuous health assessment using a single hidden Markov model.* Paper presented at the Control Automation Robotics & Vision (ICARCV), 2010 11th International Conference on.

Hastie, T., Tibshirani, R., & Friedman, J. (2009). *The elements of statistical learning: Data mining, inference, and prediction* (2nd ed.). Springer. doi:10.1007/BF02985802

Jie, S., Hong, G. S., Rahman, M., & Wong, Y. S. (2002). *Feature extraction and selection in tool condition monitoring system.* Paper presented at the 15th Australian Joint Conference on Artificial Intelligence: Advances in Artificial Intelligence.

Jun-Hong, Z., Chee Khiang, P., Lewis, F. L., & Zhao-Wei, Z. (2009). Intelligent diagnosis and prognosis of tool wear using dominant feature identification. *IEEE Transactions on Industrial Informatics*, *5*(4), 454–464. doi:10.1109/TII.2009.2023318

Mallat, S. (2008). *A wavelet tour of signal processing, The sparse way* (3rd ed.). Academic Press.

Neapolitan, R. (2003). *Learning Bayesian networks*. Prentice Hall.

Rao, B. K. N. (1996). *Handbook of condition monitoring*. Elsevier Advanced Technology.

Russell, S., & Norvig, P. (2009). *Artificial intelligence: A modern approach*. Prentice Hall.

Sansom, J., & Thomson, P. (2001). Fitting hidden semi-Markov models to breakpoint rainfall data. *Journal of Applied Probability*, *38*, 142–157. doi:10.1239/jap/1085496598

Yu, S. Z. (2010). Hidden semi Markov models. *Artificial Intelligence*, *174*(2), 215–243. doi:10.1016/j.artint.2009.11.011

Yu, S. Z., & Kobayashi, H. (2003, January). An efficient forward-backward algorithm for an explicit duration hidden Markov model. *IEEE Signal Processing Letters*, *10*(1), 11–14. doi:10.1109/LSP.2002.806705

Zhu, K. P., Hong, G. S., & Wong, Y. S. (2008). A comparative study of feature selection for hidden Markov model-based micro-milling tool wear monitoring. *Mining Science and Technology*, *12*(3), 348–369.

## KEY TERMS AND DEFINITIONS

**Bayesian Information Criterion:** A measure to assess goodness of fit in an estimated statistical model taking into account the complexity of the model using a penalizing term to penalize the complex models that may overfit the data.

**Condition Based Maintenance:** Doing the maintenance whenever it is needed considering the health condition of the system.

**Diagnostics:** Task of predicting the health state at the current time given all the observations from the beginning up to the current time.

**Feature Extraction and Selection:** Transforming data sequences into sets of features (extraction) and adopting a subset of features based on their explanatory capability to form a learning model.

**Gaussian Mixture Model:** A parametric density function which model the given data by a mixture of Gaussian functions.

**Hard-Discretization:** A discretization approach in which, each sample will be only assigned to one of the clusters (classes) and thus does not support multiple memberships.

**Prognostics:** Task of predicting the future health states while the observation data is only available up to the current time.

**Soft-Discretization:** A discretization approach in which, samples may belong to multiple clusters (classes) with various degrees of membership.

**Temporal Probabilistic Models:** A group of probabilistic graphical models which can be unrolled over time. They take into account the dependencies among the states as well as observations over the time, and use the conditional independencies to make inferences.

**Tool Condition Monitoring:** Monitoring the health status of a tool through time.

# Section 7
# Integration of Control and Prognostics

# Chapter 12
# Combining Health Monitoring and Control

**Teresa Escobet**
*Universitat Politècnica de Catalunya, Spain*

**Joseba Quevedo**
*Universitat Politècnica de Catalunya, Spain*

**Vicenç Puig**
*Universitat Politècnica de Catalunya, Spain*

**Fatiha Nejjari**
*Universitat Politècnica de Catalunya, Spain*

## ABSTRACT

*This chapter proposes the combination of system health monitoring with control and prognosis creating a new paradigm, the health-aware control (HAC) of systems. In this paradigm, the information provided by the prognosis module about the component system health should allow the modification of the controller such that the control objectives will consider the system's health. In this way, the control actions will be generated to fulfill the control objectives, and, at the same time, to extend the life of the system components. HAC control, contrarily to fault-tolerant control (FTC), adjusts the controller even when the system is still in a non-faulty situation. The prognosis module, with the main feature system characteristics provided by condition monitoring, will estimate on-line the component aging for the specific operating conditions. In the non-faulty situation, the control efforts are distributed to the system based on the proposed health indicator. An example is used throughout the chapter to illustrate the ideas and concepts introduced.*

## INTRODUCTION

The safe and reliable operation of technological systems (cars, planes, trains,...) and processes (energy, gas or water networks, chemical factories,...) is of great significance for the protection of human life and health, the environment, and the invested economic value. The correct operation of those systems has an important impact also on production cost and product quality in manufacturing.

DOI: 10.4018/978-1-4666-2095-7.ch012

The emergence of new sensors/actuators in the complex technological systems and processes made possible the development of several sophisticated monitoring and control applications where a large amount of real-time data about the monitored environment is collected and processed to activate the appropriate actuators and to achieve the desired control objectives. However, the probability of failure of some of these components will increase exponentially with their number. Thus, the safe and reliable operation must take into account mechanisms for early detection of faults to avoid performance degradation and damage to the machinery or human life.

Maintaining the health of a complex system is a difficult task that requires the in-depth analysis of the target system, principles involved and their applicability and implementation strategies. According to Ofsthun (2002), a System Health Monitoring (SHM) module implemented in the target system will be able to:

- Diagnose the root cause of a system failure,
- Furnish data and recommend solutions in real time,
- Provide prognostic capability to identify potential issues before they become critical, and
- Capture and retain knowledge for predictive maintenance and new designs.

To this aim, a SHM system consists of instrumentation components, a fault detection, isolation and response module, diagnostic and prognostic software, as well as processes and procedures responsible for information gathering about system's health and corresponding decision-making.

The scope of this chapter tries to combine two concepts. On one hand, the new concept of System Health Management (SHM) that integrates the tasks of diagnostic and prognostic modules, as well as processes and procedures responsible for information gathering about the system health enabling to make the right decisions on emergency actions and repairs (Jennions, 2010), is presented. On the other hand, the modern concept of Reliable Control (RC) that tries to design control strategies to allow a safe and a reliable operation in spite of faulty situations (Tang et al, 2008), is introduced. As a result of this combination a new paradigm appears: the Health-Aware Control (HAC) of the systems. In this paradigm, the information provided by the prognosis module about the component system health should allow the modification of the controller such that the system health is considered in the control objectives. In this way, the control actions will be generated to fulfill these objectives and, at the same time, to extend the life of the system components. HAC control contrarily to FTC adjusts the controller even when the system is still in non-faulty situation. The prognosis module will estimate on-line the component aging for the specific operating conditions. In the non-faulty situation, the control efforts are distributed to the system based on the proposed health indicator. An example will be used throughout the chapter to illustrate the ideas and concepts introduced.

The structure of the chapter is as following: first a background reviews the state of the art of the main concepts introduced in the chapter. Then, in the next section the integration of prognosis and control is discussed and the new concept of health aware control is introduced. The chapter concludes by presenting the conclusions and the main lines of future research.

## BACKGROUND

Currently in the field of security, availability and reliability of automated systems and processes we can find two different working strategies. On the one hand, there is the so-called Prognosis & Health Management (PHM) which originated in system engineering and considers aspects such as quality, reliability and maintenance. On the other hand, we have the Fault Tolerant Control (FTC) related to control engineering and that concerns issues such as safety and performance specifications. These strategies work with the fact that system components can fail either by an early damage or by accidental/fortuitous causes, respectively.

PHM technologies essentially take observations of the system and provide indications of abnormalities and, in the best cases, make predictions of future failures.

Over the last decade, research has been conducted in PHM of automotive industry, aeronautics defense and space programs and heavy industry (Mahulkar, 2009), (Tang, 2008), as a means to provide advance warnings of failure, enable forecasted maintenance, improve system qualification, extend system life, and diagnose intermittent failures.

The major part of the papers that include both control and prognosis topics are related with aircraft subsystems. As said in (Mahulkar, 2009), *"new technologies must not only provide maintainers with information regarding the condition of the aircraft subsystems (power plant, avionics, structure, etc), but must also enable operators to effectively conduct missions even when the aircraft is in a degraded state"*. Also, Tang (2008) stated that the development of PHM and FTC systems improve the reliability and survivability of safety-critical aerospace system. But, this idea can be applied to other processes and systems to meet increased safety and performance demands.

## Fault Diagnosis Overview

Fault diagnosis aims at deciding whether a fault has occurred and the time of its occurrence (fault detection), determining the fault location, indicating for example in which component the fault has occurred (fault isolation), and providing a fault identification and estimation by identifying the causes of the fault and estimating its magnitude.

Two different communities (fault detection and isolation (FDI) community in the Automatic Control area and diagnosis (DX) community in

the Artificial Intelligence area) have developed their own methodologies for fault diagnosis, one independently of the other. The FDI community has its roots in the classical theory of systems and automatic control (Gertler, 1998). On the other hand, the DX community has its roots in consistency-based diagnosis developed by Reiter (1987). The type of models used to describe the dynamic process is mainly based on qualitative models described by qualitative differential equations (Kuipers, 1994). Recently, the Bridge community (Biswas, 2004), based on the work of Cordier (2004) has provided researchers of both fields with a common framework for sharing results and techniques.

Using different techniques, both communities have developed several algorithms that combine the measurement model with the process model and obtain a set of analytical redundancy relations (ARR) (Staroswiecki, 2000) (Blanke, 2006) (Krysander, 2008) or potential conflicts (Pulido, 2004), which are defined as relations between known variables. These relations will be used in the fault diagnosis procedure to check the consistency between the observed and the predicted process behavior. Fault detection checking is based on computing, using an ARR, the difference between the predicted value from the model and the real value measured by the sensors. Next, this difference, known as the residual, is compared with a threshold value (zero in the ideal case). While a single residual is sufficient to detect faults, a set (or a vector) of residuals is required for fault isolation (Gertler, 1998). Another important issue in fault diagnosis is the optimal sensor placement problem to guarantee fault detectability and isolability (Bagajewicz, 2000) (Travé-Massuyès, 2004) (Nejjari, 2006) (Rosich, 2007) (Sarrate, 2007).

## PHM Overview

PHM permits the reliability of a system to be evaluated in its actual life-cycle conditions, to determine the advent of a failure and mitigate the system risks. In recent years, PHM has emerged as one of the most promising discipline of technologies and methods with the potential of solving reliability, survivability, and maintainability problems. *Reliability* is defined as the ability to complete a task satisfactorily and the period of time over which that ability is retained; *Maintainability* concerns the need of repairing and the cases in which the repairs can be made; and *Survivability* relates to the likelihood of conducting an operation safety whether or not the task is completed.

The aim of PHM system is to provide users with an integrated view of the health state of the overall system. For that, generally PHM systems incorporate functions of condition monitoring, state assessment, fault or failure diagnostics, failure progression analysis, predictive diagnostics (or prognostics), and maintenance or operational decision support (Sheppard, 2008).

When building a PHM system, three components are necessary for prognostics to be effective: the ability to estimate the current state of the system, the ability to predict future state, and thereby time to fail, and the ability to determine the impact of the assessment on system performance and the need for corrective or mitigating action.

PHM system can be implemented using model-based (Jata et al., 2006), data-driven (Luo et al., 2005), reliability-based (Groer, 2000) or probability-based (Sheppard et al., 2007) methods. The model-based approach takes into account the physical process and the interactions between components in the system. Prognosis of the remaining useful life (RUL) is carried out based on the knowledge of the processes causing degradation and leading to the failure of the system. The data-driven approach uses statistical pattern recognition and machine-learning to detect changes in parameter data, enabling diagnostic and prognostic measures to be calculated. Anomalies and trends or patterns are detected in the data collected to determine the state of the health of a system. The trends are then used to estimate the time-to-failure of the system. Methods such

as regression models, time series analysis, and neural networks are being applied. The reliability-based approach to predicting failure is based on statistical reliability models of component failure. Reliability predictions are used to estimate future failure based on current test results by applying a probability distribution such as the exponential distribution. The probability-based methods such as dynamic Bayesian network (DBN) architectures like hidden Markov models (HMM) and Kalman filters have been suggested as methods for using historical, sequential data to predict future failure.

Mainly, PHM tries to solve two problems. The first consists in determining the time window over which the maintenance must be performed without compromising the system's operational integrity. The second consists in estimating RUL and providing information to the operator whether it is possible to continue operating or an immediate shut-down for maintenance is required. The first shows if it is possible to reduce the needed maintenance, through optimized maintenance intervals, and at the same time avoid unplanned maintenance and associated costs as well as to improve safety and reduce environmental impacts. The second produces advice on how to change operations (speed, load, stress) that can be applied in the next maintenance opportunity as well information about whether the equipment have high probability of safe operation for the planned mission. The concept and framework of PHM have been developed based mainly on maintenance methodologies.

Maintenance can be defined by two categories: reactive or proactive (Ly, 2009). *Reactive maintenance* addresses the process by which a system is operated until failure and then corrective action is taken. *Proactive maintenance* is related with the process where "preventive" actions are performed in the hope of preventing the occurrence of a fault in the operation of the system. The second one is known as "predictive" maintenance or Condition Based Maintenance (CBM). CBM is the base of PHM (Lee, 2011). Related with maintenance, other research lines are emerging such as e-maintenance (Muller, 2008) or self-maintenance (Lee, 2011).

A PHM system will allow condition-based maintenance instead of scheduled and predictive maintenance to enable effective cost versus performance decisions.

The PHM system architecture can be formalized by using the OSA–CBM (Open System Architecture for Condition Based Monitoring) (Thurston, 2001) standards which includes six layers: Data acquisition, Data Manipulation or Signal Processing, Condition Monitoring, Health Assessment, Prognostics Generation and Advisory Generation or Decision Support and Presentation.

## Reliable and Fault-Tolerant Control Overview

Conventional feedback control designs for complex systems may result in unsatisfactory control system performance, or even instability, in the event of faults, even though it may be possible to control the plant using only the surviving inputs and outputs. It is therefore of interest to develop feedback control designs which will guarantee satisfactory closed-loop behavior despite faults. A control system designed to tolerate faults within a pre-established subset of possible faults, while retaining desired control system properties with a certain degree of reliability, will be called a "*reliable*" control system. Several approaches for the design of the reliable controllers have been proposed; however, most of those efforts are focused on linear control systems (Veillette, 1992; 1995) (Vidyasagar, 1985) rather than non-linear

ones. Although, broadly speaking, reliable control is equivalent to fault-tolerant control, there are some differences, as pointed out by Stengel (1991). In particular, reliability deals with the ability to complete a task satisfactorily and within a period of time over which that ability is retained. A reliable control system allows normal completion of tasks after component fault, while fault-tolerant is a broader concept.

According to Patton (1997) and Blanke (2006), a *fault tolerant control system* (FTCS) is a control system that can maintain system performance close to the desirable one and preserves stability conditions not only when the system is in a fault-free case but also in the presence of faulty components in the system, or at least can ensure expected degraded performances that can be accepted as a trade-off.

The general approaches to FTC can be classified into two types: *passive* and *active* (Zhang, 2008). Passive FTC is based on designing control laws to make the closed loop system robust against system uncertainties and some restrictive faults; see for example (Chen, 1999) (Zhou, 2001) (Liao, 2002). In contrast, active FTC systems use the information given by a fault diagnosis module for reconfiguring control actions in order to guarantee the stability and acceptable performance of the global system (Puig, 2001) (Zhang, 2003) (Blanke, 2006). Active FTC systems are also referred to as self-repairing, reconfigurable or self-designing control systems, among others see (Zhang, 2008). Figure 1 and Figure 2 show a representation of each one of these configurations.

*Figure 1. A general structure of passive FTC*

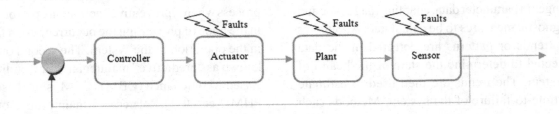

*Figure 2. A general structure of active FTC*

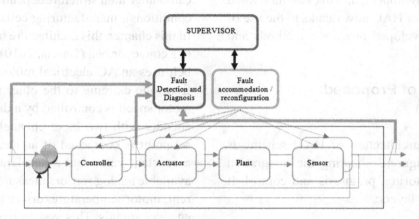

The drawbacks and advantages of passive versus active FTC have been pointed out by Benosman (2010). Passive FTC has as drawbacks the fact that the control is reliable only for the class of faults that have been taken into account in the design of the passive FTC and the performances of the closed loop are not optimized for each fault scenario. The advantage of passive FTC is related to the fact that it is able to avoid the delay due to on-line fault diagnosis and reconfiguration of the controller required in active FTC. In this paper, the author proposes an integrated strategy. First, passive FTC is applied to ensure at least the stability of the faulty system during the time period when the fault detection and diagnosis is estimating the fault. Then, active FTC takes over the passive FTC and, using the estimated faulty model the performances of the faulty system are optimized. This mixed strategy is named as hybrid in (Jiang, 2005) and (Su, 2006).

## CONTROL AND PROGNOSIS INTEGRATION

### Control and Prognosis

FTC is concerned with the control of the system when some component is already in fault. This implies the controller is adapted when the fault is detected, isolated and estimated. These tasks are done on-line by using a fault diagnosis system. However, the use of a prognosis system on-line can be used to adapt the control law even when the system is not in fault situation. In the proposed HAC approach, the information provided by the prognosis module about the component system health should allow modifying the controller accordingly. In this way, the control actions will be generated to fulfill the control objectives and, at the same time, to extend the life of the system components. HAC control, contrarily to FTC, adjusts the controller even when the system is still in non-faulty situation. The prognosis module will estimate on-line the component aging for the specific operating conditions. In the non-faulty situation, the control efforts are distributed to the system based on the proposed health indicator.

As discussed in the background section, seminal ideas about reliable control have already been suggested by (Stengel, 1991) and (Veillette, 1992; 1995). However, since at that time prognosis was still not a mature area, the HAC approach has not been developed. The main reason is that a lot of emphasis was put on diagnosing the fault and reconfiguring/accommodating the control leading to the development of the FTC. However, industry was more concerned about managing appropriately the system health before the fault appears than in applying FTC methods once the fault has already appeared. This has motivated the appearance of HAC now thanks to the use of the recently developed prognosis methods and techniques.

## Description of Proposed Architecture

The proposed architecture of HAC scheme is presented in Figure 3 showing the integration of health monitoring, prognosis and control. It consists of four layers:

**Layer 1:** Data acquisition and pre-processing.
**Layer 2:** Condition monitoring and diagnosis.
**Layer 3:** Prognosis using effect-based approaches.
**Layer 4:** Decision-making: control reconfiguration and/or maintenance actions.

These layers are not autonomous and information is transferred from one to the other. In Figure 3, the main interactions between the different layers are shown. In the following sections, the function of each layer will be explained in detail as well as how it can be implemented.

The proposed approach and architecture can be applied to different kind of systems such as batch systems, hybrid systems (i.e., systems that can change their structure depending on operating conditions), manufacturing cells, among others. In this chapter, this architecture has been applied to a conveyor belt (Garcia, 2010). This conveyor belt uses an AC electrical motor to move a cart from one extreme to the other (Figure 4). The motor speed is controlled by a driver that has an encoder with two input channels. The velocity set-point is composed of an acceleration ramp, a constant value and a deceleration ramp. The available analog sensors measure AC motor current, motor temperature, driver temperature and encoder signals. This system has also four logical sensors located in the belt and in the cart for security purposes and two control signals to move the motor to the right or to the left. The SHM has been implemented in the integrated development environment LabWindows CVI 9.0 working on a Windows platform. The system has been designed using the *multithreading* methodology and the CompactDAQ-9172 system acquisition of National Instruments (Figure 4).

## Data Acquisition and Pre-Processing

The first great challenge of SHM is to ensure the integrity of the data before using them for condition monitoring, diagnostic and prognostic

*Figure 3. SHM architecture proposal*

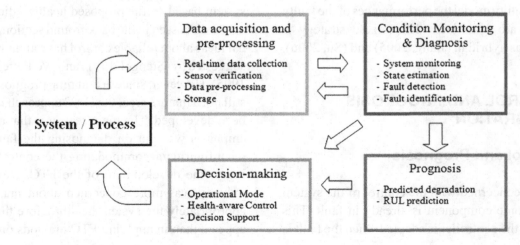

*Figure 4. Experimental test-bed*

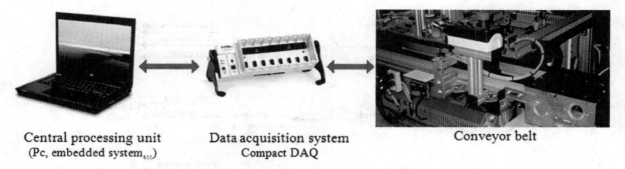

Central processing unit     Data acquisition system     Conveyor belt
(Pc, embedded system,...)       Compact DAQ

purposes. The data generally come from the sensors or intermediate equipment, using a bus of communication, either through a wireless (in the case of remote sensing and measuring) or standards buses. This can cause the data received by the SHM to suffer from a number of problems: non-constant sampling time, missing data, incorrect values and so on. These problems may be caused by communication problems or by sensor faults. Therefore, the first task is to validate the data received.

An important aspect as emphasized and reported in (Palmer, 2007), is that no single pre-processing techniques by itself can effectively handle all the types of data and sensor faults. Sensors are evaluated both individually and with respect to their roles in the monitored system. Taking into account these aspects, the internal architecture of sensor validation system is depicted in Figure 5. Three types of inputs have been considered. The first type corresponds to analog variables with a high bandwidth data and a high speed analog-digital (AD) conversion (or high speed sensor). The second one corresponds to analog variables with low bandwidth data and lower hardware requirements than the first one. And, the third type is logical variables with only two states, which can be mechanical or electrical devices informing about discrete events commanded by the user interface.

As an example, the tasks carried out for each module of the SHM system of the conveyor belt are described in the following:

- **High speed signal pre-processing:** This module works cyclically while the management system allows it. At each sampling time, it receives the data stored in AD card buffer; the data are processed and the cycle starts again. In this application, signal pre-processing includes the following tasks: on-line validation, data reconstruction, digital filtering and feature extraction. It should be remarked that not all the tasks are applied to all the variables. This module has as outputs some filtered and reconstructed data and some estimated data computed with the available measurements. A data quality index is also provided. For example, the encoder data is used to compute the cart position and the speed (Figure 6).
- **Generic signal pre-processing:** This module receives data when it is required by the SHM system. In this case, the tasks are: on-line validation, average value and noise evaluation. This module has as outputs: the average data value, the standard deviation and a quality index.

*Figure 5. Data acquisition and pre-processing*

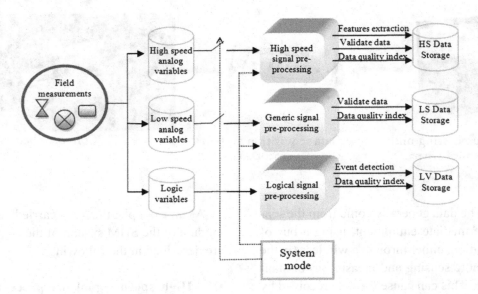

• **Logical signal pre-processing:** The variables provided by this module are always active. At each sampling time, stored data in the card buffer are received and processed. In this case, the pre-processing consists of a feature extraction which allows determining events in the logical variables. As outputs of this module, there are estimated events and, for each event, a quality index.

The on-line validation is done in the two pre-processing modules and allows calculating the quality index. Each variable is evaluated at two levels:

• **Level 1:** It checks whether the data stored in the buffer have the expected number of measurements. The relationship between actual and expected measurements indicates the quality of the data. If there are no data, a communication problem is reported.
• **Level 2:** It verifies if each data is between a maximum and minimum value. Any data out of this range is considered invalid.

Data is considered valid once passed the two validation levels. If level 2 or level 1 are not passed the quality of the data is marked as invalid because of sensor or communication problems, respectively.

## Condition Monitoring and Fault Diagnosis

Condition monitoring and fault diagnosis is an essential task for SHM. It involves monitoring system state or parameters over an operational period, when significant changes or abnormal values indicate a fault. The use of hybrid system theory for building a condition monitoring system that integrates information from multiple sources allows for determining indicators for component fault detection and degradation. In this approach, the combined use of hybrid models (Vento, 2010; Meseguer, 2010) with signal processing and parameter estimation techniques is suggested. Hybrid models combine continuous dynamics described in general by discrete-time equations with discrete-event dynamics described by finite state machines. Such heterogeneous models

*Figure 6. Belt speed and cart position extracting from encoder signals*

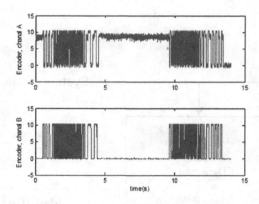 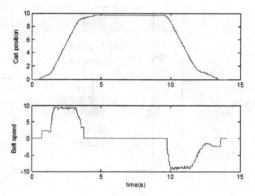

switch among many operating modes, where each mode is associated with a different dynamic law (characterized by some signals evolution, model parameters,...), and mode transitions are triggered by events. When dealing with hybrids systems, the FDI algorithm should take into account, when generating the residual, which is the current operation mode of the hybrid system to adapt the model used to generate the predicted output. Thus, fault diagnosis in hybrid systems can be separated in two stages: in the first stage, the system mode is identified while in the second the fault is detected and isolated, as displayed in Figure 7.

A physical system controlled by a supervisory controller introduces discrete switching behaviors between several operating modes (Bayoud et al. 2008).

Following the notation giving in (Benazera, 2009), a hybrid system is described by a hybrid automaton defined as a tuple $HA = (X, E, Q, T, L, \Theta)$ where $X = \{X_d, X_c\}$ is the set of discrete and continuous variables respectively, $E$ is the set of equations that describe the continuous dynamics of the system, $Q$ is a set of propositional formulas, $T$ is the set of transitions, $L$ is the set of continuous mapping functions associated to transitions, and $\Theta$ is the set of initial values.

Transitions switch $HA$ from mode $i$ to mode $j$. These transitions are due to controlled and autonomous events. Controlled transitions are triggered by logical variables which are known since they come from the supervisory controller (or operator). Autonomous transitions are triggered by conditions over the continuous state which depends on the measured (or estimated) continuous system variables or the occurrence of system faults. Because sensors could be faulty, transitions are activated after some signal processing. Mode change detection is inferred when a transition is activated, and then, $HA$ evolves from the current to the new mode.

This architecture has been applied to the conveyor belt example. In this case of study, faults could be abrupt or incipient. Abrupt faults are due to a malfunction of one of the mechanical parts preventing the cart movement or due to sensor/communication faults. Incipient faults are due to a friction increase of the mechanical parts and motor degradation or/and stress increasing.

Conveyor belt works as a discrete system and has been modeled by an $HA$ with nine modes characterized by the speed motor and the belt position described, respectively, by $X_c$ and $E$. Faults could affect variables $X$ and propositional formulas $Q$, but not the transition sequence $T$.

The faults that affect continuous dynamics can be detected using residuals computed using the general formula:

*Figure 7. Condition monitoring and fault diagnosis structure*

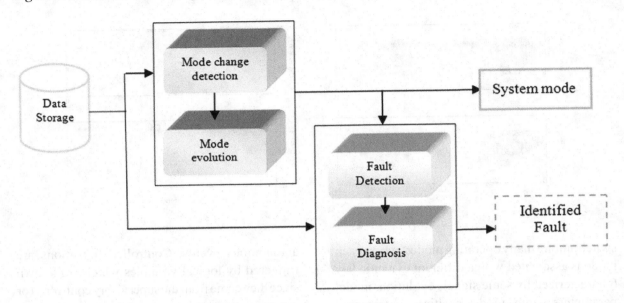

$$r(k) = x(k) - \hat{x}(k)$$

where $x(k)$ corresponds to the physical variables while $\hat{x}(k)$ corresponds to their estimation using the system model. Once the $j^{th}$ residual has been generated, it is evaluated against a threshold to detect normal or abnormal behaviors as follows:

$$s_j(k) = \begin{cases} 0 & \text{if } |r_j(k)| \le l_j & \text{normal behavior} \\ 1 & \text{if } |r_j(k)| > l_j & \text{abnormal behavior} \end{cases}$$

where $l_j$ is the threshold associated to the residual $r_j(k)$.

The module of mode recognition has four sub-modules (Figure 8):

- **Signal processing:** This module works at the defined sampling time. The set of variables $X_c$ is evaluated using the set of equations $E^i$. If the variables have, in the instant $k$, a good quality index, the associated residuals are calculated and evaluated, giv-

ing an indication of normal or abnormal behavior. This information is sent to the failure identification module. If some of the measured data does not present a good quality index, it is invalidated and an estimation provided by model equations is used to replace the invalid data. A set of actual and rebuilt data $X_c^*$ is obtained as an output of this module.

- **Variable selection:** This module selects the set of variables $X_c^*$ mapping the functions associated to transitions, $X_c^{*L} = \mathrm{P}(X_c^*) \subset L$.

- **Transition computation and checking:** This module uses the information described in $Q$ to calculate the transitions. Generally, a transition may coincide with the change of more than one event. If some discrepancies are observed the information is sent to the diagnosis system. In the logic prepositions from which transitions are calculated, the decision to take, in the presence of conflicting events, is taken into account.

*Figure 8. Condition monitoring architecture*

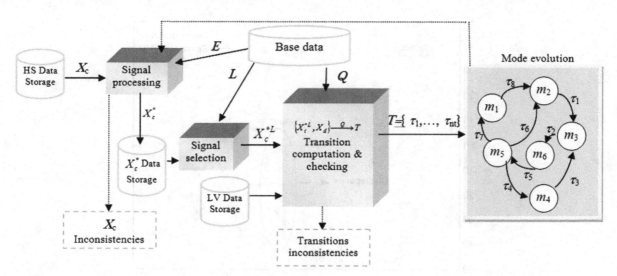

•   **Mode evolution:** This module has been implemented with a finite state machine.

Figure 9 shows the results of the signal processing module in the case of the conveyor belt system. From this figure, the validation of the belt speed can be seen. When the speed is out of the range of normal velocities calculated using the model equations, a discrepancy is indicated.

The diagnosis module has two sub-modules:

•   **Fault detection:** This module is implemented with a timed automaton (Lunze, 2002). The main motivation for dealing with a timed discrete-event representation is that the temporal distance between events includes important information for diagnosis. In the conveyor belt case of study, the mechanical degradation of the components changes the temporal behavior between events mainly during the AC motor boot and braking. This timed automaton has the same discrete system as used for mode evolution. However, it checks if the mode change $m_i \xrightarrow{t_z} m_j$ is performed with a finite time $t_z < t_{z,\max}$ and if

that is true, the algorithm saves the transition duration and proceeds through another transition evaluation. Otherwise, a mode failure, $mf_i$, is indicated.

Because degradation also affects continuous system states in each mode, the variables, $X_c^*$, are used to extract information about some state features that are sent to the prognostics module. For example, in the conveyor belt application, information about the mean speed of the AC motor in steady state and the maximum amplitude of the current in transitory states are extracted. Both transition duration and mode feature are the signals characteristics of the system.

•   **Fault diagnosis:** As shown in Figure 10, the diagnosis module is responsible for processing information from different modules and generating a diagnosis. The diagnosis model is implemented using a timed labeled transition system (TLTS) approach that recognizes the occurrence of a fault by identifying a unique sequence of observable events (Messeguer, 2010) which is derived from the model system.

*Figure 9. Detection of an inconsistency in signal processing module*

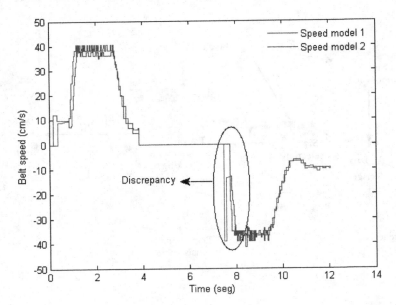

The events taken into account are: mode failure, $X_c$ and $T$ inconsistencies (provided by the mode estimation module), and the data quality index (provided by the pre-processing module). Figure 11 shows an example of the diagnostic system.

In this example, three faults have been considered: $f_1$ corresponds with a fault in the AC motor current sensor; $f_2$ refers to a stuck in the mechanical blocking systems of the cart at the end of the conveyor belt; and $f_3$ is a fault in the mechanical transmission. Table 1 shows the labeled transitions associated with the described faults, where $Ix_i$ and $I\tau_i$ mean inconsistencies in some of the variables $X_c$ and $T$, respectively.

## Prognosis

The ability to forecast machinery failure is vital to reduce maintenance costs, operation down time and safety hazards. To realize the greatest economic and social benefits, it is important to design every aspect of a prognostics system by

*Table 1. Example of fault signatures and relative indicator ordering*

| Fault | Fault labels ordering |
|---|---|
| Current sensor $f_1$ | $Ix_1$ |
| Mechanical interlocking $f_2$ | $mf_2 \rightarrow I\tau_1 \rightarrow Ix_2$ |
| Mechanical transmission $f_3$ | $(mf_2 \vee mf_3 \vee mf_4) \rightarrow Ix_3$ |

considering the asset manager's perspective. The occurrences of machinery failures are difficult to predict due to the inherent structural and operational complexities of real life systems because of the failure interaction between components, the probabilistic nature of fault symptoms, the varying operating conditions and duty parameters.

While a variety of methods are being developed using both model-based and data-driven approaches for the estimation of RUL, one of the major challenges is dealing with prediction uncertainties. Long-term prediction of RUL or time to failure increases the uncertainty bounds due to various sources, such as measurement or sensor errors, future load and usage uncertainty, model assumptions and inaccuracies, loss of in-

*Figure 10. Fault diagnosis structure*

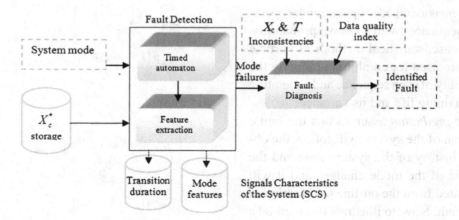

formation due to data reduction, prediction under conditions that are different from the training data, and so on (Petch, 2008). Decisions regarding the system state (maintenance activities such as repair and replacement) should therefore take into account these uncertainties. Hence, development of methods that can be used to describe the uncertainty bounds (lower and upper limits) and confidence levels for the values falling within the confidence bounds are required. Another research area is uncertainty management, in which methods to reduce the uncertainty bounds by using system data as more data becomes available are being investigated.

*Figure 11. Example of fault isolation based TLTS implementation*

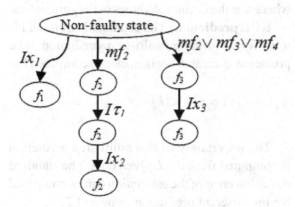

The objectives of the prognostic methods are to extract the main features of the system (one or several measures of system current damage states), to track the degradation measure of the system and to predict the remaining life. The value of degradation measure $\zeta$ can be arbitrary, but generally to facilitate the analysis, the scaled measure $\zeta$ takes values in the interval [0,1]. The $RUL$ is given by:

$$RUL = t\left(end\right) - t\left(k\right)$$

where $t\left(end\right)$ is the estimated time to component/system is no longer in specification (corresponding to $\xi$=1) and $t\left(k\right)$ is the current time.

Prediction of the remaining life can be obtained using three type of prior knowledge (Luo, 2005):

1.  *Deterministic prediction* is useful if the system will be operated according to a known sequence of mode operations (opening, closing,..) and mode durations. In this case, Monte Carlo simulations are performed for this "a priori" known operational sequence and from these simulations the estimated $RUL$ (mean and variance) can be obtained from these results.

2. *Probabilistic prediction* could be used when the system is assumed to operate under operational sequences with a known probability. In this case, statistical methods as particle filters are used for combining properly the probability of each sequence and estimating the remaining life and its variance.

3. *On-line prediction* assumes that the future operation of the system will follow the observed history of the system state and the dynamic of the mode changes and it will be updated from the on-line information of the system. Now, to illustrate this method a simple example will be presented using an autoregressive model of the observed history data of one variable of the system.

In the considered conveyor belt case study, the architecture proposed to compute RUL is presented in Figure 12. The prognosis module has two submodules: an autoregressive model (AR) estimator and a RUL predictor. These modules only work when new data have been acquired.

**Autoregressive model (AR):** The main objective of this step is to estimate a parametric time series to fit the data and analyze the trends. In the considered application, an autoregressive (AR) model is used. An AR model of order *na* is expressed by:

$$y(k) = \sum_{i=1}^{na} a_i y(k-i) + \varepsilon(k)$$

where $y(k)$ is the data at time $k$ of the observed variable, $a_i$ are model coefficients, and $\varepsilon$ is the white noise associated with $y$. This AR model can be written in state space form (de Jong, 2004)

$$\mathbf{x}(k+1) = A\mathbf{x}(k) + K\varepsilon(k)$$
$$y(k) = C\mathbf{x}(k)$$

by considering:

$$\mathbf{x}(k) = \begin{bmatrix} y(k) \\ y(k-1) \\ \vdots \\ y(k-na+1) \end{bmatrix}$$

and where

$$A = \begin{bmatrix} a_1 & \cdots & a_{na-1} & a_{na} \\ 1 & \cdots & 0 & 0 \\ \vdots & \ddots & \vdots & \vdots \\ 0 & 0 & 1 & 0 \end{bmatrix},$$

$$K = \begin{bmatrix} 1 \\ 0 \\ 0 \\ 0 \end{bmatrix}, C = \begin{bmatrix} 1 & 0 & \cdots & 0 \end{bmatrix}$$

The parameters of the system are estimated in batch using $L$ stored data and the least-square (LS) method, which seeks the minimization of:

$$V(k) = \frac{1}{L} \sum_{k=1}^{L} \left( y(k) - \hat{y}(k) \right)^2,$$

where $\hat{y}$ is the estimated value computed by:

$$\hat{y}(k+1) = \sum_{i=0}^{na} \hat{a}_i y(k-i),$$

where $\hat{a}$ is the estimated values of the parameters.

**RUL prediction:** The estimated AR model is used for a variable multi-step prediction. The predicted $\hat{y}$ on an $h$ horizon can be computed by:

$$\hat{y}(k+h|k) = CA^h \mathbf{x}(k)$$

The uncertainty of this multi-step prediction is computed from the $L$ given data. The standard deviation error of the estimation can be computed for the different prediction steps $h=1,2,3\ldots$

*Figure 12. Prognosis module*

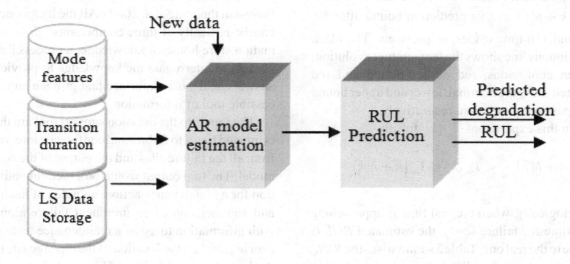

$$V(k) = \frac{1}{(L-h)} \sum_{k=1}^{L-h} \left( y(k+h) - \hat{y}(k+h) \right)^2$$

$$\sigma_j(h) = \sqrt{V_j(h)}$$

By considering a Gaussian distribution of the estimation error, a bounded envelope of the estimation of 95% can be computed on-line by using the following equations at time $k$:

$$\hat{y}(k+h|k)_{\max} = \hat{y}(k+h|k) + 2\sigma(h)$$
$$\hat{y}(k+h|k)_{\min} = \hat{y}(k+h|k) - 2\sigma(h)$$

Using minimum and maximum parameter values, a *RUL* bounds can be computed as follows:

$$RUL_{j,\max} = t_{\max}(end) - t(k)$$
$$RUL_{j,\min} = t_{\min}(end) - t(k)$$

where $t_{\max}(end)$ and $t_{\min}(end)$ are the estimated time to failure for $\xi=1$ from the upper and lower envelopes (trajectories of $y_{max}$ and $y_{min}$) respectively, or equivalently

$$t_{\max}(end) = \left\{ h \mid \hat{y}(k+h|k)_{\min} = \xi \right\}$$

and

$$t_{\min}(end) = \left\{ h \mid \hat{y}(k+h|k)_{\max} = \xi_j \right\},$$

as can be seen in Figure 13.

Motor temperature, $T_m$, is one of the variables stored in LS Data Storage (see Figure 12). The temperature is a critical variable, which must be kept below 80°C for a motor safe operation. In order to analyse the motor temperature evolution, an experiment has been carried out consisting of maintaining the motor in operation during 200 cycles at a constant working frequency. Figure 14.a shows the evolution of the scaled temperature of the motor (real temperature divided by 80°C).

From these historic data, a linear AR model of order five has been estimated using a recursive least square method. To select the order both the minimization of the unexplained output variance and the minimum number of model parameters have been taken into account.

The estimated model has been used to predict the minimum and maximun value of $\hat{T}_m(k+h|k)$

$h$ steps ahead. Figures 14.b, 14.c and 14.d show $\hat{T}_m\left(k+h\middle|k\right)$ and its prediction bound after 80, 97 and 116 time cycles, respectively. The black continuous line shows the temperature evolution; green continuous, red dashed-dotted and red dashed show the nominal, lower and upper bound temperature prediction, respectively.

In this experiment $T_m\left(k\right)$ fulfills

$$\hat{T}_m\left(k+h\middle|k\right)_{\min} < T_m\left(k\right) < \hat{T}_m\left(k+h\middle|k\right)_{\max}$$

and logically, when the real time is approaching the time for failure ($\xi$=1), the estimated *RUL* is close to the real one. Table 2 summarises the $RUL_T$ values estimated at different prediction starting points (i.e., number of cycles). $RUL_T$ values are measured in cycles.

## Decision-Making

A decision support system (DSS) is a computerized information system which contains domain-specific knowledge and analytical decision models to assist the decision maker by presenting information and the interpretation of various alternatives (Wang, 1997). A DSS is intended to enhance individual decision making by providing easier access to problem recognition, problem structure, information management, statistical tools, and the application of knowledge. Such a system is designed to enable the easier and faster generation of alternative, and to increase the awareness of deficiencies in the decision-making process. It can help the decision maker to make more effective and efficient decisions in complex situations and particularly where information is uncertain or incomplete.

A number of computational tools and techniques have been developed for DSS from simple information reporting tools to sophisticated approaches such as knowledge base (Liberatore, 1993), neural networks (Hurson, 1994), fuzzy logic and fuzzy networks (Ishikara, 1993), and Bayesian theory (Saha, 2007). All the techniques consist basically of three components: an information store house of knowledge, a process that permits to interrogate the knowledge to provide answers, and a user interface that provide an accessible tool of information.

The inputs to the decision support tool are the estimated times to failure and confidence interval from all the failure sites and the output of the cost model. The tool output would be a recommendation for a maintenance action, an index of health and annunciation of an imminent failure along with information to assist a maintenance technician in isolating the location of the damage site to the lowest repairable item. This can help reduce the risk of human errors.

Component failures and degradations lead to safety critical situations in many processes. In the proposed architecture, system health monitoring methodologies have been integrated with reliable control with the goal of the controller maintaining the health status of the system and activating remedial actions (as, f.e., virtual sensor or actuators) when problems (faults or degradations) in sensor or actuator health are detected (Montes de Oca, 2010). Recently many authors have been proposed similar works (Ginart, 2007) (Tang, 2008) (Balanban, 2011) (Meyer, 2011) (Brown, 2011) and (Guenab, 2011). In these works, suggestions as mission replanning, control reconfiguration or fault accommodation are given based on health diagnostic or prognostic information.

In this module, there is a fact to keep in mind: the quality of a decision made is directly proportional to the reliability of the information (Vanier, 2001). In other words, high-quality decisions cannot be expected without high-quality information.

A global overview of how decision-making interacts with the process is depicted in Figure 15. The outputs of the condition monitoring (system mode and identified fault) and prognosis system (predicted degradation and RUL) are used for both

*Figure 13. RUL prediction*

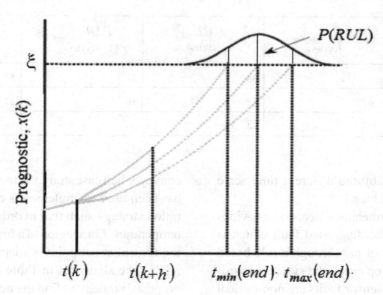

maintenance and HAC decisions. Taking into account this information, it is possible to compute the time window over which maintenance must be performed without compromising the system's operational integrity. Moreover, considering safety issue, it is able to provide information to the operator whether it is possible to continue operating or an immediate shutdown for maintenance is required. And finally, indicators provided by prognosis module are used either for control reconfiguration or for command modifications.

In the conveyor belt case study, in addition to diagnose faults and to evaluate $RUL_i$ using the analyzed data, the support maintenance decision delivers a final $RUL*$ estimation which provides information about the system health. Since it is very difficult to compute a single $RUL$ for the whole system, a maximum and minimum value is computed for each single $RUL_i$ but weighted with the confidence of the AR models estimated at the prognosis state $wm_j$ and with a weight $wf_j$ that takes into account the fault effect in the corresponding $RUL_i$

$$RUL^*_{max} = \max\left(wm_j \cdot wf_j \cdot RUL_{j,max}\right)$$

$$RUL^*_{min} = \max\left(wm_j \cdot wf_j \cdot RUL_{j,min}\right)$$

where given $m$ $RULs$, the confidence weights are computed as follows:

$$wm_j = \frac{V_j\left(t_{j,max}\left(end\right)\right)}{\sum_{j=1}^{m} V_j\left(t_{j,max}\left(end\right)\right)}$$

and $wf_j$ are computed considering the available experimental data. For example, in Figure 16, the real evolution of the variable, $Td_8$, and the prediction of its bounds after several cycles are shown; the black continuous line shows $Td_8$ evolution, the green continuous, the red dashed-dotted and the red dashed show the nominal, down and upper bounds of $Td_8$ prediction, respectively. It can be noticed that $Td_8$ evolves faster than the temperature (Figure 14), being the estimated RUL close to the real one only at 13 cycles before the critical value. The weights $wf_j$ are used to compensate the

*Table 2. Estimated RUL values*

| Cycles, $k$ | $RUL_{Tm,min}$ (cycles) | $RUL_{Tm}$ (cycles) | $RUL_{Tm,max}$ (cycles) | $RUL_{actual}$ (cycles) |
|---|---|---|---|---|
| 80 | 26 | 36 | 60 | 54 |
| 97 | 20 | 31 | 59 | 37 |
| 116 | 7 | 14 | 39 | 18 |

problem introduced by the different time scale behavior of the variables.

The support maintenance decision provides information about the diagnosed fault which is classified as critical or not through a rule based system in order to stop or not the system accordingly. For example, sensor faults are non-critical faults if they can be replaced using some estimation.

From the variables evaluated in the prognosis module, there are variables as the case of temperature that are critical and should be below some pre-established thresholds for safe operation. These critical variables are used in HAC in order to extend the system actuator life. In the conveyor belt case study, the temperature forecast has been used to implement a control reconfiguration strategy such that in order to maintain the temperature $T$ in a region of admissible behaviors, the controller set-point is adapted in function of $RUL_T$. The algorithm in Table 3 summarizes the proposed strategy to find the new set point value, $SP_{new}$, depending on corrective action, $ca$, that increases or decreases with a prefixed slope, $\gamma_0$, if the estimated $RUL_T$, is bigger or lower than de critical $RUL_T$ value, $RUL_{T,crit}$, with a security band $b$, being usually $b = 0.1\ RUL_{T,crit}$. Thus, the proposed algorithm reduces the set-point if $RUL_T$ is close to its critical value $RUL_{T,crit}$. Similar results could have been obtained keeping the original

*Figure 14. (a) $T_m$ evolution data ; (b) prognosis of $RUL_{Tm}$ after 80 cycles; (c) prognosis of $RUL_{Tm}$ after 97 cycles;(d) prognosis of $RUL_{Tm}$ after 116 cycles*

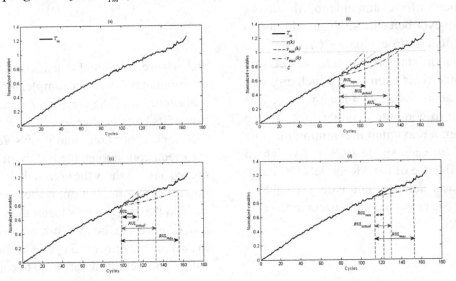

*Figure 15. A general structure of health control*

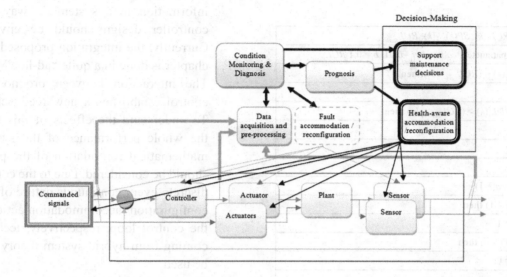

set-point but setting the integral action to zero ($K_I$ = 0) and decreasing the proportional gain $K_P$ of the proportional-integral controller. Some control strategies, such as optimal or predictive control, could easily introduce in the problem formulation, the computation of the control law that takes into account the prediction of critical variables in order to prevent components degradation.

## Solutions and Recommendations

The solutions and recommendations to be considered to address the problems presented in the previous sections are the following:

- Since the combined used of prognosis and control involves several areas of knowledge, an interdisciplinary team with deep

*Figure 16. (a) prognosis of $RUL_{Td}$ after 101 cycles; (b) prognosis of $RUL_{Td}$ after 104 cycles; (c) prognosis of $RUL_{Td}$ after 111 cycles*

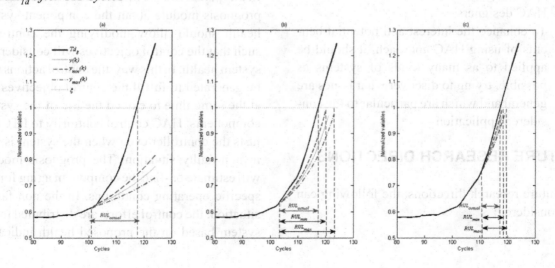

*Table 3. HAC implementation based on SP modification*

| |
|---|
| **Require**: $RUL_T(k)$, $SP(k)$, $\gamma(k)$, $RUL_{T,crit}$ |
| 1: Default parameters $ca(0)=1$, $\gamma_0=0.01$, $b = 0.1\,RUL_{T,crit}$. |
| 2: **if** $(RUL_T(k) < (RUL_{T,crit} - b))$ **then** |
| 3: $\gamma = -\gamma_0$. |
| 4: **elseif** $(RUL_T(k) > (RUL_{T,crit} - b))$ **then** |
| 5: $\gamma = \gamma_0$. |
| 6: **else** |
| 7: $\gamma = 0$. |
| 8: $ca(k) = ca(k\text{-}1) + \gamma$. |
| 9: **if** $(ca(k) > 1)$ **then** |
| 10: $ca(k) = 1$. |
| 11: **if** $(ca(k) < 0)$ **then** |
| 12: $ca(k) = 0$. |
| 13: $SP_{new}(k) = SP(k) \cdot ca(k)$. |
| 14: **return** $SP_{new}(k)$. |

knowledge of both areas should always be involved in the integration process.

- The design process of the HAC accommodation/reconfiguration strategies involves as well a deep knowledge of the system where it has to be applied. So, at this stage of the development of the HAC approach, a black box approach is not desirable. Again, the expert of the system should be working with a highly coupled interaction with the HAC designers.

- To enhance the interest and potential benefits of using HAC approach, it should be applied to as many kinds of systems as possible, trying to discover what issues are general and which are particular to the considered application.

## FUTURE RESEARCH DIRECTIONS

As future research directions, the following can be considered:

- The problem of how to include the health information in a systematic way in the controller design should be envisaged. Currently, the integration proposed in this chapter is done in a quite "ad-hoc" way.

- The interaction between prognosis and control establishes a new feedback loop. To understand the effects of this loop in the whole performance of the system, a mathematical formulation of the problem should be considered. Due to the combined discrete-event/continuous nature of the reconfiguration/ accommodation actions and the control loop, respectively, techniques coming from hybrid system theory should be used.

- The right health indicator to be used for reconfiguring/accommodating the controller is also a key issue. Some methodologies should be provided in order to facilitate the design of the HAC reconfiguration/accommodation strategy.

## CONCLUSION

This chapter has proposed the combination of system health monitoring with control and prognosis by the introduction of the HAC paradigm. In this paradigm, the information provided by the prognosis module about the component system health should allow modifying the controller such that the control objectives will consider the system health. In this way, the control actions will be generated to fulfill the control objectives but at the same time to extend the life of the system components. HAC control contrarily to FTC adjusts the controller even when the system is still in non-faulty situation. The prognosis module will estimate on-line the component aging for the specific operating conditions. In the non-faulty situation, the control efforts are distributed to the system based on the proposed health indicator.

An example has been used along the chapter to illustrate the ideas and concepts introduced.

## ACRONYMS USED

**AR:** Autoregressive model
**DBN:** Dynamic Bayesian Network
**DSS:** Decision Support System
**DX:** Principles of Diagnosis
**FDI:** Fault Detection and Isolation
**FTC:** Fault Tolerant Control
**HAC:** Health-Aware Control
**HMM:** Hidden Markov Models
**PHM:** Prognosis & Health Management
**RUL:** Remaining Useful Life
**SHM:** System Health Monitoring
**TLTS:** Timed Labeled Transition System

## ACKNOWLEDGMENT

This work has been funded by the Spanish Ministry of Science and Technology through the following project CYCYT SHERECS DPI2011-26243 and by European Comission through contract i-Sense FP7-ICT-2009-6-270428.

## REFERENCES

Bagajewicz, M. (2000). *Design and upgrade of process plant instrumentation*. Lancanster, PA: Technomic Publishers. doi:10.1002/cjce.5450800101

Balaban, E., Saxena, A., Narasimhan, S., Roychoudhury, I., & Goebel, K. (2009, September). Experimental *validation of a prognostic health management system for electro-mechanical actuators*. Paper presented at the Annual Conference of the Prognostics and Health Management Society, San Diego, CA.

Bayoudh, M., Travé-Massuyès, L., & Olive, X. (2008). Hybrid systems diagnosis by coupling continuous and discrete event techniques, In *Proceedings of the 17th World Congress*, (pp.7265-7270). Seoul, Korea.

Benazera, E., & Travé-Massuyès, L. (2009). Set-theoretic estimation of hybrid system configurations. *IEEE Transactions on Systems, Man, and Cybernetics. Part B, 39*(5), 1277–1291.

Benosman, M. (2010). A survey of some recent results on nonlinear fault tolerant control. *Mathematical Problems in Engineering*, article ID 586169, 1-25.

Biswas, G., Cordier, M.-O., Lunze, J., Travé-Massuyès, L., & Staroswiecki, M. (2004). Diagnosis of complex systems: Bridging the methodologies of FDI and DX communities. *IEEE Transactions on Systems, Man and Cybernetics, Part B: Cybernetics. Special section on Diagnosis of Complex Systems. Bridging the Methodologies of FDI and DX Communities, 34*(5), 2159–2162.

Blanke, M., Kinnaert, M., Lunze, J., & Staroswiecki, M. (2006). *Diagnosis and fault-tolerant control* (2nd ed.). Springer Publisher.

Brown, D. W., & Vachtsevanos, G. J. (2011). *A prognostic health management based framework for fault-tolerant control*. Paper presented at the Annual Conference of the Prognostics and Health Management Society, Montreal, Canada.

Chen, J., & Patton, R. J. (1999). *Robust model-based fault diagnosis for dynamic systems*. Norwell, MA: Kluwer Academic Publishers.

Cordier, M. O., Dague, P., Dumas, M., Lévy, M., Montmain, J., Staroswiecki, M., & Travé-Massuyès, L. (2004). Conflicts vs. analytical redundancy relations: A comparative analysis of the model-based diagnosis from the artificial intelligence and automatic control perspectives. *IEEE Transactions on Systems, Man and Cybernetics, Part B: Cybernetics. Special Section on Diagnosis of Complex Systems: Bridging the Methodologies of FDI and DX Communities, 34*(5), 2163–2177.

de Jong, P., & Penzer, J. (2004). The ARMA model in state space form. *Statistics & Probability Letters, 40*(1), 119–125. doi:10.1016/j.spl.2004.08.006

García, C., Escobet, T., & Quevedo, J. (2010, October). *PHM techniques for condition-based maintenance based on hybrid system model representation.* Paper presented at Annual Conference of the Prognostics and Health Management Society, Portland, US.

Gertler, J. J. (1998). *Fault detection and diagnosis in engineering systems.* New York, NY: Marcel Dekker.

Ginart, A., Barlas, I., Dorrity, J. L., Kalgren, P., & Roemer, M. J. (2006). Self-healing from a PHM perspective. In *Proceedings of the IEEE Systems Readiness Technology Conference.* New York, NY: IEEE.

Groer, P. G. (2000). Analysis of time-to-failure with a Weibull model. In *Proceedings of the Maintenance and Reliability Conference MARCON,* Knoxville, TN, USA.

Guenab, F., Weber, P., Theilliol, D., & Zhang, Y. (2011). Design of a fault tolerant control system incorporating reliability analysis under dynamic behaviour constraints. *International Journal of Systems Science, 42*(1), 219–223. doi:10.1080/00207720903513319

Hurson, A. R., Pakzad, S., & Lin, B. (1994). Automated knowledge acquisition in a neural network based decision support system for incomplete database systems. *Microcomputers in Civil Engineering, 9*(2), 129–143. doi:10.1111/j.1467-8667.1994.tb00368.x

Ishikawa, A., Amagasa, M., Tomizawa, G., Tatsuta, R., & Mieno, H. (1993). The max-min Delphi method and fuzzy Delphi method via fuzzy integration. *Fuzzy Sets and Systems, 55*(3), 241–253. doi:10.1016/0165-0114(93)90251-C

Jata, K. V., & Parthasarathy, T. A. (2005). Physics of failure. In *Proceedings of the 1st International Forum on Integrated System Health Engineering and Management in Aerospace,* Napa Valley, CA, USA.

Jennions, I. K. (2011). *IVHM Centre.* Cranfield University, Retrieved September 28, 2011, from http://www.cranfield.ac.uk/ivhm

Jiang, B., & Chowdhury, F. N. (2005). Fault estimation and accommodation for linear MIMO discrete-time systems. *IEEE Transactions on Control Systems Technology, 13*(3), 493–499. doi:10.1109/TCST.2004.839569

Krysander, M., Åslund, J., & Nyberg, M. (2008). An efficient algorithm for finding minimal overconstrained subsystems for model-based diagnosis systems. *IEEE Transactions on Man and Cybernetics. Part A, 38*(1), 197–206.

Kuipers, B. (1994). *Qualitative reasoning – Modelling and simulation with incomplete knowledge.* Cambridge, MA: MIT Press. doi:10.1016/0005-1098(89)90099-X

Lee, J., Ghaffari, M., & Elmeligy, S. (2011). Self-maintenance and engineering immune systems: Towards smarter machines and manufacturing systems. *Annual Reviews in Control, 35*(1), 111–122. doi:10.1016/j.arcontrol.2011.03.007

Liao, F., Wang, J. L., & Yang, G.-H. (2002). Reliable robust flight tracking control: An LMI approach. *IEEE Transactions on Control Systems Technology, 10*(1), 76–89. doi:10.1109/87.974340

Liberatore, M. T., & Stylianou, A. C. (1994). Using knowledge-based systems for strategic market assessment. *Information & Management, 27*(4), 221–232. doi:10.1016/0378-7206(94)90050-7

Liu, J., Djurdjanovic, D., Marko, K. A., & Ni, J. (2009). A divide and conquer approach to anomaly detection, localization and diagnosis. *Mechanical Systems and Signal Processing, 23*(8), 2488–2499. doi:10.1016/j.ymssp.2009.05.016

Lunze, J., & Supavatanakul, P. (2002). Diagnosis of discrete event systems described by timed automata. In *Proceeding of IFAC Congress*, Barcelona, Spain.

Luo, J., Pattipati, K. R., Qiao, L., & Chigusa, S. (2005). Integrated model-based and data-driven diagnostic strategies applied to an anti-lock brake system. In *Proceedings IEEE Aerospace Conference* (pp. 3702-2308). Big Sky, MT.

Ly, C., Tmo, K., Byington, C. S., Patrick, R., & Vachtsevanos, G. J. (2009). Fault diagnosis and failure prognosis for engineering systems: A global perspective. In *Proceedings Annual IEEE Conference on Automation Science and Engineering* (pp. 108-115). Bangole, India.

Mahulkar, V., Adams, D. E., & Derriso, M. (2009). Minimization of degradation through prognosis based control for a damaged aircraft actuator. In *Prooceedings of Dynamic Systems and Control Conference* (pp. 669-676). Hollywood, California, USA.

Meseguer, J., Puig, V., & Escobet, T. (2010). Fault diagnosis using a timed discrete-event approach based on interval observers: Application to sewer networks. *IEEE Transactions on Systems, Man, and Cybernetics. Part A, Systems and Humans, 40*(5), 900–916. doi:10.1109/TSMCA.2010.2052036

Meyer, R. M., Ramuhalli, P., Bond, L. J., & Cumblidge, S. E. (2011). Developing effective continuous on-line monitoring technologies to manage service degradation of nuclear power plants. In *Proceedings of IEEE Conference on Prognostics and Health Management* (PHM), Denver, CO, USA.

Montes de Oca, S., & Puig, V. (2010). Fault-tolerant control design using a virtual sensor for LPV systems. In *Proceedings of Conference on Control and Fault Tolerant Systems* (pp. 88-93). Nice, France.

Muller, A., Crespo Marquez, A., & Iung, B. (2008). On the concept of e-maintenance: Review and current research. *Reliability Engineering & System Safety, 93*(8), 1165–1187. doi:10.1016/j.ress.2007.08.006

Nejjari, F., Pérez, R., Escobet, T., & Travé-Massuyès, L. (2006). Fault diagnosability utilizing quasi-static and structural modeling. *Mathematical and Computer Modelling, 45*, 606–616. doi:10.1016/j.mcm.2006.06.008

Ofsthun, S. (2002). Integrated vehicle health management for aerospace platforms. *IEEE Instrumentation and Measurement Magazine*, September.

Patton, R. J. (1997). Fault-tolerant control: The 1997 situation. In *Proceedings of IFAC Symposium on Fault Detection Supervision and Safety for Technical Processes* (pp. 1033-1054). Hull, UK.

Puig, V., & Quevedo, J. (2001). Fault-tolerant PID controllers using a passive robust fault diagnosis approach. *Control Engineering Practice, 9*(11), 1221–1234. doi:10.1016/S0967-0661(01)00068-5

Pulido, B., & Alonso, C. (2004). Possible conflicts: A compilation technique for consistency-based diagnosis. *IEEE Transactions on Systems, Man and Cybernetics, Part B: Cybernetics. Special Section on Diagnosis of Complex Systems: Bridging the Methodologies of FDI and DX Communities, 34*(5), 2192–2206.

Reiter, R. (1987). A theory of diagnosis from first principles. *Artificial Intelligence, 32*, 57–95. doi:10.1016/0004-3702(87)90062-2

Rosich, A., Sarrate, R., Puig, V., & Escobet, T. (2007). Efficient optimal sensor placement for model-based FDI using an incremental algorithm. In *Proceedings of IEEE Conference Decision and Control* (pp. 2590-2595). New Orleans, USA.

Saha, B., Goebel, K., Poll, S., & Christopherson, J. (2007). An integrated approach to battery health monitoring using Bayesian regression, classification and state estimation. In *Proceedings of IEEE Autotestcon*. New York, NY: IEEE.

Sarrate, R., Puig, V., Escobet, T., & Rosich, A. (2007). Optimal sensor placement for model-based fault detection and isolation. In *Proceedings of IEEE Conference Decision and Control* (pp. 2584-2598). New Orleans, USA.

Sheppard, J. W., Butcher, S. G. W., & Ramendra, R. (2007, September). *Electronic systems Bayesian stochastic prognosis: Algorithm development.* Technical Report JHU-NISL-07-002.

Sheppard, J. W., Wilmering, T. J., & Kaufman, M. A. (2008). IEEE standards for prognostics and health management. In *Proceedings of* (pp. 243–248). Anaheim, California: IEEE AUTOTESTCON.

Staroswiecki, M., Cassar, J. P., & Declerk, P. (2000). A structural framework for the design of FDI system in large scale industrial plants. In Patton, R. J., Frank, P. M., & Clark, R. N. (Eds.), *Issues of fault diagnosis for dynamic systems*. Springer.

Stengel, R. F. (1991). Intelligent failure-tolerant control. *IEEE Control Systems Magazine, 11*(4), 14–23. doi:10.1109/37.88586

Tang, L., Kacprzynski, G. J., Goebel, K., Saxena, A., Saha, B., & Vachtsevanos, G. (2008). Prognostics-enhanced automated contingency management for advanced autonomous systems. In Proceeding of *International Conference on Prognostics and Health Management*, Denver, CO.

Thurston, M. G. (2001). An open standard for Web-based condition-based maintenance systems. In *Proceedings of IEEE Systems Readiness Technology Conference* (pp. 401 – 415). Valley Forge, PA, USA.

Travé-Massuyès, L., Escobet, T., & Olive, X. (2006). Diagnosability analysis based on component supported analytical redundancy relations. *IEEE Transactions on Systems, Man. Cybernetics: Part A, 36*(6), 1146–1160. doi:10.1109/TSMCA.2006.878984

Vanier, D. J. (2001). Why industry needs management tools. *Journal of Computing in Civil Engineering, 15*(1), 35–43. doi:10.1061/(ASCE)0887-3801(2001)15:1(35)

Veillette, R. J. (1995). Reliable linear-quadratic state-feedback control. *Automatica, 31*(1), 137–143. doi:10.1016/0005-1098(94)E0045-J

Veillette, R. J., Medanic, J. V., & Perkins, W. R. (1992). Design of reliable control systems. *IEEE Transactions on Automatic Control, 37*(3), 290–304. doi:10.1109/9.119629

Vento, J., Puig, V., & Sarrate, R. (2010). Fault detection and isolation of hybrid systems using diagnosers that combine discrete and continuous dynamics. In *Proceedings of Conference on Control and Fault Tolerant Systems* (pp. 149-154). Nice, France.

Vidyasagar, M., & Viswanadham, N. (1985). Reliable stabilization using a multi-controller configuration. *Automatica, 21*(5), 599–602. doi:10.1016/0005-1098(85)90008-1

Wang, H., & Wang, C. (1997). Intelligent agents in the nuclear industry. *IEEE Computer Magazine, 30*(11), 28–34.

Zhang, Y., & Jiang, J. (2008). Bibliographical review on reconfigurable fault-tolerant control systems. *Annual Reviews in Control, 32*(2), 229–252. doi:10.1016/j.arcontrol.2008.03.008

Zhang, Y. M., & Jiang, J. (2003). Fault tolerant control system design with explicit consideration of performance degradation. *IEEE Transactions on Aerospace and Electronic Systems, 39*(3), 838–848. doi:10.1109/TAES.2003.1238740

Zhou, K., & Ren, Z. (2001). A new controller architecture for high performance, robust, and fault-tolerant control. *IEEE Transactions on Automatic Control, 46*(10), 1613–1618. doi:10.1109/9.956059

## ADDITIONAL READING

Isermann, R. (2006). *Fault-diagnosis systems: An introduction from fault detection to fault tolerance.* Berlin, Germany: Springer-Verlag.

Isermann, R. (2011). *Fault-diagnosis applications.* Berlin, Germany: Springer-Verlag. doi:10.1007/978-3-642-12767-0

Johnson, S. B., Gormley, T. J., Kessler, S. S., Mott, C. D., Patterson-Hine, A., Reichard, K. M., & Scandura, P. A. (Eds.). (2011). *System health management: With aerospace applications.* London, UK: John Wiley & Sons. doi:10.1002/9781119994053

## KEY TERMS AND DEFINITIONS

**Control Accommodation:** It is an active fault-tolerant corrective action based on changing control law despite losses in closed loop performances.

**Control Reconfiguration:** It is an active fault-tolerant control approach which tries to maintain system performances despite faults changing the control structure.

**Health-Aware Control:** Control reconfiguration or accommodation based on the information provided by prognosis module.

**Hybrid Models:** It is a modeling approach which allows representing system behaviors that combine continuous and discrete-event dynamics that change with operation mode.

**Reliable Control:** Control design strategies that take into account the component reliability.

# Section 8
# Integrated Prognostics

# Chapter 13
# A Particle Filtering Based Approach for Gear Prognostics

**David He**
*The University of Illinois-Chicago, USA*

**Eric Bechhoefer**
*NRG Systems, USA*

**Jinghua Ma**
*The University of Illinois-Chicago, USA*

**Junda Zhu**
*The University of Illinois-Chicago, USA*

## ABSTRACT

*In this chapter, a particle filtering based gear prognostics method using a one-dimensional health index for spiral bevel gear subject to pitting failure mode is presented. The presented method effectively addresses the issues in applying particle filtering to mechanical component remaining useful life (RUL) prognostics by integrating a couple of new components into particle filtering: (1) data mining based techniques to effectively define the degradation state transition and measurement functions using a one-dimensional health index obtained by a whitening transform; and (2) an unbiased l-step ahead RUL estimator updated with measurement errors. The presented prognostics method is validated using data from a spiral bevel gear case study.*

## INTRODUCTION

Recently, applications of particle filtering to prognostics have been reported in the literature, for example, remaining useful life (RUL) predication of a mechanical component subject to fatigue crack growth (Zio and Peloni, 2011), on-line failure prognosis of UH-60 planetary carrier plate subject to axial crack growth (Orchard and Vachtsevanos, 2011), degradation prediction of thermal processing unit in semiconductor manufacturing (Butler and Ringwood, 2010), and prediction of lithium-ion battery capacity depletion (Saba *et al.*, 2009). The reported application results have shown that particle filtering represents a potentially powerful prognostics tool due to its capabil-

DOI: 10.4018/978-1-4666-2095-7.ch013

ity in handling non-linear dynamic systems and non-Gaussian noises using efficient sequential importance sampling to approximate the future state probability distributions. Particle filtering was developed as an effective on-line state estimation tool (see Doucet *et al.*, 2000; Arulampalam *et al.*, 2002). In order to apply particle filtering to RUL prediction of a mechanical component such as gears, a few practical implementation problems have to be solved: (1) define a state transition function $\mathbf{f}_k$ that represents the degradation evolution in time of the component; (2) select the most sensitive health monitoring measures or condition indicators (CIs) and define a measurement function $\mathbf{h}_k$ that represents the relationship between the degradation state of the component and the CIs; (3) define an effective *l*-step ahead RUL estimator. In solving the first problem, research on using particle filtering for mechanical component RUL prognostics has used Paris' law to define the state transition function $\mathbf{f}_k$ (Zio and Peloni, 2011; Orchard and Vachtsevanos, 2011). As an empirical model, Paris' law can be effective for defining a state transition function that represents a degradation state subject to fatigue crack growth. For other type of failure modes such as pitting and corrosion, effective alternatives for defining the state transition function should be explored. Regarding the second problem, on the surface, it doesn't seem to be a problem to use multiple CIs to define a measurement function for particle filtering as it allows information from multiple measurement sources to be fused in a logical manner (Zio and Peloni, 2011). In particle filtering, measurements are collected and used to update the prior state distribution via Bayes rule so as to obtain the required posterior state distribution. Subsequently, various kinds of uncertainties arise from different sources that are correlated. In most real applications, no single CI is sensitive to every failure mode of a component. This suggests that defining the measurement function $\mathbf{h}_k$ will have some form of de-correlat-

ed sensor fusion. In order to apply particle filtering to estimate the RUL, an *l*-step ahead estimator has to be defined. Both biased and unbiased *l*-step ahead estimators have been reported by Zio and Peloni (2011) and Orchard and Vachtsevanos (2011). However, as pointed out by Zio and Peloni (2011), one issue related to these estimators is that state estimation and prediction must be accompanied by a measure of the associated error.

In this chapter, a particle filtering based gear prognostics using one-dimensional health index method for spiral bevel gear subject to pitting failure mode is presented. In particular, in presenting the method, the three particle filtering prognostics implementation related issues will be addressed: (1) define the state transition function using data mining approach; (2) use an one-dimensional health index (HI) obtained by a whitening transform to define the measurement function; (3) an *l*-step ahead RUL estimator incorporated with a measure of the associated error. The presented method is validated using fatigue testing data from a spiral bevel gear case study performed in the NASA Glenn Spiral Bevel Gear Test Facility.

## THE APPROACH

The general framework of the particle filtering based gear prognostics method for spiral bevel gear subject to pitting failure mode is shown in Figure 1.

As shown in Figure 1, to predict the RUL of the spiral bevel gear subject to pitting failure mode, the oil debris mass (ODM) is used to represent the degradation state of the gear. Therefore, the state transition function $\mathbf{f}_k$ is defined by an ODM ARIMA model established using data mining based approach. The one-dimensional HI obtained by the applying a Cholesky decomposition based whitening transform and statistical generation models is used to define the measurement function $\mathbf{h}_k$ by double exponential smooth-

*Figure 1. Particle filtering based gear prognostics framework*

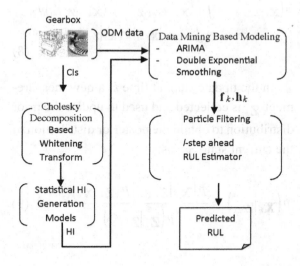

ing. Based on the defined functions $\mathbf{f}_k$ and $\mathbf{h}_k$, an *l*-step ahead RUL estimator incorporated with measurement error is used in particle filtering to provide accurate prediction of RUL. The generation of the one-dimensional HI and the *l*-step ahead RUL estimator used in particle filtering are explained in the next two sections.

## The One-Dimensional Health Index

The concept of using Cholesky decomposition to develop a one-dimensional gear health index and its threshold setting based on a probability of false alarm was first reported in Bechhoefer *et al.* (2011). To develop the one-dimensional health index, a set of correlated CIs are first de-correlated by applying the Cholesky decomposition. The Cholesky decomposition of Hermitian, a positive definite matrix results in $\mathbf{A} = \mathbf{LL}^*$, where $\mathbf{L}$ is a lower triangular, and $\mathbf{L}^*$ is its conjugate transpose. By definition, the inverse covariance is positive definite Hermitian. Let $\mathbf{F}$ be a set of correlated CIs. It then follows that:

$$\mathbf{LL}^* = \mathbf{\Sigma}^{-1} \tag{1}$$

and

$$\mathbf{Y} = \mathbf{L} \times \mathbf{F}^T \tag{2}$$

where $\mathbf{Y}$ is a vector of *n* independent CIs with unit variance and $correlation(\mathbf{Y}) = 0$. The Cholesky decomposition, in effect, creates the square root of the inverse covariance. This in turn is analogous to dividing feature by its standard deviation (the trivial case of one feature). In turn, Equation (2) creates the necessary independent and identical distributions required to define the health index for a function of distributions.

Assuming that the distributions of the CIs follow a Gaussian distribution, then three statistical HI generation models can be developed: (1) the Gaussian order statistic; (2) the sum of *n* Gaussian; and (3) the total energy of *n* Gaussian. These three models are explained as follows (Bechhoefer *et al.* 2011):

1. When the HI is defined as the Gaussian order statistic, it can be generated as following:

$$\mathbf{Y} = \mathbf{L} \times \left(\mathbf{F}^T - \mathbf{m}\right)$$
$$HI = \left(\max\{\mathbf{Y}\} + .34\right) \times {0.5}\Big/{(3.41 + 0.34)} \tag{3}$$

where $\mathbf{m}$ is the mean of $\mathbf{F}$. Subtracting the mean and multiplying by $\mathbf{L}$ transforms the features into *n*, *Z* distributions (zero mean, IID Gaussian distributions).

2. When the HI is defined as the sum of *n* Gaussian, it can be generated as following:

$$\mathbf{Y} = \mathbf{L} \times \mathbf{F}^T$$
$$HI = {0.5}\Big/{(8.352 - 0.15)}\left(-0.15 + \sum_{i=1}^{n} \mathbf{Y}_i\right) \tag{4}$$

3. When the HI is defined as the total energy of *n* Gaussian, it can be generated as following:

$$\mathbf{Y} = \mathbf{L} \times \mathbf{F}^T$$

$$HI = {0.5}\big/{3.368} \sqrt{\sum_{i=1}^{n} \mathbf{Y}_i^2} \tag{5}$$

## RUL Prediction Using Particle Filtering

### Particle Filtering for State Estimation

Consider a system described by the discrete time state space model:

$$\mathbf{x}_k = \mathbf{f}_k(\mathbf{x}_{k-1}, \omega_{k-1}) \tag{6}$$

$$\mathbf{z}_k = \mathbf{h}_k(\mathbf{x}_k, \upsilon_k) \tag{7}$$

where:

$\mathbf{f}_k : \Re^{n_x} \times \Re^{n_\omega} \to \Re^{n_x}$ is the state transition function

$\omega_k$ is an independently and identically distributed (*iid*) state noise vector of known distribution

$\mathbf{h}_k : \Re^{n_x} \times \Re^{n_v} \to \Re^{n_z}$ is the measurement function

$\upsilon_k$ is an *iid* measurement noise vector

The problem of state estimation is to estimate the dynamic state $\mathbf{x}_k$ in terms of probability density function (pdf) $p\left(\mathbf{x}_k | \mathbf{z}_{0:k}\right)$, given the measurement up to time $k$. The initial distribution of the state $p\left(\mathbf{x}_0\right)$ is assumed known.

The Bayesian solution to the state estimation problem normally consists of two steps: prediction and update. In the prediction step, the prior probability distribution of the state $\mathbf{x}_k$ at time $k$, starting from the probability distribution $p\left(\mathbf{x}_{k-1} | \mathbf{z}_{0:k-1}\right)$ at time $k$-1, is obtained as:

$$p\left(\mathbf{x}_k | \mathbf{z}_{0:k-1}\right) = \int p\left(\mathbf{x}_k | \mathbf{x}_{k-1}, \mathbf{z}_{0:k-1}\right) p\left(\mathbf{x}_{k-1} | \mathbf{z}_{0:k-1}\right) d\mathbf{x}_{k-1}$$
$$= \int p\left(\mathbf{x}_k | \mathbf{x}_{k-1}\right) p\left(\mathbf{x}_k | \mathbf{z}_{0:k-1}\right) d\mathbf{x}_{k-1} \tag{8}$$

In the update step, at time $k$, a new measurement $\mathbf{z}_k$ is collected and used to update the prior distribution to obtain the posterior distribution of the current state $\mathbf{x}_k$ as:

$$p\left(\mathbf{x}_k | \mathbf{z}_{0:k}\right) = \frac{p\left(\mathbf{x}_k | \mathbf{z}_{0:k-1}\right) p\left(\mathbf{z}_k | \mathbf{x}_k\right)}{p\left(\mathbf{z}_k | \mathbf{z}_{0:k-1}\right)} \tag{9}$$

where the normalizing constant is:

$$p\left(\mathbf{z}_k | \mathbf{z}_{0:k-1}\right) = \int p\left(\mathbf{x}_k | \mathbf{z}_{0:k-1}\right) p\left(\mathbf{z}_k | \mathbf{x}_k\right) d\mathbf{x}_k \tag{10}$$

Obtaining exact state estimation solutions for Equation (8) and Equation (9) is not realistic for most cases. Therefore, particle filtering is used to obtain the heuristic solutions. The prediction at time $k$ can be accomplished by particle filtering to perform the following two tasks: (1) sampling $N$ number of random samples (particles) $x_{k-1}^i, i = 1, ..., N$ from the probability distribution of the state noise $\omega_{k-1}$ and (2) generating new set of samples $x_k^i, i = 1, ..., N$ using Equation (6). In the update step, each new sampled particle $x_k^i$ is assigned a weight $w_k^i$ based on the likelihood of the new measurement $\mathbf{z}_k$ at time $k$ as:

$$w_k^i = \frac{p\left(\mathbf{z}_k | x_k^i\right)}{\sum_{i=1}^{N} p\left(\mathbf{z}_k | x_k^i\right)} \tag{11}$$

Then the approximation of the posterior distribution $p\left(\mathbf{X}_k\middle|\mathbf{Z}_{0:k}\right)$ can be obtained from the weighted samples $\left\{x_k^i, w_k^i, i = 1, ..., N\right\}$ (Doucet et al., 2000)

## Particle Filtering for State Estimation

In order to apply particle filtering to estimate the RUL, an *l*-step ahead estimator has to be developed. An *l*-step ahead estimator will provide a long term prediction of the state pdf $p\left(\mathbf{X}_{k+l}\middle|\mathbf{Z}_{0:k}\right)$ for $l = 1, ..., T - k$, where $T$ is the time horizon of interest. In making an *l*-step ahead prediction, it is necessary to assume that no information is available for estimating the likelihood of the state following the future *l*-step path $\mathbf{X}_{k+1:k+l}$, that is, future measurements $\mathbf{Z}_{k+l}$, $l = 1, ..., T - k$ cannot be used for making the prediction. Therefore, one can only project the initial condition $p\left(\mathbf{X}_k\middle|\mathbf{Z}_{0:k}\right)$ using state transition pdf $p\left(\mathbf{X}_j\middle|\mathbf{X}_{j-1}\right), j = k+1, ..., k+l$ along all possible future paths weighted by their probability $\prod_{j=k+1}^{k+l} p\left(\mathbf{X}_j\middle|\mathbf{X}_{j-1}\right)d\mathbf{X}_{j-1}$. By combining Equation (6) and Equation (9), an unbiased *l*-step ahead estimator can be obtained (Zio and Peloni, 2011; Orchard and Vachtsevanos, 2011):

$$p\left(\mathbf{X}_{k+l}\middle|\mathbf{Z}_{0:k}\right) =$$
$$\int \cdots \int \prod_{j=k+1}^{k+l} p\left(\mathbf{X}_j\middle|\mathbf{X}_{j-1}\right) p\left(\mathbf{X}_k\middle|\mathbf{Z}_{0:k}\right) \prod_{j=k}^{k+l-1} d\mathbf{X}_j$$

$$(12)$$

In theory, an unbiased estimator would give the minimum variance estimation. However, solving Equation (12) can be either difficult or computationally expensive. A particle filtering approximation procedure of the *l*-step ahead estimator is provided in (Zio and Peloni, 2011).

Assume that the state $\mathbf{X}_k$ represents a mono-dimensional health indicator and RUL is the time remained before its crossing of a pre-specified critical value $\lambda$. At each time $k + l$ projected $l$ steps from current time $k$, estimating $\hat{p}\left(RUL \le l\middle|\mathbf{Z}_{0:r}\right)$ is equivalent to estimating $\hat{p}\left(\mathbf{X}_{k+l} \ge \lambda\middle|\mathbf{Z}_{0:r}\right)$.

Note that in computing the *l*-step ahead RUL estimator using particle filtering, at each updating step, a weight is computed according to Equation (11) without considering any measurement of the associated errors. Define $\hat{\mathbf{Z}}_k$ the estimated measurement at time $k$ computed from Equation (7). Then a weighting process in particle filtering that takes into account the measurement errors can be defined as:

$$w_k^i = \frac{p\left(\left(\mathbf{Z}_k - \hat{\mathbf{Z}}_k\right)\middle|x_k^i\right)}{\sum_{i=1}^N p\left(\left(\mathbf{Z}_k - \hat{\mathbf{Z}}_k\right)\middle|x_k^i\right)} \qquad (13)$$

In the particle filtering based gear prognostics method presented in this chapter, the *l*-step ahead RUL estimator is computed using the weights defined by Equation (13).

## SPIRAL BEVEL GEAR CASE STUDY

In this chapter, data from a spiral bevel gear case study conducted in the NASA Glenn Spiral Bevel Gear Test Facility at are used to validate the presented method.

### Experimental Setup and Data Collection

A detailed description of the test rig and test procedure is given in (Dempsey et al., 2002). The rig (as shown in Figure 2) was used to quantify the performance of gear material, gear tooth design and lubrication additives on the fatigue strength of gears. During the testing, vibration CIs and oil debris data were used to detect pitting damage on spiral bevel gears.

The tests consisted of running the gears under load through a "back to back" configuration, with vibration data collected once per minute using a sampling rate of 100 kHz for 2 seconds duration, generating time synchronous averages (TSA) on the gear shaft (36 teeth). The pinion, on which the damage occurred, has 12 teeth. The tests were performed for a specific number of hours or until surface fatigue occurs. In this chapter, data collected from experiments 5 and 6 are used. At test completion of the experiments, destructive pitting could be observed on the teeth of the pinions (see Figure 3).

TSA data was re-processed with gear CI algorithms presented in (Zakrajsek *et al.*, 1993) and (Wemhoff *et al.*, 2007). A total of 6 CIs were used for the HI calculation: residual RMS, energy operator RMS, FM0, narrowband kurtosis, amplitude modulation kurtosis, and frequency modulation RMS.

## The Results

The ODM and HI data from experiments 5 and 6 are shown in Figure 4 and 5, respectively.

In order to define the state transition function using the ODM data, various ARIMA models were fitted into the ODM data of experiment 6. The best fitted ARIMA model was: ARIMA(1,1,1). Let: $x_k$ = true ODM value at time $k$, $\hat{x}_k$ = pre-

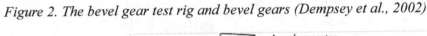

*Figure 2. The bevel gear test rig and bevel gears (Dempsey et al., 2002)*

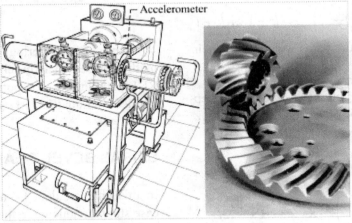

*Figure 3. Spiral gear destructive pitting: (1) experiment 5, (2) experiment 6*

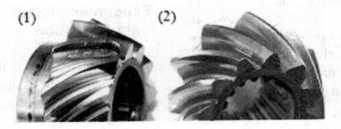

*Figure 4. ODM and HI of experiment 5*

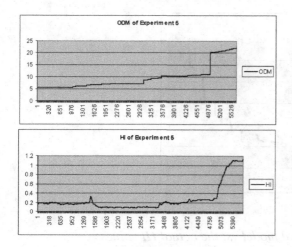

*Figure 5. The ODM and HI of experiment 6*

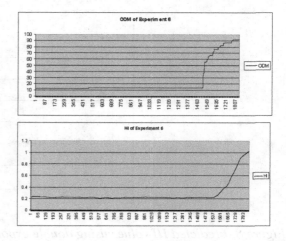

dicted ODM value at time $k$. Then the state transition function $\mathbf{f}_k$ can be defined as:

$$x_k = 0.0165 + 1.1415 x_{k-1} - 0.415 x_{k-2}$$
$$- 0.1032 \left( x_{k-1} - \hat{x}_{k-1} \right) + \omega_k$$

The plot of actual ODM values against the predicted ODM values is shown in Figure 6. From Figure 6, it is obvious that the ARIMA(1,1,1) model is almost a perfect fit to the ODM data.

The measurement function $\mathbf{h}_k$ was defined using a double exponential smoothing model with $\alpha = 0.05$ to fit the relation between HI and ODM. Figure 7 shows the plot of HI against ODM for experiment 6.

Figure 8 shows the predicted HI values using the double exponential smoothing model against the actual HI values.

Using the station transition function $\mathbf{f}_k$ and the measurement function $\mathbf{h}_k$ defined by the ODM and HI data from experiment 6, the particle filter-

*Figure 6. The actual ODM vs. the predicted ODM using ARIMA(1,1,1) model*

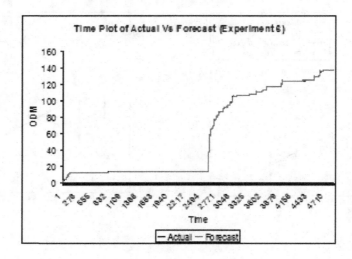

*Figure 7. Plot of HI against ODM for experiment 6*

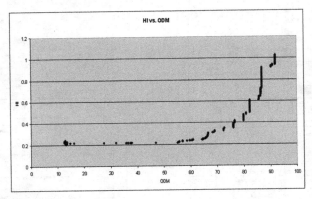

*Figure 8. Predicted HI values using double exponential model vs. the actual HI values*

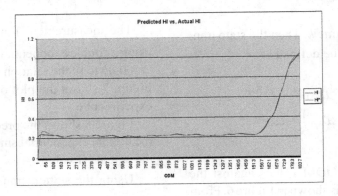

*Figure 9. The predicted mean RUL and corresponding 90% confidence intervals using estimator updated with error measurement*

*Figure 10. The predicted mean RUL and corresponding 90% confidence intervals using estimator updated without error measurement*

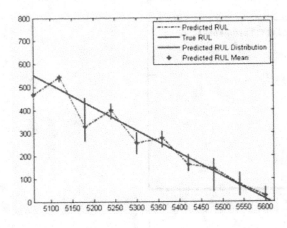
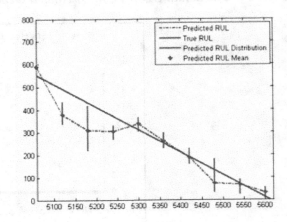

ing based *l*-step ahead RUL estimator was run on the data from experiment 5 using $N = 1000$ particles. To compute the RUL, the critical value $\lambda$ was set to be the level of ODM = 22 mg. Updating the estimated pdf on the basis of the measurements collected very 100 temporal steps, the estimated mean RUL and corresponding 90% confidence intervals are shown in Figure 9.

To make a comparison, the estimated mean RUL and corresponding 90% confidence intervals using the estimator without error measurement update are shown in Figure 10.

From Figure 9 and Figure 10, one can see that the *l*-step ahead RUL estimator updated with the error measurement give a better performance.

## SUMMARY

A particle filtering based gear prognostics method using a one-dimensional health index for spiral bevel gear subject to pitting failure mode is presented in this chapter. The presented method effectively addresses the issues in applying particle filtering to mechanical component remaining useful life prognostics by integrating a couple of new components into particle filtering: (1) data mining based techniques to effectively define the degradation state transition and measurement functions using a one-dimensional health index obtained by a whitening transform; (2) an unbiased *l*-step ahead RUL estimator updated with measurement errors. The presented prognostics method is validated using data from a spiral bevel gear case study. The validation results have shown the effectiveness of the presented method.

## ACKNOWLEDGMENT

The authors would like to thank Dr. Paula Dempsey at NASA Glenn for providing the spiral bevel gear case study data to support the research presented in this chapter.

## REFERENCES

Arulampalam, S. M., Maskell, S., Gordon, N., & Clapp, T. (2002). A tutorial on particle filters for online nonlinear/non-Gaussian Bayesian tracking. *IEEE Transactions on Signal Processing, 50*(2), 174–188. doi:10.1109/78.978374

Bechhoefer, E., He, D., & Dempsey, P. (2011). Gear health threshold setting based on a probability of false alarm. *Proceedings of the 2012 Prognostics and Health Management Conference*, Montreal, Canada, September 23 – 28, 2011.

Butler, S., & Ringwood, J. (2010). Particle filters for remaining useful life estimation of abatement equipment used in semiconductor manufacturing. *Proceedings of the First Conference on Control and Fault-Tolerant Systems*, Nice, France, (pp. 436–441).

Dempsey, P., Handschuh, R., & Afjeh, A. (2002). *Spiral bevel gear damage detection using decision fusion analysis*. NASA/TM-2002-211814.

Doucet, A., Godsill, S., & Andrieu, C. (2000). On sequential Monte Carlo sampling methods for Bayesian filtering. *Statistics and Computing, 10*(3), 197–208. doi:10.1023/A:1008935410038

Orchard, M. E., & Vachtsevanos, G. J. (2011). A particle-filtering approach for on-line fault diagnosis and failure prognosis. *Transactions of the Institute of Measurement and Control, 31*(3-4), 221–246. doi:10.1177/0142331208092026

Saha, B., Goebel, K., Poll, S., & Christophersen, J. (2009). Prognostics methods for battery health monitoring using a Bayesian framework. *IEEE Transactions on Instrumentation and Measurement, 58*(2), 291–296. doi:10.1109/TIM.2008.2005965

Wemhoff, E., Chin, H., & Begin, M. (2007). Gearbox diagnostics development using dynamic modeling. *Proceedings of the AHS 63rd Annual Forum*, Virginia Beach, VA, 2007.

Zakrajsek, J., Townsend, D., & Decker, H. (1993). *An analysis of gear fault detection method as applied to pitting fatigue failure data.* NASA Technical Memorandum 105950.

Zio, E., & Peloni, G. (2011). Particle filtering prognostic estimation of the remaining useful life of nonlinear components. *Reliability Engineering & System Safety, 96*(3), 403–409. doi:10.1016/j.ress.2010.08.009

## Section 9
# Life Cycle Cost and Return on Investment for Prognostics and Health Management

# Chapter 14
# Supporting Business Cases for PHM:
## Return on Investment and Availability Impacts

**Peter Sandborn**
*CALCE, University of Maryland, USA*

**Taoufik Jazouli**
*CALCE, University of Maryland, USA*

**Gilbert Haddad**
*Schlumberger Technology Center, USA*

## ABSTRACT

*The development and demonstration of innovative prognostics and health management (PHM) technology is necessary but not sufficient for widespread adoption of PHM concepts within systems. Without the ability to create viable business cases for the insertion of the new technology into systems and associated management processes, PHM will remain a novelty that is not widely disseminated. This chapter addresses two key capabilities necessary for supporting business cases for the inclusion and optimization of PHM within systems. First, the chapter describes the construction of life-cycle cost models that enable return on investment estimations for the inclusion of PHM within systems and the valuation of maintenance options. Second, the chapter addresses the support of availability-centric requirements (e.g., availability contracts) for critical systems that incorporate PHM, and the resulting value that can be realized. Examples associated with avionics, wind turbines, and wind farms are provided.*

## INTRODUCTION

Prognostics and health management (PHM) is a discipline consisting of technologies and methods to assess the reliability of a system in its actual life-cycle conditions, to determine the advent of failure, and mitigate system risks (Pecht, 2008; Cheng et al. 2010). The objective of PHM is to avoid unanticipated system failure while concurrently minimizing the amount of remaining useful life that is thrown away by maintenance actions. As with any maintenance strategy, the incorporation

DOI: 10.4018/978-1-4666-2095-7.ch014

of PHM within a system is a tradeoff between the value realized and the cost of creating, incorporating, and supporting the associated management activities. Quantification of this value is extremely important in order to justify the large investments of money, time and resources required to create and support PHM in a system.

## Maintenance Paradigms

Maintenance refers to the measures taken to keep a product in operable condition or repair it to operable condition (Dhillon, 1999). Mission, safety, and infrastructure critical systems are examples of sustainment-dominated systems (Sandborn and Myers, 2008) where the costs of sustainment (a significant component of which is maintenance) exceeds the cost of system procurement. Examples of sustainment-dominated systems include: aircraft, military systems, industrial controls, medical equipment and communications infrastructure. It is rarely practical to throwaway these types of systems when they fail, so they nearly always have well developed maintenance strategies. Maintenance paradigms are constantly evolving. The following definitions capture a view of the primary maintenance approaches used for complex systems:

Corrective Maintenance – The simplest approach to maintenance is to run a system to failure and fix it when it breaks. The advantage of corrective maintenance is that all the life in the system is consumed (none is thrown away), however in this case failures are unpredictable and may have expensive or catastrophic consequences. In the case of corrective maintenance, the failure threshold is fixed and defined by the user's definition of a failure. The scope of the maintenance activities is also limited to individual system instances. Corrective maintenance is also referred to as "reactive" or "unscheduled maintenance" and is the primary approach used for most electronic systems today.

Preventative Maintenance – Preventative maintenance (also called fixed-interval maintenance or scheduled maintenance) introduces the idea of performing maintenance (repair or replacement) at fixed intervals measured by some relevant usage parameter (time, miles, cycles, etc.). While inspection may be used to initiate maintenance actions, it may not be possible for many types of systems, e.g., electronics. Preventative maintenance allows the avoidance of unscheduled failures, at the cost of large amounts of wasted remaining useful life (life that is disposed of) – preventative maintenance is usually quite conservative. An example for preventative maintenance is changing the oil in your car every 3000 miles – whether it needs to be changed or not.

Reliability Centered Maintenance (RCM) – Reliability centered maintenance uses knowledge of the failure rate of the system components as a function of key usage parameters to determine viable maintenance intervals. In RCM the failure modes of the system components are analyzed to identify optimal failure management strategies usually through the use of a Failure Mode, Effects and Criticality Analysis (FMECA). RCM targets failure avoidance.

Condition Based Maintenance (CBM) - CBM is based on using real-time data from the system to observe the actual state of the system (condition monitoring) and thus determine the system's health, and then acting only when maintenance is necessary. CBM allows one to minimize the remaining useful life of the system component that is thrown away. CBM includes logistics-driven maintenance thresholds and allows fault isolation and diagnostics. RCM and CBM are similar and sometimes used interchangeably. The major difference between RCM and CBM is that RCM employs a rigorous quantitative evaluation of equipment criticality based on identified failure modes and effects, i.e., an FMECA.

Prognostics and Health Management (PHM) Enabled Maintenance - PHM is a broader concept than CBM that combines current system health with expected future use conditions to predict remaining useful life (or remaining useful perfor-

*Figure 1. The value of maintenance (Haddad et al. 2012a)*

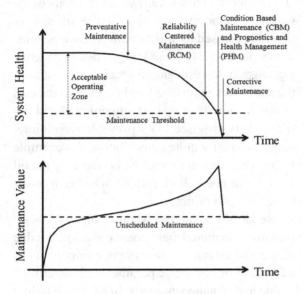

mance) for a system. PHM enabled maintenance is outcome/requirements driven and enables (and optimizes) the use of outcome contracts. The distinction between PHM and CBM in Figure 1 is that PHM supports a variable maintenance threshold that depends on the expected future conditions and the availability of maintenance resources.

The goal of all maintenance approaches is to either: 1) Perform maintenance such that the remaining useful life (RUL) and the remaining useful performance (RUP) that are thrown away are minimized, while avoiding failures; or 2) Find the optimum mix of scheduled and unscheduled maintenance that minimizes the life-cycle cost. The fundamental maintenance/value tradeoff associated with putting PHM into systems in shown in Figure 1.

## System and Enterprise Level Maintenance Value

The value of a maintenance approach, and specifically the value of PHM, can be realized at two distinctly different levels: individual system instances and the enterprise (fleet).

System-level maintenance value means taking action based on a fixed interval, condition or prognostics to manage one specific instance of a system, e.g., one truck or one airplane. Often these actions are "real-time" and fall into the following classifications:

- Modify the system (e.g., adaptive re-configuration, redundancy)
- Modify the mission (e.g., reduce speed, take a different route)
- Modify the sustainment (e.g., call ahead to arrange for a maintenance action)

Most of today's value propositions for PHM are focused at the system level with a primary interest in failure avoidance in order to protect investments in assets from the expensive or catastrophic consequences of failure.

Enterprise-level (or fleet-level) value means taking action based on a fixed interval, condition or prognostics to manage an enterprise, e.g., a whole fleet of trucks or a whole airline. In this case the actions tend to be longer-term strategic planning things (usually not real-time):

- Optimizing the logistics
- Adapting business models

While there is certainly value in maintenance strategies at both the system and enterprise levels, business cases are usually made at the enterprise level. It is also possible that the optimum maintenance solution for an individual member of a population of systems does not lead to the optimum for the whole population (the enterprise). Consider a wind farm consisting of many individual wind turbines. Optimizing the maintenance for each individual turbine may not represent the optimum for the entire farm of turbines if the farm must meet specific availability (electricity generation) requirements.

This chapter discusses the determination of value. First the chapter describes the construction of cost models that enable return on investment estimations for the inclusion of PHM within systems, and enable the valuation of maintenance options. Second, we will address the support of availability-centric requirements (e.g., availability contracts) for critical systems that incorporate PHM; which will include the process of performing design for availability.

## RETURN ON INVESTMENT (ROI)

When management considers spending money they usually want to formulate a business case that not only describes the process they wish to follow, but also the value that they expect to gain as a result of the investment. For the manufacturing and life-cycle support of a complex system, business cases could be required for the following activities: spending money to modify a manufacturing line, refresh the design of a system, add or expand product or system management activities, or adopt a new technology. One common way to quantify the value is to compute a return on investment (ROI) for a given use of money.

While the formulation of a ROI associated with investing money in a financial instrument (e.g., the stock market) is straightforward, the calculation of a ROI associated with the generation of an increase in the customer base, a cost savings, or a future cost avoidance is not as simple to perform, especially when the investment involves new technology.

## ROI Definition

A rate of return is the benefit received from an investment over a period of time. Generally returns are ratios relating the amount of money that is gained or lost to the amount of money risked. Return on Investment (ROI) is the monetary benefit derived from having spent money on developing, changing, or managing a product or system. ROI is

a common performance measure used to evaluate the efficiency of an investment or to compare the efficiency of a number of different investments opportunities. To calculate ROI, the benefit or gain associated with an investment is divided by the cost of the investment and the result is expressed as a percentage or a ratio,

$$ROI = \frac{Return - Investment}{Investment} = \frac{V_f - V_i}{V_i} \tag{1}$$

where $V_f$ and $V_i$ are the final and initial values of an investment respectively. The quantity expressed in (1) is the true rate of return on an investment that generates a single payoff after one period (where the period is the length of time over which one wishes to measure the value).

A key to using (1) is to realize that the *Return* (or final value, $V_f$) includes the *Investment* (or initial value, $V_i$) and that the difference of the two is the gain realized by making the investment. For the formulation in (1), an *ROI* of 0 represents a break-even situation, i.e., the value you receive in return exactly equals the value you invested. If *ROI* is > 0 then there is a gain and if *ROI* is < 0 there is a loss. Constructing a business case for a product does not necessarily require that the *ROI* be greater than zero; in some cases, the value of a product is not fully quantifiable in monetary terms, or the product is necessary in order to meet a system requirement that could not otherwise be attained such as an availability requirement. However, ROIs are still an important part of business cases even if they are < 0.

ROIs are easy to calculate but deceivingly difficult to get right. Financial investment ROIs are straightforward, but when one is evaluating the ROI of a cost savings, market share increase or cost avoidance, the difference between costs that are investments and those that are returns is blurred. A detailed treatment of ROI as it applied to technology and systems is given in Sandborn (2012).

## Cost Avoidance ROI

The type of return associated with maintenance activities and in particular PHM is generally a cost avoidance. Cost avoidance is a "metric" that results from a "spend" that is lower than the spend that would have otherwise been required if the cost avoidance exercise had not been undertaken (Ashenbaum, 2006). Restated, cost avoidance is a reduction in costs that have to be paid in the future. Cost avoidance is commonly used as a metric by organizations that have to support and maintain systems, to quantify the value of the services that they provide and the actions that they take. The reason that cost avoidance is used rather than cost savings, is that if the value of an action is characterized as a cost savings, then someone wants the saved money back. In the case of system sustainment activities there is no money to give back. Requesting resources to create a cost avoidance is not as persuasive as making a cost savings or a return on investment argument.

Consider the maintenance of an electronic system. Electronics is almost always managed via an unscheduled maintenance policy, i.e., only fix it when it breaks. To formulate the ROI for adding PHM to an electronic system we first have to decide what we are measuring the ROI relative to. In the case of electronics we will measure the ROI of PHM relative to the unscheduled maintenance case since this is the commonly used default maintenance policy. The ROI from (1) becomes,

$$ ROI = \frac{V_f - V_i}{V_i} = \frac{C_u - C_{PHM}}{I_{PHM} - I_u} \tag{2} $$

where,

$C_u$ = life-cycle cost of the system when managed using unscheduled maintenance

$C_{PHM}$ = life-cycle cost of the system when managed using a PHM approach

$I_{PHM}$ = investment in PHM when managing the system using a PHM approach

$I_u$ = investment in PHM when managing the system using unscheduled maintenance.

To form (2), replace $V_f$-$V_i$ with $C_u$-$C_{PHM}$ (which assumes $C_u > C_{PHM}$) and $V_i$ with $I_{PHM}$-$I_u$. Note, $C_u$ and $C_{PHM}$ are total life-cycle costs that include their respective investment costs, $I_u$ and $I_{PHM}$. The denominator is the investment (relative to the unscheduled maintenance case). By definition, $I_u =$ 0 (contains no investment in PHM because there is no PHM), therefore, (2) simplifies to,

$$ ROI = \frac{C_u - C_{PHM}}{I_{PHM}} \tag{3} $$

In (3) ($C_u - C_{PHM}$) excludes all the costs that are a "wash" (i.e., the same independent of the maintenance approach). The formulation of the ROI in this manner solves the problem of splitting up the costs, i.e., we never need to address which particular life-cycle costs are due to the maintenance policy. In (3), if $C_u = C_{PHM}$ the *ROI* = 0 implying that the cost avoidance that results from PHM exactly equals the investment made (which is correct, again note that $C_{PHM}$ includes $I_{PHM}$ within it). In (2) and (3) the PHM investment cost is given by,

$$ I_{PHM} = C_{NRE} + C_{REC} + C_{INF} \tag{4} $$

where,

$C_{NRE}$ = PHM management non-recurring costs

$C_{REC}$ = PHM management recurring costs (cost of putting PHM hardware into each instance of the system)

$C_{INF}$ = PHM management infrastructure costs.

PHM management NRE costs are the costs of designing (hardware and software) to perform the PHM. PHM infrastructure cost are the costs of acquiring and keeping PHM management resources in place (equipment, people, training, software, databases, plan development, etc.). One question that arises is, is $I_{PHM}$ complete? Are there other investment costs that are not captured in (4)? This can be a difficult question to answer. For example:

- What if my PHM approach results in more maintenance actions (the need for more spare parts) than an unscheduled maintenance approach (generally it will, since it will cause maintenance to be performed prior to failure)? Is the cost of the extra spare parts accounted for in the investment ($I_{PHM}$)?
- What if (for simplicity) my PHM management approach resulted in buying the exact same number of spare parts for exactly the same price per part as my unscheduled maintenance approach, but I buy them at different times. Due to the cost of money (a non-zero discount rate), this does not end up costing the same. Is the cost of money part of $I_{PHM}$?

In general, the costs in the two examples above would not be included in the investment cost because they are considered the result of the PHM management approach (i.e., the result of the investment) and are reflected in the life-cycle cost $C_{PHM}$.

Performing the calculation of the life-cycle costs in (3) and investments in (4) is beyond the scope of this chapter, however, it is useful to look qualitatively at an example result (see Feldman et al. (2009) for the details of the model that was used to analyze this example). The ROI as a function of time for the application of a data-driven PHM approach[1] to an electronic display unit in the cockpit of a Boeing 737 is shown in Figure 2. Unscheduled maintenance, in this case means that the display unit will run until failure (i.e.,

no remaining useful life will be left) and then an unscheduled maintenance activity will take place. In the case of an airline, an unscheduled maintenance activity will generally be more costly to resolve than a scheduled maintenance activity because, depending on the time of the day that it occurs, it may involve delaying or canceling a flight. Alternatively, an impending failure that is detected by the PHM approach ahead of time will allow maintenance to be performed at a time and place of the airline's choosing, thus not disrupting flights and therefore being less expensive to resolve. These effects can be qualitatively seen in Figure 2.

Figure 2 was generated by simulating the life cycle of one instance of the "socket" that the display unit resides in managed using unscheduled maintenance and the data-driven PHM approach and applying (3) and (4). A socket is the location in a system that a module or LRU (Line Replaceable Unit) resides in. Sockets are tracked instead of modules because a socket could be occupied by one or more modules during its lifetime and socket cost and availability are more relevant to systems than the cost and availability of the modules. The ROI starts at a value of -1 at time 0, this represents the initial investment to put the PHM technology into the unit with no return ($C_u - C_{PHM}$ $= -I_{PHM}$). After time 0, the ROI starts to step down.[2] In this analysis the inventory cost (the cost of holding spares in the inventory) is 10% of the spare purchase price per year. Since spares cost more for PHM (due to higher recurring costs of the display unit due to the inclusion of PHM), inventory costs more. In the period from year 0 to 4, $C_{PHM}$ is increasing while $C_u$ and $I_{PHM}$ are constant (inventory costs are considered to be a result of the PHM investment, not part of the PHM investment). The step size decreases as time increases in part due to a non-zero cost of money (the discount rate in this example is 7%). If there was no inventory charge (or if the inventory charge was not a function of the spare purchase price) and there was no annual infrastructure cost, then

*Figure 2. ROI as a function of time for the application of a data-driven PHM approach to an electronic display unit in the cockpit of a Boeing 737 ((Sandborwn, 2012). © 2012 World Scientific Oublishing. Used with permission.)*

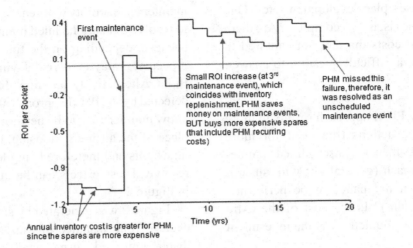

the ROI would be a constant -1 until the first maintenance event. The first maintenance event occurs in year 4 and is less expensive to resolve for PHM than for unscheduled maintenance (PHM successfully detected the failure ahead of time) – as a result the ROI increases to above zero. During the period from 4 to 8 years the decreases in ROI are inventory charges and annual PHM infrastructure costs – note, even though PHM infrastructure costs are an investment, they still affect the ROI ratio. A second maintenance event that was successfully detected by PHM occurs at year 8. In year 11 a third maintenance event oc-

curs and more spares are purchased. In year 18 there is a system failure that was not detected by PHM.

## The ROI of Putting Health Management into Wind Turbines

This section presents a simple case study associated with the insertion of health management hardware into turbine blades in a wind farm.

TRIADE (Figure 3) is a technology that was developed for aeronautic applications by EADS. The architecture encompasses temperature, pres-

*Figure 3. Architecture and photograph of the TRIADE health monitoring solution, (Haddad et al., 2011). © 2011 Taylor & Francis. Used with permission.*

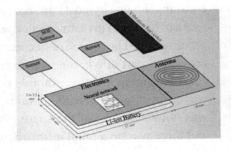 

*Table 1. Description of the sensors of TRIADE (Haddad et al., 2011)*

| Sensor | Quantity | Range | |
|---|---|---|---|
| Temperature | 1 1 | -40/+125°C | < 0.04°C/year drift |
| Pressure | 1 | 10/1100 mbar | -1 mbar/year drift |
| Vibration | 1 (3D) | 2-1000 Hz | Max sampling rate 12 kHz |
| Strain gauge | 6 1 | ± 5000 με | Self-compensated |
| Acoustic | 3 channels of 8 sensors each | Impact detection | |

sure, vibration, strain and acoustic sensors that can be used to monitor the health of a turbine (see Table 1).

The TRIADE solution is designed to be autonomous, and for ultra-low-power applications, having flexible clocking system, multiple low-power modes, instant wake-up and intelligent autonomous peripherals, features that enable true ultra-low-power optimization and dramatically extend battery life. The TRIADE Smart Tag acts like a passive RFID tag during communications. Passive tags are powered (during the communication phase) from the electromagnetic field generated by the RFID reader antenna. All these characteristics mean that it may be implemented in places that are not readily accessible, e.g. to monitor gearboxes in wind turbines.

The data used in this example, representing turbines in a wind farm, was adopted from Andrawus et al. (2006) and Arabian-Hoseynabadi et al. (2010), Table 2. The ROI analysis was performed using the life-cycle cost model from Sandborn and Wilkinson (2007).

To enable the calculation of ROI, the analysis first determines the optimal prognostic distance when using a data-driven PHM approach (see Figure 4). Due to uncertainties in the RUL predicted by the PHM approach, waiting for the whole predicted RUL before taking maintenance

action will result in a significant number of unscheduled failures. Prognostic distance is the amount of time before the forecasted failure (end of the RUL) that maintenance action should be taken. Small prognostics distances cause PHM to miss failures, while large distances are overly conservative and throw away lots of life. For the combination of PHM approach, implementation costs, reliability information, and operational profile assumed in this example, a prognostic distance of 470 hours yielded the minimum life-cycle cost over the support life of the turbine.

*Table 2. Inputs to ROI model*

| Inputs | Values |
|---|---|
| Failure rate per year (exponential distribution assumed) | 0.308 |
| PHM acquisition cost (Euros) | 300 |
| Operational time per blade per year (hours) | 2891 |
| Support life (years) | 20 |
| PHM annual infrastructure costs (Euros) per blade | 1282 |
| PHM recurring costs per blade (Euros) | 1820 |
| Blade base cost (Euros) | 37736 |
| PHM non-recurring costs per fleet (Euros) | 4520 |
| Number of blades per fleet | 78 |
| PHM non-recurring costs per blade (Euros) | 57.95 |
| Time to replace for scheduled maintenance (calendar hours) | 168 |
| Fraction of maintenance events requiring replacement (%) | 100 |
| Materials/logistics cost per maintenance (replacement) event | 1606 |
| Labor cost per maintenance (replacement) event (Euros) | 2753 |
| Total materials and labor | 4359 |
| Value added tax (%) | 17.5 |
| Discount rate/cost of money (%) | 8.2 |
| Cost per hour out of service (Euros) | 11 |
| Time to replace for scheduled maintenance (calendar days) | 7 |
| Time to replace for unscheduled maintenance (calendar hours) | (Variable) |
| Number of turbines | 26 |
| Number of blades per turbine | 3 |

*Figure 4. Variation of mean life-cycle cost with a prognostic distance for a data-driven PHM approach (1000-socket population). © 2011 Taylor & Francis. Used with permission.*

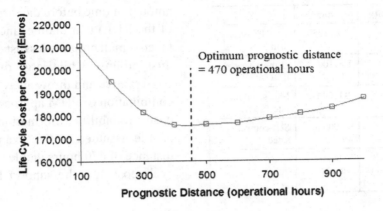

Similar analysis was conducted to determine the optimum fixed-interval scheduled maintenance interval. A fixed maintenance interval of 8,000 hours yielded the minimum life-cycle cost over the support life. Larger fixed maintenance intervals miss failures, while small intervals are overly conservative.

Figure 5 shows the life-cycle cost per socket for both data-driven PHM and fixed-interval scheduled maintenance case. A socket is a location in a system (in the wind turbine) where a single instance of the item being maintained (a blade) is installed. The socket may be occupied by one or more items during the lifetime of the system.

The time history of costs for each of 1000 sockets is shown in Figure 5. The data-driven PHM case resulted in an overall lower life-cycle cost (mean = €173,213) compared to the fixed-interval scheduled maintenance case (mean = €356,999). The data-driven PHM case requires fewer spares throughout the support life of the system. This is primarily due to maximizing the useful life of the blades, i.e., early warning of failures in the data-driven PHM case provided an opportunity to schedule and perform maintenance events closer to the actual failures, thus, avoiding failures while maximizing the useful life. Alternatively, the fixed-interval scheduled maintenance case re-

*Figure 5. (a) Life-cycle cost per socket for a fixed-interval scheduled maintenance approach. (b) Life-cycle cost per socket for a data-driven PHM approach. 1000-socket population. © 2011 Taylor & Francis. Used with permission.*

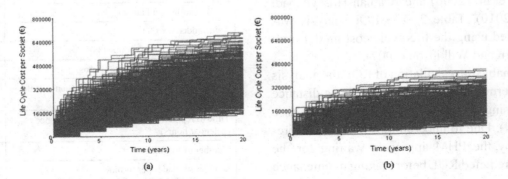

sulted in either throwing-away more useful-life (early intervention). In both cases, some unscheduled maintenance events (intervention that is too late) occurred. Intuitively the advantage of PHM over fixed-interval maintenance for this case is shown in Figure 5 where the life-cycle cost was minimized in the PHM case when the prognostics distance was 470 operational hours versus 8000 operational hours in the fixed-interval case.

We now wish to estimate the return on investment (ROI) of the data-driven PHM approach relative to a fixed-interval scheduled maintenance approach. The mean total life-cycle cost per blade, for a data-driven PHM approach, was €173,213 (mean), with an effective investment cost per blade of €25,408 (mean), representing the cost of developing, supporting, and installing PHM in the blade. This cost was compared to the fixed-interval scheduled maintenance approach, where the total life-cycle cost per blade was €356,999 (mean). Note that the investment cost for the fixed-interval scheduled maintenance policy is by definition zero.

Figure 6 shows the histogram of the computed ROIs for 1000-socket population (due to uncertainties in all quantities, each socket in a population will have a unique ROI) relative to

*Figure 6. Histogram of ROI for a data-driven PHM solution for a 1000-socket population. © 2011 Taylor & Francis. Used with permission.*

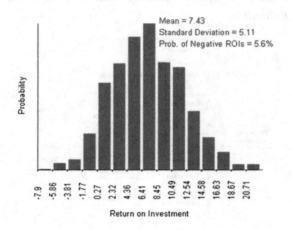

the fixed-interval maintenance solution. In this example, the computed mean ROI of investing in a data-driven PHM approach for the population of blades was 7.43. Notice that some of the ROI values in Figure 6 are negative. This means that that implementing a data-driven PHM approach for the blades will result in an economic loss for some of the blades, i.e., you will end up being worse off than fixed-interval scheduled maintenance. Based on Figure 6, this example predicts that a data-driven PHM approach would result in a positive ROI (cost benefit) with a 94.4% confidence.[3]

## SYSTEM-LEVEL MAINTENANCE VALUE

One view of the fundamental value of advanced maintenance paradigms discussed earlier (RCM, CBM, and PHM) is that they create flexibility for managing a system. The scope of the flexibility that is created differentiates RCM from CBM and CBM from PHM. RUL is the remaining useful life that a system has and it effectively represents the lead time (subject to appropriate uncertainties) for the decision-maker or other maintenance entities to take preventive actions prior to a failure. This can be described as a flexibility phenomenon whereby entities involved with the operation, management, and maintenance of a system have the flexibility to take actions at any time up to the end of the RUL. Therefore, assessing the value of the using the RUL is of primary importance and provides the decision-maker with the true value of the cost avoidance possible when using PHM.

In a system that incorporates PHM, after a prognostic indication (the information provided by the PHM system that indicates that the RUL is limited), several different actions can be taken by the decision-maker to manage the health of the system. Examples of the actions that can be taken are fault accommodation, changing loads, and tactical control. Hence the decision-maker has a set of *options* among which they can choose. The term *options* will be used in the remainder of the section to denote a choice or action the decision maker can take after a prognostic indication. The flexibility that PHM allows increases the value of the system if the decision maker can take advantage of the options it provides. A real option (RO) is an alternative or choice that becomes available with a business investment opportunity, (Investopidia, 2011); it is a right, but not an obligation, to take action within a period of time. In our case, the investment opportunity is the investment in putting PHM into the system.[4]

Real options have been used for maintenance applications, e.g., the comparison of different maintenance strategies and their effects on the total costs for the maintenance and management of an existing bridge for thirty years (Koide et al., 2001). ROs have also been applied in the maintenance, repair, and overhaul industry (Miller and Park, 2004). Jin et al. (2009) used an option-based cost model for scheduling joint production and preventive maintenance for a manufacturing industry when demand was uncertain. The Jin et al. option-based mathematical model provides recommendations for maintenance decisions in the environment of uncertain demand. Mappings from financial options have been proposed in the literature, e.g., (Kodukula and Papudesu, 2006). In order to value maintenance options, Haddad et al., (2012a) propose the extension to maintenance options to address the options created by PHM.

Figure 7 shows the general categories of options that can arise after a prognostic indication. For instance, maintenance can be carried out immediately after the prognostic indication, or it can be delayed in order to use up the RUL. Alternatively, the mission can be abandoned completely if it is judicious to do so. The option to abandon is the right the decision-maker has to quit the mission or stop the operation of the system. This may be the case when the consequences of the predicted failure are large. The most general option relevant to PHM, and the subject of the remainder of the

discussion in this section, is the option to *wait to perform maintenance* (Haddad et al., 2012b). The option to wait proposes a solution to the fundamental problem for systems with prognostics: when is the optimal time to maintain while maximizing the use of RUL and minimizing the risk of failure.

Numerous benefits or cost-avoidance opportunities are derived from the knowledge of when the system will fail. Hence assessing the value of waiting represents a means for monetizing the true value of having PHM in a system. The option to wait is the right that the decision-maker has to delay maintenance actions thus using up the RUL. If the system is used until failure, then unscheduled maintenance will need to be performed and in this case the option expires, and the cost invested in PHM did not produce any cost avoidance. However, sometimes choosing unscheduled maintenance may be a better policy. The value of waiting (and related options such as abandoning) are the key to applying RO theory to the PHM valuation problem.

*Figure 7. Options arising after a prognostic indication*

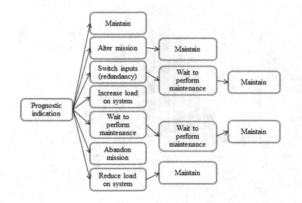

## The NPV of Scheduled Maintenance versus PHM without Maintenance Flexibility

The cost data used in this section is based on Andrawus et al. (2006) who present a cost-benefit analysis for implementing PHM on a wind farm of 26 land-based turbines, and compare the net present value (NPV) of scheduled maintenance and PHM. We reproduce the NPV analysis for 7 turbines[5] using the costs from Andrawus et al. (2006), rescaling the farm to 7 turbines, and discounting over a period of 18 years (support life of the turbines after they go out of warranty) with a discount rate of 8.2%. This analysis results in a NPV of 32,869£ for scheduled maintenance, and 64,374£ for PHM. This result indicates that scheduled maintenance is more beneficial than PHM.

Andrawus et al. (2006) consider that the PHM system avoids failures of the following subsystems: blade, bearing, main shaft, gearbox, and generator. The frequency of failure $\alpha$, indicates the probability of failure. When the PHM system avoids failure a cost avoidance, called failure consequence ($F_C$) results. Andrawus et al. (2006) identify these cost avoidance opportunities as an annual cost reservation, $A_{CR}$,[6] which is the product of the frequency of failure $\alpha$, the failure consequence ($F_C$), and the number of turbines in the farm $N_T$:

$$A_{CR} = \alpha F_C N_T \qquad (5)$$

The failure consequence ($F_C$) consists of production lost when turbine is not operating ($TC_{PL}$), the cost of material to maintain the turbines ($TC_{MT}$), the cost of labor to maintain the turbines ($TC_{LB}$), and cost of accessing the turbines ($TC_{AS}$):

$$F_C = TC_{PL} + TC_{MT} + TC_{LB} + TC_{AS} \qquad (6)$$

The cost of production loss in (6) is given by,

$$TC_{PL} = 24 N_{dy} WT_{PR} C_{EH} C_f \qquad (7)$$

where $N_{dy}$ is the number of days of downtime per failure, $WT_{PR}$ is the turbine power rating, $C_{EH}$ is the cost of energy, and $C_f$ = capacity factor.

The other costs appearing in (6) are computed by Andrawus et al. (2006) for various turbine failure modes. These costs appear in Table 3 and are not varied in this analysis. Table 3 shows the failure consequence that results from an assumed capacity factor of 0.33.

For the case of a generator, $N_{dy}$ = 7, $WT_{PR}$ = 600 KW, $C_{EH}$ = 0.5£/KW hour, and $C_f$ = 0.33, then $TC_{PL}$ is 1663.20£ from (7). Considering the case of the generator with a frequency of failure $\alpha$ of 0.00641, from the last row of Table 3 and (5) the annual cost reservation $A_{CR}$ is 1,613£. This result can be seen in the last row of Table 4. Note, only gearbox and generator failures are included in Table 4 since no blade, main bearing or main shaft failures were observed in the 7 turbine sample used in this study.

*Table 3. Failure consequence (£)*

| Failure mode | $TC_{MT}$ | $TC_{LB}$ | $TC_{AS}$ | $TC_{PL}$ | $F_C$ |
|---|---|---|---|---|---|
| Blade failure | 34,545 | 2,400 | 8,460 | 1,663.20 | 47,068 |
| Main bearings failure | 9,851.49 | 2,400 | 8,460 | 1,663.20 | 22,375 |
| Main shaft failure | 11,133.36 | 4,800 | 11,280 | 1,900.80 | 29,114 |
| Gearbox failure | 61,687.50 | 3,600 | 11,280 | 1,900.80 | 78,468 |
| Generator failure | 23,441.25 | 2,400 | 8,460 | 1,663.20 | 35,964 |

*Table 4. Calculation of annual cost reservation (£)*

|  | Number of events | $\alpha$ | $F_C$ | $A_{CR}$ | $A_{CR}$ scaled to 7 turbines |
| --- | --- | --- | --- | --- | --- |
| Gearbox failures | 2 | 0.01282 | 78,468 | 26,156.10 | 7,042 |
| Generator failures | 1 | 0.00641 | 35,964 | 5,994.08 | 1,613 |

Similar results are obtained for the gearbox and these are also shown in Tables 3 and 4. The other turbine subsystems are not included in the analysis as they are assumed to have a frequency of failure of 0. The annual cost reservation is realized over the 18 years of support life. Using

$$NPV_{PHM} = \sum_{y=1}^{N} \frac{\text{Total } A_{CR}}{\left(1+i\right)^y} \qquad (8)$$

which is cost is discounted to year 0 and results in a net present value of 80,004£. This value is added to the net present value of inspection (or time based maintenance (TBM), and results in a total of 112,873£ ($NPV_{TBM}$), which is the real cost of scheduled maintenance. To assess the value of PHM, the difference in NPV of scheduled maintenance and PHM is calculated and results in 48,498£. This value compares the net present value accounting for the cost and cost avoidance derived from scheduled maintenance and PHM over the life cycle of the turbine.

## Maintenance Options Analysis

The Andrawus et al. (2006) analysis presented in the last subsection and other traditional methods used to quantify the benefits of PHM do not account for the value of the options that the decision-maker can take after a prognostic indication. Furthermore, it is necessary to account for uncertainties in the capacity factor and other parameters. For example, if there are forecasts for wind speed, then the decision-maker may decide to continue running the system (post the prognostic indication) while there is probability of high wind speeds in order to

harness the upside effect of uncertainty. Another uncertainty that needs to be accounted for is the uncertainty within the PHM system (i.e., its RUL predictive capability is not perfect).

For example, to estimate the uncertainty in capacity factor, we consider the power from a healthy turbine. The power output from this turbine is averaged every day, and divided by 600KW, the maximum theoretical power the turbine can produce to form a capacity factor ($C_f$). The resulting distribution of capacity factors is shown in Figure 8. This distribution is fit with a beta distribution (also shown in Figure 8) that has the following form (the best fit parameters are: $\gamma = 0.734$ and $\beta = 2$):

$$f(x;\gamma,\beta) = \frac{x^{\gamma-1}(1-x)^{\beta-1}}{\int_0^1 u^{\gamma-1}(1-u)^{\beta-1}\,du} \qquad (9)$$

*Figure 8. Distribution fitting for the capacity factor over a year for healthy turbine*

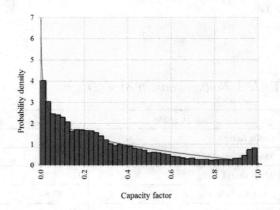

To incorporate flexibility in the model, we assume that the decision-maker can turn the turbine off when there is a prognostic indication for the gearbox. The cost of PHM for the gearbox is assumed to be 40% of its cost of failure (i.e., if PHM detects the problem before failure, the cost to resolve it is 40% of the cost of unscheduled failure) and the cost of resolving a problem detected during scheduled maintenance is 70% of the cost of resolving a failure that was unanticipated (Walford and Roberts, 2006). The costs of production loss from Table 3 for a gearbox and generator are 1900.80£ and 1663.20£ respectively. Using (7) we can calculate the number of days the turbine is down for each system: 8 for gearbox and 7 for generator. Using (7) and the capacity factor distribution, we can update the production loss accounting for the uncertainty in the capacity factor. As a result, the net present value of the difference in the two maintenance paradigms demonstrating the value of PHM can be now represented as a distribution.

Now assume that there is a misclassification rate of 0.05 associated with the PHM system. This implies that the PHM system will not provide a prediction of failure 5% of the time. Figure 9 shows a decision tree for uncertainty associated with PHM.

We will assume that the turbines considered here have the same number of failure events (same $\alpha$) that that Andrawus et al. (2006) assumed. However, we will assume that when predicted by PHM, the decision-maker has the option to aban-

don and halt the operation of the wind turbine. This will result in a 30% cost saving in supporting maintenance (under the assumptions listed earlier: the cost of PHM for the gearbox is 40% of its cost of failure and the cost of scheduled maintenance is 70% of the cost unscheduled maintenance).

The expected cost avoidance value is obtained by multiplying the probability of occurrence by the outcome of each branch in the decision tree shown in Figure 6 as given by,

$$Cost\ Avoidance = 0.95(NPV_{TBM} - NPV_{PHM})$$
$$+ 0.05(NPV_{TBM} - NPV_{PHM})_{no\ failure\ prediction}$$
$$(10)$$

where $NPV_{TBM} = (7/26)117,021 = 31,505£$ (117,021£ is the total cost savings of the 18 year life cycle for time based maintenance given by Andrawus et al. 2006) and $NPV_{PHM}$ is obtained from Monte Carlo simulation. The net present value comparing PHM and TBM implies that the true cost of supporting TBM is higher than 117,021£ (or 31,505£ for 7 turbines). The frequency of failure is used in conjunction with the failure consequence in (5) to obtain the annual cost reservation. The annual cost reservation is the cost that must be added to the cost of supporting TBM to obtain its true cost. The difference between the true cost of TBM and the cost of supporting PHM is the value of PHM in the net present value analysis.

*Figure 9. Decision tree for uncertainty within the PHM system*

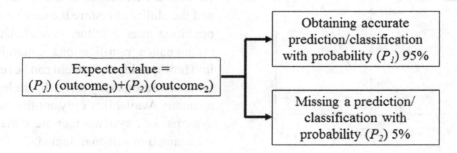

The second term in (10) reflects that fact that when the PHM system fails to predict the failures, the annual cost reservation is less than it is for a perfect PHM system. The fraction of the prediction when the PHM system does not predict the failure should not be added to the annual cost reservation of TBM since neither TBM nor PHM would have predicted the failure and avoided it. To reflect this in our analysis, we consider that there is a 5% chance the PHM system does not predict the failure (as shown in Figure 9 and in (10)). This is a modeled by assuming that the frequency of failure is 0 (this results in a cost savings of only 117,021£ for 26 turbines or 31,505£ for 7 turbines when using TBM, no additional cost reservation because no failures were avoided).

Now we complete the steps of the methodology and represent them in a distribution of net present value (called value at risk and gain diagram - VARG diagram). 500 Monte Carlo simulations were run and combined with the decision trees are represented in Figure 10. We assume that the decision-maker can exercises the abandon option 50 days prior to the failure.[7] Figure 10 shows both the results from the analysis with flexibility and the analysis that does not account for flexibility or the uncertainties. In other words, the use of

PHM on systems results in a cost avoidance that is the amount of value the PHM system is providing the user.

Figure 10 shows that the 18% value at risk is 48,850£. This result indicates that there is an 18% chance that the value of PHM on the gearbox and generator is smaller than the value obtained from the analysis that does not account for flexibility and uncertainties. The 50% value at gain is 56,096£. This result means that there is a 50% chance that the value of PHM will be greater than 56,096£. This additional value is a result of accounting for the uncertainty in the capacity factor, and the cost-avoidance that is possible by turning the turbine off to avoid failure (leading to a lower maintenance cost).

This section included uncertainties and flexibility in the quantification of the benefit of PHM. We consider the support life and show that exercising flexibility can result in a higher value from PHM. This can potentially reduce the operation and maintenance cost further than analysis that do not consider maintenance options arising from implementing PHM. Other options arising from PHM also exist (e.g., alter mission, reduce load, etc.), the valuation of these options is also key to understanding the benefit of PHM during system operation.

## AVAILABILITY REQUIREMENTS

Availability is the ability of a service or a system to be functional when it is requested for use or operation. The concept of availability accounts for both the frequency of the failure (reliability) and the ability to restore the service or system to operation after a failure (maintainability). The maintenance ramifications generally translate into how quickly the system can be repaired upon failure and are usually driven by the logistics management. Availability only applies to maintained systems, i.e., systems that are either externally maintained or self-maintained.

*Figure 10. VARG diagram for model accounting for real capacity factor and option to turn the turbine off 50 days before failure*

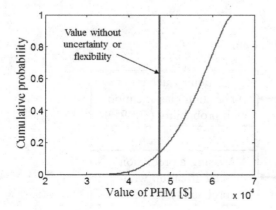

Availability has been a critical design parameter for the aerospace and defense communities for many years, but recently it is beginning to be recognized, quantified, and studied by other types of systems. Many real world systems are significantly impacted by availability. A failure, i.e., a decrease of availability, of an ATM machine causes inconvenience to customers; poor availability of wind farms can make them non-viable; the unavailability of a point-of-sale system to retail outlets can generate a huge financial loss; and the failure of a medical device or of hospital equipment can result in loss of life. In these example systems, insuring the availability of the system becomes the primary interest and the owners of the systems are often willing to pay a premium (purchase price and/or support) for higher availability.

## Availability Contracting

Customers of avionics, large scale production lines, servers, and infrastructure service providers with high availability requirements are increasingly interested in buying the availability of a system, instead of actually buying the system itself; resulting in the introduction of "availability-based contracting." Availability-based contracts are a subset of outcome-based contracts (Ng et al., 2009); where the customer pays for the delivered outcome, instead of paying for specific logistics activities, system reliability management or other tasks. Basically, in this type of contract, the customer pays the service or system provider to ensure a specific availability requirement. For example the Availability Transformation: Tornado Aircraft Contract–ATTAC (BAE, 2008), is an availability contract, where BAE Systems has agreed to support the Tornado GR4 aircraft fleet at a specified availability level throughout the fleet service life for the UK Ministry of Defense. The agreement aims to implement a new approach to improve the availability of the fleet while minimizing the life-cycle cost (BAE, 2008). Another form of outcome-based contracting that is used by the U.S.

Department of Defense is called performance-based contracting (or PBL – Performance Based Logistics). In PBL contracts the contractor is paid based on the results achieved, not on the methods used to perform the tasks (Beanum, 2006, and Hyman, 2009). Availability contracts, and most outcome-based contracting, include cost penalties that could be assessed for failing to fulfill a specified availability requirement within a defined time frame (or a contract payment schedule that is based on the achieved availability).

The evaluation of an availability requirement is a challenging task for both suppliers and customers. From a suppliers' perspective, it is not trivial to estimate the cost of delivering a specific availability. Entering into an availability contract is a non-traditional way of doing business for the suppliers of many types of safety- and mission-critical systems. For example, the traditional avionics supply chain business model is to sell the system; and then separately provide the sustainment of the system. As a result, the avionics suppliers may sell the system for whatever they have to in order to obtain the business, knowing that they will make their money on the long-term sustainment of the system. From a customers' perspective, the amount of money that should be spent on a specific availability contract is also a mysterious quantity – if a choice has to be made between two offers of availability contracts where the value of the promised availabilities are similar (e.g., one contract offers an availability of 95%, and the other one offers 97%), then how much money should the customer be willing to spend for a specific availability improvement?

Two common mechanisms that may include elements of availability contracting are Product Service Systems (PSS) and leasing models. Figure 11 shows an example PSS spectrum that indicates the concept of outcome-based contracting models (of which availability contracting is an example). PSS provide both the product and its service/support based on the customer's requirements (Bankole et al., 2009), which could include an

*Figure 11. Example PSS spectrum for a car*

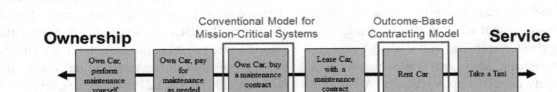

availability requirement. Lease contracts (Yeh and Chang, 2007) are use-oriented PSS, where the ownership of the product is usually retained by the service provider. A lease contract may indicate not only the basic product and service provided but also other use and operation constraints such as the failure rate threshold. In leasing agreements the customer has an implicit expectation of a minimum availability, but the availability is generally not quantified contractually.[8]

## Design for Availability

Availability optimization and requirements-based availability optimization (e.g., Kiureghiana et al. 2007, Sharma and Misra, 1988) have the objective of deriving the optimal system parameters to maximize availability, while satisfying other system management requirements (e.g., demand requirements, reliability requirements, inventory management requirements, etc.). Availability optimization problems are usually approached using constrained-optimization methods. The availability can also be maximized while concurrently reducing the costs associated with supporting the system; which usually requires performing an availability versus cost tradeoff analysis. Previous work has been done in the area of maximizing a cost-benefit function that combines the accumulated life-cycle costs associated with a specific system's management (e.g., logistics, maintenance, reliability, etc.) and the availability achieved, e.g., Naikan and Rao, (2005), and

Reineke et al. (1999). These approaches maximize the return by choosing the optimal system design and support parameters. All the work described so far is focused on solving problems where availability is computed and maximized (e.g., Dadhich and Roy, 2010; De Castro and Cavalca, 2003), by representing the system management as closed-form analytical equations, which can be used to derive closed-form analytical solutions for the availability. These solutions provide value when system management can be represented by a relatively simple set of design and support parameters. Existing availability optimization approaches described previously in this section all seek to determine system parameters that maximize availability at a specific point in time rather than meeting a minimum availability requirement at all times during the entire support life of the system. Also, existing availability optimization approaches usually require closed-form equations to compute and optimize the availability. Alternatively, in discrete event simulation solutions, the quantities are accumulated throughout the timeline; and the availability can be computed based on the accumulated event sequence up to any point in time during the system support life.

Discrete event simulation is usually the preferred approach to model and predict the life-cycle characteristics (cost and availability) of large populations of complex real systems managed for long periods of time. However, while using discrete event simulation to predict the availability of a system or a population of systems based on known or predicted system parameters is relatively straightforward; determining the system parameters that result in a desired availability using discrete event simulation is not, and is generally performed using "brute force" search-based methods that become quickly impractical for designing systems with more than a few variables and when uncertainties are present.

A discrete-event simulation approach to design for availability was presented by Jazouli and Sandborn, (2011). In this approach, an availability requirement is used to compute and impose the uptimes and downtimes required to meet a specified availability throughout the system's life using a discrete event simulator. Then, it uses these imposed uptime and downtime values to solve for the unknown system parameter. The methodology is summarized in Figure 12. Unlike existing availability optimization approaches that seek to determine system parameters that maximize availability at a specific point in time, the

design for availability methodology in Figure 12 insures that a minimum availability requirement is met at all times during the entire support life of the system.

As an example of the design for availability approach, the maximum allowable inventory lead time (*ILT*) necessary to fulfill a specific availability requirement for unscheduled maintenance versus a data-driven PHM approach are compared. *ILT* is the duration of time between ordering spares and receiving them. *ILT* can be an important differentiator between sources for critical parts. Determining the maximum allowable *ILT* for a specific availability requirement could be used to improve logistics management and potentially reduce life-cycle cost. If the availability drops below a specified threshold value, a cost penalty could be assessed; determining upfront the necessary *ILT* could avoid these potential cost penalties. Also, knowing the maximum allowable *ILT*,

*Figure 12. Design for availability methodology, Jazouli and Sandborn (2011). © 2011 IEEE. Used with permission.*

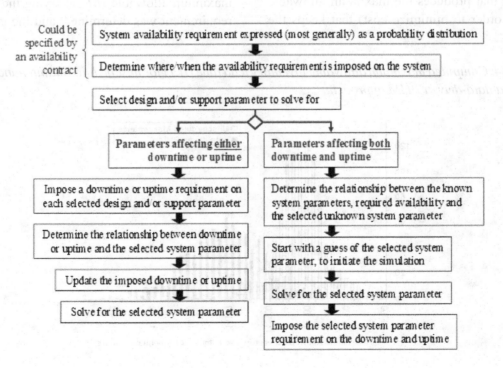

*Figure 13. Required availability distribution (input to the model). © 2011 IEEE. Used with permission.*

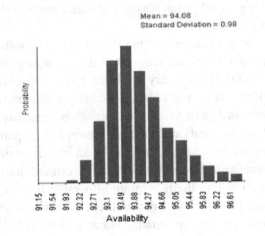

availability value at or above the availability requirement.

By running the simulation (all inputs are summarized in Jazouli and Sandborn (2011)) with the imposed availability (input) requirement shown in Figure 13, the *ILT* (output) satisfying this requirement was determined for the unscheduled maintenance policy (the light gray histogram bars in Figure 14).

The methodology has also been applied to the same system managed using a data-driven PHM approach. Using the input data (provided in the Jazouli and Sandborn (2011)) with the data-driven PHM approach, an optimal prognostic distance of 600 operational hours results in the minimum life-cycle cost over the entire support life (determined similarly to Figure 4. Also, a symmetric triangular distribution with a width of 500 hours was assumed to represent the effectiveness of the data-driven PHM approach (see Sandborn and Wilkinson (2007) for details of modeling the PHM uncertainty).

customers could require their suppliers to deliver within a specific lead time.

Implicitly, reducing the delivery time, i.e., *ILT* (considered as the only variable input in this example) would increase the availability. However, we also want to maximize the *ILT* to reduce the cost. Basically, we want to generate an optimal solution that produces the maximum allowable *ILT* (in order to minimize cost) that keeps the

After running the simulation with the imposed availability requirement shown in Figure 13, the maximum allowable *ILT* satisfying the contract requirement was determined, and the generated

*Figure 14. Computed maximum allowable inventory lead time (ILT) for unscheduled maintenance policy, and for a data-driven PHM approach.*

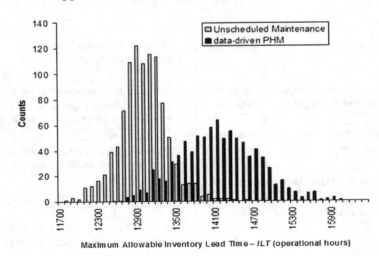

*ILT* probability distribution is shown in Figure 14 (black histogram bars).

In Figure 14, both of the maintenance approaches satisfy the same availability requirement shown in Figure 13. In this example, the data-driven PHM approach allows for a larger *ILT* (mean = 13,936 operational hours), compared to the unscheduled maintenance case (mean = 12,961 operational hours). In other words, using a data-driven PHM approach allows a given availability requirement to be met if *ILTs* are longer, or alternatively stated, the use of PHM would allow a supply chain with longer *ILTs* to be used. The use of a data-driven PHM approach has shifted the maximum allowable *ILT* distribution by approximately 1000 hours to the right. This result is due the fact that data-driven PHM has provided early warning of failures; therefore converting maintenance from unscheduled to scheduled reducing the accumulated operational downtime. For a fixed *ILT*, this would result in an improved operational availability of the system. However, in this example the same availability requirement was imposed on both cases (unscheduled maintenance and data-driven PHM), thus the accumulated operational downtime was used as a fixed quantity (imposed by the contract availability requirement); therefore, the avoided

unscheduled maintenance downtime was added to the inventory downtime, resulting in a larger allowed *ILT* in the data-driven PHM case.

Figure 15 shows how the maximum allowable *ILT* results could be practically interpreted. For example, if the *ILT* of each spares replenishment order is equal or greater than 12,700 operational hours, then the system manager would be 95% confident to meet the availability requirement using a data-driven PHM approach, and only 78% confident to meet the same availability requirement under an unscheduled maintenance approach.

Notice that, the application of the design for availability methodology doesn't only determine the unknown system parameter (e.g., *ILT*) but also illustrates the effect of each adopted sustainment approach on the selected system parameter. In other words, for a specific PHM approach or any other maintenance policy, one could forecast the variation of the selected system parameter, and react accordingly to maintain the required availability.

To summarize, a design for availability methodology was applied to an avionics case study for two different maintenance approaches, to satisfy a specific contract availability requirement. For both approaches, unscheduled maintenance and data-driven PHM, we were able to determine the

*Figure 15. Computed TTF cumulative distribution function for unscheduled maintenance and data-driven PHM*

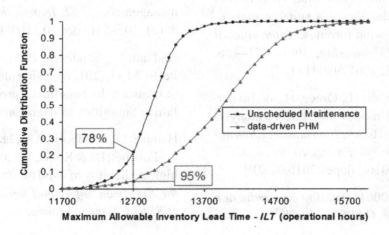

unknown system parameter (the maximum allowable *ILT* in this case) satisfying the availability requirement. Then a comparison of the results showed that data-driven PHM allows larger *ILTs* compared to the unscheduled maintenance policy case. See Jazouli and Sandborn (2011) for a more detailed discussion.

## CONCLUSION

The insertion of prognostics and health management strategies within systems has the potential to create transformative business and system management practices, through: 1) logistics transformation—the development of logistics processes necessary for the enterprise-level management; 2) implementation impacts—the development of methodologies to accurately measure the value of prognostics and associate risk mitigation in ways that are useful to system integrators; 3) business process changes—innovative ways of doing business based on the adoption PHM technologies; and 4) business enablers—developing business models that enable system integrators to successfully implement prognostics and system health management into their systems.

## REFERENCES

Andrawus, J. A., Watson, J., Kishk, M., & Adam, A. (2006). The selection of a suitable maintenance strategy for wind turbines. *International Journal of Wind Engineering*, 30(6), 471–486. doi:10.1260/030952406779994141

Arabian-Hoseynabadi, H., Oraee, H., & Tavner, P. (2010). Failure modes and effects analysis (FMEA) for wind turbines. *International Journal of Electrical Power & Energy Systems*, 32(7), 817–824. doi:10.1016/j.ijepes.2010.01.019

Ashenbaum, B. (2006). *Defining cost reduction and cost avoidance*. CAPS Research.

BAE. (2008). *BAE 61972 BAE annual report*.

Bankole, O. O., Roy, R., Shehab, E., & Wardle, P. (2009). Affordability assessment of industrial product-service system in the aerospace defense industry. *Proceedings of the CIRP Industrial Product-Service Systems (IPS2) Conference* (p. 230).

Beanum, R. L. (2006). Performance-based logistics and contractor support methods. *Proceedings of the IEEE Systems Readiness Technology Conference (AUTOTESTCON)*.

Cheng, S., Azarian, M., & Pecht, M. (2010). Sensor systems for prognostics and health management. *Sensors (Basel, Switzerland)*, 10, 5774–5797. doi:10.3390/s100605774

Dadhich, V., & Roy, S. (2010). Availability-based optimization of preventive maintenance schedule: A parametric study. *Chemical Engineering Transactions*, 21, 949–954.

De Castro, H., & Cavalca, K. (2003). Availability optimization with genetic algorithm. *International Journal of Quality & Reliability Management*, 20(7), 847–863. doi:10.1108/02656710310491258

Dhillon, B. S. (1999). *Engineering maintainability*. Houston, TX: Gulf Publishing Company.

Feldman, K., Jazouli, T., & Sandborn, P. (2009). A methodology for determining the return on investment associated with prognostics and health management. *IEEE Transactions on Reliability*, 58(2), 305–316. doi:10.1109/TR.2009.2020133

Haddad, G., Sandborn, P. A., Eklund, N. H., & Pecht, M. G. (2012b). Using maintenance options to measure the benefits of prognostics for wind farms. Submitted to *Wind Energy*.

Haddad, G., Sandborn, P. A., Jazouli, T., Pecht, M. G., Foucher, B., & Rouet, V. (2011). Guaranteeing high availability of wind turbines. *Proceedings of the European Safety and Reliability Conference (ESREL)*, Troyes, France.

Haddad, G., Sandborn, P. A., & Pecht, M. G. (2012a). An options approach for decision support of systems with prognostic capabilities. To be published *IEEE Transactions on Reliability.*

Hyman, W. A. (2009). Performance-based contracting for maintenance. *NCHRP Synthesis 389.*

Investopedia (2011). Real option. *Investopedica.* Retrieved September 1, 2011, from http://www.investopedia.com/terms/r/realoption.asp#axzz1WiKKDwgK

Jazouli, T., & Sandborn, P. (2011). Using PHM to meet availability-based contracting requirements. *Proceedings of the IEEE International Conference on Prognostics and Health Management,* Denver, CO.

Jin, X., Li, L., & Ni, J. (2009). Options model for joint production and preventive maintenance system. *International Journal of Production Economics, 119*(2), 347–353. doi:10.1016/j.ijpe.2009.03.005

Kiureghiana, A., Ditlevsenb, D., & Song, J. (2007). Availability, reliability and downtime of systems with repairable components. *Reliability Engineering & System Safety, 92,* 231–242. doi:10.1016/j.ress.2005.12.003

Kodukula, P., & Papudesu, C. (2006). *Project valuation using real options: A practitioner's guide.* Fort Lauderdale, FL: J. Ross Publishing.

Koide, Y., Kaito, K., & Abe, M. (2001). Life cycle cost analysis of bridges where the real options are considered. In Das, P. C., Frangopol, D. M., & Nowak, A. S. (Eds.), *Current and future trends in bridge design, construction and maintenance* (*Vol. 2*, pp. 387–394). London, UK: Thomas Telford Publishing.

Miller, L. T., & Park, C. S. (2004). Economic analysis in the maintenance, repair, and overhaul industry: an options approach. *The Engineering Economist, 49,* 21–41. doi:10.1080/00137910490432593

Naikan, V., & Rao, P. (2005). Maintenance cost benefit analysis of production shops – An availability based approach. *Proceedings of Reliability and Maintainability Symposium,* (pp. 404-409). January 24-27.

Ng, I. C. L., Maull, R., & Yip, N. (2009). Outcome-based contracts as a driver for systems thinking and service-dominant logic in service science: Evidence from the defence industry. *European Management Journal, 27*(6), 377–387. doi:10.1016/j.emj.2009.05.002

Pecht, M. (2008). *Prognostics and health management of electronics.* Hoboken, NJ: Wiley. doi:10.1002/9780470385845

Reineke, D., Murdock, W., Pohl, E., & Rehmert, I. (1999). Improving availability and cost performance for complex systems with preventive maintenance. *Proceedings of Reliability and Maintainability Symposium,* (pp. 383-388).

Sandborn, P. (2012). *Cost analysis of electronic systems.* Singapore: World Scientific.

Sandborn, P., & Myers, J. (2008). Designing engineering systems for sustainment. In Misra, K. B. (Ed.), *Handbook of performability engineering* (pp. 81–103). London, UK: Springer. doi:10.1007/978-1-84800-131-2_7

Sandborn, P., & Wilkinson, C. (2007). A maintenance planning and business case development model for the application of prognostics and health management (PHM) to electronic systems. *Microelectronics and Reliability, 47*(12), 1889–1901. doi:10.1016/j.microrel.2007.02.016

Sharma, U., & Misra, K. (1988). Optimal availability design of a maintained system. *Reliability Engineering & System Safety, 20*(2), 147–159. doi:10.1016/0951-8320(88)90094-4

Walford, C., & Roberts, D. (2006). Condition monitoring of wind turbines: Technology overview, seeded-fault testing, and cost benefit analysis. Tech. Rep. 1010419. Palo Alto, CA EPRI.

Yeh, R. H., & Chang, W. L. (2007). Optimal threshold value of failure-rate for leased products with preventive maintenance actions. *Mathematical and Computer Modelling, 46*, 730–737. doi:10.1016/j.mcm.2006.12.001

## KEY TERMS AND DEFINITIONS

**Availability:** The ability of a service or a system to be functional when it is requested for use or operation.

**Availability Contracts:** A subset of outcome-based contracts; where the customer pays for the delivered outcome, instead of paying for specific logistics activities, system reliability management or other tasks.

**Cost Avoidance:** A reduction in costs that have to be paid in the future.

**Design for Availability:** Methodologies for determining parameters (system and logistics) that guarantee the ability to satisfy a minimum availability requirement at all times during the support life of the system.

**Prognostics and Health Management (PHM):** A discipline consisting of technologies and methods to assess the reliability of a system in its actual life-cycle conditions, to determine the advent of failure, and mitigate system risks.

**Real Options:** An alternative or choice that becomes available with a business investment opportunity; it is a right, but not an obligation, to take action within a period of time.

**Return on Investment (ROI):** The monetary benefit derived from having spent money on developing, changing, or managing a product or system.

## ENDNOTES

1    Data-driven PHM means that you are directly observing the system and deciding that it looks unhealthy (e.g., monitoring for precursors to failure, use of canaries, anomaly detection). See Pecht (2008) for details.

2    In Figure 2 all the accounting is done on an annual basis, so ROI is only recalculated once per year.

3    Some blades will never fail and in these cases, the investment in PHM is wasted (negative ROI).

4    Real options (RO) are different than financial options. Financial options are derivative instruments, whereas real options are choices that an organization can make that pertain to tangible assets.

5    We rescaled the wind farm from 26 to 7 turbines due to the availability of actual failure data for 7 different turbines in a farm.

6    The cost reservation is the amount of money that could have been saved if the failures could have been detected and prevented.

7    The analysis in (10) can be used to assess the value of PHM associated with halting the operation of the turbine as any point prior to the predicted end of the RUL. Figure 10 is the specific result for 50 days prior to the end of the predicted RUL.

8    Leasing contracts often have availability-like requirements; however, the primary difference is that the requirement is usually imposed by the owner of the system upon the customer, rather than the other way around. For example, a copy machine lease may require that the customer to make 1000 copies per month or less (if they make more they pay a penalty) - similarly there may be a maximum number of minutes you can use per month on your mobile phone plan. Alternatively, if this was an availability contract, the customer of the copy machine would tell the owner of the machine that they must be able to successfully make at least 1000 copies per month or they will pay the owner of the machine less for the lease.

# Section 10
# Physics Based Diagnostics

292

# Chapter 15
# Remote Fault Diagnosis System for Marine Power Machinery System

**Chengqing Yuan**
*Reliability Engineering Institute, Wuhan University of Technology, China*

**Yuelei Zhang**
*Reliability Engineering Institute, Wuhan University of Technology, China*

**Xinping Yan**
*Reliability Engineering Institute, Wuhan University of Technology, China*

**Chenxing Sheng**
*Reliability Engineering Institute, Wuhan University of Technology, China*

**Zhixiong Li**
*Reliability Engineering Institute, Wuhan University of Technology, China*

**Jiangbin Zhao**
*Reliability Engineering Institute, Wuhan University of Technology, China*


## ABSTRACT

*Marine power machinery parts are key equipments in ships. Ships always work in rigorous conditions such as offshore, heavy load, et cetera. Therefore, the failures in marine power machinery would badly threaten the safety of voyages. Keeping marine power machineries running reliably is the guarantee of voyage safety. For the condition monitoring and fault diagnosis of marine power machinery system, this study established the systemic condition identification approach for the tribo-system of marine power machinery and developed integrated diagnosis method by combining on-line and off-line ways for marine power machinery. Lastly, the remote fault diagnosis system was developed for practical application in marine power machinery, which consists of monitoring system in the ship, diagnosis system in laboratory centre, and maintenance management & maintenance decision support system.*


## INTRODUCTION

Marine power machinery systems provide power supply for ships. Any failures in the system may induce terrible marine accidents. Hence, the normal operation of the marine power machinery system is essential for a safe trip. However, exposed to hostile environment, the marine power machinery systems are readily to break down (Jones & Li, 2000; Yan, 2005a; Li, Z., 2010a, 2011a & 2011b; Li, W. 2001). It is therefore imperative to diagnose

DOI: 10.4018/978-1-4666-2095-7.ch015

Copyright © 2013, IGI Global. Copying or distributing in print or electronic forms without written permission of IGI Global is prohibited.

impending faults of marine power machinery systems to prevent malfunctions.

*Machinery condition monitoring and fault diagnosis* (CMFD) technique initially emerged at the end of 1960s, and the research work published in the journal of Automatica can be regarded as the milestone of CMFD (Mehra & Peschon, 1971), which led to a series of scientific and industrial activities in the field of large-scale machinery and equipment condition monitoring projects. As for the field of marine engineering, the marine industry over the world has made strict instructions for the CMFD of marine power machinery. American Bureau of Shipping (ABS) has drafted the "Test Guide of Preventive Maintenance" in 1987 (An aligned, 2006; Leontopoulos, 2005; Low & Lim, 2004). Det Norske Veritas (DNV) has emphasized the CMFD of main engine and shaft line in their "Test Handbook of marine main engine and shaft line" (An aligned, 2006; Leontopoulos, 2005; Low & Lim, 2004). Nippon Kaiji Kyokai (NK) has developed new CMFD technologies for the marine power machinery (An aligned, 2006; Leontopoulos, 2005; Low & Lim, 2004). China Classification Society (CCS) has compiled the "Guide of Diesel Engine grease condition monitoring" and "Guide of Propeller condition monitoring".

Benefited from decades of development, numerous CMFD methodologies have been put forward. These methods can be divided into several major categories, such as the performance parameter monitoring, vibration analysis, oil analysis, and instantaneous speed monitoring etc. The information flow for machinery and equipment condition monitoring is shown in Figure 1 (Liu & Yan, 2010). The performance parameter monitoring is usually used to warn of abnormal operation of the machines under the condition that the concerned specific parameters (such as the temperature, pressure, etc.) have exceeded the 'baseline'. The oil analysis has been now used for marine equipment condition monitoring and fault diagnosis, and the commercial services have already been provided by the Mobil Oil Company and Lloyd's Register. It mainly concerns the geometrical characteristics and chemical characteristics of the wear particles in the lubricant. As for the vibration analysis and the instantaneous speed monitoring, they are the most used and simplest methods for the CMFD of marines. The recent advancements on the smart sensor technique and signal processing have made the vibration analysis and the instantaneous speed monitoring very efficient and easy to realize for industrial application.

*Figure 1. The information flow for marine machinery and equipment using CMFD technique (Liu & Yan, 2010)*

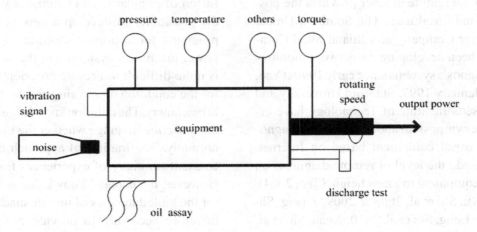

Although many fault diagnostic methods have been developed and proven to be effective in laboratory for the condition monitoring and fault diagnosis of marine power machinery systems, their industrial applications have been seldom reported. This is because the industrial machines running in a complex condition that is very different to the simulated one in the laboratory. In addition, the advanced data analysis equipments are always unavailable in the ship. As a result, the fault diagnosis procedure is difficult to implement. Fortunately, recent progress in advanced techniques has been greatly contributing to improvement of reliability and efficiency of the health monitoring and fault diagnosis for the marine power machinery systems. These techniques include the new concept of Internet of Things (IOT), fiber sensor technique, wireless sensor network (WSN), advanced signal and image processing methods, etc. One of the primary trends of today and tomorrow's Marine Power Machinery System Fault Diagnosis is the development of IOT based automatic intelligent Prognostics and Health Management (PHM) system, which consequently poses significant challenges on multidisciplinary studies for fault diagnosis. IOT based diagnosis system can be simply regarded as the remote diagnosis system, which is very suitable for the offshore marines. Inspired by the success of remote diagnosis system in medical practice, to monitor the marine power machinery systems in the remote manner provides the possibility of industrial uses. The Southern United States Power Company and Inland Steel Company have been developing the network monitoring and diagnosis systems since early 1990s (Yao, 2007). In January 1997, Stanford University and Massachusetts Institute of Technology have organized the symposium about the remote diagnosis of industrial equipment based on Internet, which upgrade the level of remote diagnosis on industrial equipment to a new height (Tao, 2004). Shi et al (Xu, Shi et al, 2008 & 2009; Zhang, Shi et al, 2009; Long, Shi et al, 2010; Xuan, Shi et al,

2009) were dedicated to the research of fault mechanism, signal analysis and intelligent fault diagnosis system, and have established the remote fault diagnostic site server for large scale mechanical systems. Other literature (Yan, 2005b & 2010) has also shown interesting results on the remote diagnosis system. The remote system can fulfill the high industrial requirements of effectiveness and accuracy of the diagnosis results. The rapid developments of computer, network communication, automatic control technologies, comprehensive research on the on-line monitoring technology (Wang, 2007) about the performance parameters, lubricant oil analysis, vibration and instantaneous speed analysis, have gradually made the remote diagnosis system of marine power machinery systems feasible and available.

However, existing remote diagnosis systems for marine power machinery have not integrated different monitoring methodologies together to enhance their fault diagnosis ability. Literature review has indicated that the oil analysis, the vibration and instantaneous speed analysis are used independently in practice. Due to the complex nature of machinery, only 30% to 40% of faults can be detected when they are used independently (Peng, 2003 & 2005). The fusion of different diagnostic techniques can extend the fault diagnosis range and increase the fault detection precision (Mathew & Stecki, 1987; Maxwell & Johnson, 1997; Akagaki, 2006; Chee, 2007; Maru, 2007). Hence, effective fusion of popular CMFD methods will provide the possibility to develop a new generation of programs for condition monitoring of marine power machinery systems. On the other hand, it is quite difficult to acquire criterion knowledge for the condition monitoring and fault diagnosis in machinery. The criterion knowledge determines the '*baseline*' to judge whether the machine runs normally. The traditional approach depends on constant summary of experiences from experts. However, the experts' knowledge is insufficient for the fault diagnosis of marine machinery. It is therefore necessary to provide comprehensive knowledge using advanced network platform. The knowledge-based remote diagnostic services system integrates comprehensive experts' knowledge and other kind information and thus can provide a new solution for condition monitoring of marine power machinery systems.

To build a remote diagnostic system that is suitable for the industrial use, this chapter proposes a new knowledge-based remote diagnostic system for the tribo-system of marine power machinery and develops integrated diagnosis method that combines performance parameter monitoring, oil analysis, vibration analysis and instantaneous speed monitoring for marine power machinery. The newly proposed system consists of three level structures, e.g. the Monitoring System in the Ship, the Diagnosis System in Laboratory Centre, and the Maintenance Management & Maintenance Decision Support System. A series of experiments have been conducted on the marine power machinery test rigs to verify the diagnosis methods employed in this new system (Yuan, 2005a, 2005b, 2007, 2008a, 2008b & 2009; Li, Z. 2010b, 2011c & in press; Sheng, Yan et al, in press; Sheng, Xie, in press). In this paper, the industrial application of the new knowledge-based remote diagnostic system is presented in the dredgers, buoy tenders, and ocean salvage vessels. The application performance shows high efficiency of the proposed remote diagnostic system.

# DESCRIPTION OF THE KNOWLEDGE-BASED REMOTE DIAGNOSTIC SYSTEM

## The Overall Design of the System

The overall view of the proposed remote diagnosis system is shown in Figure 2 (Yan, 2010). It consists of the Monitoring System in the Ship (MSS), the Diagnosis System in Laboratory Centre (DSLC), and the Maintenance Management & Maintenance Decision Support System (MMDS). The 3G (B3G)

wireless communication system has been adopted to connect the MMS to the DSLC and MMDS, and the local area network (LAN) has been used to share information between the DSLC and MMDS. Figure 3 shows the framework of the system. The LAN hub has been adopted to connect the fault diagnosis servers and engine room server, the display terminals of chief engineer room and marine cab in the ship. The fault diagnosis servers are used to analyze the monitoring data. The diagnosis result is displayed on the engineer room server, the chief engineer room and the marine cab. At the same time the result is transferred to MMDS through 3G (B3G) wireless network for the on-line monitoring purpose. In addition, the off-line monitoring is carried out in the DSLC in order to deeply mine the monitoring data.

## Design of the MSS

The two-criteria and four-dimension condition monitoring and fault diagnosis strategy has been proposed in the MSS design. The two criteria are the abnormal alarm level and the fault detection level. The four dimensions include the performance parameter monitoring, oil analysis, vibration analysis and instantaneous speed monitoring. The abnormal alarm level means to initially identify whether damages happen. It mainly uses the performance parameter monitoring to realize the abnormal alarm. The temperature, pressure, rotational speed, etc. of the marine power machinery are compared with their baseline value, and then determine the machinery states. If the performance parameters exceed the safe baseline values, the compared result is "1", then alarm; else, the result is "0", and the monitor stay calm. The fault detection level means to analyze the sensor signals using advanced signal process techniques to detect impending faults. Based on the abnormal alarm level, the oil analysis, vibration analysis and instantaneous speed monitoring have been integrated to provide more powerful fault diag-

*Figure 2. The overall design of the remote diagnosis system (Yan, 2010)*

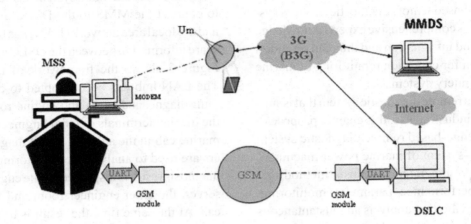

nosis ability. Figure 4 gives the workflow of the integrated diagnosis methodology.

The abnormal alarm level firstly figures out the damaged components in the power system. Then, these components will be the focus of the fault detection level. The Fuzzy neural network (FNN) has been employed in the intelligent fault diagnosis system (Li, Z. in press). The intelligent fault diagnosis system based on oil analysis is given by Figure 5 and the principle and expected

results of the intelligent fault diagnosis system based on vibration analysis is shown in Figure 6. The combined fault diagnosis process can select suitable fusion techniques to relate the diagnosis results of the oil analysis and vibration analysis. The information fusion methods available in the combined analysis include the D-S theory (Fan & Zuo, 2006), rough set theory (Tay & Shen, 2003), fuzzy logic (Escobet, 2011), etc.

*Figure 3. The framework of the proposed remote fault diagnosis system*

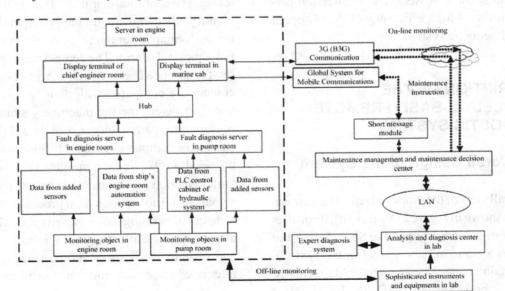

*Figure 4. The framework of the proposed remote fault diagnosis system*

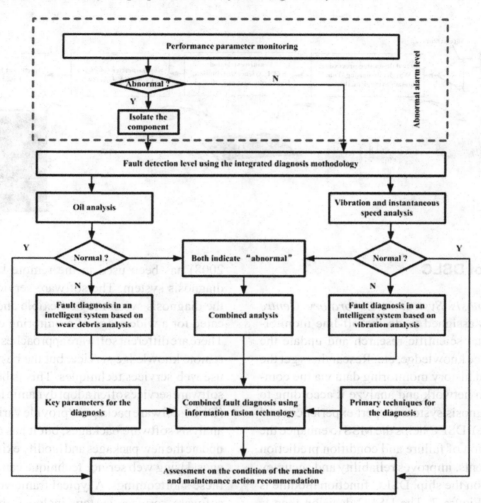

*Figure 5. The intelligent fault diagnosis system based on oil analysis*

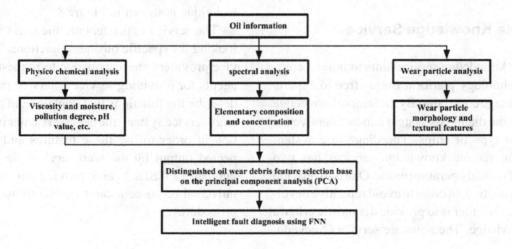

*Figure 6. The principle and expected results of the intelligent fault diagnosis system based on vibration analysis*

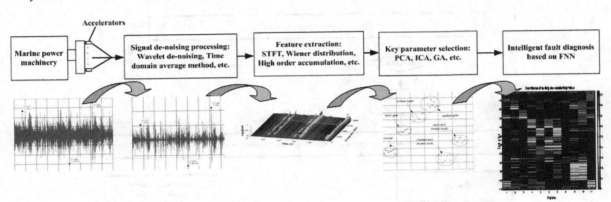

## Design of DSLC

The *Diagnosis System in Laboratory Centre* (DSLC) is assigned to do the off-line monitoring, conduct scientific research and update the maintenance knowledge, etc. Researchers get the current and history monitoring data via the communication network and analyze it according to expert diagnosis system (expert experiences and knowledge). DSLC helps the MSS to enhance the identification of failure and condition prediction of equipments, improves reliability and accuracy of CMFD on the ship. DSLC function module is shown in Figure 7. The DSLC decision then is transferred to the MMDS and MSS via wired and wireless communication networks.

## Remote Knowledge Service

Remote knowledge service aims to build a common technology platform that is free to experts and language constrains. By the remote knowledge service, the diagnosis system can be adaptive for different type of ships. The diagnostic system based on remote knowledge services has two kinds of typical operating mode. One is to provide certain function, in order to avoid repetitive development; the other is to provide diagnostic criteria for knowledge. The software service (Acevedo, 2008) has been used in the remote knowledge diagnosis system. The software service enables the diagnosis system to be flexible and sophisticated for a wide range of monitoring objectives. There are different software approaches to provide remote knowledge service, but the best way is to use web services techniques. This is because the software services often adopt dynamic link library (DLL) software package to provide various signal analysis software packages, but it has difficult to update the new packages and modify existing packages. Using web service technique can overcome these shortcomings. A typical framework of web software services system includes three parts: the service provider, service agent and services requester. The architecture using web services technique is shown in Figure 8.

The service requesters are the users who are looking for specific business functions. The service providers are responsible to requesters and agents for providing services. Service providers describe the functions that contained in the software service system, the necessary enter information in order to use these features and the expected output by the web services description language (WSDL), and provide the uniform/ universal resource locator (URL) to the service procedures.

*Figure 7. Function modules of the DSLC*

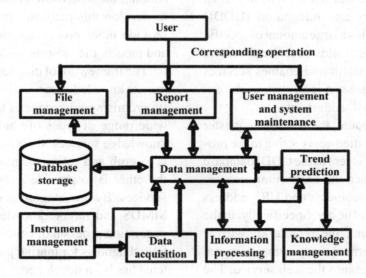

WSDL is one of the cores of Web Services technique, and it is used to describe the call interface of Web Services. WSDL is an XML format for describing network services as a set of endpoints operating on messages containing either document-oriented or procedure-oriented information. The operations and messages are described abstractly, and then they are bound to a concrete network protocol and message format to define an endpoint. Related concrete endpoints are combined into abstract endpoints (services). WSDL is extensible to allow description of endpoints and their messages regardless of what message formats or network protocols used to communicate. A client program connecting to a web service can read the WSDL to determine what functions are available on the server.

*Figure 8. Typical architecture of Web services*

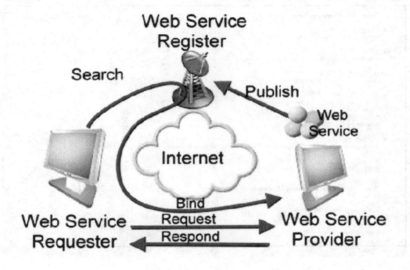

The core of the service agent is the universal description discovery and integration (UDDI) database which contains a large amount of specific commercial functions. It allows service providers to bulletin service content and enables services requesters to find these services, such as the expected output and URL address etc.

In a service procedure, the services requester firstly sends the formatted query string to the provider. Then the query is sent to the UDDI database to search expected functions. Lastly, the necessary input data, WSDL documents and URL address are prepared for the re1uester. Specifically, in the fault diagnosis system, the remote knowledge services have transformed many common signal processing techniques into the web service. The services can not only process oil information, but vibration signals. For the distribution use, we do not have to repeat the developments of signal analysis functions, but directly use these functions through the remote signal analysis services. The functions listed in the remote signal processing services which have been incorporated in our diagnosis system are shown in Figure 9. We have

adopted the Microsoft Visual Studio NET 2003 to develop this platform. In the future, we can provide more processing functions to upgrade and modify the existing system.

The framework of diagnostic system based on remote knowledge service is shown in Figure 10. Its promising advantage is that it can process a wide range of types of ships using the remote knowledge services. It consists of two parts, one is to grub and refine diagnosis knowledge, and the other is to provide knowledge for external services. By the integration of the MSS, DSLC, MMDS and remote knowledge service, a new generation program of condition monitoring and fault diagnosis for marine power machinery systems has been developed.

## CASE STUDY: GEAR PUMP DAMAGE DETECTION AND DIAGNOSIS

The knowledge-based remote fault diagnosis system has been developed for practical applica-

*Figure 9. The function list of remote signal processing services (Liu & Yan, 2010)*

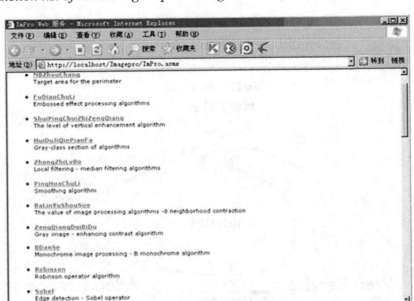

tion of condition monitoring and fault diagnosis for marine power machinery. The hardware and software have been installed on the marine engine room and the diagnostic reports can be read by the captain and the engineers from the display equipments.

The knowledge-based remote fault diagnosis system has already used in a series of real ships and has been proven to be useful for the fault detection and equipment maintenance. The ship types include the dredger, buoy tender, and the ocean salvage vessel.

Hereby, the application of the proposed CMFD system on the dredger named "Changjing 2" has been taken as a case study to illustrate the CMFD processing of the proposed knowledge-based remote fault diagnosis system. Figure 11 gives the controller cabinet of the knowledge-based remote fault diagnosis system. It uses the personal computer (PC) to conduct the intelligent CMFD procedure. The PC connects the marine power machinery's monitoring data acquired from multi-dimensional sensors. The intelligent diagnosis software has been installed in the PC, and it can be operated in both the automatic and manual manners. In this chapter, we will discuss the diagnosis of the gear pump damage to illustrate how the proposed system works.

## On-Line Abnormal Alarm

As mentioned in Section 2, the abnormal alarm is firstly used to determine whether there is a fault in the marine power system. By doing so, it can accelerate the fault diagnosis speed and enhance the detection rate. As for the abnormal alarm level, the performance parameter monitoring has been adopted. Figure 12 shows the software interface of the abnormal alarm configuration for gearboxes. The software design of the marine diesel engines and other components is similar to the gearbox abnormal alarm configuration. Several typical performance parameters have been monitored for abnormal alarm. These performance parameters include the gearbox grease pressure at the inlet and outlet, the grease, gear and bearing temperature, and the rotational speed of the gearbox shaft. The function of the abnormal alarm is to inform the users the potential danger of the equipment if any of the above parameters behaves singularly.

The key issue of the abnormal alarm is to determine the baseline value to judge the performance parameters according to the theoretic results and empirical values. A large number of qualitative rules/criteria need to be specified in the actual diagnosis procedure. In order to determine the component state, it is inevitable to figure out

*Figure 10. The diagnostic system framework based on remote knowledge service*

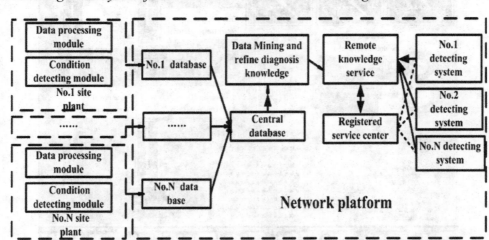

*Figure 11. The controller cabinet of the diagnosis system*

that what is a "big vibration", what phase belongs to the "stability" and how much is "instability", and so on. These rules or criteria are the key link that determines whether the alarm is right.

To make the abnormal alarm effective, in our research and practice projects, a large amount of investigations have been implemented on the experimental and industrial marine power machineries. Many experts in the field have also considered in these investigations and their experiences and knowledge have been taken into account. As a result, effective baseline values have been obtained for most of the performance parameters and hence reliable abnormal alarm can be put into force.

In addition to the above performance parameters, the on-line ferrographic analysis (Wu, 2009) has been incorporated into the abnormal alarm. The on-line ferrographic analysis aims to count the wear debris number to check the pollution lever of the lubricant oil. Then, the health condition of the diesel engine can be informed. Figure 13 gives

*Figure 12. The software interface of the abnormal alarm configuration for gearboxes*

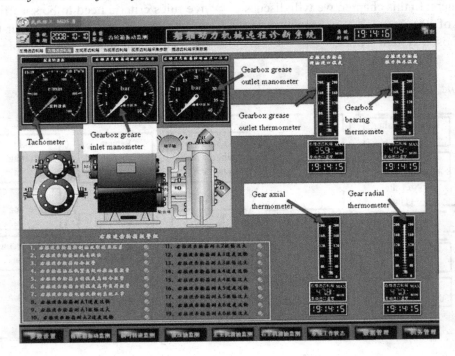

the principle of the on-line ferrographic device. The ferrographic device is composed of a PC, a ferrographic sensor, an electromagnetic valve, and an oil tank, etc. In the working processing, it firstly uses the PC to control the electromagnetic valve in the "through" state, and the oil then can inflow into the ferrographic sensor oil circuit. After that, the PC lights the electromagnet on so as to deposit the wear debris contained in the oil and thus the ferrographic image can be gotten by the ferrographic sensor. An image processing program is lastly used to analyze the ferrographic image to calculate the wear debris number and size.

The on-line ferrographic device has been installed near the scavenge pipe of the diesel engine so that the ISO standards can be met. Figure 14 shows the location of the on-line ferrographic device. The analysis method for on-line monitoring has been reported in (Wu, 2011), where the index of particle covered area (IPCA) has been proposed to measure the wear debris size.

In the case study, the diesel engine has been tested for 100 hours. the gear pump temperature increased gradually and exceeded the alarm value after 50 hours. The IPCA curve is shown in Figure 15 (Sheng, Yan et al, in press). It can be seen that abnormal wear appears after 50 testing hours, which indicates that a terrible fault has occurred in the diesel engine.

Figures 16 and 17 show the on-line ferrographic images of the diesel engine from the start to 70 testing hours. It is noticeable in the figures that very large wear particles are observed after 50 testing hours. Hence, the on-line abnormal alarm sounds a warning to inform the users that the engine is under faulty operation condition. However, at this stage, it is difficult to determine where the damage located and what the fault pattern is. Hence, the off-line analysis has been adopted to diagnose the abnormality.

## Off-Line Fault Diagnosis

The wear debris texture characteristics and chemical characteristics of the lubricant oil were analyzed by off-line wear particle analysis. These

*Figure 13. The principle of the on-line ferrographic device (Wu, 2009)*

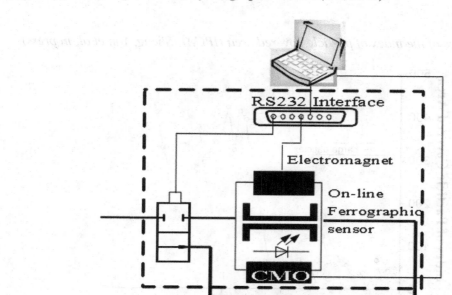

*Figure 14. The on-line wear particle monitoring device: (left) right profile, and (right) left profile*

wear information may help us to find the position and severity of the engine damages. Figure 18 shows the sampling location of the lubricant oil. The sampling location is installed at the pipeline following the hydraulic pump. By doing so, the oil sample is out of pollution.

In practice, the wear debris analysis requires a large amount of professional knowledge, and engineers need to consult many experts to learn the wear failure mechanism and fault patterns. To overcome this problem, the remote knowledge services system was used in the wear debris analysis. After received the wear debris photos transferred from MSS, the system will get the parameters of these wears debris and feedback to MSS. The ferrographic image shown in Figure 19 is presented via the web-page manner. The knowledge service system can quickly processing the image and give the conclusions that this wear failure type may indicate a fault has appeared in the cylinder, bearings or gears.

To enhance the diagnosis result of the remote knowledge services system, the spectral analysis was furthermore carried out. The lubricant oil samples were analyzed by the Spectroil M oil analysis spectrometers. The contents of the element Fe, Cu, Al and Cr in the lubricant oil are list in Table 1. It can be seen that the contents of the

*Figure 15. The curves of the index of particle covered area (IPCA) (Sheng, Yan et al, in press)*

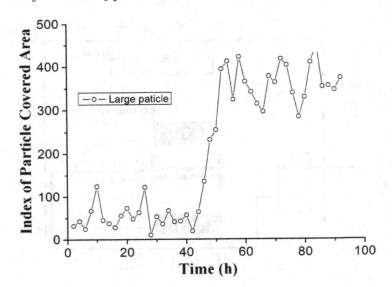

*Figure 16. The on-line ferrographic images: (left) scale indication, and (right) two testing hours*

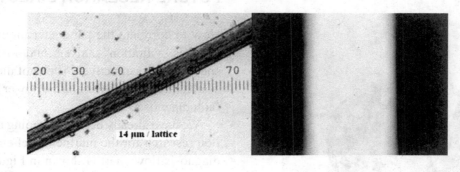

*Figure 17. The ferrographic images of the diesel engine at different testing time*

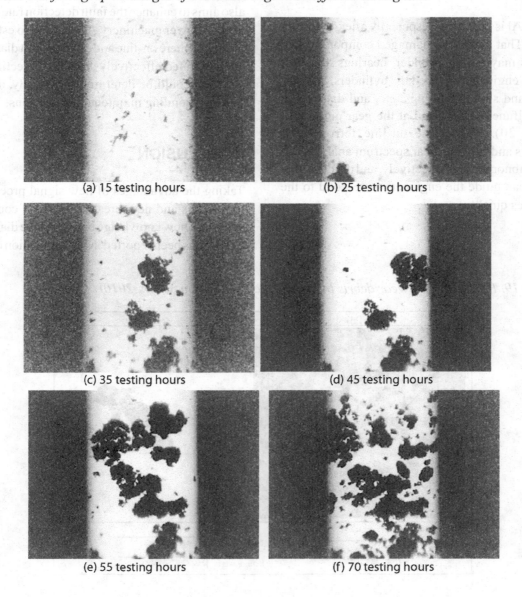

(a) 15 testing hours

(b) 25 testing hours

(c) 35 testing hours

(d) 45 testing hours

(e) 55 testing hours

(f) 70 testing hours

*Figure 18. The sampling location of the lubricant oil*

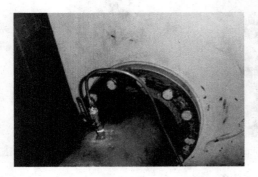

Fe and Al is very large, especially after 45 testing hours. That means the damaged component suspicions may be the cylinder, bearings or gears. Hence, engineers checked the cylinders, the gear pump and sliding bearings, etc, and discovered the malfunction occurred at the gear pump (see Figure 20). Hence, the on-line ferrographic analysis and off-line wear spectrum analysis can detect abnormalities effectively, and the detection result can guide the engineers to respond to the damages quickly.

## FUTURE RESEARCH DIRECTIONS

How to correlate the parameter monitoring, oil analysis, vibration analysis and instantaneous speed analysis effectively is one of the future research directions in the marine power machinery fault diagnosis.

We have already been undergoing the correlation research for the marine diesel engines. The diagnosis flow chart is shown in Figure 21. The integrated diagnosis aims to identify these faults that using individual method is hard to detect. It also aims to enhance the fault detection rate for the marine power machinery. Our goal is to establish a system where on-line and off-line fault diagnosis are integrated effectively and fault detection and isolation could be determined in timely, as well as corresponding maintenance decisions.

## CONCLUSION

Taking the advantages of new signal processing techniques and new breakthroughs in computer science, a new knowledge based remote diagnosis system has been reported in the application of con-

*Figure 19. Processing the wear debris by knowledge service (Liu & Yan, 2010)*

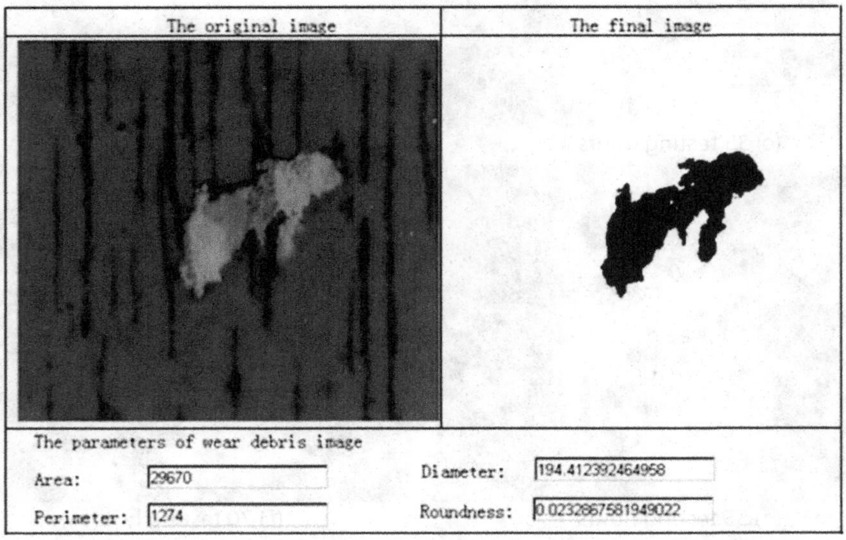

| The original image | The final image |
|---|---|

The parameters of wear debris image

| Area: | 29670 | Diameter: | 194.412392464958 |
|---|---|---|---|
| Perimeter: | 1274 | Roundness: | 0.0232867581949022 |

*Table 1. The element contents using the spectral analysis*

| Testing time (hours) | Element content (mg/L) | | | |
|---|---|---|---|---|
| | Fe | Cu | Al | Mo |
| 2 | 2.11 | 0.51 | 2.01 | 0.08 |
| 15 | 2.24 | 0.56 | 1.56 | 0.01 |
| 25 | 2.35 | 0.62 | 1.01 | 0.01 |
| 35 | 2.46 | 0.67 | 1.74 | 0.02 |
| 45 | 2.57 | 0.72 | 1.42 | 0.03 |
| 55 | 2.79 | 0.77 | 1.65 | 0.23 |
| 70 | 2.84 | 0.81 | 2.02 | 0.32 |

dition monitoring and fault diagnosis (CMFD) for marine power machinery systems. The constructed two lever diagnosis system integrates the performance parameters, lubricant oil analysis, vibration and instantaneous speed analysis to make the remote diagnosis system of marine power machinery systems feasible and available. The gear pump test shows that the proposed system is competent for fault detection. The proposed knowledge based remote diagnosis system has been proven to be feasible in engineering practice, and efficient for failure detection for diesel engines.

*Figure 20. The damaged gear pump (Sheng, Yan et al, in press)*

## ACKNOWLEDGMENT

This project is sponsored by the grants from the State Key Program of National Natural Science of China (NSFC) (No. 51139005), the National Natural Sciences Foundation of China (NSFC) (No. 50975213 and No. 50705070), and the Program of Introducing Talents of Discipline to Universities (No. B08031).

*Figure 21. The integrated fault diagnosis flow chart*

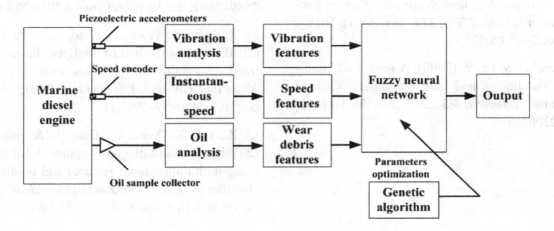

# REFERENCES

Acevedo, P., Martinez, J., & Gonzalez, M. (2008). Case-based reasoning and system identification for control engineering learning. *IEEE Transactions on Education, 51*(2), 271–281. doi:10.1109/TE.2007.909361

Akagaki, T., Nakamur, M., Monzen, T., & Kawabata, M. (2006). Analysis of the behaviour of rolling bearings in contaminated oil using some condition monitoring techniques. *Proceedings of the Institution of Mechanical Engineers, Part J, Journal of Engineering Tribology, 220*(5), 447–453. doi:10.1243/13506501J00605

An aligned view on shafts? (2006). *Marine Engineers Review, 5,* 44-47.

Chee, T., Phil, I., & David, M. (2007). A comparative experimental study on the diagnostic and prognostic capabilities of acoustics emission. vibration and spectrometric oil analysis for spur gears. *Mechanical Systems and Signal Processing, 21*(1), 208–233. doi:10.1016/j.ymssp.2005.09.015

Escobet, A., Nebot, À., & Cellier, F. (2011). Fault diagnosis system based on fuzzy logic: Application to a valve actuator benchmark. *Journal of Intelligent and Fuzzy Systems, 22*(2), 155–171.

Fan, X., & Zuo, M. (2006). Fault diagnosis of machines based on D-S evidence theory. Part 2: Application of the improved D-S evidence theory in gearbox fault diagnosis. *Pattern Recognition Letters, 27*(5), 377–385. doi:10.1016/j.patrec.2005.08.024

Jones, N., & Li, Y. (2000). A review of condition monitoring and fault diagnosis for diesel engines. *Tribotest, 6*(3), 267–291. doi:10.1002/tt.3020060305

Lanktree, C., & Briere, J. (1991). *Early data on the trauma symptom checklist for children (TSC-C).* Paper presented at the meeting of the American Professional Society on the Abuse of Children, San Diego, CA.

Leontopoulos, C., Dovies, P., & Park, K. (2005). Shaft alignment analysis: Solving the reverse problem. *Proceedings of the Institute of Marine Engineering, Science and Technology Part B. Journal of Marine Design and Operations, 8B,* 3–12.

Li, W., Gu, F., Ball, A., Leung, A., & Phipps, C. (2001). A study of the noise from diesel engines using the independent component analysis. *Mechanical Systems and Signal Processing, 15*(6), 1165–1184. doi:10.1006/mssp.2000.1366

Li, Z., & Yan, X. (2011a). Independent component analysis and manifold learning with applications to fault diagnosis of VSC-HVDC systems. *Hsi-An Chiao Tung Ta Hsueh, 45*(1), 44–48.

Li, Z., Yan, X., Wu, J., & Peng, Z. (in press). Study on the fusion of multi-dimensional sensors for health monitoring of rolling bearings. [in press]. *Sensor Letters.*

Li, Z., Yan, X., Yuan, C., Peng, Z., & Li, L. (2011b). Virtual prototype and experimental research gear multi-fault diagnosis using wavelet-autoregressive model and principal component analysis method. *Mechanical Systems and Signal Processing, 25*(7), 2589–2607. doi:10.1016/j.ymssp.2011.02.017

Li, Z., Yan, X., Yuan, C., Zhao, J., & Peng, Z. (2010a). A new method of nonlinear feature extraction for multi-fault diagnosis of rotor systems. *Noise and Vibration Worldwide, 41*(10), 29–37. doi:10.1260/0957-4565.41.10.29

Li, Z., Yan, X., Yuan, C., Zhao, J., & Peng, Z. (2010b). The fault diagnosis approach for gears using multidimensional features and intelligent classifier. *Noise and Vibration Worldwide, 41*(10), 76–86. doi:10.1260/0957-4565.41.10.76

Li, Z., Yan, X., Yuan, C., Zhao, J., & Peng, Z. (2011c). Fault detection and diagnosis of the gearbox in marine propulsion system based on bispectrum analysis and artificial neural networks. *Journal of Marine Science and Application, 10*(1), 17–24. doi:10.1007/s11804-011-1036-7

Liu, J., & Yan, X. (2010). Study of diagnosis system framework using remote knowledge service. *Proceedings of the 2010 Prognostics and System Health Management Conference*, Macau, China, January 12, 2010 - January 14.

Long, Y., Shi, T., & Xiong, L. (2010). Excimer laser electrochemical etching n-Si in the KOH solution. *Optics and Lasers in Engineering, 48*(5), 570–574. doi:10.1016/j.optlaseng.2009.12.001

Low, K., & Lim, S. (2004). Propulsion shaft alignment method and analysis for surface crafts. *Advances in Engineering Software, 35*, 45–58. doi:10.1016/S0965-9978(03)00082-6

Maru, M., Castillo, R., & Padovese, L. (2007). Study of solid contamination in ball bearings through vibration and wear analyses. *Tribology International, 40*(3), 433–440. doi:10.1016/j.triboint.2006.04.007

Mathew, J., & Stecki, J. (1987). Comparison of vibration and direct reading ferrographic techniques in application to high-speed gears operating under steady and varying load conditions. *Lubrication Engineers, 43*(8), 646–653.

Maxwell, H., & Johnson, B. (1997). Vibration and lube oil analysis in an integrated predictive maintenance program. In *Proceedings of the 21st Annual Meeting of the Vibration Institute*, USA.

Mehra, R., & Peschon, J. (1971). An innovations approach to fault detection and diagnosis in dynamic systems. *Automatica, 7*(5), 637–640. doi:10.1016/0005-1098(71)90028-8

Peng, Z., & Kessissoglou, N. (2003). An integrated approach to fault diagnosis of machinery using wear debris and vibration analysis. *Wear, 255*(7-12), 1221-1232.

Peng, Z., Kessissoglou, N., & Cox, M. (2005). A study of the effect of contaminant particles in lubricants using wear debris and vibration condition monitoring techniques. *Wear, 258*(11-12), 1651–1662. doi:10.1016/j.wear.2004.11.020

Sheng, C., Xie, H., Zhang, Y., Hui, F., & Wu, J. (in press). Nonlinear blind source separation of multi-sensor signals for marine diesel engine fault diagnosis. [in press]. *Sensor Letters*.

Sheng, C., Yan, X., Wu, T., Zhang, Y., & Wu, J. (in press). Wear properties of marine diesel engine under abnormal operation state using Ferrographic analysis and Spectrum analysis. [in press]. *Mechanika*.

Tao, Y., Huang, S., & Wei, G. (2004). Research on remote network monitoring and fault diagnosis system of steam turbine sets. *Power Engineering, 24*, 840–844.

Tay, F., & Shen, L. (2003). Fault diagnosis based on rough set theory. *Engineering Applications of Artificial Intelligence, 16*(1), 39–43. doi:10.1016/S0952-1976(03)00022-8

Wang, W., Tse, P., & Lee, J. (2007). Remote machine maintenance system through Internet and mobile communication. *International Journal of Advanced Manufacturing Technology, 31*(5), 783–789.

Wu, T., Mao, J., Wang, J., Wu, J., & Xie, Y. (2009). A new on-line visual ferrograph. *Tribology Transactions, 52*(5), 623–631. doi:10.1080/10402000902825762

Wu, T., Wang, J., Wu, J., Mao, J., & Xie, Y. (2011). Wear characterization by on-line ferrograph image. *Proceedings of the Institution of Mechanical Engineers, Part J, Journal of Engineering Tribology, 225*(1), 23–24.

Xu, Z., Xuan, J., & Shi, T. (2008). Fault diagnosis of bearings based on the wavelet grey moment vector and CHMM. *China Mechanical Engineering, 19*(15), 1858–1862.

Xu, Z., Xuan, J., & Shi, T. (2009). Application of a modified fuzzy ARTMAP with feature-weight learning for the fault diagnosis of bearing. *Expert Systems with Applications, 36*(6), 9961–9968. doi:10.1016/j.eswa.2009.01.063

Xuan, J., Shi, T., Liao, G., & Liu, S. (2009). Statistical analysis of frequency estimation methods of vibration signal. *Key Engineering Materials, 413-414*, 195–200. doi:10.4028/www.scientific.net/KEM.413-414.195

Yan, X., Zhang, Y., Sheng, C., & Yuan, C. (2010). Study on remote fault diagnosis system using multi monitoring methods on dredger. In *Proceedings of the 2010 Prognostics and System Health Management Conference*, Macau, China, January 12, 2010 - January 14.

Yan, X., Zhao, C., Lu, Z., Zhou, X., & Xiao, H. (2005a). A study of information technology used in oil monitoring. *Tribology International, 38*(10), 879–886. doi:10.1016/j.triboint.2005.03.012

Yan, X., Zhao, C., Lv, Z., & Xiao, H. (2005b). The study on information technology used in oil monitoring. *Tribology International, 38*(10), 879–886. doi:10.1016/j.triboint.2005.03.012

Yao, H., Bin, L., & Liu, G. (2007). Plan and design of remote service center for steam turbine. *Turbine Technology, 49*, 81–84.

Yuan, C., Jin, Z., Tipper, J., & Yan, X. (2009). Numerical surface characterization of wear debris from artificial joints using atomic force microscopy. *Chinese Science Bulletin, 54*(24), 4583–4588. doi:10.1007/s11434-009-0588-2

Yuan, C., Peng, Z., & Yan, X. (2005b). Surface characterisation using wavelet theory and confocal laser scanning microscopy. *Journal of Tribology, ASME, 127*(2), 394–404. doi:10.1115/1.1866161

Yuan, C., Peng, Z., Yan, X., & Zhou, X. (2008b). Surface roughness evolutions in sliding wear process. *Wear, 265*(3-4), 341–348. doi:10.1016/j.wear.2007.11.002

Yuan, C., Peng, Z., Zhou, X., & Yan, X. (2005a). The surface roughness evolutions of wear particles and wear components under lubricated rolling wear condition. *Wear, 259*(1-6), 512–518.

Yuan, C., Yan, X., & Peng, Z. (2007). Prediction of surface features of wear components based on surface characteristics of wear debris. *Wear, 263*(7-12), 1513-1517.

Yuan, C., Yan, X., & Zhou, X. (2008a). Effects of Fe powder on sliding wear process under lubricants with different viscosity. *Proceedings of the Institution of Mechanical Engineers, Part J, Journal of Engineering Tribology, 222*(4), 611–616. doi:10.1243/13506501JET316

Zhang, C., Liu, S., Shi, T., & Tang, Z. (2009). Improved model-based infrared reflectrometry for measuring deep trench structures. *Journal of the Optical Society of America. A, Optics, Image Science, and Vision, 26*(11), 2327–2335. doi:10.1364/JOSAA.26.002327

## KEY TERMS AND DEFINITIONS

**Condition Monitoring:** Monitoring the operation condition of the machinery/system.

**Fault Diagnosis:** Detect faults, isolate them and find the fault roots.

**Marine Power Machinery:** Machinery to provide power to ships, such as diesel engines, generators, shaft line, gearboxes, etc.

**Oil Analysis:** Monitoring the system by analyzing the lubricant oil.

**Remote:** A device that can be used to control a machine or apparatus from a distance.

**Tribo-System:** Typical rubbing pairs, such as cylinder-piston pairs, gear meshing pairs, etc.

**Vibration Analysis:** Using the machinery vibration signal to determine the system' health states.

# Section 11
# Prognostics

# Chapter 16
# Prognostics and Health Management of Choke Valves Subject to Erosion:
## A Diagnostic–Prognostic Frame for Optimal Maintenance Scheduling

**Giulio Gola**
*Institute for Energy Technology, & IO-center for Integrated Operations, Norway*

**Bent H. Nystad**
*Institute for Energy Technology*

## ABSTRACT

*Oil and gas industries are constantly aiming at improving the efficiency of their operations. In this respect, focus is on the development of technology, methods, and work processes related to equipment condition and performance monitoring in order to achieve the highest standards in terms of safety and productivity. To this aim, a key issue is represented by maintenance optimization of critical structures, systems, and components. A way towards this goal is offered by Condition-Based Maintenance (CBM) strategies. CBM aims at regulating maintenance scheduling based on data analyses and system condition monitoring and bears the potential advantage of obtaining relevant cost savings and improved operational safety and availability. A critical aspect of CBM is its integration with condition monitoring technologies for handling a wide range of information sources and eventually making optimal decisions on when and what to repair. In this chapter, a CBM case study concerning choke valves utilized in Norwegian offshore oil and gas platforms is proposed and investigated. The objective is to define a procedure for optimizing maintenance of choke valves by on-line monitoring their condition and determining their Remaining Useful Life (RUL). Choke valves undergo erosion caused by sand grains transported by the oil-water-gas mixture extracted from the well. Erosion is a critical problem which can affect the correct valve functioning, resulting in revenue losses and cause environmental hazards.*

DOI: 10.4018/978-1-4666-2095-7.ch016

# 1 INTRODUCTION

A diagnostic-prognostic scheme for assessing the actual choke valve health state and eventually estimating its RUL is here proposed. In particular, focus has been on the identification of those parameters which contribute to the actual erosion of the choke valve, the development of a model-based approach for calculating a reliable indicator of the choke valve health state, the actual estimation of the choke RUL based on that indicator and, finally, the investigation of methods to reduce the uncertainty of the RUL estimation. On July 6, 1988 an explosion and a resulting fire on the North Sea oil production platform Piper Alpha caused nearly two hundreds fatalities and a £1.7 billion estimated loss. According to the Cullen Inquiry (Ross, 2008), the disaster was caused by an initial condensate leak as a result of maintenance work carried out simultaneously on a pump and the related safety valve. The safety valve was taken out of a production line for repair and replaced with end flanges. Pressure was put back in the system, which later resulted in a gas leakage and combustion. Piper Alpha's operator was found guilty of having inadequate maintenance and safety procedures.

More recently, on April 20, 2010 an explosion on the Deepwater Horizon platform in the Gulf of Mexico caused many fatalities and a 4.9-million-barrel crude oil spill into the water which resulted in extensive damages to marine and wildlife habitats and to the Gulf's fishing and tourism industries. A critical factor in the causal chain of events that contributed to this accident was the failure of a blow-out preventer to seal the leaking reservoir (Nomack, 2011). The British Petroleum team conducted an audit on the disaster including the maintenance management system for the blow-out preventer. The findings made it clear that control-related equipment maintenance were manually documented on separate spread sheets and in the daily logbook, but had not been recorded in the Transocean maintenance management system (RMS-ll). This made it difficult to track blow-out preventer maintenance actions. The fact that maintenance records were not accurately reported in the maintenance management system was identified as a potential cause of the failure of the blow-out preventer system.

Although these disasters resulted from a complex mix of mechanical failures, human judgement, engineering design, operational implementation and team interfaces, they illustrate the importance of valves as safety barriers in a process system (Meland, 2011; Andrews *et al.*, 2005; Haugen *et al.*, 1995; Ngkleberg & Sontvedt, 1995; Hovda & Lejon, 2010).

Failure to close on command and leakages through the valve when in closed position are so-called dangerous undetected failure modes and represent serious hazards for the safe operation of the system.

To reveal and repair dangerous undetected failure modes, valves must be periodically tested. In the offshore oil and gas industry, tests are normally performed during annual revisions and involve stopping the hydrocarbons flow and consequently a process shut-down. On average, operation tests for a single valve last between three and four hours (Meland, 2011; Haugen *et al.*, 1995). Production downtime can be reduced either by increasing the length of the test interval or by performing the test during unplanned system shutdowns. Another approach would be removing tests without compromising safety by resorting to methods for monitoring the valve condition with an accuracy that would allow avoiding testing (Haugen *et al.*, 1995). In this view, it is critical to devise systems which can early detect, identify, quantify and accurately predict the degradation processes directly related to the dangerous undetected failure modes.

A viable way to achieve this is offered by Condition-Based Maintenance (CBM) strategies (Baraldi *et al.*, 2011a; Baraldi *et al.*, 2011b; Grall *et al.*, 2002; Tsang, 1995; Williams *et al.*, 1994). Components' CBM relies on an adequate component condition monitoring, high quality documentation of tests and maintenance events, complete and accurate data acquisition of the system health state and effective data cleaning to increase the robustness of the models. Given the component's current condition and past operation profile, the output of CBM is the knowledge of the component's time-to-failure expressed in terms of Remaining Useful Life (RUL), a stochastic variable that can be used to plan future inspections, repairs or replacements of the degraded components (Grall *et al.*, 2002; van Noortwijk & Pandey, 2003). Major advantages of performing CBM are the increase of plant availability and therefore economic revenues, a more efficient use of human and technical resources during optimally scheduled maintenance operations and a shift towards higher safety standards.

As previously mentioned, valves are critical components in oil- and gas-related processes. They serve different functions such as regulating the flow rate, preventing back-flow (check valves), stopping the flow and relieving pressure.

The current Chapter, which is the aggregation and extension of the research partly presented in Nystad *et al.* (2010) and Gola & Nystad (2011), focuses on choke valves which are especially used to regulate the oil-water-gas mixture extracted from the well and to control the pressure in the extraction pipeline. Choke valves suffer from erosion due to sand production in mature fields. Erosion management is vital to avoid failures that may result in loss of containment, production being held back, and increased maintenance costs. Moreover, several chokes are located subsea,

where the replacement cost is high (Haugen *et al.*, 1995; Ngkleberg & Sontvedt, 1995; Andrews *et al.*, 2005; Bringedal *et al.*, 2010; Hovda & Lejon, 2010; Wallace *et al.*, 2004).

Currently, fixed maintenance is the most common way to manage choke replacement; a CBM strategy is here investigated to optimize choke valves' maintenance based on the indications of their actual condition (i.e. health state) and on the estimations of their RUL (Nystad *et al.*, 2010; Kiddy, 2003; Gola & Nystad, 2011; van Noortwijk & Pandey, 2003).

The integration of condition monitoring systems with CBM strategies is critical for handling a wide range of information sources and providing accurate indications of the component's health state based on which reliable estimates of the component's RUL can be provided. In this respect, the measurements of those parameters considered relevant to assess the erosion state of the choke valve are first processed by an empirical model called Virtual Sensor (PCT/NO2008/00293, 2008; Roverso, 2009) to obtain an accurate indication of the choke erosion (component diagnostics); a stochastic method based on the gamma process is then used to effectively estimate the choke remaining useful life (component prognostics) (Figure 1).

The Chapter is organized as follows. Section 2 briefly illustrates the general guidelines and standards for CBM. Section 3 describes the problem of maintenance of choke valves subject to sand-driven erosion. The case study hereby investigated is reported in Section 4. Sections 5 and 6 illustrate the results of the condition monitoring

*Figure 1. General diagnostic-prognostic frame*

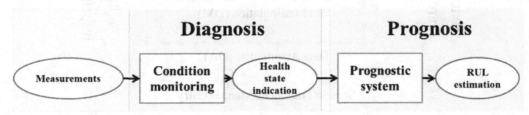

system used to assess the choke erosion state and the methods applied to estimate its remaining useful life, respectively. Conclusions are drawn in the last Section.

## 2 GENERAL GUIDELINES AND STANDARDS FOR CBM

General guidelines for condition monitoring and diagnostics of machines can be found in ISO 13374-1 (2003) and ISO 13374-2 (2007). For the prognostic part of machine health, which requires foreknowledge of the probable failure modes, the ISO standard ISO 13381-1 (2004) can be used. Interesting alliances of organisations are the Machinery Management Open Systems Alliance (MIMOSA, www.mimosa.org) and the open system alliance CBM (OSA-CBM).

The primary objective of MIMOSA is to project maintenance management as a business function that operates on business objects with well-defined properties, methods and information interfaces. In order to implement condition monitoring successfully, ISO has made a general model (Figure 2) of the recommended data processing and information flow. Within this frame, the efforts of OSA-CBM were aimed at developing open and interchangeable systems that cater to the maintenance arena (i.e. CBM systems). OSA-CBM has developed a seven-layered architecture similar to the one illustrated in Figure 2 that encompasses the typical stages of the development, deployment and integration of maintenance solutions under the CBM framework (Mitchell *et al.*, 1998).

A description of the technology-specific blocks and the functions they are expected to provide follows:

1.   Data Acquisition (DA) converts an output from the transducer to a digital parameter representing the physical quantity and the related information (such as time, calibration, data quality, data collector utilized and sensor configuration).

2.   Data Manipulation (DM) performs signal analysis, computes meaningful descriptors and derives virtual sensor readings from raw measurements.

*Figure 2. ISO data-processing and information-flow blocks*

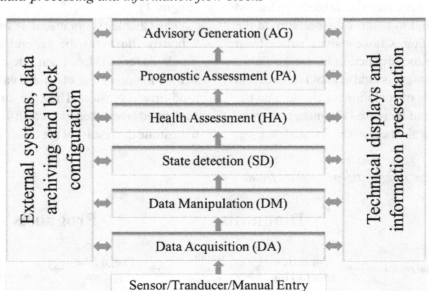

3. State Detection (SD) facilitates the creation and maintenance of normal baseline profiles, searches for abnormalities whenever new data are required and determines in which state data belong.

The final three blocks normally attempt to combine monitoring technologies in order to assess the current health of the component:

4. Health Assessment (HA) diagnoses any faults and rates the current health of the component, considering all state information.
5. Prognostic Assessment (PA) determines future health states and failure modes based on the current health assessment and projected usage loads on the component and predicts its remaining useful life.

Advisory Generation (AG) provides actionable information regarding maintenance or operational changes required to optimize the life of the component.

# 3 CONDITION-BASED MAINTENANCE OF CHOKE VALVES

As a common practice in the oil industry, liquid and gas mixtures are flown through chokes to control flow rates and protect equipment from unusual pressure fluctuations. Both surface and subsurface chokes are utilized. At the surface all chokes are of the rotating disk type. Here the throttle mechanism consists of two circular disks each with a pair of circular openings to create variable flow areas. One of the disks is fixed in the valve body, the other being rotated either by manual operation or by actuator, to vary or close the aperture. The mating surfaces of both disks are precision lapped to ensure tight sealing and prevention of ingress of grained grit, sand and scale between the disks. Multi orifice valve (MOV) chokes (Figure 3) with trim sizes ranging from 2x½" to 2x2" are used on Gullfax and Statfjord platforms in the North Sea.

For large pressure drops, the well stream containing gas, liquid and sand particles can reach 400-500 m/s and produce heavy metal loss mainly due to solids, liquid droplets, cavitation and combined mechanisms of erosion-corrosion,

*Figure 3. Typical MOV choke valve of rotating disk type: by rotating the disk the flow is throttled (Metso Automation, 2005)*

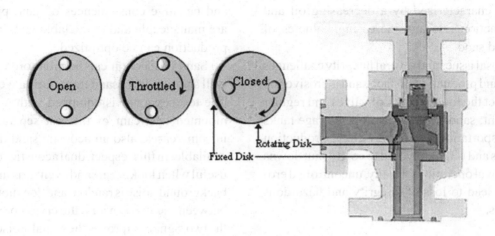

resulting in choke lifetimes of less than a year. Prolonged exposure to abrasive flow conditions will eventually lead to erosion on sealing surfaces and the valve body. The rotating disk valves are designed to produce near linear flow characteristics making them ideal for e.g. control applications. Erosion is a slow process and will produce a gradual change from the design characteristics. For a specific opening the valve flow coefficient will increase as the valve circular openings erode (change in geometry).

As previously mentioned, erosion management is fundamental due to the relevant safety and economic issues involved. Nevertheless, calculation of the erosion rate is a complex task and is in general dependent on factors like the design and build quality (design geometry, materials used), operational history (particle impact angle, velocity, fraction of abrasion particles and change in geometry) and the current state of the valve as measured by a set of indicators.

The following Sections illustrate the procedures currently adopted to manage the problem of sand extraction from oil fields in the North Sea and the method used to calculate the erosion in choke valves.

## 3.1 Sand Management and Detection

Many oil fields in the North Sea are mature and so-called brown fields. Production from these fields is characterized by a decreasing oil and gas extraction rate and increasing volumes of water and sand.

Disposal issues and several integrity challenges follow sand production. In fact, sand is erosive and may affect the functionality of valves and regular equipment; sand fill in separators, storage tanks and transportation vessel may cause production problems and ultimately lead to costly shut downs and removal operations; finally, uncontrolled erosion can lead to loss of integrity and hazardous situations.

A sand management strategy has been implemented by the operators on the fields. For several years, the Maximum Sand-Free Rate (MSFR) strategy was widely accepted. It focuses on finding the maximum production rate that would not result also in sand production. This conservative, on-site disposal strategy was a safe approach that also reduced problems related to erosion of equipment. However, it also resulted in production being unnecessarily held back and limited the economic reservoir lifetime and development during tail production.

A strategy that overcomes these limitations is known as the Acceptable Sand Rate (ASR) strategy. This is a risk management approach based on the idea of handling topside the sand extracted from the well before proceeding to the down-hole, on-site disposal (Andrews *et al.*, 2005; Bringedal *et al.*, 2010). The idea behind ASR is to accept sand production at some level together with the fact that sand enters the production system, causes damage to the equipment and eventually introduces disposal issues. In return, the increased profit coming from extra production is expected to exceed the costs of equipment maintenance and other costs. This strategy requires monitoring the erosion well by well, thus enabling wells with low risk of sand production to increase the production and choking back those (few) from which too much sand is extracted. By doing so, sand is produced in a safe and controlled way and negative consequences of sand production are manageable and predictable and oil and gas production can be optimized.

Sand production can be monitored by taking well samples from sand traps during well testing. The test separator is equipped with a sand trap mounted upstream of the test separator itself; in some cases, also an acoustic sand detector is available. In this respect, dual acoustic sensors are useful when looking at sand events: assuming that background noise is random and does not correlate between the two sensors, the cross correlation of the two signals improves the signal-noise ratio and also provides a direct determination of sand flow velocity between the position of the two sensors. Also raw data from acoustic sensors are useful: in fact, when this information is correlated to the well behaviour (rates, pressures, choke movements etc.) in a real time system, a good picture of the well sand tendency can be obtained. In-line erosion probes are useful in the higher velocity range. Nevertheless, accurate measurements of sand production in multiphase environments are inherently difficult and sensors cannot provide exact measurements of sand quantities. Once the erosion rate is accurately predicted for given geometries and operating conditions, oil and gas production rates from individual wells can be optimized within reasonable erosion levels.

Finally, models for sand production monitoring can be used as frames for risk-based inspections in order to prioritize and plan actions by assessing the probability (based on degradation models) and consequences of component failures given a risk measure.

### 3.2 Erosion Monitoring

Production experience has shown that these MOV chokes are prone to erosion in the disks and in the outlet sleeve. An example of MOV choke valve replaced after being highly eroded in shown in Figure 4. Erosion has also been observed in the blinded tee immediately downstream the choke. Recently, Tungsten Carbide (TC) outlet sleeves have been installed in these chokes due to its superior erosion resistance compared to steel.

From the physical point of view, extensive studies have shown that the particle erosion rate is highly dependent on the particle impact velocity and the impact angle of the sand grains through the choke discs. The erosion rate in steel is proportional to the particle impact velocity with a power law from 2 to 3. As such, high velocity and not large quantities of sand is the primary cause for erosion, e.g. doubling velocity leads to a nearly six-fold erosion rate increase, whereas

doubling sand volume equal to doubling the erosion rate. The angle through the choke is determined by the trim size and the choke opening.

From the mathematical point of view, the generic choke valve fluid dynamic model states that the total flow $w$ through the choke is proportional to the pressure $\Delta p$ drop through the choke:

$$w = C_V \sqrt{\frac{\Delta p}{\rho}} \tag{1}$$

where $\rho$ is the average mixture density and $C_V$ is called valve flow coefficient. $C_V$ depends on the valve type. It is related to the effective flow cross-section of the valve and is proportional to the choke opening according to a function depending on the type of choke valve and given by the valve constructors, i.e. for a given choke opening, $C_V$ is expected to be constant (Meland, 2011; Hovda & Andrews, 2007).

When erosion occurs, a gradual increase of the valve area available for flow transit is observed (as shown in Figure 4) even at constant pressure drop. Such phenomenon is therefore related to an abnormal increase of the valve flow coefficient

*Figure 4. Damage caused by sand erosion. In the picture the original circular holes have a major wear on the upper side of the left hole and lower side of the right hole. The choke is no longer able to close and stop a flow.*

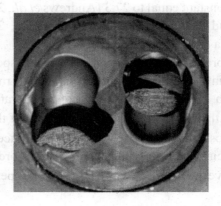

with respect to its expected theoretical value, hereby denoted as $C_V^{th}$.

For this reason, for a given choke opening the difference $\delta_{C_V}$ between the actual value of the valve flow coefficient, hereby simply denoted as $C_V$, and its theoretical value $C_V^{th}$ is retained as an indication of the choke erosion. The difference $\delta_{C_V} = C_V - C_V^{th}$ is expected to monotonically increase throughout the choke life since it should reflect the physical behavior of the erosion process. When $\delta_{C_V}$ eventually reaches a pre-defined failure threshold, the choke must be replaced.

The actual valve flow coefficient $C_V$ cannot be directly measured, but it can be calculated from the following analytical expression which accounts for the physical parameters involved in the process:

$$C_V = \frac{w_g + w_o + w_w}{N_6 F_p \sqrt{p_{in} - p_{out}}} \sqrt{\frac{f_g}{\rho_g J^2} + \frac{f_o}{\rho_o} + \frac{f_w}{\rho_w}}$$

(2)

where $p_{in}$ and $p_{out}$ are the pressures upstream and downstream of the choke, $w_o$, $w_w$ and $w_g$ are the flow rates of oil, water and gas, $f_o$, $f_w$ and $f_g$ the corresponding fractions with respect to the total flow rate and $\rho_o$, $\rho_w$ and $\rho_g$ the corresponding densities, $J$ is the gas expansion factor, $F_p$ is the piping geometry factor and $N_6$ is a constant equal to 27.3 (Andrews *et al.*, 2005; Nystad *et al.*, 2010; Metso Automation, 2005; Hovda & Andrews, 2007).

Correct choke operation is a critical aspect to account for minimizing erosion. A good rule of thumb is that chokes should not be operated at more than 50% of their maximum $C_V$. If this is not possible, the erosion rate can be reduced by replacing the trim with smaller discs in order to achieve the same $C_V$ with a larger choke opening.

## 4 CHOKE VALVE EROSION: THE CASE STUDY

A case study on a choke valve located top side on the Norwegian continental shelf is here considered.

Measurements and calculations related to the physical parameters involved in the process are available as daily values. In particular, the pressures upstream and downstream of the choke are directly measured, whereas oil, gas and water flow rates are calculated based on the daily production rates of other wells of the same field. Pressure measurements are considered reliable since they are directly related to the well under analysis, whereas the calculations of oil, gas and water flow rates expected form that well might not be realistic and therefore might not reflect the actual physical composition of the extracted mixture.

In addition to the daily measurements and calculations, seven well tests are carried out throughout the valve life at regular intervals, during which oil, gas and water flow rates are accurately measured using a multi-phase fluid separator. The valve choke opening is also provided as a parameter.

Since oil, gas and water flow rates are used to compute the actual $C_V$ (Equation 2), inaccuracies in their calculation might negatively affect the $C_V$ calculation itself and thus the quality of the erosion indication $\delta_{C_V}$.

Figures 5 and 6 illustrate the parameters used to compute the actual $C_V$ and the resulting erosion indication $\delta_{C_V}$, respectively.

The mismatch between the values of oil, water and gas flow rates daily calculated accounting for the other wells and the values of the same three parameters measured during the well tests is evident in the bottom graphs in Figure 5. Notice that there is instead no mismatch for the pressure drop and, obviously, for the choke opening indication (top graphs in Figure 5).

*Figure 5. Choke opening and pressure drop (top graphs) and oil, water and gas flow rates (bottom graphs) during daily measurements (black line) and well tests (stars)*

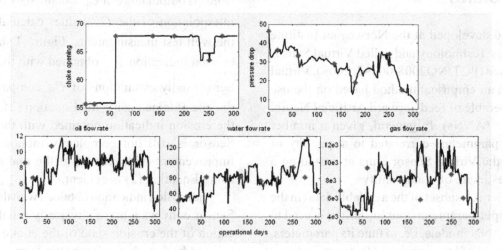

As a consequence of the inaccurate daily calculations of oil, water and gas flow rates, the daily erosion indication $\delta_{C_V}$ (black line in Figure 6) results non-monotonic and very noisy, generally showing an unphysical behaviour. On the other hand, when $\delta_{C_V}$ is computed using the well test measurements of oil, water and gas flow rates, its behaviour results monotonic and provide a reliable information on the physical erosion process.

Nevertheless, a diagnostic assessment on the erosion state of the valve and a prognostic estimation of its remaining useful life cannot be made based on the daily erosion indications (black line in Figure 6), whilst this is currently the sole erosion indicator available to the technicians at oil platforms. In the next Section, an empirical model-based approach is used to produce a reliable daily calculation of the erosion state which is then fed to a prognostic system for estimating the choke remaining useful life.

*Figure 6. Erosion indication $(\delta_{C_V})$ obtained with $C_V$ calculations based on daily measurements and calculations (black line) and computed using the measurements of the well tests (stars)*

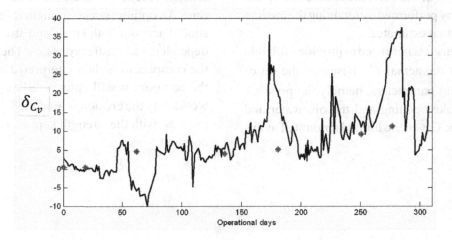

## 5 CHOKE VALVE CONDITION MONITORING

A method developed at the Norwegian Institute for Energy Technology and called Virtual Sensor is here used (PCT/NO2008/00293, 2008). Virtual Sensor is an empirical method based on the use of an ensemble of feed-forward Artificial Neural Networks (ANNs). In general, given a number of input parameters correlated to a quantity of interest, the Virtual Sensor aims at providing a reliable estimate of that quantity.

In general, a subset of the available data (in the format input-parameters/output-target) is used to train the ANN models, i.e. to tune its parameters, with the goal of learning the ANN to estimate the output target. Once the model is trained, it can be used on-line by providing a stream of input measurements in order to obtain an estimate of the (unknown) output (Rumelhart & McClelland, 1986).

Virtual Sensor exploits the concepts of ensemble modelling which bear the advantages of ensuring high accuracy and robustness of the estimation without the need of developing one single optimal model (Bauer & Kohavi, 1999; Breiman, 1996; Breiman, 1999; Hansen & Salamon, 1990; Opitz & Shavlik,1996). Critical aspects of ensemble modelling are the diversity of the individual models, hereby ensured by randomizing the training initial conditions of the ANNs, and the aggregation of the outcomes of the individual models, hereby performed by retaining the median of the individual estimates.

Virtual Sensor is here used to provide a reliable estimation of the actual $C_V$ based on the set of available input parameters, namely the pressure drop, the choke opening and the oil, water and gas flow rates. Given the limited amount of available data, the Virtual Sensor has been trained by using as output target a $C_V$ obtained by the linear interpolation of the $C_V$ values calculated with the well test measurements. Figure 4 shows the erosion indication $\delta_{C_V}$ obtained with the Virtual Sensor daily estimations of $C_V$ compared with the one obtained using the Equation (2). Despite the erosion indication obtained with the Virtual Sensor is still not completely monotonic, the improvement with respect to the one obtained using Equation (2) is evident.

The erosion indication obtained with the Virtual Sensor conveys a more physically reliable indication of the erosion state of the choke and can be used both within a diagnostic frame to assess the valve performance in the present and within a prognostic system for predicting the temporal evolution of the erosion, eventually estimating when the erosion will cross the failure threshold and the valve needs to be replaced. The choke valve RUL estimation based on the erosion indication provided by the Virtual Sensor is illustrated in the next Section.

## 6. CHOKE VALVE REMAINING USEFUL LIFE ESTIMATION

The remaining useful life of the choke is calculated based on the estimate of the time at which erosion reaches a threshold value for the first time. An optimal stress-strength reliability model should account both stress and strength as time-dependent stochastic processes. The reliability of the component is then interpreted as how likely the component will enter a failure state. In this work, only the erosion is modelled as a stochastic process, with the strength kept as a deterministic constant.

## 6.1 Gamma Process Modeling of the Erosion Indicator

The challenge in modelling time-dependent reliability is that the erosion process indicated by $\delta_{C_V}$ is uncertain during the lifetime of the component. Good prognostics are able to quantify and eventually reduce such uncertainty.

The gamma process (van Noortwijk & Pandey, 2003) of the cumulative, time-dependent $\delta_{C_V}$ at time $t$ will follow a gamma distribution with shape parameter $v(t) > 0$ and constant scale parameter $u$. The shape function $v(t)$ must be an increasing function to reflect the physical monotonic nature of the erosion. Further requirements are that it must be right-continuous, real-valued, and equal to zero at the beginning of the calculation, i.e. $v(0) = 0$.

Let $t_0$ be the time at which operation starts after a required maintenance action has been performed, $t$ be the specified point in time starting at operational time $t_0$, i.e. $t = 0$ at operational time $t_0$ and $\delta_{C_V}(t)$ be the $\delta_{C_V}$ value at time

$t$. Formally, the gamma process with shape function $v(t) > 0$ and scale parameter $u > 0$ is defined as a continuous-time stochastic process $\{Y(t), t \geq 0\}$ with the following properties:

1. $Y(0) = 0$ with probability one;
2. $Y(t_2) - Y(t_1) \sim Ga(v(t_2) - v(t_1), u)$, $\forall t_2 > t_1 > 0$;
3. $Y(t)$ has independent increments.

In this case study, the stochastic process $Y(t)$ represents the erosion indicator $\delta_{C_V}(t)$. The probability density function is then given by:

$$f_{Y(t)}(t) = \frac{u^{v(t)}}{\Gamma(v(t))} y^{v(t)-1} e^{uy} = Ga(y; v(t), u)$$

(3)

where $\Gamma(a) = \int_0^\infty z^{a-1} e^{-z} dz$ is the gamma function defined for $a > 0$.

The expected value and the variance of $Y(t)$ at time $t$ are

*Figure 7. Erosion indication $(\delta_{C_V})$ obtained with $C_V$ calculated with Equation (2) (black line) and with the Virtual Sensor (light blue line)*

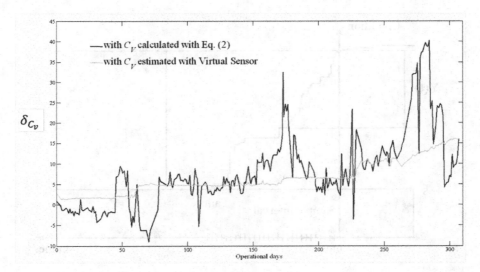

$$E(Y(t)) = \frac{v(t)}{u} \qquad (4)$$

$$Var(Y(t)) = \frac{v(t)}{u^2} \qquad (5)$$

Figure 8 illustrates the generic gamma process model of a health indicator $Y(t)$. The faulty state is defined as the down-crossing of $R(t) = r_0 - Y(t)$ below the stress threshold $s$. The first assumption claims that the stress $s$ and the initial health condition $r_0 = Y(0)$ are known a priori with no uncertainty.

Let $\rho = r_0 - s$ be the design fault margin and $T_\rho$ the time at which a fault occurs. The cumulative probability distribution function of the lifetime $T_\rho$ can be derived from Equation (3):

$$F_{T_p} = \Pr(T_p \le t) = \Pr(Y(t) \ge \rho)$$
$$= \int_{y=\rho}^{\infty} f_{Y(t)}(y)dy = \frac{\Gamma(v(t), \rho u)}{\Gamma(v(t))} \qquad (6)$$

where $\Gamma(a, x) = \int_{z=x}^{\infty} z^{a-1} e^{-z} dz$ is the incomplete gamma function for $x \ge 0$ and $a > 0$.

## 6.2 Estimation of the Gamma Process Parameters

In order to apply the gamma process to the indicator $\delta_{C_V}(t)$, statistical methods are used to estimate the process parameters.

Data sets consist of daily calculations of $\delta_{C_V}(t)$ based on daily measurements of the five parameters shown in Figures 5 and 6 at times (i.e. days) $t_i, i = 1, 2, ..., n$. The observations of $\delta_{C_V}(t)$ must be increasing in order to be modelled by a gamma process, i.e. $0 = t_0 < t_1 < t_2 < ... < t_n$ and $0 = y_0 < y_1 < y_2 < ... < y_n$.

After a pre-analysis of the $\delta_{C_V}$ trends, a general power law is assumed to be a flexible candidate and is therefore used to calculate the expected value:

$$E(Y(t)) = \frac{v(t)}{u} = \frac{ct^b}{u} \propto t^b \qquad (7)$$

*Figure 8. Gamma process model of erosion indicator (van Noortwijk & Pandey, 2003)*

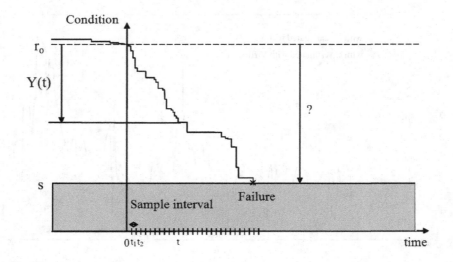

The estimation of the parameters $b$, $c$ and $u$ can be carried out in different ways. Two methods, i.e. maximum likelihood and method of moments (van Noortwijk & Pandey, 2003), are applied in this case study to derive the estimators of $\hat{b}$, $\hat{c}$ and $\hat{u}$.

The maximum-likelihood estimators are obtained by maximizing the logarithm of the likelihood function of the increments. The likelihood of the observed erosion increments $\delta_i = y_i - y_{i-1}$, $i = 1,...,n$, is given as the product of independent gamma densities:

$$L = \prod_{i=1}^{n} f_{Y(t_i)-Y(t_{i-1})}(\delta_i)$$
$$= \prod_{i=1}^{n} \frac{u^{c(t_i^b - t_{i-1}^b)}}{\Gamma(c(t_i^b - t_{i-1}^b))} \delta_i^{c(t_i^b - t_{i-1}^b)-1} e^{-u\delta_i} \qquad (8)$$

By computing the first partial derivatives of the log-likelihood function of the increments with respect to parameters $b$, $c$ and $u$, the maximum-likelihood estimates can be numerically solved from the followings equations:

$$\hat{u} = \frac{\hat{c}t_n^b}{y_n} \qquad (9)$$

$$\sum_{i=1}^{n} \left(t_i^b - t_{i-1}^b\right)\left(\psi\left(c\left(t_i^b - t_{i-1}^b\right)\right) - \ln \delta_i\right)$$
$$= t_n^b \ln\left(\frac{\hat{c}t_n^b}{y_n}\right) \qquad (10)$$

$$\sum_{i=1}^{n} \left(t_i^b \ln t_i - t_{i-1}^b \ln t_{i-1}\right)\left(\psi\left(c\left(t_i^b - t_{i-1}^b\right)\right) - \ln \delta_i\right)$$
$$= t_n^b \ln t_n \ln\left(\frac{\hat{c}t_n^b}{y_n}\right) \qquad (11)$$

A more computationally attractive solution is offered by the method of moments. In fact, in case the parameter $b$ is known a priori (an uncommon, yet possible situation) the non-stationary gamma

process can be easily transformed into a stationary gamma process. The resulting equations for the estimators $\hat{c}$ and $\hat{u}$ become:

$$\frac{\hat{c}}{\hat{u}} = \frac{y_n}{t_n^b} = \overline{\delta} \qquad (12)$$

$$\frac{y_n}{\hat{u}}\left(1 - \frac{\sum_{i=1}^{n} \omega_i^2}{\left(\sum_{i=1}^{n} \omega_i\right)^2}\right) = \sum_{i=1}^{n}\left(\delta_i - \overline{\delta}\omega_i\right)^2 \qquad (13)$$

where $\omega_i = t_i^b - t_{i-1}^b$ is the transformed time corresponding to a $\delta_{C_V}(t)$ increment and $\delta_i = y_i - y_{i-1}$ is the size of the erosion increment.

Given the estimator $\hat{u}$ (Equation 12), the expected erosion at time $t$ can be written as:

$$E(Y(t)) = \frac{\hat{c}t^b}{\hat{u}} = \frac{\hat{c}t^b}{\dfrac{\hat{c}t_n^b}{y_n}} = y_n\left(\frac{t}{t_n}\right)^b \qquad (14)$$

Since cumulative amounts of $\delta_{C_V}(t)$ are available, the last value of $\delta_{C_V}(t)$ received at timestamp $t_n$ is conjectured to contain the most relevant information and therefore the expected value of the gamma process at $t_n$ is equal to the to the last available value of $\delta_{C_V}(t)$, i.e.:

$$E(Y(t_n)) = y_n\left(\frac{t_n}{t_n}\right)^b = y_n \qquad (15)$$

A drawback of estimating parameter $b$ with the maximum likelihood method is that, due to the gamma process property of having independent increments, the $b$ estimate is not sensitive to a change in the new upcoming values of $\delta_{C_V}(t)$.

A viable way to overcome this limitation is to increase the memory of its estimator by resorting to the least square fit of the general power law assumed in Equation (7) and by adding a constraint to the estimator. In this view, the expected value of the least square estimator at timestamp $t_n$ must be equal to the last available value $y_n$. The objective is to adjust the parameters of the model to best fit the data by minimizing the sum of squared residuals:

$$S(b, c, u) = \sum_{i=1}^{n-1} \left( y_i - \frac{ct_i^b}{u} \right)^2 \wedge E(Y(t_n))$$
$$= y_n = \frac{ct_n^b}{u}$$

(16)

Since there is no closed-form solution to this non-linear least square problem, the approximated logarithmic transformation of the least-square fit model for the expected value is used, i.e.

$$\ln E(Y(t)) = \ln \frac{c}{u} + b \ln t_i$$

Finally, to allow assigning different weights $w_i$ to the data, the weighted least-squares objective becomes:

$$Q(b, c, u) = \sum_{i=1}^{n-1} \left( w_i \left( \ln y_i - \ln \frac{c}{u} - b \ln t_i \right) \right)^2$$
$$\wedge \ln E(Y(t_n)) = \ln y_n = \ln \frac{c}{u} + b \ln t_n$$

(17)

The solution to the constrained Equation (17) is equal to the maximum-likelihood Equation (9) and the method of moments Equation (12) which is used in Equation (17) to find its minimum by setting equal to zero its partial derivative with respect to $b$ (Equation 18).

Equation (18) is an attractive closed-form solution of the estimator $\hat{b}$. For example, to give a better fit to the most recent values of $\delta_{C_V}(t)$ larger values of the corresponding weights $w_i$ are assigned, thus making the estimator biased. A closed-form solution for all parameters ($b$, $c$ and $u$) in the power-law assumption for the non-stationary gamma process is then available by combining Equation (18), Equation (12) and Equation (13).

## 6.3 Remaining Useful Life as a Conditional Lifetime Distribution

Assume that a component with time to failure $T$ is put into operation at time $t = t_0 = 0$ and is still functioning at time $t = s$. The cumulative probability distribution function (lifetime distribution) at timestamp $t = s$ is $\Pr(T \leq s) = F_T(s)$ and the expression for the survival function at timestamp $t = s$ is $\Pr(T > s) = 1 - F_T(s) = R_T(s)$.

In the traditional lifetime approach to reliability modelling, the remaining useful life distribution of a component which has survived up to time $s$ is given as the conditional lifetime distribution:

*Equation 18*

$$\hat{b} = \frac{\sum_{i=1}^{n-1} w_i^2 (\ln y_i)(\ln t_i - \ln t_n) - \ln y_n \sum_{i=1}^{n-1} w_i^2 (\ln t_i - \ln t_n)}{\sum_{i=1}^{n-1} w_i^2 (\ln t_i - \ln t_n)^2}$$

(18)

$$F_T\left(t\middle|s\right) = \Pr(T \le t \middle| T > s) = \frac{F_T(t) - F_T(s)}{1 - F_T(s)}, \forall t > s \tag{19}$$

The conditional probability density function called remaining useful life density function $f_T\left(t\middle|s\right)$ is then written as:

$$f_T\left(t\middle|s\right) = \frac{\partial}{\partial t}\left(F_T\left(t\middle|s\right)\right) = \frac{f_T(t)}{R_T(s)} \tag{20}$$

Equation (20) shows that the distribution of the time-based remaining useful life $f_T\left(t\middle|s\right)$ is the original lifetime distribution $f_T(t)$ left truncated at time $t = s$.

An alternative approach, hereby called state-based approach, is here adopted, which accounts for the knowledge of the actual component state rather than its survival time. In fact, the remaining useful life distribution can differ a lot from the left-truncated lifetime, depending on the actual level of state $y_s$ at time $t = s$.

The state-based remaining useful life is therefore defined as the remaining time to the first

passage past the threshold $\rho$ given the knowledge of its value at time $t = s$, i.e. $Y(s) = y_s < \rho$. The state-based remaining useful life is defined as:

$$F_{T_\rho}\left(t\middle|s\right) = \Pr\left(T \le t \middle| Y(s) = y_s\right), \forall t \ge s \tag{21}$$

This approach exploits information of noticeably higher quality, given that a pre-defined list of discrete health states for a component is available based on expert analysis (Kiddy, 2003; Gola & Nystad, 2011).

## 6.4 Results

In this case study, measurements corresponding to 305 operational days are available. Approximately 235 operational days of measurements are collected and processed with the Virtual Sensor to produce reliable erosion state indications $\delta_{C_V}$ before the gamma process is devised to estimate the choke remaining useful life. This amount of measurements is conjectured to be sufficient to achieve reliable calculations of the three parameters.

*Figure 9. RUL estimation and uncertainty obtained with the gamma process when parameter $b$ is calculated with the weighted least-square optimization and when it is fixed to 2.2. The actual RUL is indicated by the dashed line.*

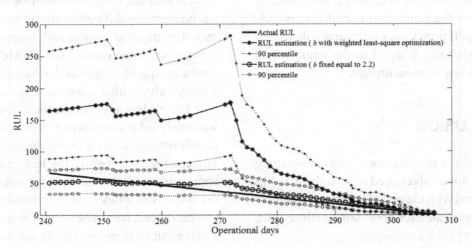

The weighted least-square optimization is done by considering $K = 1000$ replicates of the last $N = 50$ erosion state indications. The estimation of the RUL and its uncertainty is then carried out every operational day until the choke is actually replaced. The failure threshold for the erosion indicator $\delta_{C_V}$ is set equal to 16. Since the gamma process requires a monotonic data series, the erosion indicator $\delta_{C_V}$ is first filtered with a combination of moving average and moving maxima.

Results of the remaining useful life estimation are shown in Figure 9 and compared to those obtained when the $b$ parameter is set constant and equal to 2.2 which is the value that best fits the last 50 available erosion state indications $\delta_{C_V}$ in terms of least-square error.

The slowly increasing values calculated for the erosion indicator $\delta_{C_V}$ up to 273 operational days (Figure 7) lead to having values of $b$ with the weighted least-square optimization smaller than 1. As a consequence, the resulting convex shape of the expected gamma function hits the failure threshold at considerably large times, thus returning an overestimated value of the choke remaining useful life. When values of the erosion indicator $\delta_{C_V}$ show a sharp increase towards the end of the choke life, the weighted least-square optimization allows to quickly update the value of $b$ with the effect of obtaining a more precise estimation of the remaining useful life, which, after 290 operational days is comparable to that obtained by fixing $b$ equal to the value which best fits the last 50 measurements.

## 7 CONCLUSION

In this chapter, a practical case study concerning erosion in choke valves used in oil industries has been analysed with the aim of defining a diagnostic-prognostic frame for optimizing maintenance scheduling of such components.

Two objectives have been identified: 1) the development of a condition monitoring system capable of providing reliable calculations of the erosion state based on collected measurements of physical parameters related to the choke erosion and 2) the development of a prognostic system to accurately estimate the remaining useful life of the choke.

An empirical, model-based approach has been used to fulfil the diagnostic objective of providing reliable calculations of the erosion state, whereas a statistical method based on the gamma probability distribution has been adopted to reach the prognostic goal of accurately estimating the remaining useful life of the choke.

Although the results obtained so far are encouraging with respect to the goal of defining a diagnostic-prognostic frame for optimizing maintenance scheduling of choke valves, a strong limitation of the proposed procedure has been envisioned in the amount and the quality of the available data.

In fact, the inaccurate daily calculations of oil, water and gas mass flow rates are hard to improve without additional instrumentation and/or a larger set of data related to similar valves. Improved instrumentation seems to be the best choice. Meland (Meland, 2011) studied internal leakage of safety critical valves. Models based on analytical relationships can be suited for understanding the basic principles. However, quantification of leakage is inherently difficult and the information from acoustic emission sensors and dynamic pressure sensors is of great value. Meland (Meland, 2011) used a comparative approach where he compared a leaky valve with a group of other leaky valves with known leakage. The frequency spectral data was analysed and nonlinear regression methods based on the support vector machine principle was implemented to quantify the leakage. The same approach could also be used for improving e.g. the mass flow rates related to erosion of choke valves. Indeed, if the leakage rate or erosion rate of a valve can be monitored more accurately, it will lead to improvements to the planning of maintenance actions, which ultimately cuts costs, by, for example, delaying the replacement of a valve, or having less functional tests thus reducing system downtime. Furthermore, an improved condition monitoring will also increase operation safety.

The reference database is very important. With large quantities of data it will be important to query the data according to failure mode, size of valve, sensor type, sensor measurement frequency and units, liquid/gas and so on. Important feedback to maintenance crew performing inspections and safety experts will be the output from all the functional blocks of the condition-based maintenance framework in Figure 2. Important questions like which data and quality of data is needed to quantify erosion of choke valves can then be given to personnel performing inspections. A baseline for making a handbook for common understanding of documentation of degradation mechanisms related to failure modes will be available.

The relation between safety instrumented systems and condition monitoring is not discussed in this chapter. More details about this extensive topic can be found in IEC 61508 (1998) and IEC 61511 (2003) and OLF (2001).

Attention has been recently steered towards the development of physical or data-driven models as well as hybrid systems to be used ahead of the diagnostic scheme to provide reliable estimates of the values of those parameters used in the diagnostic model.

Finally, it is evident that having data corresponding to one single valve considerably affect the general applicability of the approach which has not been yet demonstrated. With a larger amount of data related to many similar valves one could in fact perform a more consistent training of the Virtual Sensor and eventually define an optimal value for the shape-parameter of the gamma function.

# REFERENCES

Andrews, J., Kjørholt, H., & Jøranson, H. (2005, June). *Production enhancement from sand management philosophy: A case study from Statfjord and Gullfaks.* Paper presented at SPE European Formation Damage Conference, Sheveningen, The Netherlands.

Baraldi, P., Compare, M., Despujols, A., & Zio, E. (2011a). Modelling of the effect of maintenance actions on the degradation of an electric power plant component. *Proceedings of the Institution of Mechanical Engineers, Part O. Journal of Risk and Reliability, 225*(2), 169–184.

Baraldi, P., Compare, M., Rossetti, G., Zio, E., & Despujols, A. (2011b). A modelling framework to assess maintenance policy performance in electrical production plants. In Andrews, J., Berenguer, C., & Jackson, L. (Eds.), *Maintenance modelling and applications* (pp. 263–282). Det Norske Veritas, AS.

Bauer, E., & Kohavi, R. (1999). An empirical comparison of voting classification algorithms: Bagging, boosting and variants. *Machine Learning, 36*, 105–139. doi:10.1023/A:1007515423169

Breiman, L. (1996). Bagging predictors. *Machine Learning, 24*(2), 123–140. doi:10.1007/BF00058655

Breiman, L. (1999). Combining predictors. In Sharkey, A. J. C. (Ed.), *Combining artificial neural nets*. Springer.

Bringedal, B., Hovda, K., Ujang, P., With, H. M., & Kjørrefjord, G. (2010, November-December). *Using on-line dynamic virtual flow metering and sand erosion monitoring for integrity management and production optimization.* Paper presented at Deep Offshore Technology Conference, Houston, Texas.

Gola, G., & Nystad, B. H. (2011, May-June). *Comparison of time- and state-space non-stationary Gamma processes for estimating the remaining useful life of choke valves undergoing erosion*. Paper presented at COMADEM Conference, Stavanger, Norway.

Grall, A., Bérenguer, C., & Dieulle, L. (2002). A condition-based maintenance policy for stochastically deteriorating systems. *Reliability Engineering & System Safety, 76*(2), 167–180. doi:10.1016/S0951-8320(01)00148-X

Hansen, L. K., & Salamon, P. (1990). Neural network ensembles. *IEEE Transactions on Pattern Analysis and Machine Intelligence, 12*(10), 993–1001. doi:10.1109/34.58871

Haugen, K., Kvernvold, O., Ronold, A., & Sandberg, R. (1995). Sand erosion of wear resistant materials: Erosion in choke valves. *Wear, 186-187*, 179–188. doi:10.1016/0043-1648(95)07158-X

Hovda, K., & Andrews, J. S. (2007, May). *Using Cv models to detect choke erosion - A case study on choke erosion at Statfjord C-39*. Paper presented at SPE Applied Technology Workshop on Sound Control, Phuket, Thailand.

Hovda, K., & Lejon, K. (2010, March). *Effective sand erosion management in chokes and pipelines - Case studies from Statoil*. Paper presented at 4th European Sand management Forum, Aberdeen, UK.

IEC 61508. (1998). *Functional safety of electrical/ electronic/programmable electronic safety-related systems*. Geneva, Switzerland: International Electrotechnical Commission.

IEC 61511. (2003). *Functional safety - safety instrumented systems for the process industry*. Geneva, Switzerland: International Electrotechnical Commission.

ISO 13374-1. (2003). *Condition monitoring and diagnostics of machines – data processing, communication and presentation – Part 1: General guidelines*. ISO, Switzerland.

ISO 13374-2. (2007). *Condition monitoring and diagnostics of machines – data processing, communication and presentation – Part 2: Data processing*. ISO, Switzerland.

ISO 13381-1. (2004). *Condition monitoring and diagnostics of machines – prognostics – Part 1: General guidelines*. ISO, Switzerland.

Kiddy, J. S. (2003, June). *Remaining useful life prediction based on known usage data*. Paper presented at the SPIE Conference, Santa Fe, NM.

Meland, E. (2011). *Condition monitoring of safety critical valves*. PhD thesis, Department of Marine Engineering, Norwegian University of Science and Technology, Trondheim, Norway.

Metso Automation. (2005). *Flow control manual* (4th ed.).

Mitchell, J., Bond, T., Bever, K., & Manning, N. (1998). Mimosa - Four years later. *Sound and Vibration*, 12–21.

Ngkleberg, L., & Sontvedt, T. (1995). Erosion in choke valves-oil and gas industry applications. *Wear, 186-187*(2), 401–412. doi:10.1016/0043-1648(95)07138-5

Nomack, M. (2011, January). *BP's Deepwater Horizon accident investigation report*. Paper presented at NCSE 11th National conference on Science, Policy, and the Environment, Washington, DC.

Nystad, B. H., Gola, G., Hulsund, J. E., & Roverso, D. (2010, October). *Technical condition assessment and remaining useful life estimation of choke valves subject to erosion*. Paper presented at the Annual Conference of the Prognostics and Health Management Society, Portland, Oregon.

OLF. (2001). *Application of IEC 61508 and IEC 61511 in the Norwegian petroleum industry.* Retrieved from http://www.itk.ntnu.no/sil/OLF-070-Rev2.pdf

Opitz, D. W., & Shavlik, J. W. (1996). Actively searching for an effective neural network ensemble. *Connection Science, 8*(3-4), 337–354. doi:10.1080/095400996116802

PCT/NO2008/00293. (2008). *System and method for empirical ensemble-based virtual sensing.*

Ross, P. (2008). *The night the sea caught fire: Remembering Piper Alpha.* Scotland on Sunday, June 15, 2008.

Roverso, D. (2009, September). *Empirical ensemble-based virtual sensing – A novel approach to oil-in-water monitoring.* Paper presented at Oil-in-Water Monitoring Workshop, Aberdeen, UK.

Rumelhart, D. E., & McClelland, J. L. (1986). *Parallel distributed processing.* Cambridge, MA: MIT Press.

Tsang, A. H. C. (1995). Condition-based maintenance: Tools and decision making. *Journal of Quality in Maintenance Engineering, 1*(3), 3–17. doi:10.1108/13552519510096350

van Noortwijk, J. M., & Pandey, M. D. (2003, November). *A stochastic deterioration process for time-dependent reliability analysis.* Paper presented at IFIP WG 7.5 Working Conference on Reliability and Optimization of Structural Systems, Banff, Canada.

Wallace, M. S., Dempster, W. M., Scanlon, T., Peters, J., & McCulloch, S. (2004). Prediction of impact erosion in valve geometries. *Wear, 256,* 927–936. doi:10.1016/j.wear.2003.06.004

Williams, J. H., Davies, A., & Drake, P. R. (1994). *Condition-based maintenance and machine diagnostics.* London, UK: Chapman & Hall.

## KEY TERMS AND DEFINITIONS

**Choke Valves:** Type of valve designed to create chocked flow in a fluid. Over a wide range of valve settings the flow through the valve can be understood by ignoring the viscosity of the fluid passing through the valve; the rate of flow is determined only by the ambient pressure on the upstream side of the valve.

**Condition-Based Maintenance (CBM):** Direction of maintenance actions based on indications of asset health as determined from non-invasive measurement of operation and condition indicators. CBM allows preventative and corrective actions to be optimized by avoiding traditional calendar or run based maintenance.

**Diagnostics:** Discipline aimed at identifying the nature and cause of anything. In system engineering and computer science, diagnosis is typically used to determine the causes of symptoms, mitigations for problems and solutions to issues.

**Empirical Modelling:** Any kind of computer-based modelling which exploits empirical observations (typically data collected in the form input-output) rather than mathematically describable relationships based on first-principle assumptions to model a system.

**Gamma Process:** Stochastic process with independent, gamma-distributed increments. In prognostics, it is often used to model the deterioration of a system or a component due to the positive-defined increments which reflect the degradation of a system or a component not subject to maintenance.

**Prognostics:** Discipline focused on predicting the time at which a system or a component will no longer perform its intended function. This lack of performance is most often a failure beyond which the system can no longer be used to meet desired performance.

**Remaining Useful Life (RUL):** Difference between the estimated time at which a system or a component will no longer be able to perform its intended function and the current time.

# Section 12
# Prognostics (Review)

# Chapter 17
# Prognostics and Health Management of Industrial Equipment

**E. Zio**
*Ecole Centrale Paris, France & Politecnico di Milano, Italy*

## ABSTRACT

*Prognostics and health management (PHM) is a field of research and application which aims at making use of past, present, and future information on the environmental, operational, and usage conditions of an equipment in order to detect its degradation, diagnose its faults, and predict and proactively manage its failures. This chapter reviews the state of knowledge on the methods for PHM, placing these in context with the different information and data which may be available for performing the task and identifying the current challenges and open issues which must be addressed for achieving reliable deployment in practice. The focus is predominantly on the prognostic part of PHM, which addresses the prediction of equipment failure occurrence and associated residual useful life (RUL).*

## INTRODUCTION

In human health care, a medical analysis is made, based on the measurements of some parameters related to health conditions; the examination of the collected measurements aims at detecting anomalies, diagnosing illnesses and predicting their evolution. By analogy, technical procedures of health management are used to capture the functional state of industrial equipment from historical recordings of measurable parameters (Vachtsevanos, Lewis, Roemer, Hess & Wu, 2006). With the term 'equipment', from now on we shall intend a System, Structure or Component (SSC).

DOI: 10.4018/978-1-4666-2095-7.ch017

The knowledge of the state of equipment and the prediction of its future evolution are at the basis of condition-based maintenance strategies (Jarrell, Sisk & Bond, 2004): according to these strategies, maintenance actions are carried out when a measurable equipment condition shows the need for corrective repair or preventive replacement. From the point of view of production performance, by identifying the problems in the equipment at their early stages of development, it is possible to allow the equipment to run as long as it is healthy and to opportunely schedule the maintenance interventions for the most convenient and inexpensive times. The driving objectives are maximum availability, minimum unscheduled shutdowns of production, economic maintenance (Jardine, Lin & Banjevic, 2006).

The condition of the equipment is usually monitored at a regular interval and once the reading of the monitored signal exceeds a threshold level, a warning is triggered and maintenance actions may be planned based on the prediction of the future evolution of the degradation process. The monitoring interval influences the equipment overall cost and performance: a shorter interval may increase the cost of monitoring, whereas a longer one increases the risk of failure. The monitoring system should be reliable in order to avoid false alarms. A decision must be taken every time an alarm is indicated; to ignore an alarm may give rise to serious consequences. A first option is to make further investigation of the alarm, without stopping the equipment; an alternative option is to stop the equipment for an overhaul. In the first option, a false alarm would result in extra cost due to the time and manpower necessary to make the diagnosis; the second option could result in greater losses, where lost production and manpower costs occur simultaneously. The greatest losses will occur when ignoring the alarm, in case of accidents with damages and loss of assets.

The dynamic scheduling of condition-based maintenance represents a challenging task, which requires the prediction of the evolution of the

monitored variables representing the equipment condition. Upon detection of failure precursors, prognostics becomes a fundamental task; this entails predicting the reliability or the probability of failure of the equipment at future times, and the residual useful life (RUL), i.e. the amount of time the equipment will continue to perform its function according to design specifications. This prediction/forecasting/extrapolation process needs to account for the current state assessment and the expected future operational conditions. The 'fortune teller' of such prognostic task is the intelligent integration of the information and data available into accurate models solved by efficient computational algorithms.

Equipment state knowledge and prediction are also central to the management of abnormal events in process plants (Venkatasubramanian, Rengaswamy, Yin & Kavuri, 2003). A single abnormal event may give rise to a catastrophic accident with significant economic, safety and environmental impacts (the accident at the Kuwait petrochemical refinery in 2000 led to an estimated 100 million dollars in damages (Venkatasubramanian et al.)). On the other hand, minor accidents occur relatively frequently and may cumulate to numerous occupational injuries and illnesses, and relevant costs to the industry (estimates of these costs in the US and UK range in the order of 20 billion dollars per year (Nimmo, 1995; Laser, 2000)). This explains the great interest in, and attention paid to the development of effective methods and procedures for abnormal event management.

Successful abnormal event management requires the timely detection of the abnormal conditions, diagnosis of the causal fault, prognosis of the process evolution; these elements feed the procedure of correction of the equipment fault and of plant control to safe conditions. From the point of view of safety, the recognition of the state of equipment (diagnostics) and prediction of its future evolution (prognostics) enable safer and more reliable operation, under a proactive approach to operations and maintenance which has the potential to improve the understanding of the (safety) margins between operating values and safety thresholds, prioritize aging factors that impact life cycle asset management in the long term, reduce unplanned outages, optimize staff utilization, reduce impacts to high value capital assets. These issues are of utmost relevance for the development and operation of technologies of safety concern, such as nuclear and process technologies.

Complete reliance on human operators for the management of abnormal events and emergencies has become increasingly difficult. In particular, the diagnostic and prognostic tasks in a complex plant are made quite difficult by the variety of equipment failure occurrences and of the related process responses, and by the large number of monitored process variables (of the order of a thousand in modern process plants) which lead to information overload. So, the grand challenge is the creation of adequate, automated methods for PHM of industrial equipment, in support to the human operators.

In the present paper, the focus is on prognostics. Due to the broad nature of the problem, it is not possible to be complete and exhaustive in its treatment and in the detailed discussion of the methods, nor in the reference to the specialized literature. The intent is to provide a general view on the state of knowledge and available methods of prognostics, placed in context with the types of information and data which may be available, while pointing at the main challenges and open issues for practical applications.

Indeed, the development of prognostic systems may rely on quite different information and data on the past, present and future behavior of the equipment; there may be situations in which sufficient and relevant statistical equipment failure data are available, others in which the equipment behavior is known in a sufficiently accurate way to allow building a sufficiently accurate model, and yet others with scarce data on the failure behavior of the equipment (this is typically the case of highly valued equipment which is rarely allowed to run to failure) but with available process data measured by sensors and related to the equipment degradation and failure processes. Correspondingly, a wide range of approaches have been developed, based on different sources of information and data, modeling and computational schemes, and data processing algorithms. A number of taxonomies have been proposed to categorize the different approaches (Pecht, 2008).

In general, a distinction can be made among first-principle model-based, reliability model-based and process sensor data-driven approaches (also referred to as white box, black box and grey box models, respectively). In the first type of approach, a mathematical model is derived from first principles to describe the degradation process leading to failure; the model is then used to predict the evolution of the equipment state and infer the (failure) time in correspondence of which the state reaches values beyond the threshold which describes loss of functionality. Where applicable, this approach leads to the most accurate prognostic results. The difficulty in its use lies in the definition of the model, which may turn out actually impossible for real complex systems subject to multiple, complicated, stochastic processes of degradation. Physic-based Markov models are a typical example. If experimental or field degradation data are available, they can be used to calibrate the parameters of the model or to provide ancillary information related to the degradation state, within the state-observer formulation typical of a filtering problem, e.g. Kalman and particle filtering (Myotyri, Pulkkinen & Simola, 2006; Pedregal & Carnero, 2006; Peel, 2008). Reliability model-based approaches use traditional reliability laws to estimate the RUL. They estimate the life of average equipment under average usage conditions. The parameters of the reliability models can be estimated on the basis of the available data related to the equipment failure behavior. In other words, time to failure data are used to tune the parameters of the failure time distribution. The most common way is by Weibull Analysis. It is also possible to include the effects of the environmental stresses (e.g. temperature, pressure, load, vibration, etc.) under which the equipment operates, to estimate the RUL of an average equipment under the given usage conditions. This can be done for example by the Proportional Hazard Models. Unfortunately, often in practice the availability of sufficiently representative data is rare, especially for very reliable equipment and for new equipment for which experience feedback data is scarce or non-existent. Finally, there are approaches which do not use any explicit model and rely exclusively on process data measured by sensors related to the degradation and failure states of the equipment. Empirical techniques like artificial neural networks, support vector machines, local Gaussian regression, pattern similarity are typical examples. The advantage of these methods lies on the direct use of the measured process data for equipment failure prognostics.

## CHARACTERISTICS OF PROGNOSTICS METHODS

Some desirable characteristics of prognostic methods for practical application are listed below. Their degree of desirability is dependent on the type of equipment and on the objective of the PHM, also in view of its different benefits for production and safety as discussed in the previous Section.

### Quick Prediction

In the application of prognostics, the estimation of the equipment state and the prediction of its future evolution must be performed in a time which is compatible with the time scale of the equipment life and the time constants of the corrective actions (inspections, repairs, overhauls). If the prediction of failure is predicted 'too early', more than needed lead time is used to verify the developing failure, monitor the related variables

and perform the preventive correction; on the contrary, if the failure is predicted too late, the time available to assess the situation and act accordingly is reduced. An even worse situation occurs if the failure prediction arrives after the failure has already occurred. In this sense, a positive time bias (early prediction) is preferable to a negative one (late prediction), and boundaries must be set on the maximum allowable late and early times of prediction. However, the timing of the prediction must be balanced with the need of collecting representative information and data sufficient for providing accurate results. Further, in fast developing situations, e.g. of accident management, the prediction task entails the use of fast-running, approximate models; in these cases, also a proper balance between computational rapidity and precision of the results must be sought, in order to avoid arriving quickly at wrong decisions of intervention.

### Robustness

It is desirable that the performance of the prognostic method does not degrade abruptly in the presence of noises, uncertainties and novelties of situations. For the latter, it is important that the novel malfunctions be promptly recognized and the method adapt to recognizing and handling them.

### Confidence Estimation

State estimations and predictions must be accompanied by a measure of the associated error, accounting for the incomplete and imprecise information available on the process. This measure is fundamental for projecting the confidence on the predictions of the RUL and the related decisions on the actions to take on the equipment to effectively and reliably control its function.

The best representation of the uncertainty associated to the RUL predictions is the complete distribution. This is usually possible to obtain when an analytical model for the prediction of the RUL is available, which also improves the understanding of the degradation behavior. However, this is not always the case in practice.

## Adaptability

Processes in general change in time, as do the functioning of equipment due to changes in external environments, structural changes, changes in the input, retrofitting etc. The prognostic method must be able to accommodate these changes and perform equally well in different working conditions. A distinction between stationary and transient conditions is also expected to be needed, as the correlations among the process variables may differ significantly in the two cases, and require different methods ('static' methods like Principal Component Analysis, Partial Least Squares or others for the stationary conditions, and simulation and trend analysis methods for the transient case) (Hussey, Lu & Bickford, 2009).

## Clarity of Interpretation

PHM methods in general are seen as constituting a complementary aid for plant operators. In this view, the prognostic indications provided must be clearly interpretable to allow the operator to act accordingly, in full conscience of the situations and confidence in the decisions. Graphical tools of representation of the outcomes of the analysis, and the associated uncertainty, can provide an added value.

## Modeling and Computational Burden

For timely prognostics and clarity of interpretation, the amount of modeling required, and the storage and computational burden associated must be properly gauged to the application.

## Multiple Faults Handling

Estimating the equipment state and predicting its evolution is an additional challenge whose solution must be provided for practical interests. A particular challenge relates to the possibility of diagnosing multiple faults simultaneously. A codebook approach would lead to the enumeration of an exponential number of possible fault combinations and their symptoms (Kliger, Yemini, Yemini, Oshie & Stolfo, 1995). Also, a combination of multiple faults may produce symptoms that can be confused with single fault symptoms while the effect of several failure mechanisms may be much higher than the sum of effects of each individual mechanism.

In general, the problem is NP-hard[1] both logically and probabilistically (de Kleer & Williams, 1987; Pearl, 1988) and the more combinations are possible, the more powerful and informative set of measurements we need for discriminating among the different occurrences and avoiding shielding effects, i.e. conditions for which the failures of certain components render impossible the retrieval of information on the state of other components. Possible effective approaches to multiple fault handling might be ones that handle the faults incrementally as they occur over time; otherwise, some kind of approximation on the fault combinations considered is necessary (Rish, Brodie & Ma, 2002; Steinder & Sethi, 2002).

## INFORMATION AND DATA FOR PHM

Different forms of information and data may be available for the prognostic assessment of the trajectory to failure of an equipment undergoing degradation:

- Equipment inherent characteristics (material, physical, chemical, geometrical etc.); these may vary from one individual equipment to another of the same type; this variability is described probabilistically by a probability distribution function of the values of the characteristic parameters.

- Time To Failure (TTF) data of a population of identical or similar equipment, e.g. as recorded for maintenance management purposes.
- External parameters (environmental, operational etc.) which may vary with time during the equipment life. They are not directly related to the equipment degradation state but may influence its evolution. Some of these parameters may be directly observable and measured by sensors, others may not; for some there may be a priori knowledge of their behavior in time (e.g. because defining the operational settings of the equipment, like in the flight plan of an airplane), while the behavior of others may be uncertain (e.g. because depending on external environmental conditions, like those occurring in the actual flight of an airplane).
- Data related to the values of observable process parameters measured by sensors during the evolutions to failure of a population of identical or similar equipment. The observable process parameters are directly or indirectly related to the equipment degradation state. It can then be assumed that a time-dependent parameter indicating the degradation state is available, either because one among the observable parameters directly measurable or because constructed/inferred from them. A threshold on such parameter is established, indicating failure of the equipment which is considered unable to function when the degradation parameter exceeds such threshold. These parameters can be called "Internal parameters" because, as explained, they are different from the "External parameters" of the first bullet above since they are related to the degradation state (for anal-

ogy, in health care the heartbeat of a human is an internal parameter).

Each of these sources of information may be available alone or in combination with other sources. The data may come from the field operation of the equipment or from experimental tests undertaken in laboratory under controlled conditions.

## APPROACHES TO PHM

As said in the previous Section, the development of a prognostic system may face quite different situations with regards to the information and data available on the past, present and future behavior of the equipment; there may be situations in which sufficient and relevant TTF data on the equipment behavior are available, others in which the equipment behavior is known in a way to allow building a sufficiently accurate model, and yet others with scarce TTF data of the equipment, e.g. because being highly valued it is rarely allowed to run to failure, but with process data measured by sensors and related to the equipment degradation and failure processes. Correspondingly, a wide range of approaches have been developed, based on different sources of information, modeling schemes and data processing algorithms.

Various categorizations of these approaches have been proposed in the literature. Perhaps the most useful ones attempt to distinguish the approaches depending on the type of information and data they use, as discussed above. Still, in practice one is often faced with various sources of information and data on the equipment degradation and failure processes, which are best exploited in frameworks of integration of different types of prognostics approaches.

For the sake of presentation, here we distinguish among first-principle model-based, reliability model-based and process sensor data-driven approaches.

## First-Principle Model-Based Approaches

In these approaches, mathematical models of the degradation process leading to failure are built and used to predict the evolution of the equipment state in the future, and infer its probability of failure and time to failure (Pecht, 2007).

The model is typically a system of equations which mathematically describe the evolution of the degradation, as known from first-principle laws of physics. Typical examples are the mathematical models describing the physical laws of evolution of cracks due to fatigue phenomena, of wearing and corrosion processes, etc. Experimental or field degradation data are used to calibrate the parameters of the model, which is then used to predict the degradation state evolution.

Most prognostic models of this type (e.g. Samanta, Vesely. Hsu & Subudly (1991), Lam & Yeh (1994), Hontelez, Burger & Wijnmalen (1996), Kopnov (1999), Bigerelle & Iost (1999), Berenguer, Grall & Castanier (2000)) track the degradation (damage) of the equipment in time and predict when it will exceed a predefined threshold of failure. Physics-based Markov models are often used to describe the degradation evolution in time (Kozin & Bogdanoff, 1989). In practice, the degradation measure may not be a directly measured parameter but it could be a function of several measured variables. If the actual equipment degradation state is not directly observable, then a state-observer formulation of the prognostic problem can be adopted in which a Markov state equation is used to represent the evolution of the hidden degradation state and an observation equation is introduced to relate the measured variables to the degradation state. This formulation sets up a typical framework of signal analysis by filtering approaches like Kalman filtering and particle filtering. Shock models can be introduced to describe the evolution of equipment subject to randomly arriving shocks which deliver a random amount of damage.

## Reliability Model-Based Approaches

A common approach to prognostics of equipment failure time is by modeling its probability distribution, based on available TTF data from laboratory testing and field operation. The distribution thereby obtained allows estimating the lifetime of the average equipment under average usage conditions. A comprehensive review of these methods is found in (Si et al 2010).

The parametric distribution type most commonly used for prognostic purposes is the Weibull distribution which in terms of the formula for the hazard rate $\lambda(t)$ reads (Abernethy, 1996):

$$\lambda(t) = \frac{\beta}{\vartheta}\left(\frac{t}{\vartheta}\right)^{\beta-1} \tag{1}$$

With a proper choice of the value of the shape parameter $\beta$ greater than unity, the hazard rate $\lambda(t)$ can model aging equipment, characterized by failure rates increasing in time.

A disadvantage of this approach is that it only provides the failure time distribution of the average equipment under average conditions of operations. On the contrary, it is expected that equipment under harsh conditions will fail at earlier times than equipment operating in mild environments. To account for the environment and operating conditions, stress data must be brought into the model. This leads to the development of degradation models including explanatory variables (covariates) that describe the operation environment. For example, the Proportional Hazard Model (PHM) is a method capable of including information on the environmental and operating conditions to modify a baseline, "average" hazard rate $\lambda_0(t)$ (Cox, 1972; Cox & Oakes, 1984; Bendell, Wightman & Walker, 2002):

$$\lambda(t;z) = \lambda_0(t)\exp\left(\sum_{j=1}^{q} \beta_j z_j\right) \qquad (2)$$

The environmental information condition is represented by the multiplicative covariates $z_j$ and failure data collected at different covariate conditions are used to estimate the coefficients $\beta_j$ by ordinary least square algorithms (Vlok, Coetzee, Banjevic, Jardine & Makis, 2002). Time-dependent covariates can also be included, within a gamma process description of the increasing degradation (Bagdonavicius & Nikulin, 2000; Deloux, Castanier & Berenguer, 2008).

An alternative approach is that of the cumulative hazard rate, which also amounts to modeling the effect of the environment by the introduction of covariates in the hazard function (Singpurwalla, 1995; Singpurwalla, 2006).

Markov models can also be used to account for the influence of the working environment on the equipment failure behavior (Samanta et al., 1991; Yeh, 1997; Grall, Berenguer & Chu, 1998; Marseguerra, Zio & Podofillini, 2002; Zhao, Fouladirad, Berenguer & Bordes, 2008). This typically entails:

- The definition of a prognostic parameter which indicates the equipment failure condition state, upon comparison with a threshold. The value of the prognostic parameter for which actual equipment failure occurs is typically uncertain and can be established via statistical analysis of failure data, if available.
- The development of a stochastic model of the random evolution of the environmental conditions, based on the physics of the process.
- The development of a stochastic functional relationship of the effect of the environmental conditions on the evolution of the prognostic parameter. Usually, the func-

tional form of this relationship is cumulative, since the environmental stressors tend to non-decreasingly deteriorate the equipment.

A disadvantage is the large number of degradation states and related probability distribution parameters to be fitted by data (Zille, Despujols, Bataldi, Rossetti & Zio, 2009). Indeed in practice, reliability model-based approaches suffer from the fact that equipment becomes more and more reliable, so that fewer data are available for fitting the reliability model distributions (Coble & Hines, 2009a) and the estimation of the characteristic parameters is quite difficult even by accelerated life testing (Bagdonavicius & Nikulin, 2000; Lehmann, 2006). For this reason, alternative hybrid approaches, based on the integration of fuzzy logic models are investigated with the aim of including in the model also qualitative information on the operational and environmental conditions based on expert knowledge of the equipment design, usage, degradation processes and failure modes (Baraldi, Zio, Compare, Rossetti & Despujols, 2009).

General Path Models (GPM) are also used to predict future degradation in time, based on measurable degradation data collected. A functional relationship is established of the damage evolution in time and the model parameters are estimated using historical data (Lu & Meeker, 1993). The underlying model describes the equipment degradation in terms of the evolution of an identified prognostic indicator of damage; the model accounts for both population (fixed) and individual (random) effects. The adaptability of the predictive model to changing scenarios of working conditions can be achieved by a dynamic Bayesian approach for updating the parameters estimates as historical degradation data become available from tests or field operation (Coble & Hines, 2009a).

## Process Sensor Data-Driven Approaches

In situations where developing physics-based models of the degradation and failure behavior of an equipment is not possible or favorable, sufficient TTF data are not available whereas process sensors are available, which collect data related to the degradation state of the equipment, one can employ a data-driven approach to prognostics (Schwabacher, 2005; Hines, Garvey, Preston & Usynin, 2008). In this case, the prognosis of the state of the equipment for RUL estimation relies on the availability of run-to-failure data. Based on these data, the RUL can be estimated either directly through a multivariate pattern matching process from the data to the remaining life, or indirectly through damage estimation followed by extrapolation to damage progression up to the failure threshold. The latter approach is closer to the typical engineering reasoning but requires the definition of both the damage parameter and the failure criterion.

Common to all data-driven approaches is the estimation of the output values, without necessarily modeling the equipment physical behavior and operation. Such approaches include conventional numerical algorithms like linear regression and time series analysis (Hines & Garvey, 2007) as well as machine learning and data mining algorithms, like artificial neural networks, fuzzy logic systems, support vector machines (Vachtsevanos & Wang, 2001; Wang, Yu & Lee, 2002; Wang, Goldnaraghi & Ismail, 2004; Yan, Koç & Lee, 2004; Schwabacher & Goebel, 2007; Sotiris & Pecht, 2007; Peng, Zhang & Pan, 2010; Sikorska, Kelly & McGrath, 2010; Heng et al., 2009). Indeed, since predicting the TTF of an equipment can be seen as a regression problem, regression analysis and time-series estimation and forecasting methods can be used to build models for the direct estimation of the TTF from the available data.

## EXAMPLES

For exemplary purposes, one first-principle model-based approach, within a filtering framework, and one data-driven approach are here illustrated with reference to their application in a prognostic task regarding a non linear fatigue crack growth process, typical of a certain class of industrial and structural equipment (Oswald & Schueller, 1984; Sobezyk & Spencer, 1992; Bolotin & Shipkov, 1998; Myotyri et al., 2006). The choice of the approaches presented is not motivated by any declaration of alleged superiority in comparison to the many other methods proposed in the literature, but by the need to rely on the experience of the author in their development and application.

The common Paris-Erdogan model is adopted for describing the evolution of the crack depth $x$ as a function of the load cycles $t$ (Pulkkinen, 1991):

$$\frac{dx}{dt} = C(\Delta S)^m \tag{3}$$

where $C$ and $m$ are constants related to the material properties (Provan, 1987; Kozin & Bogdanoff, 1989), which can be estimated from experimental data (Bigerelle & Iost, 1999) and $\Delta S$ is the stress intensity amplitude, roughly proportional to the square root of $x$ (Provan, 1987):

$$\Delta S = \gamma \sqrt{x} \tag{4}$$

where $\gamma$ is again a constant which may be determined from experimental data.

The intrinsic stochasticity of the process may be inserted in the model by modifying Equation (3) as follows (Provan, 1987):

$$\frac{dx}{dt} = e^\omega C \left( \gamma \sqrt{x} \right)^m \tag{5}$$

where $\omega \sim N(0, \sigma_\omega^2)$ is a white Gaussian noise. For $\Delta t$ sufficiently small, the state-space model (5) can be discretized to give:

$$x\left(t_k\right) = x\left(t_{k-1}\right) + e^{\omega_j} C(\Delta S)^m \Delta t \tag{6}$$

which represents a non-linear Markov process with independent, non-stationary degradation increments of the degradation state $x$.

At the generic inspection time $T_j$, the degradation state $x\left(T_j\right)$ is generally not directly measurable. In the case of non-destructive ultrasonic inspections a logit model for the observation $f\left(T_j\right)$ can be introduced (Simola & Pulkkinen, 1998):

$$\ln \frac{f\left(T_j\right)}{d - f\left(T_j\right)} = \gamma_0 + \gamma_1 \ln \frac{x\left(T_j\right)}{d - x\left(T_j\right)} + v_k \tag{7}$$

where d is the equipment material thickness, $\gamma_0 \in (-\infty, \infty)$ and $\gamma_1 > 0$ are parameters to be

estimated from experimental data and $v$ is a white Gaussian noise such that $v \sim N(0, \sigma_v^2)$.

The following standard transformations are introduced:

$$y\left(T_j\right) = \ln \frac{f\left(T_j\right)}{d - f\left(T_j\right)} \tag{8}$$

$$\mu\left(T_j\right) = \gamma_0 + \gamma_1 \ln \frac{x\left(T_j\right)}{d - x\left(T_j\right)} \tag{9}$$

In the case study here considered (taken from Myotyri et al. (2006)), the parameters of the state Equation (6) are $C = 0.005$, $m = 1.3$ and $\gamma = 1$, whereas those in the measurement Equation (7) are $\gamma_0 = 0.06$, and $\gamma_1 = 1.25$. The process and measurement noise variances are $\sigma_\omega^2 = 2.89$ and $\sigma_v^2 = 0.22$, respectively. The equipment is assumed failed when the crack depth $x \geq d = 100$, in arbitrary units.

Figure 1 shows the degradation-to-failure pattern that in the following will be used as test

*Figure 1. Crack growth pattern used as test pattern*

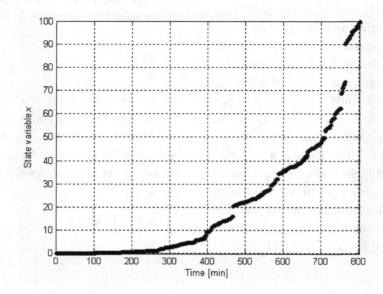

pattern in the procedure for predicting the equipment RUL. The crack depth $x$ reaches the full material thickness d=100 at 802 [min].

In the case study, the interval between two successive inspections is equal to 100 [min], if the estimated RUL>200 [min] or otherwise it is equal to 10 [min], reflecting a more frequent inspection of the equipment integrity as the equipment is approaching the end of life.

## A First-Principle Model-Based Approach by Particle Filtering

In general terms, under a filtering formulation of the first-principle model-based approach the problem of estimating the degradation state of an equipment is carried out in a discrete time domain, considering both a set of measurements and a model linking the equipment states among themselves and with the measurements.

In correspondence of a sequence of equidistant discrete times $t$, where $t$ stands for $\tau_t = t \cdot \Delta t, (t = 0, 1, 2, ....)$, it is desired to infer the unknown (hidden) state $x_t \equiv x(\tau_t)$ on the basis of all the previously estimated state values $x_{0:t-1} \equiv (x_0, x_1, ..., x_{t-1})$ and of all the measurements $z_{0:t} \equiv (z_0, z_1, ..., z_t)$, $z_i \equiv z(\tau_i)$, collected up to time $t$ by a set of sensors. Both the equipment states and the measurements, which may be multidimensional variables, are affected by inherent noises.

In a Bayesian context, the associated filtering problem amounts to evaluating the posterior distribution $p(x_t \mid z_{0:t})$. This can be done by sampling a large number $N_s$ of time sequences $\{x_{0:t}^i\}_{i=1}^{N_s}$ from a suitably introduced importance

function $q(x_{0:t} \mid z_{0:t})$ (Doucet, de Freitas & Gordon, 2001). In the state space, this sample of sequences represents an ensemble of trajectories of state evolution similar to those simulated in particle transport phenomena: the problem is then that of utilizing the ensemble of $N_s$ simulated trajectories for filtering out the unobserved trajectory of the real process.

The filtering distribution $p(x_t \mid z_{0:t})$ is the marginal of the probability $p(x_{0:t} \mid z_{0:t})$, i.e. the multiple integral of this latter with respect to $x_{t_0}, x_{t_1}, ..., x_{t-1}$ in $[-\infty, \infty]^t$ viz., $p(x_t \mid z_{0:t}) = \int p(x_{0:t} \mid z_{0:t}) dx_{0:t-1}$. The integration may be formally extended to include also the variable $x_t$ by means of a $\delta$-function, i.e.

$$p(x_t \mid z_{0:t}) = \int p(x_{0:t-1}, u \mid z_{0:t}) \delta(x_t - u) dx_{0:t-1} du.$$

By sampling a large number $N_s$ of trajectories $\{x_{0:t}^i\}_{i=1}^{N_s}$ from the importance function $q(x_{0:t} \mid z_{0:t})$, the integral is approximated as Equation 10 where the weights $w_t^i$ of the estimation are

$$w_t^i = \frac{p(x_{0:t}^i \mid z_{0:t})}{q(x_{0:t}^i \mid z_{0:t})} \tag{11}$$

which can be recursively computed as

$$w_t^i = w_{t-1}^i \frac{p(z_t \mid x_t^i) p(x_t^i \mid x_{t-1}^i)}{q(x_t^i \mid x_{t-1}^i)} \tag{12}$$

Unfortunately, the trajectories sampled according to the procedure illustrated suffer from

*Equation 10*

$$p(x_t \mid z_{0:t}) = \int \left[ \frac{p(x_{0:t-1}, u \mid z_{0:t})}{q(x_{0:t-1}, u \mid z_{0:t})} \delta(x_t - u) \right] q(x_{0:t-1}, u \mid z_{0:t}) dx_{0:t-1} du \approx \sum_{i=1}^{N_s} w_t^i \delta(x_t - x_t^i) \tag{10}$$

the so called degeneracy phenomenon: after few samplings, most of the $N_s$ weights in (12) become negligible so that the corresponding trajectories do not contribute to the estimate of the probability density function (pdf) of interest (Doucet et al., 2001).

A possible remedy to this problem is to resort to the so called resampling method (Arulampalam, Maskell, Gordon & Clapp, 2002), based on the bootstrap technique which essentially consists in sampling balls from an urn with replacement (Efron, 1979; Efron & Tibshirani, 1993). At each time $t$, $N_s$ samplings with replacement are effectuated from an urn containing $N_s$ balls; the i-th ball is labelled with the pair of known numbers $\left\{ w_t^i, x_t^i \right\}$ and it will be sampled with a probability proportional to the weight value $w_t^i$; a record of the sampled pairs is maintained; at the end of these $N$ multinomial samplings, there is a good chance that the recorded sample will contain several replicas of the balls with larger weights (in other words, that the final record will contain several identical copies of the same label), whereas a corresponding number of balls with smaller weights will not appear in the sample (in other words, a corresponding number of labels is lost from the sample).

In the described bootstrap procedure, it is evident that the sampled weights are i.i.d. so that the same weight $1/N_s$ may be assigned to all sampled pairs. Then, the filtering procedure continues with the original pairs $\left\{ w_t^i, x_t^i \right\}_{i=1}^{N_s}$ replaced by new pairs $\left\{ 1/N_s, x_t^{i^*} \right\}_{i^*=1}^{N_s}$ in which several $i^*$ may correspond to the same i in the original pairs. Equation (10) then becomes

$$p(x_t \mid z_{0:t}) \approx \sum_{i=1}^{N_s} \frac{p(z_t \mid x_t^i) p(x_t^i \mid x_{t-1}^i)}{q(x_t^i \mid x_{t-1}^i)} \delta(x_t - x_t^i)$$
$$\approx \sum_{i=1}^{N_s} \frac{1}{N_s} \delta(x_t - x_t^{i^*})$$

(13)

A pseudo-code describing the basic steps of the procedure is:

- At $t=0$, a sequence $\left\{ x_0^i \right\}_{i=1}^{N_s}$ is sampled from $p(x_0)$;
- At the generic time $t>0$:
  - a value $z_t$ is measured (or simulated if we are dealing with a case study);
  - a sequence $\left\{ x_t^i \right\}_{i=1}^{N_s}$ is sampled from the given $q(x \mid x_{t-1}^i)$;
  - the $N_s$ likelihoods $\left\{ p(z_t \mid x_t^i) \right\}_{i=1}^{N_s}$ are evaluated;
  - the weights $w_t^i$ required by the described resampling procedure are evaluated from (12) in which $w_{t-1}^i = 1$;
  - the resampling procedure is performed and the obtained $x_t^{i^*}$ yield the resampled realizations of the states at time $t$;
  - the $x_t^{i^*}$-range, $X_t = \max_{i^*}(x_t^{i^*}) - \min_{i^*}(x_t^{i^*})$, is divided in a given number of intervals and the mean probability values in these intervals are given by the histogram of the $x_t^{i^*}$.

The particle filtering estimation method has been applied to the case study of literature previously illustrated, concerning the nonlinear crack propagation due to fatigue as modelled by the Paris-Erdogan law.

The application of particle filtering for RUL estimation entails the evaluation of the conditional cumulative distribution function (cdf) of the stochastic observable variable related to the degradation state. For more details on the procedure, the interested reader may refer to Cadini, Zio & Avram (2009).

From Equation (8) it follows that the transformed observation $Y\left(T_j\right) \sim N\left(\mu\left(T_j\right), \sigma_v^2\right)$ is a Gaussian random variable with cdf:

$$cdf_{Y\left(T_j\right)}\left(y\left(T_j\right)\middle|x\left(T_j\right)\right)$$
$$= P\left(Y\left(T_j\right) < y\left(T_j\right)\middle|x\left(T_j\right)\right) \qquad (14)$$
$$= \Phi\left(\frac{y\left(T_j\right) - \mu\left(T_j\right)}{\sigma_v}\right)$$

where $\Phi\left(u\right)$ is the cdf of the standard normal distribution $N\left(0,1\right)$.

The conditional cdf of the stochastic measurement variable $F\left(T_j\right)$ related to the stochastic degradation state $X\left(T_j\right)$ is then:

$$cdf_{F\left(T_j\right)}\left(f\left(T_j\right)\middle|x\left(T_j\right)\right)$$
$$= cdf_{Y\left(T_j\right)}\left(\ln\frac{f\left(T_j\right)}{d - f\left(T_j\right)}\middle|x\left(T_j\right)\right) \qquad (15)$$
$$= \Phi\left(\frac{1}{\sigma_v}\left(\ln\frac{f\left(T_j\right)}{d - f\left(T_j\right)} - \mu\left(T_j\right)\right)\right)$$

with corresponding pdf:

$$pdf_{F\left(T_j\right)}\left(f\left(T_j\right)\middle|x\left(T_j\right)\right)$$
$$= \frac{1}{\sqrt{2\pi}\sigma_v}e^{-\frac{1}{2}\left(\frac{\ln\frac{f\left(T_j\right)}{d-f\left(T_j\right)} - \mu\left(T_j\right)}{\sigma_v}\right)^2} \frac{d}{f\left(T_j\right) \cdot \left(d - f\left(T_j\right)\right)} \qquad (16)$$

The estimates of the RUL obtained resorting to particle filtering are plotted in Figure 2 in thin continuous lines with the bars of one standard

*Figure 2. Comparison of the RUL estimations for the crack propagation pattern of Figure 1 provided by the particle filtering and similarity-based approaches*

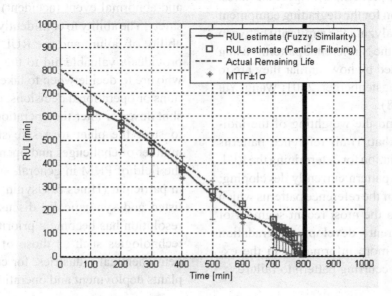

deviation of the samples; the $R\hat{U}L\left(T_j\right)$ estimates at the inspection times of Figure 2, are indicated in squares. After fault detection, the particle filtering estimates $R\hat{U}L\left(T_j\right)$ are shown to move away from the Mean Time To Failure $MTTF\left(T_j\right)$ values towards the real RUL (dashed thick line). In the Figure, the bold vertical line indicates the time of crack depth exceedance of the limit on the material thickness.

## A Data-Driven Approach: Pattern Fuzzy Similarity

In general terms, in a similarity-based approach to RUL estimation, it is assumed that J measurements taken at predefined inspection times are available for N degradation-to-failure trajectories (reference patterns) of equipment of the type of interest; these trajectories last all the way to equipment failure, i.e., to the instance when the degradation state reaches the threshold value beyond which the equipment loses its functionality (Dubois, Prade & Testemale, 1988; Joentgen, Mikenina, Weber & Zimmermann, 1999).

A degradation trajectory (test pattern) is developing in the equipment under analysis, which is monitored at the predefined inspection times. The RUL estimation for the degrading equipment is performed by analyzing the similarity between the test pattern and the N reference patterns, using their RULs weighted by how similar they are to the test pattern (Angstenberger, 2001; Wang, Yu, Siegel & Lee, 2008).

The ideas behind the weighting of the individual RULs are that: i) all reference patterns bring useful information for determining the RUL of the degradation pattern currently developing; ii) those segments of the reference patterns which are most similar to the most recent segment of length $n$ of the currently developing degradation pattern should be more informative in the extrapolation of the occurring pattern to failure.

For more details on the procedure, the interested reader may refer to Zio & Di Maio (2010).

The procedure has been applied to the fatigue crack propagation case study. A database of $N = 50$ reference crack propagation patterns of differing initial conditions has been used. These are compared for similarity with the test pattern containing the values of the measured signal of the developing degradation pattern of Figure 1. The RUL estimates of the individual reference patterns are computed and then aggregated in a weighted sum, with the weights opportunely calculated (Zio & Di Maio, 2010).

The estimates of the $MTTF\left(T_j\right)$ are plotted in Figure 2 in circles linked by a thick continuous line with the bars of one standard deviation Again, the estimates $R\hat{U}L\left(T_j\right)$ at the beginning match the $MTTF\left(T_j\right)$; then, upon fault detection, the $R\hat{U}L\left(T_j\right)$ estimates move away from the $MTTF\left(T_j\right)$ values towards the real RUL value.

## CHALLENGES AND WAYS FORWARD

PHM is the pinnacle of reliability engineering and abnormal event (accident) management for safety. The ability to confidently predict the probability of failure and the RUL of an equipment provides a valuable aid to the human operators who are to decide when to take maintenance actions or operational decisions, e.g. as to whether shut down or alter the operational configuration to steer the plant to a safe state. However, a number of challenges and open issues remain in the field of PHM in general, and of prognostics in particular (Hines & Usynin, 2008). Some are listed and synthetically discussed below: their resolution has become a priority in high-valued technologies such as those of the nuclear and petrochemical industries, for current and future plants deployment and operation.

## Hybrid Information and Data

The efficacy of a prognostic model relies on the information and data available. These often come from multiple sources (e.g. equipment parameters measured by multiple, different sensors) of different types (including measured equipment parameters values and time-to-failure data from maintenance records): their fusion into an effective prognostic algorithm represents a fascinating challenge (Goebel & Bonissone, 2005). The data must be integrated across different collection platforms, with all issues associated to different interfaces, resolutions and coverages (US Department of Energy, 2002).

Attempts in this direction are being made (Muller, Suhner & Iung, 2008). In Aumeier, Alpay, Lee & Akcasu (2006), a probabilistic dynamics framework which combines a Bayesian formulation with the solution of the Chapman-Kolomogorov equation has been developed for fault diagnosis. The generalized system transport equation which is formulated makes explicit use of the equipment reliability data, the process data and plant measurements, thus with the hybridization of the continuous process variables and discrete equipment states within a proper mathematical model of the transport of the system among its reachable states. The probability density functions of state transitions are then obtained via an adaptive Kalman filtering approach. The application of the method in practice and its extension to prognosis would be an attractive challenge, with significant potential. In Gebraeel, Elwany & Pan (2009), equipment failure times from historical maintenance records are first fitted to a Bernstein distribution by maximum likelihood estimation; the estimated parameters are used to estimate the prior distributions of the parameters of a linear or exponential degradation model; a bayesian approach is adopted to update the degradation model parameters, based on the real-time observation of the degradation signals evolution.

Furthermore, even in cases where no failures have been explicitly documented or observed there may exist a way to exploit the data available on the equipment behavior for PHM purposes (Sotiris, Tse & Pecht, 2010).

In effects, a wide range of situations may be encountered in PHM, with respect to information and data availability. One may deal with equipment time-to-failure data only, with the current historical pattern of degradation developing in the equipment, with the historical patterns of degradation of similar equipments under similar operating conditions, with information on exogenous operational and environmental parameters, with any combination of these: depending on the situation, different methods, or their combination thereof, may be applied with more or less success. A structured procedure for guiding the choice of the approach to follow in the different situations is needed.

## Definition of Prognostic Indicators

The efficacy of a prognostic method relies on the representativeness of the prognostic indicators chosen. A number of desirable characteristics are expected to be look at in the choice (Vachtsevanos, 2003; Coble & Hines, 2009a):

- **Monotonicity:** The indicators are wished to present an overall positive or negative trend in time, excluding possible self-healing situations.
- **Prognosability:** The distribution of the final value that an indicator takes at failure is wished to be 'peaked', i.e. not too wide-spread.
- **Trendability:** The entire histories of evolution of the indicator towards failure are wished to have quite similar underlying shapes, describable with a common underlying functional form.

Other characteristics may be desirable. For any characteristic sought, a metric must be introduced to allow comparing the different potential prognostic indicators on the different characteristics. A detailed list of possible metrics, and their meaning, is given in Saxena et al. (2008) with the distinction among accuracy-based, precision-based and robustness-based metrics. Furthermore, in the manipulation of prognostic indicators for the tasks of state estimation and prediction it is often convenient to reduce the multivariate problem into a single-variable one, by opportunely combining the multiple indicators, e.g. by weighted average (Coble & Hines, 2009b). Multiobjective optimization problems may arise from these issues.

## Ensemble and Hybrid Methods

Increasing interest is arising towards the use of ensembles of diagnostic and prognostic models for PHM. These ensembles build their state estimation and prediction from a combination of the estimates and predictions of a set of individual models. The individual models perform well and make errors in different regions of the parameters space; the errors are balanced out in the combination and as a result the performance of the ensemble is superior to that of the single best model of the ensemble (Polikar, 2006; Baraldi, Razavi-Far & Zio, 2011). Furthermore, by exploiting the nature of the ensemble itself, it is possible to provide measures of confidence in the ensemble outcomes (Baraldi, Razavi-Far & Zio, 2010).

More so, given the variety of information and data sources and types, and of prognostic indicators and their characteristics, it is becoming more and more attractive to ensemble-combine first-principle model-based and process sensor data-driven PHM methods (Penha & Hines, 2002; Peel, 2008; Yan & Lee, 2008). This way of proceeding is aimed at augmenting the robustness and interpretability of first-principle model-based methods with the sensitivity of process sensor data-driven methods. The modeling framework underpinning

hybrid methods is certainly more complicated, but offers clear advantages on the reliability of the predictions. Purely process sensor data-driven methods cannot guarantee effective performance in their extrapolation to new regions of operation determined by plant configuration changes and/or external factors: steering the predictions towards the first-principle model-based ones when new operating conditions are encountered, and to process sensor data-driven ones when in the familiar operating conditions allows to improve the reliability of the prognosis; the investigation on efficient methods of steering hybrid methods need to be continued.

## RUL and Reliability Estimation with Uncertainty Quantification

A prognostic method should provide the estimation of the equipment RUL as well as of its reliability, with the related uncertainties. The proper assessment of the uncertainties in the prognostic outcomes is fundamental for their effective and reliable use. Uncertainty in inputs, model parameters and structure, algorithms, operational modes must all be properly represented and propagated. Efforts are being made in this direction, both for first-principle and reliability model-based (Orchard, Kacprzynski, Goebel, Saha & Vachtsevanos, 2008), as well as for process sensor data-driven methods (Chryssolouris, Lee & Ramsey, 1996; Zio, 2006).

## Validation and Verification of Prognostic Methods

At the level of research and development, it seems desirable that benchmarks on common datasets and with agreed evaluation criteria be established, to allow the evaluation of the technical and economical feasibility of the different prognostic methods proposed. From the practical application point of view, proof-of concept experiments and proof-of-practice applications are needed in order to be able to license a prognostic method with given measures of accuracy, stability and reliability (Byington, Roemer & Kalgren, 2005; Dzakovic & Valentine, 2007). In the case of high-value equipment (e.g. those employed in the nuclear, aerospace and process industries), this poses the problem of dimensional scaling and accelerated testing of integral testing facilities (Coble & Hines, 2009b).

## Instrumentation Design

Although not in the scope of the present paper, it seems in order to also mention the fundamental role played by the instrumentation for data measurement. The possibility of efficiently collecting and managing field data on equipment degradation and failure processes is the cornerstone which PHM is founded upon. The design of the required instrumentation must then be planned harmoniously with the planning of the PHM methods, so as to avoid the difficult, cumbersome and less efficient work of retrofitting the instruments to the methods or vice versa. Also, datasets should be shared among the researchers in the field, so as to allow consistent evaluation of the different methods proposed.

## PHM Integration in Control, Operation, and Maintenance Procedures

The final challenge for practical application of PHM is its integration within the autonomous, intelligent plant control and information systems which are nowadays deployed on one side, and with the inspection and testing procedures which support preventive (on condition) maintenance on the other side. This will provide a complete and valuable decision-making asset for the operators during abnormal events and emergency accidents.

## SUMMARY AND CONCLUSION

PHM is destined to play a more and more relevant role in modern maintenance practice and accident management, supported by the maturity of condition monitoring technology. The promises of reducing downtime, spares inventory, maintenance costs and risk exposure are very attractive for the Industry. However, to date most prognostic studies have been carried out in research laboratories, where simplifications of certain practical aspects are often adopted.

This paper has reviewed the state of knowledge on the different methods for performing PHM, placing them in context with the different information and data which may be available for capturing and predicting the health state of industrial equipment. The practical requisites of PHM methods have been laid out and the types of data and information upon which these methods may rely have been specified. This has guided the illustration of the different approaches for PHM, under a classification into first-principle model-based, reliability model-based and process sensor data-driven. Examples have been given with reference to the prognostics of a non linear fatigue crack growth process typical of a certain class of industrial and structural equipment. In particular, particle filtering and pattern fuzzy similarity have been illustrated and compared with respect to their strengths and similarities. Finally, a discussion has been provided on the main challenges and open issues of PHM in general, and of prognostics in particular, with the hope of stimulating researchers to contributing the developments necessary for its full maturity and application. The ultimate goal indeed remains to be achieved, that of developing reliable prognostic frameworks for application to real-life situations.

# REFERENCES

Abernethy, R. B. (1996). *The new Weibull handbook*, 2nd ed. North Palm Beach, FL: Robert B. Abernethy. ISBN 0 9653062

Angstenberger, L. (2001). Dynamic fuzzy pattern recognition. In Zimmermann, H. J. (Ed.), *International series in intelligent technologies, 17*. Dordrecht, The Netherlands: Kluwer Academic Publishers.

Arulampalam, M. S., Maskell, S., Gordon, N., & Clapp, T. (2002). A tutorial on particle filters for online nonlinear/non-Gaussian Bayesian tracking. *IEEE Transactions on Signal Processing, 50*(2), 174–188. doi:10.1109/78.978374

Aumeier, S. E., Alpay, B., Lee, J. C., & Akcasu, A. Z. (2006). Probabilistic techniques for diagnosis of multiple component degradations. *Nuclear Science and Engineering, 153*, 101–123.

Bagdonavicius, V., & Nikulin, M. (2000). Estimation in degradation models with explanatory variables. *Lifetime Data Analysis, 7*, 85–103. doi:10.1023/A:1009629311100

Baraldi, P., Razavi-Far, R., & Zio, E. (2010, November). *A method for estimating the confidence in the identification of nuclear transients by a bagged ensemble of FCM classifiers*. Paper presented at the 7th American Nuclear Society International Topical Meeting on Nuclear Plant Instrumentation, Controls and Human Machine Interface Technology, Las Vegas, Nevada, USA.

Baraldi, P., Razavi-Far, R., & Zio, E. (2011). Bagged ensemble of FCM classifier for nuclear transient identification. *Annals of Nuclear Energy, 38*(5), 1161–1171. doi:10.1016/j.anucene.2010.12.009

Baraldi, P., Zio, E., Compare, M., Rossetti, G., & Despujols, A. (2009, September). *A novel approach to model the degradation of components in electrical production plants*. Paper presented at the European Safety and Reliability Conference, Prague, Czech Republic.

Bendell, D., Wightman, D., & Walker, E. (1991). Applying proportional hazard modeling in reliability. *Reliability Engineering & System Safety, 34*, 35–53. doi:10.1016/0951-8320(91)90098-R

Bérenguer, C., Grall, A., & Castanier, B. (2000). Simulation and evaluation of condition-based maintenance policies for multi-component continuous-state deteriorating systems. In M. Cottam, D. Harvey, R. Pape, & J. Tait (Eds.), *Proceedings of the ESREL 2000 Foresight and Precaution Conference* (pp. 275-82). Rotterdam, The Netherlands: Balkema Publishers.

Bigerelle, M., & Iost, A. (1999). Bootstrap analysis of FCGR, application to the Paris relationship and to lifetime prediction. *International Journal of Fatigue, 21*, 299–307. doi:10.1016/S0142-1123(98)00076-0

Bolotin, V. V., & Shipkov, A. A. (1998). A model of the environmentally affected growth of fatigue cracks. *Journal of Applied Mathematics and Mechanics, 62*, 313. doi:10.1016/S0021-8928(98)00037-9

Byington, C. S., Roemer, M. J., & Kalgren, P. W. (2005, April). *Verification and validation of diagnostic/prognostic algorithms*. Paper presented at Machinery Fault Prevention Technology Conference (MFPT 59). Virginia Beach, Virginia, USA.

Cadini, F., Zio, E., & Avram, D. (2009). Monte Carlo-based filtering for fatigue crack growth estimation. *Probabilistic Engineering Mechanics, 24*, 367–373. doi:10.1016/j.probengmech.2008.10.002

Chryssolouris, G., Lee, M., & Ramsey, A. (1996). Confidence interval prediction for neural network models. *IEEE Transactions on Neural Networks*, *7*(1), 229–232. doi:10.1109/72.478409

Coble, J., & Hines, J. W. (2009a, April). *Fusing data sources for optimal prognostic parameter selection*. Paper presented at the 6th American Nuclear Society International Topical Meeting on Nuclear Plant Instrumentation, Controls and Human Machine Interface Technology, Knoxville, TN, US.

Coble, J., & Hines, J. W. (2009b, April). *Analysis of prognostic opportunities in power industry with demonstration*. Paper presented at the 6th American Nuclear Society International Topical Meeting on Nuclear Plant Instrumentation, Controls and Human Machine Interface Technology, Knoxville, Tennessee, USA.

Cox, D. (1972). Regression models and life tables. *Journal of the Royal Statistics B*, *34*, 187–202.

Cox, D. R., & Oakes, D. (1984). *Analysis of survival data*. USA: Chapman and Hall.

de Kleer, J., & Williams, B. C. (1987). Diagnosing multiple faults. *Artificial Intelligence*, *32*(1), 97–130. doi:10.1016/0004-3702(87)90063-4

Deloux, E., Castanier, B., & Bérenguer, C. (2008, September). *Condition-based maintenance approaches for deteriorating system influenced by environmental conditions*. Paper presented at the European Safety and Reliability Conference, Valencia, Spain.

Doucet, A., de Freitas, J. F. G., & Gordon, N. J. (2001). An introduction to sequential Monte Carlo methods. In Doucet, A., de Freitas, J. F. G., & Gordon, N. J. (Eds.), *Sequential Monte Carlo in practice*. New York, NY: Springer-Verlag.

Dubois, D., Prade, H., & Testemale, C. (1988). Weighted fuzzy pattern matching. *Fuzzy Sets and Systems*, *28*, 313–331. doi:10.1016/0165-0114(88)90038-3

Dzakowic, J. E., & Valentine, G. S. (2007, April). *Advanced techniques for the verification and validation of prognostics and health management capabilities*. Paper presented at Machinery Failure Prevention Technology (MFPT 60). Virginia Beach, Virginia, USA.

Efron, B. (1979). Bootstrap methods: Another look at the Jacknife. *Annals of Statistics*, *7*, 1–26. doi:10.1214/aos/1176344552

Efron, B., & Tibshirani, R. J. (1993). *An introduction to the bootstrap*. New York, NY: Chapman and Hall.

Gebraeel, N., Elwany, A., & Pan, J. (2009). Residual life predictions in the absence of prior degradation knowledge. *IEEE Transactions on Reliability*, *58*(1), 106–117. doi:10.1109/TR.2008.2011659

Goebel, K., & Bonissone, P. (2005). Prognostic information fusion for constant load systems. In *Proceedings of the 7th Annual Conference on Information Fusion*, Vol. 2 (pp. 1247-1255).

Grall, A., Bérenguer, C., & Chu, C. (1998). Optimal dynamic inspection/replacement planning in condition-based maintenance. In S. Lydersen, G. Hansen, & H. Sandtorv (Eds.), *Proceedings of the European Safety and Reliability Conference ESREL'98* (pp. 381-388). Rotterdam, The Netherlands: Balkema Publishers.

Heng, A., Tan, A., Mathew, J., Montgomery, N., Banjevic, D., & Jardine, A. (2009). Intelligent condition-based prediction of machinery reliability. *Mechanical Systems and Signal Processing*, *23*(5), 1600–1614. doi:10.1016/j.ymssp.2008.12.006

Hines, J. W., & Garvey, D. R. (2007, October). *Data based fault detection, diagnosis and prognosis of oil drill steering systems.* Paper presented at the Maintenance and Reliability Conference. Knoxville, Tennessee, USA.

Hines, J. W., Garvey, J., Preston, J., & Usynin, A. (2008, January). *Empirical methods for process and equipment prognostics.* Paper presented at 53$^{rd}$ Annual Reliability and Maintainability Symposium. Las Vegas, Nevada, USA.

Hines, J. W., & Usynin, A. (2008). Current computational trends in equipment prognostics. *International Journal of Computational Intelligence Systems, 1*(1), 94–102. doi:10.2991/ijcis.2008.1.1.7

Hontelez, J. A. M., Burger, H. H., & Wijnmalen, D. J. D. (1996). Optimum condition-based maintenance policies for deteriorating systems with partial information. *Reliability Engineering & System Safety, 51*, 267–274. doi:10.1016/0951-8320(95)00087-9

Hussey, A., Lu, B., & Bickford, R. (2009, April). *Performance enhancement of on-line monitoring through plant data classification.* Paper presented at the 6th American Nuclear Society International Topical Meeting on Nuclear Plant Instrumentation, Controls and Human Machine Interface Technology, Knoxville, TN, US.

Jardine, A. K. S., Lin, D., & Banjevic, D. (2006). A review on machinery diagnostics and prognostics implementing condition based maintenance. *Mechanical Systems and Signal Processing, 20*, 1483–1510. doi:10.1016/j.ymssp.2005.09.012

Jarrell, D. B., Sisk, D. R., & Bond, L. J. (2004). Prognostics and condition-based maintenance: A new approach to precursive metrics. *Nuclear Technology, 145*, 275–286.

Joentgen, A., Mikenina, L., Weber, R., & Zimmermann, H. J. (1999). Dynamic fuzzy data analysis based on similarity between functions. *Fuzzy Sets and Systems, 105*(1), 81–90. doi:10.1016/S0165-0114(98)00337-6

Kliger, S., Yemini, S., Yemini, Y., Oshie, D., & Stolfo, S. (1995). A coding approach to event correlation. In *Proceedings of the 4$^{th}$ International Symposium on Intelligent Network Management* (pp. 266–277). London, UK: Chapman and Hall.

Kopnov, V. A. (1999). Optimal degradation process control by two-level policies. *Reliability Engineering & System Safety, 66*, 1–11. doi:10.1016/S0951-8320(99)00006-X

Kozin, F., & Bogdanoff, J. L. (1989). Probabilistic models of fatigue crack growth: Results and speculations. *Nuclear Engineering and Design, 115*, 143–171. doi:10.1016/0029-5493(89)90267-7

Lam, C., & Yeh, R. (1994). Optimal maintenance policies for deteriorating systems under various maintenance strategies. *IEEE Transactions on Reliability, 43*, 423–430. doi:10.1109/24.326439

Laser, M. (2000). Recent safety and environmental legislation. *Transactions of the Institution of Chemical Engineers, 78*(B), 419-422.

Lehmann, A. (2006). Joint modeling of degradation and failure time data. *Journal of Statistical Planning and Inference, 139*(5), 1693–1706. doi:10.1016/j.jspi.2008.05.027

Lu, C. J., & Meeker, W. Q. (1993). Using degradation measures to estimate time-to-failure distribution. *Technometrics, 35*(2), 161–173.

Marseguerra, M., Zio, E., & Podofillini, L. (2002). Condition-based maintenance optimization by means of genetic algorithms and Monte Carlo simulation. *Reliability Engineering & System Safety, 77*, 151–165. doi:10.1016/S0951-8320(02)00043-1

Muller, A., Suhner, M. C., & Iung, B. (2008). Formalisation of a new prognosis model for supporting proactive maintenance implementation on industrial system. *Reliability Engineering & System Safety*, *93*, 234–253. doi:10.1016/j.ress.2006.12.004

Myotyri, E., Pulkkinen, U., & Simola, K. (2006). Application of stochastic filtering for lifetime prediction. *Reliability Engineering & System Safety*, *91*, 200–208. doi:10.1016/j.ress.2005.01.002

Nimmo, I. (1995). Adequately address abnormal situation operations. *Chemical Engineering Progress*, *91*(9), 36–45.

Orchard, M., Kacprzynski, G., Goebel, K., Saha, B., & Vachtsevanos, G. (2008, October). *Advances in uncertainty representation and management for particle filtering applied to prognostics*. Paper presented at International Conference on Prognostics and Health Management, Denver, USA.

Oswald, G. F., & Schueller, G. I. (1984). Reliability of deteriorating structures. *Engineering Fracture Mechanics*, *20*(1), 479–488. doi:10.1016/0013-7944(84)90053-5

Pearl, J. (1988). *Probabilistic reasoning in intelligent systems*. San Mateo, CA: Morgan Kaufmann.

Pecht, M. (2008). *Prognostics and health management of electronics*. New Jersey: John Wiley. doi:10.1002/9780470385845

Pedregal, D. J., & Carnero, M. C. (2006). State space models for condition monitoring: A case study. *Reliability Engineering & System Safety*, *91*, 171–180. doi:10.1016/j.ress.2004.12.001

Peel, L. (2008, October). *Data driven prognostics using a Kalman filter ensemble of neural network models*. Paper presented at the International Conference on Prognostics and Health Management, Denver, CO.

Peng, Y., Zhang, S., & Pan, R. (2010). Bayesian network reasoning with uncertain evidences. *International Journal of Uncertainty. Fuzziness and Knowledge-Based Systems*, *18*(5), 539–564. doi:10.1142/S0218488510006696

Penha, R. L., & Hines, J. W. (2002). Hybrid system modeling for process diagnostics. *Proceedings of the Maintenance and Reliability Conference MARCON*, Knoxville, USA.

Polikar, R. (2006). Ensemble based systems in decision making. *IEEE Circuits and Systems Magazine*, *6*(3), 21–45. doi:10.1109/MCAS.2006.1688199

Provan, J. W. (1987). *Probabilistic fracture mechanics and reliability*. Dordrecht, The Netherlands: Martinus Nijhoff.

Pulkkinen, U. (1991). A stochastic model for wear prediction through condition monitoring. In Holmberg, K., & Folkeson, A. (Eds.), *Operational reliability and systematic maintenance* (pp. 223–243). London, UK: Elsevier.

Rish, I., Brodie, M., & Ma, S. (2002). Accuracy vs. efficiency trade-offs in probabilistic diagnosis. In *Proceedings of the 18th National Conference on Artificial Intelligence,* Edmonton, Canada.

Samanta, P. K., Vesely, W. E., Hsu, F., & Subudly, M. (1991). *Degradation modeling with application to ageing and maintenance effectiveness evaluations. NUREG/CR-5612*. US Nuclear Regulatory Commission.

Saxena, A., Celaya, J., Balaban, E., Goebel, K., Saha, B., Saha, S., & Schwabacher, M. (2008, October). *Metrics for evaluating performance of prognostic techniques*. Paper presented at International Conference on Prognostics and Health Management. Denver, USA.

Schwabacher, M. (2005, September). *A survey of data driven prognostics*. Paper presented at AIAAA Infotech@Aerospace Conference, Atlanta, USA.

Schwabacher, M., & Geobel, K. (2007, November). *A survey of artificial intelligence for prognostics*. Paper presented at the AAAI Fall Symposium, Arlington, Virginia, USA.

Sikorska, J. Z., Kelly, P. J., & McGrath, J. (2010, June). *Maximizing the remaining life of cranes*. Paper presented at the ICOMS Asset Management Conference, Adelaide, Australia.

Simola, K., & Pulkkinen, U. (1998). Models for non-destructive inspection data. *Reliability Engineering & System Safety, 60*, 1–12. doi:10.1016/S0951-8320(97)00087-2

Singpurwalla, N. (1995). Survival in dynamic environments. *Statistical Science, 10*(1), 86–103. doi:10.1214/ss/1177010132

Singpurwalla, N. (2006). *Reliability and risk: A Bayesian perspective*. West Sussex, UK: Wiley & Sons.

Sobezyk, K., & Spencer, B. F. (1992). *Random fatigue: From data to theory*. Boston, MA: Academic Press.

Sotiris, V., & Pecht, M. (2007, July). *Support vector prognostics analysis of electronic products and systems*. Paper presented at the AAAI Conference on Artificial Intelligence, Vancouver, Canada.

Sotiris, V., Tse, P., & Pecht, M. (2010, June). Anomaly detection through a Bayesian support vector machine. *IEEE Transactions on Reliability, 59*(2), 277–286. doi:10.1109/TR.2010.2048740

Steinder, M., & Sethi, A. S. (2002). End-to-end service failure diagnosis using belief networks, *Proceedings of Network Operations and Management Symposium*. IEEE Conference Publications.

US Department of Energy. (2002, May). *Instrumentation, controls and human-machine interface*. Presented at the IC&HMI Technology Workshop. Gaithersburg, Maryland, PA.

Vachtsevanos, G. (2003). Performance metrics for fault prognosis of complex systems. In *AUTOTESTCON 2003, IEEE Systems Readiness Technology Conference* (pp. 341-345). IEEE Conference Publications.

Vachtsevanos, G., Lewis, F. L., Roemer, M., Hess, A., & Wu, B. (2006). *Intelligent fault diagnosis and prognosis for engineering systems* (1st ed.). Hoboken, NJ: John Wiley & Sons. doi:10.1002/9780470117842

Vachtsevanos, G., & Wang, P. (2001). Fault prognosis using dynamic wavelet neural networks. In *AUTOTESTCON Proceedings, IEEE Systems Readiness Technology Conference* (pp.857-870). IEEE Conference Publications.

Venkatasubramanian, V., Rengaswamy, R., Yin, K., & Kavuri, S. N. (2003). A review of process fault detection and diagnosis: Part I and III. *Computers & Chemical Engineering, 27*, 293–311, 327–346. doi:10.1016/S0098-1354(02)00160-6

Vlok, P. J., Coetzee, J. L., Banjevic, D., Jardine, A. K. S., & Makis, V. (2002). An application of vibration monitoring in proportional hazards models for optimal component replacement decisions. *The Journal of the Operational Research Society, 53*(2), 193–202. doi:10.1057/palgrave.jors.2601261

Wang, T., Yu, J., Siegel, D., & Lee, J. (2008, October). *A similarity based prognostic approach for remaining useful life estimation of engineered systems*. Paper presented at the International Conference on Prognostics and Health Management. Denver, USA.

Wang, W. Q., Goldnaraghi, M. F., & Ismail, F. (2004). Prognosis of Machine health condition using neuro-fuzzy systems. *Mechanical Systems and Signal Processing, 18*, 813–831. doi:10.1016/ S0888-3270(03)00079-7

Wang, X., Yu, G., & Lee, J. (2002, September). *Wavelet neural network for machining performance assessment and its implication to machinery prognostic.* Paper presented at 5th International Conference on Managing Innovations in Manufacturing (MIM), Milwaukee, Wisconsin, USA.

Yan, J., Koç, M., & Lee, J. (2004). A prognostic algorithm for machine performance assessment and its application. *Production Planning and Control, 15*(8), 796–801. doi:10.1080/0953728 0412331309208

Yan, J., & Lee, J. (2008, October). *A hybrid method for on-line performance assessment and life prediction in drilling operations.* Paper presented at International Conference on Prognostics and Health Management, Denver, USA.

Yeh, R. H. (1997). State-age-dependent maintenance policies for deteriorating systems with Erlang sojourn time distributions. *Reliability Engineering & System Safety, 58*, 55–60. doi:10.1016/ S0951-8320(97)00049-5

Zhao, X., Fouladirad, M., Berenguer, C., & Bordes, L. (2008, September). *Optimal periodic inspection/replacement policy of deteriorating systems with explanatory variables.* Paper presented at the European Safety and Reliability Conference, Valencia, Spain.

Zille, V., Despujols, A., Baraldi, P., Rossetti, G., & Zio, E. (2009, September). *A framework for the Monte Carlo simulation of degradation and failure processes in the assessment of maintenance programs performance.* Paper presented at the European Safety and Reliability Conference, Prague, Czech Republic.

Zio, E. (2006). A study of the bootstrap method for estimating the accuracy of artificial neural networks in predicting nuclear transient processes. *IEEE Transactions on Nuclear Science, 53*(3), 1460–1478. doi:10.1109/TNS.2006.871662

Zio, E., & Di Maio, F. (2010). A data-driven fuzzy approach for predicting the remaining useful life in dynamic failure scenarios of a nuclear power plant. *Reliability Engineering & System Safety, 95*(1), 49–57. doi:10.1016/j.ress.2009.08.001

## KEY TERMS AND DEFINITIONS

**Condition Monitoring:** Process of keeping the state of a system, structure or component under check.

**Condition-Based Maintenance:** Process of maintenance intervention on a system, structure or component, upon information of its state of condition.

**Degradation Modeling:** Description of the evolution of change in the performance and functionality of a system, structure or component.

**Failure Prediction:** Task of anticipating when a failure may occur in the future.

**Prognostics:** Prediction of the state of a system, structure or component.

**Residual Useful Life:** Usage time remaining before a system, structure or component reaches a state beyond which it cannot longer function as designed.

## ENDNOTE

[1] NP-hard (non-deterministic polynomial-time hard), in computational complexity theory, is a class of problems that are, informally, "at least as hard as the hardest problems in NP". A problem *H* is NP-hard if and only if there is an NP-complete problem L that is polynomial time Turing-reducible

to H (i.e., $L \leq_T H$). In other words, $L$ can be solved in polynomial time by an oracle machine with an oracle for $H$. Intuitively, we can think of an algorithm that can call such an oracle machine as a subroutine for solving $H$, and solves $L$ in polynomial time, if the subroutine call takes only one step to compute. (Taken from http://en.wikipedia.org/wiki/NP-hard).

# Section 13
# Prognostics (Structure)

# Chapter 18
# Structure Reliability and Response Prognostics under Uncertainty Using Bayesian Analysis and Analytical Approximations

**Xuefei Guan**
*Clarkson University, USA*

**Ratneshwar Jha**
*Clarkson University, USA*

**Jingjing He**
*Clarkson University, USA*

**Yongming Liu**
*Clarkson University, USA*

## ABSTRACT

*This study presents an efficient method for system reliability and response prognostics based on Bayesian analysis and analytical approximations. Uncertainties are explicitly included using probabilistic modeling. Usage and health monitoring information is used to perform the Bayesian updating. To improve the computational efficiency, an analytical computation procedure is proposed and formulated to avoid time-consuming simulations in classical methods. Two realistic problems are presented for demonstrations. One is a composite beam reliability analysis, and the other is the structural frame dynamic property estimation with sensor measurement data. The overall efficiency and accuracy of the proposed method is compared with the traditional simulation-based method.*

## INTRODUCTION

Diagnostics and prognostics of modern engineering systems have drawn extensive attentions in the past decade due to their increasing complexities (Brauer & Brauer, 2009; Melchers, 1999). In particular, time-dependent reliability estimate for high reliability demanding systems such as aircraft and nuclear facilities must be quantified in order to prevent system failures. The central idea of reliability analysis involves computation of a multi-dimensional integral over the failure domain of the performance function (Ditlevsen & Madsen, 1996; Madsen, Krenk, & Lind, 1986; Rackwitz,

DOI: 10.4018/978-1-4666-2095-7.ch018

2001). For problems with high-dimensional parameters, the exact evaluation of this integral is either analytically intractable or computationally prohibitive (Yuen, 2010). Analytical approximations and numerical simulations are two major computational methods to solve such problems.

The simulation-based method includes direct Monte Carlo (MC) (Kalos & Whitlock, 2008), Importance Sampling (IS) (Gelman & Meng, 1998; Liu, 1996), and other MC simulations with different sampling techniques. Analytical approximation methods, such as first- and second- order reliability methods (FORM/SORM) have been developed to estimate the reliability without large numbers of MC simulations. FORM and SORM computations are based on linear (first-order) and quadratic (second-order) approximations of the limit state surface at the most probable point (MPP) in the standard normal space (Ditlevsen & Madsen, 1996; Madsen, Krenk, & Lind, 1986; Rackwitz, 2001). Under the condition that the limit state surface at the MPP is close to its linear or quadratic approximation and that no multiple MPPs exist on the limit state surface, FORM/SORM are sufficiently accurate for engineering purposes (Bucher & Bourgund, 1990; Cai & Elishakoff, 1994; Zhao & Ono, 1999). If the final objective is to calculate the system response given a reliability index, the *inverse reliability method* can be used. The most well-known approach is inverse FORM method proposed in (Der Kiureghian & Dakessian, 1998; Der Kiureghian, Zhang, & Li, 1994; Li & Foschi, 1998). Du, Sudjianto, and Chen (2004) proposed an inverse reliability strategy and applied it to the integrated robust and reliability design of a vehicle combustion engine piston. Saranyasoontorn and Manuel (2004) developed an inverse reliability procedure for wind turbine components. Lee, Choi, Du, and Gorsich (2008) used the inverse reliability analysis for reliability-based design optimization of nonlinear multi-dimensional systems. Cheng, Zhang, Cai, and Xiao (2007) presented an artificial neural network based inverse FORM method

for solving problems with complex and implicit performance functions.

Conventional forward and inverse reliability analyses are based on existing knowledge about the system (e.g., underlying physics, distributions of input variables). Time-dependent reliability degradation and system response changes are not reflected. For many engineering problems, usage monitoring or inspection data are usually available at a regular time interval either via structural health monitoring system or non-destructive inspections. The new information can be used to update the initial estimate of system reliability and responses. The critical issue is how to incorporate the existing knowledge and new information into the estimation. Bayesian updating is the most common approach to incorporate these additional data. By continuous Bayesian updating, all the variables of interest are updated and the inference uncertainty can be significantly reduced, provided the additional data are relevant to the problem and they are informative (Guan, Jha, & Liu, 2011). Hong (1997) presented the idea of reliability updating using inspection data. Papadimitriou, Beck, and Katafygiotis (2001) reported a reliability updating procedure using structural testing data. Graves, Hamada, Klamann, Koehler, and Martz (2008) applied the Bayesian analysis for reliability updating. Wang, Rabiei, Hurtado, Modarres, and Hoffman (2009) used Bayesian reliability updating for aging airframe. A similar updating approach using Maximum relative Entropy principles has also been proposed in (Guan, Giffin, Jha, & Liu, 2011). In those studies, Markov chain Monte Carlo (MCMC) simulations have been extensively used. For practical problems with complicated performance functions, simulations are time-consuming and efficient computations are critical for time constrained reliability evaluation and system response prognostics. In structural health management, simulation-based method may be not feasible because updating is frequently performed upon the arrival of sensor data. All these applications require efficient and accurate computations. However, very few studies are available on the investigation of complete analytical updating and estimation procedure without using simulations.

The objective of this study is to develop an efficient analytical computational framework for system reliability and response updating without using simulations. Computational components evolved in this approach are Bayesian updating, reliability estimation, and system response estimation given a reliability index. For Bayesian updating, Laplace method (Tierney & Kadane, 1986) is proposed to obtain an analytical representation of the posterior distribution and avoid using simulations. Once the analytical posterior distribution is obtained, the FORM/SORM method can be applied to estimate the updated system reliability or probability. In addition, system response predictions given a reliability index or confidence interval can also be updated using the inverse FORM method.

The chapter is organized as follows. First, a general Bayesian posterior model for uncertain variables is formulated and uncertainties are explicitly included. Relevant information such as response measures and usage monitoring data are used for model updating. Then an analytical approximation to the posterior distribution is derived based on the *Laplace method*. Next, FORM/SORM methods are introduced to evaluate the reliability and a simplified algorithm based on inverse FORM method is formulated to calculate system response given a reliability index or confidence interval. Following this, two realistic structural scale problems are presented to demonstrate the overall methodology and computational procedure. One is a composite beam reliability analysis and the other is the structural frame dynamic property estimation with sensor measurement data. The efficiency and accuracy of the proposed method are compared with traditional simulation-based methods.

## BAYESIAN MODELING AND LAPLACE APPROXIMATION

Consider a general parameterized model $M(x)$ describing an observable event $d$, where $x$ is an uncertain model parameter vector that needs to be updated. If the model is perfect, one obtains $M(x) = d$. In reality, such a perfect model is rarely available due to uncertainties from the simplification of the actual complex physical mechanisms, statistical identification of model parameter $x$, and the measurement noise in $d$. Given a prior probability distribution of $x$, $p(x)$, and the known relationship between $d$ and $x$ (e.g., conditional probability distribution or likelihood function), $p(d \mid x, M)$, the posterior probability distribution of $x$ given $d$ can be expressed as

$$p(x \mid d, M)$$
$$= p(x \mid M)p(d \mid x, M)\frac{1}{Z} \propto p(x \mid M)p(d \mid x, M),$$

(1)

where $Z = \int_x p(x \mid M)p(d \mid x, M)\mathrm{d}x$ is the normalizing constant.

Assume there is only one model $M$ available. The notation of $M$ is omitted hereafter for simplicity. For problems with high-dimensional parameters, the evaluation of Equation (1) is difficult because the exact normalizing constant $Z$, which is a multi-dimensional integral, is either analytically intractable or computationally prohibitive. Instead of evaluating this equation directly, the most common approach is to draw samples from $p(x \mid d)$ using MCMC simulations. For applications where performance functions are computationally expensive to evaluate, MCMC simulations are time-consuming and hence not suitable for on-line updating and prognostics. To improve the computational efficiency, Laplace method is proposed to obtain a normalized approximation to the posterior distribution of $p(x \mid d)$. The derivation of Laplace approximation is presented below.

Consider the unnormalized distribution (i.e., $Z$ is not known to a constant) $p(x \mid d)$ in Equation (1) and its natural logarithm $\ln p(x \mid d)$. Expand $\ln p(x \mid d)$ using Taylor series around an arbitrary point $x^*$ and obtain,

$$\ln p(x \mid d) = \ln p(x^* \mid d) + (x - x^*)^T \nabla \ln p(x^* \mid d)$$
$$+ \frac{1}{2}(x - x^*)^T \nabla^2 \ln p(x^* \mid d)(x - x^*) + O((x - x^*)^3),$$
(2)

where $\nabla \ln p(x^* \mid d)$ is the gradient of $\ln p(x \mid d)$ evaluated at $x^*$, $\nabla^2 \ln p(x^* \mid d)$ is the Hessian matrix evaluated at $x^*$, and $O(\cdot)$ are higher-order terms. Assume that the higher-order terms are negligible in computation with respect to the other terms to obtain,

$$\ln p(x \mid d) \approx \ln p(x^* \mid d) + \underbrace{(x - x^*)^T \nabla \ln p(x^* \mid d)}_{(*)}$$
$$+ \frac{1}{2}(x - x^*)^T \nabla^2 \ln p(x^* \mid d)(x - x^*).$$
(3)

Term (*) is zero at local maxima (denoted as $x_0$) of the distribution since the term $\nabla \ln p(x^* \mid d) = 0$ at $x_0$. Therefore, expanding $\ln p(x \mid d)$ around the distribution local maxima $x_0$, which is usually the mode of the distribution, can eliminate term (*) in Equation (3) and yield,

$$\ln p(x \mid d) \approx \ln p(x_0 \mid d)$$
$$+ \frac{1}{2}(x - x_0)^T \nabla^2 \ln p(x_0 \mid d)(x - x_0)$$
(4)

Exponentiation of Equation (4) yields

$$\exp\{\ln p(x \mid d)\} = p(x \mid d)$$
$$\approx p(x_0 \mid d) \exp\left\{\frac{1}{2}(x - x_0)^T \nabla^2 \ln p(x_0 \mid d)(x - x_0)\right\}.$$
(5)

Let $\Sigma = [\nabla^2 \ln p(x_0 \mid d)]^{-1}$, then Equation (5) can be conveniently expressed as

$$p(x \mid d) \approx p(x_0 \mid d) \exp\left\{-\frac{1}{2}(x - x_0)^T \Sigma^{-1}(x - x_0)\right\},$$
(6)

which is a multivariate normal distribution with a covariance matrix of $\Sigma = [\nabla^2 \ln p(x_0 \mid d)]^{-1}$, and its normalizing constant can be easily obtained as

$$Z = \int_x p(x \mid d)\mathrm{d}x = p(x_0 \mid d)\sqrt{(2\pi)^k \mid \Sigma \mid}.$$
(7)

Therefore, the posterior distribution of Equation (1) can now be approximately represented by a multivariate normal distribution, i.e.,

$$p(x \mid d) \approx \frac{1}{\sqrt{(2\pi)^k \mid \Sigma \mid}} \exp\left\{-\frac{1}{2}(x - x_0)^T \Sigma^{-1}(x - x_0)\right\}.$$
(8)

To find the mean vector $x_0$ and the covariance matrix $\Sigma$, the first step is to compute the local maxima of $\ln p(x \mid d)$. Numerical root-finding algorithms can be used to obtain $x_0$, such as Gauss—Newton algorithm (Dennis Jr, Gay, & Walsh, 1981), Levenberg—Marquardt algorithm (More, 1978), trust-region dogleg algorithm (Powell, 1970), and so on. In many engineering problems uncertain variables are considered as normal variables (Gregory, 2005) and Laplace method can yield accurate results. With the analytical representation of the posterior distribution $p(x \mid d)$, the reliability index can be calculated using the FORM method. In addition, system response predictions given a reliability index or confidence intervals can also be calculated using the inverse FORM method.

## RELIABILITY ESTIMATE USING FIRST- AND SECOND-ORDER RELIABILITY METHOD

Reliability analysis entails computation of a multi-dimensional integral over the failure domain of the performance function.

$$P_F \equiv P[g(x) < 0] = \int_{g(x)<0} p(x)\mathrm{d}x , \qquad (9)$$

where $x \in R^n$ is a real-valued $n$-dimensional random variable, $g(x)$ is the performance function, such that $g(x) < 0$ represents the failure domain, $P_F$ is the probability of failure, and $p(x)$ is the probability distribution of $x$. For example, it can be the prior or posterior distribution in Bayesian analysis. The surface $g(x) = 0$ is usually called the limit state surface. In FORM/SORM methods, computations are based on standard normal variables in the standard normal space. The variable $x$ needs to be transformed from the original probability space to a variable $z$ in the standard normal space. Denote the transformed performance function as $g(z)$, where $z \in R^n$ is an $n$-dimensional standard normal variable. The minimal distance between the limit state surface $g(z) = 0$ to the origin in the standard normal space is the Hasofer—Lind reliability index (Madsen, Krenk, & Lind, 1986), denoted as $\beta_{HL}$ in Figure 1. The corresponding point on the limit state surface is referred to as the most probable point (MPP).

The computation of $\beta_{HL}$ and the MPP is a standard constrained optimization problem which is defined as

minimize: $\| z \|$   subject to $g(z) = 0$,

$$\qquad (10)$$

where $\| z \|$ denotes the distance between a point $z$ on $g(z) = 0$ and the origin in the standard

normal space. The MPP, denoted as $z^*$, is generally not known a priori. An iterative process is usually involved to solve the above optimization problem. If the distribution of a random variable is not a standard normal distribution, Rackwitz—Fiessler (Madsen, 1977) procedure is usually employed to transform the random variable from its original probability space to the standard normal space. The idea requires that the cumulative density function (CDF) and the probability density function (PDF) of the target distribution be equal to a normal CDF and PDF at the value of variable $x$ on the limit state surface. This procedure finds the mean $\mu_{eq}$ and standard deviation $\sigma_{eq}$ of the equivalent normal distribution and thus the variable $x$ can be reduced to a standard normal variable $z = (x - \mu_{eq}) / \sigma_{eq}$. After the necessary transformation, numerical minimization algorithms can be used to obtain $z^*$. The Hasofer & Lind—Rackwitz & Fiessler (HL—RF) algorithm (Hasofer & Lind, 1974; Rackwitz & Flessler, 1978) is a widely-used algorithm to find $z^*$. With an initial guess of $z_0$ on the limit state surface,

*Figure 1. Linear (FORM) and quadratic (SORM) approximations of the performance function at MPP on the limit state surface*

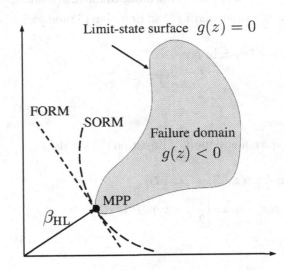

the algorithm computes a new location for $z^*$ iteratively according to

$$z_{k+1} = \frac{1}{\left|\nabla g(z_k)\right|^2}\left[\nabla g(z_k)z_k - z_k\right]\nabla g(z_k)^T, \tag{11}$$

A reasonable guess can be fixing the first $n-1$ components of $z_0$ to its distribution means and solving for the last component on the limit state surface. The iterative procedure terminates based on some criteria such as $\left|\beta_{k+1} - \beta_k\right| < \varepsilon_\beta$, where $\varepsilon_\beta$ is a small control parameter assigned by users. Usually a value of $\varepsilon_\beta = 10^{-4}$ to $10^{-3}$ yields accurate results for $\beta_{HL}$ and the MPP (Cheng, Zhang, Cai, & Xiao, 2007). After obtaining $z^*$ and $\beta_{HL}$ by solving Equation (10) using the iterative formula of Equation (11), FORM or SORM can be used to calculate the probability of failure using a linear or quadratic approximation of the performance function, respectively. Both of them are based on the Taylor series expansion of the performance function around the MPP truncated to linear and quadratic terms. For example, FORM method estimates the probability of failure as

$$P_F^{\text{FORM}} \approx \Phi(-\beta_{HL}), \tag{12}$$

where $\Phi$ is the standard normal CDF. Experience has shown that the FORM method can produce accurate results for general engineering purposes (Cheng, Zhang, Cai, & Xiao, 2007).

Given that $\beta$ is large enough, the estimate of the probability of failure using SORM method can be asymptotically expressed as (Der Kiureghian, Lin, & Hwang, 1987)

$$P_F^{\text{SORM}} \approx \Phi(-\beta_{HL})\prod_{i=1}^{n-1}\sqrt{(1+\beta\kappa_i)}, \tag{13}$$

where $\kappa_i$ denotes the $i$th main curvature of the performance function $g(z)$ at the MPP.

## SYSTEM RESPONSE ESTIMATE USING INVERSE FORM METHOD

The idea of inverse reliability analysis is to locate the MPP given a target reliability index $\beta_t$. The inverse reliability analysis can also be expressed as an optimization problem such that

$$\text{minimize: } g(z) \quad \text{subject to } \|z\| = \beta_t. \tag{14}$$

In inverse reliability analysis, among the different values of performance function $g(z)$ taking on $z$ that pass through the $\beta_t$ curve in the reduced variable space, the one $z^*$ that minimizes the performance function is sought. Figure 2 illustrates the inverse reliability analysis. The point $z^*$ is also called MPP and the corresponding minimal value of $g(z^*)$ is called probabilistic performance measure (PPM). Both reliability analysis and inverse reliability analysis search for MPPs. The difference is that the former searches for the MPP on the limit state surface $g(z) = 0$ while the latter searches for MPP on a given $\beta_t$ curve. Based on the idea of inverse FORM procedure proposed in (Der Kiureghian, Zhang, & Li, 1994), an efficient and simplified iterative formula in the reduced variable space is formulated as

$$z_{k+1} = z_k + \lambda\left[-\beta_t\frac{\nabla g(z_k)}{|\nabla g(z_k)|} - z_k\right], \tag{15}$$

where $\nabla$ is the gradient vector with respect to $z$ and $\lambda$ is the step size at the $k$th iteration (a small constant is used in this formula instead of an adaptive value). The initial value $z_0$ is usually assigned to the distribution mean value. The it-

*Figure 2. Inverse reliability analysis and MPP for target reliability index $\beta_t$ and the linear approximation of the performance function at MPP labeled as FORM. Values of $g(z)$ are for illustration purposes only.*

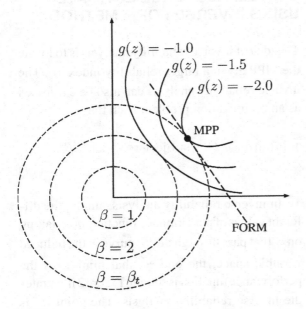

erative procedure proceeds until a convergence is achieved, i.e., when

$$|z_{k+1} - z_k| / |z_{k+1}| \le \varepsilon, \qquad (16)$$

where $\varepsilon$ is a small quantity assigned by the user. For practical problems, $\varepsilon_t = 10^{-4}$ to $10^{-3}$ usually yields satisfactory estimates (Cheng, Zhang, Cai, & Xiao, 2007).

Since the MPP in the inverse FORM method corresponds to the target reliability index $\beta_t$, the inverse FORM method can be used to calculate the confidence intervals of the system response estimate. It takes three steps. The first step converts confidence intervals to reliability indexes. For example, a 95% confidence interval consists of an upper level of 97.5% and a lower level of 2.5%. The reliability indexes associated with the two levels can be calculated using the normal inverse

CDF, i.e., $\beta_l = \Phi^{-1}(0.025) = -1.96$ and $\beta_u = \Phi^{-1}(0.975) = 1.96$. The next step calculates MPPs $z_l^*$ and $z_u^*$ for $\beta_l = -1.96$ and $\beta_u = 1.96$, respectively. The final step computes system responses by evaluating $g(z_l^*)$ and $g(z_u^*)$. The resulting $[g(z_l^*), g(z_u^*)]$ are the 95% confidence intervals of the system response.

Both iterative formulae in Equation (11) and Equation (15) implicitly assume that the components of $z$ are uncorrelated. For correlated component variables in $z$, the correlated components need to be transformed into uncorrelated components via the orthogonal transformation of $\tilde{z} = L^{-1}z$, where $L$ is the lower triangular matrix obtained by Cholesky factorization of the correlation matrix $R$ such that $LL^T = R$, where $L^T$ is the transpose of $L$. General methodologies in handling correlated random variables for reliability analysis can be found in (Haldar & Mahadevan, 2000). The overall computational procedure of the proposed method and the traditional simulation-based method are shown in Figure 3. The traditional simulation-based method needs $2k$ function evaluations for model $M(\cdot)$, where $k$ is the number of simulation samples. In order to ensure results are reliable, $k$ is usually large. The number of function evaluations required by the proposed method is usually much smaller to reach the stop criteria in Equation (3), Equation (11), and Equation(15) for Laplace approximation, FORM, and inverse FORM computations, respectively. To demonstrate the proposed method, examples are presented in the following section.

## EXAMPLES

A numerical example with twenty random variables is given first to illustrate the use and the effectiveness of the proposed method. Following this, a structural scale problem is used to demon-

*Figure 3. Overall computational procedure of the proposed analytical method and the traditional simulation-based method. (a) proposed, and (b) traditional*

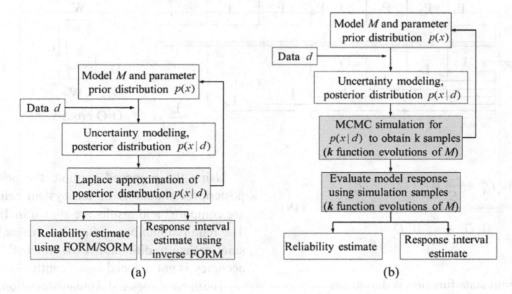

(a)　　　　　　　　　　　　　　(b)

strate the capability of the proposed method for realistic structures. Comparisons with traditional simulation-based methods (MCMC simulations with standard Metropolis—Hastings algorithm (Hastings, 1970) are made to investigate the accuracy and computational efficiency of the proposed method.

## A Composite Beam Reliability Example

A composite beam example from (Sudjianto, Du, & Chen, 2005) is used to demonstrate the overall procedure of the proposed method. The problem is described as follows. A wood beam with Young's modulus $E_1$ has an aluminum plate with Young's modulus $E_2$ securely fastened to its bottom face.

The beam is $L$ mm long and the cross-section of the wood beam is $W_1$ mm wide by $D_1$ mm deep. The cross-section of the aluminum plate is $W_2$ mm wide by $D_2$ mm deep. Six external vertical forces, $P_1$, $P_2$, $P_3$, $P_4$, $P_5$ and $P_6$ are applied at six different locations ($L_1$, $L_2$, $L_3$, $L_4$, $L_5$ and $L_6$) along the beam. The allowable tensile stress is $S$. Figure 4 shows the diagram of the beam and applied forces. The problem has twenty random variables defined in Equation 17.

The twenty random variables are given in Table 1. The maximum stress of the beam occurs in the middle cross-section $O - O$ of the beam and the stress can be obtained using the Equation 18 where

*Equation 17*

$$\mathbf{x} = [W_1, D_1, W_2, D_2, L_1, L_2, L_3, L_4, L_5, L_6, L, P_1, P_2, P_3, P_4, P_5, P_6, E_1, E_2, S] \tag{17}$$

*Figure 4. Diagram of the composite beam and applied forces*

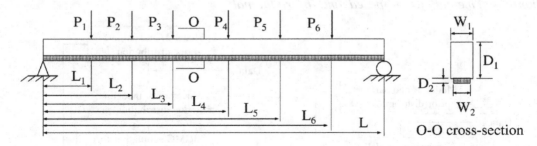

O-O cross-section

$$\Gamma = \frac{0.5 W_1 D_1^2 + \dfrac{E_2}{E_1} W_2 D_2 (D_1 + D_2)}{W_1 D_1 + \dfrac{E_2}{E_1} W_2 D_2}. \qquad (19)$$

The limit state function is defined as

$$\sigma_o(\mathbf{x}) - S. \qquad (20)$$

For the purpose of illustration, a measurement of the response variable $\sigma_{o'} = 23.5 \text{MPa}$ is used to perform Bayesian updating and the likelihood is assumed to be a zero-mean normal distribution with a standard deviation of $e_o = 0.5 \text{MPa}$. Based on the information given, the posterior distribution of $\mathbf{x}$ can be expressed as,

$$p(\mathbf{x} \mid \sigma_{o'}) = p(\mathbf{x}) \frac{1}{\sqrt{2\pi} e_o} \exp \left\{ -\frac{1}{2} \left[ \frac{\sigma_o(\mathbf{x}) - \sigma_{o'}}{e_o} \right]^2 \right\}. \qquad (21)$$

Using the proposed method, the prior and posterior estimations of the system reliability are computed and results are shown in Table 2. The required number of function evaluations is smaller than traditional MCMC method but the accuracy is not reduced significantly.

The mean values and standard deviation values obtained using the proposed method and traditional simulation-based method are shown in Table 2, where the proposed method yields close results with traditional method. It is also observed that the difference (between the proposed and traditional methods) for non-symmetric distributions is larger than symmetric ones. For example, differences of mean values and standard deviations for variables $P_1$, $P_2$, $P_3$, $P_4$, $P_5$ and $P_6$ (following extreme value type I distributions) are larger than other variables (following normal distributions). This is due to the fact that Laplace approximation uses multivariate normal distribution to approximate the posterior distribution in Equation (21) The approximation will introduce nu-

*Equation 18.*

$$\sigma_o(\mathbf{x}) = \frac{\left[ \dfrac{\sum_{i=1}^{6} P_i (L - L_i) L_3}{L} - P_1 (L_2 - L) - P_2 (L_3 - L_2) \right] \Gamma}{\dfrac{W_1 D_1^3}{12} + W_1 D_1 \left( \Gamma - 0.5 D_1 \right)^2 + \dfrac{E_2 W_2 D_2^3}{12 E_1} + \dfrac{E_2 W_2 D_2}{E_1} \left( 0.5 D_2 + D_1 - \Gamma \right)^2} \qquad (18)$$

*Table 1. Random variables of the composite beam problem*

| No. | Variable | Mean | Standard deviation | Distribution type |
|-----|----------|------|--------------------|-------------------|
| 1 | $W_1$ | 100mm | 0.2mm | Normal |
| 2 | $D_1$ | 100mm | 0.2mm | Normal |
| 3 | $W_2$ | 80mm | 0.2mm | Normal |
| 4 | $D_2$ | 20mm | 0.2mm | Normal |
| 5 | $L_1$ | 200mm | 1mm | Normal |
| 6 | $L_2$ | 400mm | 1mm | Normal |
| 7 | $L_3$ | 600mm | 1mm | Normal |
| 8 | $L_4$ | 800mm | 1mm | Normal |
| 9 | $L_5$ | 1000mm | 1mm | Normal |
| 10 | $L_6$ | 1200mm | 1mm | Normal |
| 11 | $L$ | 1400mm | 1mm | Normal |
| 12 | $P_1$ | 20kN | 4kN | Extreme type I |
| 13 | $P_2$ | 20kN | 4kN | Extreme type I |
| 14 | $P_3$ | 15kN | 2kN | Extreme type I |
| 15 | $P_4$ | 15kN | 2kN | Extreme type I |
| 16 | $P_5$ | 15kN | 2kN | Extreme type I |
| 17 | $P_6$ | 15kN | 2kN | Extreme type I |
| 18 | $E_1$ | 70GPa | 7GPa | Normal |
| 19 | $E_2$ | 8.75GPa | 1GPa | Normal |
| 20 | $S$ | 27MPa | 2.78MPa | Normal |

*Table 2. Prior and updated estimates of probability of failure (PoF), the number of function evaluations (NoF), and time of computations (based on MATLAB® 2008a) for the composite beam problem*

| Distribution | Method | PoF | NoF | Time (s) |
|--------------|--------|-----|-----|----------|
| Prior | Proposed | 0.03 | 398 | 0.2 |
| | MCMC | 0.02 | $2\times10^5$ | 21.3 |
| Posterior | Proposed | 0.09 | 2433 | 1.1 |
| | MCMC | 0.09 | $2\times10^5$ | 23.7 |

merical errors because $P_{1,...,6}$ are not normal random variables.

This numerical example demonstrates the overall process of the proposed method and its capability of solving a medium size problem. A practical structural example is presented next to further investigate the method.

## A Structural Scale Updating Example Using Health Monitoring Data

A realistic structure from ASCE structural health monitoring (SHM) benchmark problems (Lam, 2004) is adopted here to investigate the performance of the proposed method in structural scale problems. The structure is a four-story, two-bay by two-bay steel frame structure. It has a 2.5m×2.5m plan and it is 3.6m tall. The material is hot rolled grade 300W steel. The columns are all oriented to be stronger in bending toward the x-direction (i.e., about the y-axis). The floor beams are oriented to be stronger in bending vertically for those oriented with longitudinal axis parallel to the x-axis. There is one floor slab per bay per floor: four 800kg slabs at the first level, four 600kg slabs at each of the second and third levels, and four 400kg slabs at the fourth floor. A finite element model based on this structure was developed to generate simulated response data (IASC-ASCE SHM Task Group, 1999), i.e. the acceleration signals. The diagram of the analytical model is shown in Figure 5, where

*Table 3. Means and standard deviations of the posterior distribution obtained by both methods*

| No. | Variable | Proposed | | MCMC | |
|---|---|---|---|---|---|
| | | Mean | Standard deviation | Mean | Standard deviation |
| 1 | $W_1$ | 99.99mm | 0.20mm | 99.99mm | 0.20mm |
| 2 | $D_1$ | 199.99mm | 0.20mm | 200.00mm | 0.20mm |
| 3 | $W_2$ | 80.00mm | 0.20mm | 80.00mm | 0.20mm |
| 4 | $D_2$ | 19.99mm | 0.20mm | 19.98mm | 0.20mm |
| 5 | $L_1$ | 200.01mm | 1.00mm | 200.00mm | 0.99mm |
| 6 | $L_2$ | 399.99mm | 1.00mm | 400.00mm | 1.00mm |
| 7 | $L_3$ | 600.02mm | 1.00mm | 600.03mm | 1.00mm |
| 8 | $L_4$ | 800.00mm | 1.00mm | 800.02mm | 1.00mm |
| 9 | $L_5$ | 1000.00mm | 1.00mm | 1000.00mm | 1.01mm |
| 10 | $L_6$ | 1200.00mm | 1.00mm | 1200.00mm | 1.00mm |
| 11 | $L$ | 1400.10mm | 2.00mm | 1400.10mm | 2.02mm |
| 12 | $P_1$ | 22.46kN | 2.44kN | 22.71kN | 2.36kN |
| 13 | $P_2$ | 21.92kN | 2.83kN | 21.90kN | 2.74kN |
| 14 | $P_3$ | 15.76kN | 1.55kN | 15.53kN | 1.63kN |
| 15 | $P_4$ | 15.60kN | 1.66kN | 15.21kN | 1.76kN |
| 16 | $P_5$ | 15.42kN | 1.77kN | 14.89kN | 1.99kN |
| 17 | $P_6$ | 15.22kN | 1.88kN | 14.33kN | 2.40kN |
| 18 | $E_1$ | 67.87GPa | 6.84GPa | 66.62GPa | 6.88GPa |
| 19 | $E_2$ | 9.07GPa | 0.94GPa | 9.21GPa | 0.94GPa |
| 20 | $S$ | 27.00MPa | 2.78MPa | 27.08MPa | 2.78MPa |

the x-direction is the strong direction due to the orientation of the columns. The finite element model is assured to be closely consistent with the actual quarter-scale model (Lam, 2004), and the simulated sensor data can be used to represent the actual sensor measurements. The properties of structural members are shown Table 4.

Due to the manufacturing induced uncertainties, the basic material properties, namely Young's modulus $E$ and the Poisson's ratio $\nu$ are considered to be uncertain variables. The two variables are commonly treated as normal variable truncated into the physically constrained region (Haldar & Mahadevan, 2000). Steel experimental data indicate that the coefficient of variation (CoV) for Young's modulus is usually between 0.0056 and 0.045 (Hess, Bruchman, Assakkaf, & Ayyub, 2002; Vrouwenvelder, 2002), and the CoV for Poisson's ratio is usually considered to be 0.05 (Haldar & Mahadevan, 2000). In practice, the geometry dimensions are generally manufactured to satisfy absolute dimensional tolerance, and thus the variability in dimension can be safely ignored compared to material properties (Schmidt &

Bartlett, 2002). In addition, independent stochastic excitations are applied to the structure at each of the floors to represent the earthquake or ambient excitations as in Figure 5(b). The excitations are modeled as independent filtered Gaussian white noise, and generated by Gaussian white noise processes passed through a sixth order low-pass Butterworth filter with a 100Hz cutoff. The deterministic accelerometer data (sensor measurements) are calculated using the equation of motion of the finite element model subject to the stochastic excitations. The 10% noise elements are added to the deterministic accelerometer data to represent realistic noisy measurements. The noise elements are Gaussian pulse processes with root mean square (RMS) 10% of the largest RMS of the acceleration responses. Details of response generation and the MATLAB code package are available in (IASC-ASCE SHM Task Group, 1999). The sampling frequency of the measurement is 1000Hz.

Frequency responses extracted from the sensor measurement are used to perform Bayesian update for $(E, \nu)$. Reliability analysis and system response

*Figure 5. Diagram of the structure. (a) quarter scale model taken from (Lam, 2004), and (b) the corresponding analytical finite element model*

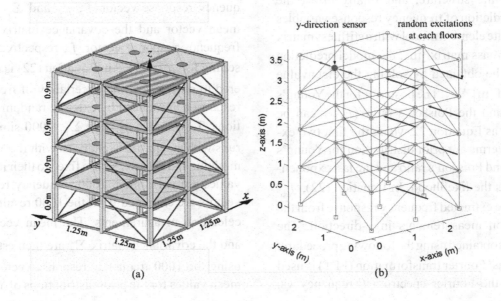

(a)　　　　　　　　　　　　　　　(b)

*Table 4. Properties of structural members (Lam, 2004)*

| Property (Lam, 2004) | Columns | Floor beams | Braces |
|---|---|---|---|
| Section type | B100×9 | S75×11 | L25×25×3 |
| Cross-sectional area $B(m^2)$ | $1.113×10^{-3}$ | $1.43×10^{-3}$ | $0.141×10^{-3}$ |
| Moment of inertia (strong direction) $I_y(m^4)$ | $1.97×10^{-6}$ | $1.22×10^{-6}$ | 0 |
| Moment of inertia (weak direction) $I_z(m^4)$ | $0.664×10^{-6}$ | $0.249×10^{-6}$ | 0 |
| St. Venant torsion constant $J(m^4)$ | $8.01×10^{-9}$ | $38.2×10^{-9}$ | 0 |
| Young's modulus (nominal) $E(GPa)$ | 200 | 200 | 200 |
| Poisson's ratio (nominal) $\nu$ | 0.3 | 0.3 | 0.3 |
| Shear modulus $G(GPa)$ | $E/(2+2\nu)$ | $E/(2+2\nu)$ | $E/(2+2\nu)$ |
| Mass per unit volume $\rho(kg/m^3)$ | 7800 | 7800 | 7800 |

estimates are then based on the updated distribution of $(E,\nu)$. The vector of first four natural frequencies in y-direction, $f=(f_1,f_2,f_3,f_4)$ is considered as the frequency response variable, and is modeled as a multivariate normal vector. Uncertain variables $(E,\nu)$ affect the stiffness matrix $[k]$ and damping matrix of the finite element model of the structure, and finally affect the model prediction of frequency response variables $f$. For finite element model with a stiffness matrix $[k]$ and a mass matrix $[m]$, the model prediction of $f$ can be obtained by solving the eigenvalue problem of $[m]^{-1}[k]$. The posterior for the Young's modulus and the Poisson's ratio $(E,\nu)$ can be expressed as Equation 22 where the first two exponential terms are prior distribution for Young's modulus and Poisson's ratio and the last exponential term is the likelihood. In Equation (22), $f_{ext}$ denotes the extracted frequency response from the acceleration measurements in y-direction. The term $f_{ext}$ is obtained using the following procedure: first, the fast Fourier transformation (FFT) is used to obtain the Fourier spectrum (Frequency vs.

Amplitude) of the acceleration measurements, then the peak picking algorithm is used to locate the frequencies $f_{ext}$ in the spectrum. $\mu_E$ and $\mu_\nu$ are the nominal values for the Young's modulus and the Poisson's ratio respectively, $\sigma_E = \text{CoV}_E \cdot \mu_E$ and $\sigma_\nu = \text{CoV}_\nu \cdot \mu_\nu$ are standard deviations of the two variables, $k=4$ is the length of the frequency response vector $f$. $\mu_f$ and $\Sigma_f$ are the mean vector and the covariance matrix of the frequency response vector $f$, respectively. The setting for $\mu_f$ and $\Sigma_f$ in Equation (22) is considered to only reflect the uncertainty of frequency responses introduced by external random excitations. To estimate $\mu_f$ and $\Sigma_f$, 1000 simulation runs with stochastic excitations with the Young's modulus and Poisson's ratio fixed to their nominal values are performed. 1000 frequency response vectors are extracted from the 1000 resulting acceleration measurements. The mean vector $\mu_f$ and the covariance matrix $\Sigma_f$ are then estimated using the 1000 frequency response vectors. The mean values for the prior distributions of Young's

*Equation 22*

$$p(E, \nu \mid A) = \frac{1}{\sqrt{2\pi}\sigma_E} \exp\left\{-\frac{(E - \mu_E)^2}{2\sigma_E^2}\right\} \frac{1}{\sqrt{2\pi}\sigma_\nu} \exp\left\{-\frac{(\nu - \mu_\nu)^2}{2\sigma_\nu^2}\right\} \times \frac{1}{(2\pi)^{k/2}\sqrt{|\Sigma_f|}} \exp\left\{-\frac{1}{2}[f_{\text{ext}} - \mu_f]'\Sigma_f^{-1}[f_{\text{ext}} - \mu_f]\right\} \quad (22)$$

modulus and Poisson's ratio are 200GPa and 0.3, respectively. The CoVs for Young's modulus and Poisson' ratio in prior distributions are taken as 0.01 and 0.05, respectively (Hess, Bruchman, Assakkaf, & Ayyub, 2002). The overall flowchart of Bayesian update and reliability analysis is shown in Figure 6.

For illustration purposes, the failure event is defined as the first natural frequency of the structure to be less than a critical value of $f_c = 9.3$ Hz. In practice, $f_c$ is a design related value. The limit state function is

$$M(E, \nu)_1 - f_c, \quad (23)$$

*Figure 6. Overall flowchart of Bayesian update and reliability analysis for the structural example*

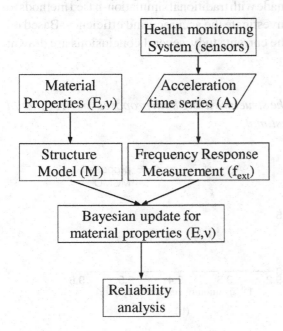

where $M(E, \nu)_1$ denote the first component of the model prediction of the frequency response $f$, When $M(E, \nu)_1 - f_c < 0$, the system is considered as failed. The sensor measurement data used to perform Bayesian updating are simulated with the Young's modulus equal to 195GPa and Poisson's ratio equal to 0.3. The sampling frequency of the sensor is 1000Hz and 10 seconds data are simulated. The frequency responses are obtained as $f_{\text{ext}} = (9.2174, 25.211, 38.518, 47.308)$ Hz.

Table 5 shows the prior and posterior results of reliability analysis using both the proposed method and traditional MCMC simulations. The prior estimate of probability of failure uses the prior distribution of $(E, \nu)$ in the limit state function of Equation (23) and the posterior estimate uses the updated distribution of $(E, \nu)$. Using the proposed method, the posterior distribution of $(E, \nu)$ has a mean vector of (195.79, 0.30) and standard deviation vector of (0.649, 0.03), respectively. $E$ and $\nu$ have no correlations. The traditional MCMC method reports a mean vector of (196.08, 0.30) and a standard deviation vector of (0.631, 0.0301), respectively. The CDF of the first frequency response obtained using the proposed and the traditional methods are shown in Figure 7, where a close agreement between the proposed method and traditional MCMC simulations is observed.

The two examples demonstrate that the proposed method and its results are comparable with the traditional simulation-based method. The computational cost in terms of number of required function evaluations and the actual time is significantly reduced and the results are still accurate

*Table 5. Prior and updated estimates of probability of failure (PoF), the number of function evaluations (NoF), and time of computations (based on MATLAB® 2008a) for the structural scale example*

| Distribution | Method | PoF | NoF | Time (s) |
|---|---|---|---|---|
| Prior | Proposed | 0.0096 | 131 | 46 |
| | MCMC | 0.0100 | $1 \times 10^5$ | 7897 |
| Posterior | Proposed | 0.1154 | 212 | 69 |
| | MCMC | 0.1085 | $2 \times 10^5$ | 9097 |

compared with the traditional method. For problems where the response function is time-consuming to evaluate, the proposed method can improve the computational efficiency significantly. For example, the response function in the structure problem involves the assembly of finite element component, matrix inversion, and FFT. The computational time of the proposed method is less than two minutes while the traditional simulation-based method takes more than two hours. It should be noted that the proposed method can yield accurate results when uncertain parameters are Gaussian-like distributed. In general, symmetric and uni-modal distributions can be considered as Gaussian-like distributions. However, results must be carefully interpreted when dealing with those distributions. The Laplace approximation relies on numerical differentiations

and gradient-based algorithms. In practice, the gradient-based optimization algorithm works well for problems with a moderate number of variables (Baldi, 1995). For problems with hundreds or thousands of random variables, simulation method would be more appropriate (Schuëller & Pradlwarter, 2007). An alternative is to apply the principal component analysis technique to retain only principal variables before using the proposed method (Jolliffe, 2002).

## CONCLUSION

This study has developed an efficient analytical Bayesian method for reliability and system response updating. The method is capable of incorporating additional information such as inspection data to reduce uncertainties and improve the estimation accuracy. One major difference between the proposed work and the traditional approach is that the proposed method performs all the calculations including Bayesian updating without using MC or MCMC simulations. A twenty-variable numerical example and a structural scale problem are presented for demonstration. Comparisons are made with traditional simulation-based methods to investigate the accuracy and efficiency. Based on the current study, several conclusions are drawn:

*Figure 7. Comparison of the first frequency response of the structure using both proposed and traditional MCMC methods. (a) prior estimate, and (b) posterior estimate*

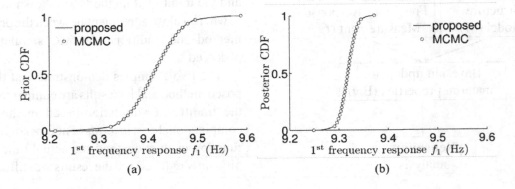

The proposed method provides an efficient analytical computational procedure for computing and updating system reliability responses. No MC or MCMC simulation is required; therefore, it provides a feasible and practical solution to time-constrained prognostics. The method is also beneficial for structural health monitoring problems where Bayesian updating and system response predictions are frequently performed upon the arrival of sensor data.

The proposed method is capable of incorporating additional information such as the inspection data and usage data from health monitoring system by way of Bayesian updating. This property is beneficial for time-dependent systems where prior estimates for reliability and system response may become unreliable at a future time. By continuous Bayesian updating, estimation uncertainties can be reduced.

The proposed method yields almost identical results to those produced by traditional simulation-based methods given that uncertain variables are Gaussian-like distributed. This is true for many engineering problems where the uncertain parameters are normal or log-normal variables (which can be transformed or truncated into normal variables). When these conditions are not assured, the results need careful interpretation. The computational efficiency the proposed method is significantly improved compared with the traditional method.

## REFERENCES

Baldi, P. (1995). Gradient descent learning algorithm overview: A general dynamical systems perspective. *IEEE Transactions on Neural Networks*, *6*(1), 182–195. doi:10.1109/72.363438

Brauer, D. C., & Brauer, G. D. (2009). Reliability-centered maintenance. *IEEE Transactions on Reliability*, *36*(1), 17–24. doi:10.1109/TR.1987.5222285

Bucher, C., & Bourgund, U. (1990). A fast and efficient response surface approach for structural reliability problems. *Structural Safety*, *7*(1), 57–66. doi:10.1016/0167-4730(90)90012-E

Cai, G., & Elishakoff, I. (1994). Refined second-order reliability analysis. *Structural Safety*, *14*(4), 267–276. doi:10.1016/0167-4730(94)90015-9

Cheng, J., Zhang, J., Cai, C., & Xiao, R. C. (2007). A new approach for solving inverse reliability problems with implicit response functions. *Engineering Structures*, *29*(1), 71–79. doi:10.1016/j.engstruct.2006.04.005

Dennis, J. E. Jr, Gay, D. M., & Walsh, R. E. (1981). An adaptive nonlinear least-squares algorithm. [TOMS]. *ACM Transactions on Mathematical Software*, *7*(3), 348–368. doi:10.1145/355958.355965

Der Kiureghian, A., & Dakessian, T. (1998). Multiple design points in first and second-order reliability. *Structural Safety*, *20*(1), 37–49. doi:10.1016/S0167-4730(97)00026-X

Der Kiureghian, A., Lin, H. Z., & Hwang, S. J. (1987). Second-order reliability approximations. *Journal of Engineering Mechanics*, *113*(8), 1208–1225. doi:10.1061/(ASCE)0733-9399(1987)113:8(1208)

Der Kiureghian, A., Zhang, Y., & Li, C. C. (1994). Inverse reliability problem. *Journal of Engineering Mechanics*, *120*(5), 1154–1159. doi:10.1061/(ASCE)0733-9399(1994)120:5(1154)

Ditlevsen, O., & Madsen, H. O. (1996). *Structural reliability methods* (*Vol. 315*). Citeseer.

Du, X., Sudjianto, A., & Chen, W. (2004). An integrated framework for optimization under uncertainty using inverse reliability strategy. *Journal of Mechanical Design*, *126*, 562. doi:10.1115/1.1759358

Gelman, A., & Meng, X. L. (1998). Simulating normalizing constants: From importance sampling to bridge sampling to path sampling. *Statistical Science*, *13*(2), 163–185. doi:10.1214/ss/1028905934

Graves, T., Hamada, M., Klamann, R., Koehler, A., & Martz, H. (2008). Using simultaneous higher-level and partial lower-level data in reliability assessments. *Reliability Engineering & System Safety*, *93*(8), 1273–1279. doi:10.1016/j.ress.2007.07.002

Gregory, P. C. (2005). *Bayesian logical data analysis for the physical sciences: A comparative approach with Mathematica support*. Cambridge, UK: Cambridge Univ Press. doi:10.1017/CBO9780511791277

Guan, X., Giffin, A., Jha, R., & Liu, Y. (2011). (in press). Maximum relative entropy-based probabilistic inference in fatigue crack damage prognostics. *Probabilistic Engineering Mechanics*. doi:doi:10.1016/j.probengmech.2011.11.006

Guan, X., Jha, R., & Liu, Y. (2011). Model selection, updating, and averaging for probabilistic fatigue damage prognosis. *Structural Safety*, *33*(3), 242–249. doi:10.1016/j.strusafe.2011.03.006

Haldar, A., & Mahadevan, S. (2000). *Probability, reliability, and statistical methods in engineering design*. John Wiley & Sons, Inc.

Hasofer, A. M., & Lind, N. C. (1974). Exact and invariant second-moment code format. *Journal of the Engineering Mechanics Division*, *100*(1), 111–121.

Hastings, W. K. (1970). Monte Carlo sampling methods using Markov chains and their applications. *Biometrika*, *57*(1), 97. doi:10.1093/biomet/57.1.97

Hess, P. E., Bruchman, D., Assakkaf, I., & Ayyub, B. M. (2002). Uncertainties in material strength, geometric, and load variables. *ASNE Naval Engineers Journal*, *114*(2), 139–165. doi:10.1111/j.1559-3584.2002.tb00128.x

Hong, H. P. (1997). Reliability analysis with nondestructive inspection. *Structural Safety*, *19*(4), 383–395. doi:10.1016/S0167-4730(97)00018-0

IASC-ASCE SHM Task Group. (1999). ASCE structural health monitoring (SHM) benchmark problem. Retrieved from http://bc029049.cityu.edu.hk/asce.shm/phase1/ascebenchmark.asp

Jolliffe, I. (2002). *Principal component analysis*. Wiley Online Library.

Kalos, M. H., & Whitlock, P. A. (2008). *Monte carlo methods*. Wiley-VCH. doi:10.1002/9783527626212

Lam, H. (2004). Phase I IASC-ASCE structural health monitoring benchmark problem using simulated data. *Journal of Engineering Mechanics*, *130*(3).

Lee, I., Choi, K., Du, L., & Gorsich, D. (2008). Inverse analysis method using MPP-based dimension reduction for reliability-based design optimization of nonlinear and multi-dimensional systems. *Computer Methods in Applied Mechanics and Engineering*, *198*(1), 14–27. doi:10.1016/j.cma.2008.03.004

Li, H., & Foschi, R. O. (1998). An inverse reliability method and its application. *Structural Safety*, *20*(3), 257–270. doi:10.1016/S0167-4730(98)00010-1

Liu, J. S. (1996). Metropolized independent sampling with comparisons to rejection sampling and importance sampling. *Statistics and Computing*, *6*(2), 113–119. doi:10.1007/BF00162521

Madsen, H. (1977). *Some experience with the Rackwitz-Fiessler algorithm for the calculation of structural reliability under combined loading*, (pp. 73-98). DIALOG-77. Lyngby, Denmark: Danish Engineering Academy.

Madsen, H., Krenk, S., & Lind, N. (1986). *Methods of structural safety*. Englewood Cliffs, NJ: Prentice-Hall, Inc.

Melchers, R. E. (1999). *Structural reliability analysis and prediction*. John Wiley & Son Ltd.

More, J. (1978). The Levenberg-Marquardt algorithm: Implementation and theory. *Numerical Analysis*, 105-116.

Papadimitriou, C., Beck, J. L., & Katafygiotis, L. (2001). Updating robust reliability using structural test data. *Probabilistic Engineering Mechanics*, *16*(2), 103–113. doi:10.1016/S0266-8920(00)00012-6

Powell, M. (1970). A FORTRAN subroutine for solving systems of nonlinear algebraic equations. *Numerical Methods for Nonlinear Algebraic Equations, 115*.

Rackwitz, R. (2001). Reliability analysis-a review and some perspectives. *Structural Safety*, *23*(4), 365–395. doi:10.1016/S0167-4730(02)00009-7

Rackwitz, R., & Flessler, B. (1978). Structural reliability under combined random load sequences. *Computers & Structures*, *9*(5), 489–494. doi:10.1016/0045-7949(78)90046-9

Saranyasoontorn, K., & Manuel, L. (2004). Efficient models for wind turbine extreme loads using inverse reliability. *Journal of Wind Engineering and Industrial Aerodynamics*, *92*(10), 789–804. doi:10.1016/j.jweia.2004.04.002

Schmidt, B., & Bartlett, F. (2002). Review of resistance factor for steel: Data collection. *Canadian Journal of Civil Engineering*, *29*(1), 98–108. doi:10.1139/l01-081

Schuëller, G., & Pradlwarter, H. (2007). Benchmark study on reliability estimation in higher dimensions of structural systems-An overview. *Structural Safety*, *29*(3), 167–182. doi:10.1016/j.strusafe.2006.07.010

Sudjianto, A., Du, X., & Chen, W. (2005)... *SAE Transactions*, *114*(6), 267–276.

Tierney, L., & Kadane, J. B. (1986). Accurate approximations for posterior moments and marginal densities. *Journal of the American Statistical Association*, *81*(393), 82–86.

Vrouwenvelder, A. (2002). Developments towards full probabilistic design codes. *Structural Safety*, *24*(2-4), 417–432. doi:10.1016/S0167-4730(02)00035-8

Wang, X., Rabiei, M., Hurtado, J., Modarres, M., & Hoffman, P. (2009). A probabilistic-based airframe integrity management model. *Reliability Engineering & System Safety*, *94*(5), 932–941. doi:10.1016/j.ress.2008.10.010

Yuen, K. V. (2010). *Bayesian methods for structural dynamics and civil engineering*. Wiley. doi:10.1002/9780470824566

Zhao, Y. G., & Ono, T. (1999). A general procedure for first/second-order reliability method (form/sorm). *Structural Safety*, *21*(2), 95–112. doi:10.1016/S0167-4730(99)00008-9

# Chapter 19
# Fatigue Damage Prognostics and Life Prediction with Dynamic Response Reconstruction Using Indirect Sensor Measurements

**Jingjing He**
*Clarkson University, USA*

**Xuefei Guan**
*Clarkson University, USA*

**Yongming Liu**
*Clarkson University, USA*

## ABSTRACT

*This study presents a general methodology for fatigue damage prognostics and life prediction integrating the structural health monitoring system. A new method for structure response reconstruction of critical locations using measurements from remote sensors is developed. The method is based on the empirical mode decomposition with intermittency criteria and transformation equations derived from finite element modeling. Dynamic responses measured from usage monitoring system or sensors at available locations are decomposed into modal responses directly in time domain. Transformation equations based on finite element modeling are used to extrapolate the modal responses from the measured locations to critical locations where direct sensor measurements are not available. The mode superposition method is employed to obtain dynamic responses at critical locations for fatigue crack propagation analysis. Fatigue analysis and life prediction can be performed given reconstructed responses at the critical location. The method is demonstrated using a multi degree-of-freedom cantilever beam problem.*

DOI: 10.4018/978-1-4666-2095-7.ch019

# INTRODUCTION

The issue of predicting the remaining useful life (RUL) of machines has attracted considerable attention in recent years. Among the various aging effects that influence the remaining useful life, fatigue is one of the most important failure modes. Fatigue prognosis is still a challenging problem for those structures suffered from the aging problem. The main objective of *fatigue prognosis* is to predict the remaining useful life of engineering materials and structures under cyclic loadings. The prognosis methods can be generally classified to two major approaches: data-driven method and physics-based method. Data-driven methods are applicable where the physics of the problem does not change much. For example, the loading spectrum of training samples needs to be similar with those of predictions. This requirement limits the applicability of the data-driven approaches. This paper uses physics-based models for damage prognosis, which are capable of different random loading spectrums. One of the most uncertain factors of a reliable fatigue RUL prediction is the loading uncertainty. Classical fatigue damage tolerance analysis and design used specified design spectrums for the entire fleet. The progress of structural health monitoring and usage monitoring systems makes it possible for the fatigue damage prognosis to use measured loading spectrums for each individual vehicle, which will significantly advance the next generation vehicle health management(Gupta, Ray, & Keller, 2007; Link & Weiland, 2009; Papazian et al., 2007). Several advanced sensor techniques have been developed for usage monitoring system. The usage monitoring system collects data for the usage information, such as mechanical load, temperature, humidity, etc. of a system (Chang, 1998). The information from usage monitoring system greatly facilitates the accurate RUL prediction of an individual structural system under service loading conditions. One of the objectives of this study is to propose a general methodology for the fatigue damage prognosis integrating usage monitoring system.

Ideally, the remaining useful life can be evaluated using physical models/ data-driven models after collecting the data from the critical spot. This straightforward methodology requires that the critical spot being monitored is readily accessible (Chang, 1998). Several methods for fatigue prognosis using direct sensor measurements are available(Adams & Nataraju, 2002; Au & Beck, 2001; Gupta, et al., 2007; Papazian et al., 2009; Papazian, et al., 2007; Yan & Gao, 2006). However, the number of sensors is always limited for a large structural system, and it is not possible to put sensors at every critical location. If critical locations are not covered by sensors or the critical location is uncertain due to the complex operational conditions, proper extrapolation process using the sensor data obtained from available locations is required in order to identify and reconstruct the state on the critical spot. This is especially true for realistic practice of large systems. The proposed study uses the dynamic system identification technique and finite element-based extrapolation to estimate the dynamic response at different critical locations in the structure. Extensive researches have been done on the system identification using sensor data. Wavelet transform (WT), which decomposes the measured signal in the frequency-time domain, is widely used for system identification(Gurley & Kareem, 1999; Haase & Widjajakusuma, 2003; Luk & Damper, 2006; Pislaru, Freeman, & Ford, 2003; Tan, Liu, Wang, & Yang, 2008). Hilbert transform (HT) has also received considerable attention in the system identification field, and has been applied to the single-degree-of-freedom (SDOF) case (M Feldman, 1985; M. Feldman, 1997). However, there are several limitations for these two methodologies. The selection of the type of the basic wavelet function is critical for the WT method and it will affect the effectiveness in identification (Yan & Gao, 2006). On the other hand, well-behaved HT requires the mono-component frequency for input data which is hardly achieved for practical applications (N.E. Huang, Shen, & Long, 1999; Norden E. Huang et al., 1998; Yang, Lei, Pan, & Huang, 2003). In comparison, Hilbert-Huang transform (HHT) does not suffer from these limitations (Bao, Hao, Li, & Zhu, 2009; N.E. Huang, et al., 1999; Yang, et al., 2003). In HHT method, the Empirical Mode Decomposition (EMD) method is employed to decompose the signal into several Intrinsic Mode Functions (IMFs) which only contains a single frequency component before the well-behaved HT transform can be performed. Recently, HHT method has been applied to identify the modal parameters of multiple-degree-of-freedom (MDOF) for both linear and nonlinear systems (Bao, et al., 2009; Norden E. Huang, et al., 1998; Poon & Chang, 2007; Yang, et al., 2003). In this paper, the HHT-based system identification method is used.

The current work focuses on concurrent fatigue prognosis using the limited sensor data collected from usage monitoring system. The flowchart for the proposed method is shown in Figure 1. EMD methodology is employed to decompose the measured signal data into a series of IMFs cooperating with intermittency criteria. Those IMFs which represent the dynamic response under mode coordinate are used to extrapolate the dynamic response at the critical spot with respect to mode coordinate. Full Mode information obtained from finite element method is used as the basis of the extrapolation. When all the mode information is extrapolated, mode superposition method is employed to reconstruct the dynamic response in time domain. After the local dynamic response is determined, fatigue prognosis can be performed by the physical fatigue crack growth model.

One major benefit of this method is that only limited numbers of sensors are required, which greatly facilitate the realistic applications. Another benefit is that the fatigue damage can be obtained concurrently with structural dynamics analysis, which is critical for real-time decision making. A novel small time scale formulation of fatigue crack growth is employed to predict the

crack growth in this paper. This physical fatigue model is fundamentally different from the traditional cycle-based approach. It describes the crack from time instant based on the geometry at the crack tip. The fatigue prognosis does not suffer from the cycle-counting requirements and stress ratio effects compared to the cycle-based fatigue prognosis models, and has great potential for real-time structural fatigue prognosis.

This chapter is organized as follows. A brief introduction about EMD methodology is given first. Then, the extrapolation and reconstruction process are proposed and verified with numerical examples. Next, the physical fatigue crack growth model used in this paper is introduced. Following

*Figure 1. Flowchart of the concurrent fatigue analysis*

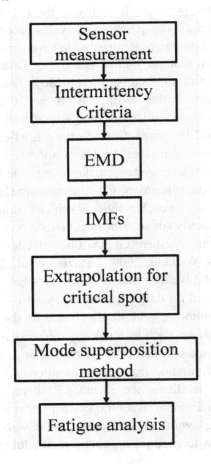

this, a cantilever beam is taken as an example to demonstrate the procedure for structural fatigue prognosis integrating usage monitoring system. Finally, some conclusions are drawn based on the current study.

## EMPIRICAL MODE DECOMPOSITION

The overall computational procedure of the proposed dynamic response reconstruction method involves two sequential steps. The first step is to decompose the measured signals into modal responses (i.e. in terms of intrinsic model functions (IMFs)) using the EMD method with intermittency criteria. The resulting IMFs are used in the second step to reconstruct the dynamic responses at critical locations which are sensor-inaccessible. For the completeness of the proposed method description, the EMD method with intermittency criteria is briefly introduced.

The EMD is a method of breaking down a signal without leaving the time domain. The idea of the EMD method is to break down a given signal into functions which form a complete and nearly orthogonal basis for the original signal. The resulting functions which have a mean value of zero and only one extreme between zero crossings, known as IMFs, are sufficient to describe the underlying dynamics of the signal. The sufficiency and completeness of the EMD method is ensured by the way the signal is decomposed (N.E. Huang & Shen, 2005). Using IMFs to represent the signal allows for varying frequency in time to be preserved, which is hidden in the Fourier domain or in wavelet coefficients (Norden E. Huang, et al., 1998). Those approximately orthogonal constituent IMFs bear close resemblance to base functions and are problem specific and data oriented (N.E. Huang, et al., 1999; Li, Wang, Tao, Wang, & Du, 2011)

## Standard Sifting Process

To filter out IMFs from given signals; the EMD method employs a so-called sifting process. Sifting process is to subtract a signal from its mean value. For a time domain signal, denoted as $y(t)$, the sifting process is described as follows (Norden E. Huang, et al., 1998; Yang, et al., 2003):

1.  Determine the upper and lower envelopes using cubic-spline interpolation of local maxima and minima of the signal. Let $m_1$ denote the mean of the upper and lower envelopes.
2.  Compute the first component $h_1 = y(t) - m_1$ and check to see whether $h_1$ is an IMF. If it is not an IMF, continue the sifting process using $h_1$ as the new signal data.
3.  Construct the envelopes for $h_1$ and denote the mean value of its envelopes as $m_{11}$. Obtain the component $h_{11} = h_1 - m_{11}$, where $m_{1k}$ is the mean of the upper and lower envelopes of the $k^{th}$ iteration.
4.  The sifting process is repeated $k$ times, until the resulting $h_{1k}$ (i.e., $h_{1k} = h_{1(k-1)} - m_{1k}$) is an IMF.
5.  Denote $f_1(t) = h_{1k}$, which is the first IMF component from the data. It contains the shortest period component of the signal.

Repeat the above steps to get the second IMF $f_2(t)$ from the rest signal of $y(t) - f_1(t)$. Continue the computation to obtain up to the nth IMF until the remaining of the signal $r(t) = y(t) - \sum_{i=1}^{n} f_i(t)$ is the main trend of the signal (Yang, et al., 2003). By applying the EMD method with the standard sifting process, the original signal $y(t)$ can be expressed as the summation of $n$ IMFs and a residue term, as shown in Equation (1).

$$y(t) = \sum_{i=1}^{n} f_i(t) + r(t), \qquad (1)$$

where $f_i(t)$ is the ith IMF and $r(t)$ is the residue. Term $r(t)$ also represents the mean trend or constant for this signal (Yang, et al., 2003).

## Sifting Process with Intermittency Criteria

IMFs obtained from the EMD method may contain several frequency components in each IMF, which is not a modal response. To obtain IMFs corresponding to the system modal responses, an intermittency frequency denoted by $\omega_{int}$ can be imposed in the sifting process to ensure that each of the IMFs contains only one frequency component. The idea is to remove all frequency components lower and larger than $\omega_{int}$ from an IMF, and this can be done during the sifting process using a band-pass filter. The process to obtain the modal response corresponding to the ith natural frequency $\omega_i$ is discussed in (Yang, et al., 2003) and is summarized here.

1. An approximate range for $\omega_i$, $\omega_{iL} < \omega_i < \omega_{iH}$ is estimated either by Fourier transform of $y(t)$ or finite element model computations.
2. Process the signal data $y(t)$ using a band-pass filter with a frequency range $\omega_{iL} < \omega_i < \omega_{iH}$. The data obtained from the filter are then processed through the sifting process.
3. Process the filtered signal using the standard sifting process.

By repeating the above procedure with different frequency ranges for different natural frequencies, all modal responses can be obtained. These IMFs have several characteristics: 1) Each IMF contains the intrinsic characteristics of the signal; 2) Once

an IMF is obtained, the next IMF will not have the same frequency at the same time instant (Bao, et al., 2009; N.E. Huang, et al., 1999); and 3) The first IMF for each IMFs series is considered to be the approximation of modal response. Using the sifting process with intermittency criteria, the original signal expression can be written as Equation (2).

$$y(t) \approx \sum_{i=1}^{m} x_i(t) + \sum_{i=1}^{n-m} f_i(t) + r(t), \qquad (2)$$

where $x_i(t)$ is the modal response (that is also an IMF) for the ith mode. Terms $f_i(t)$ $(i = 1, \cdots n - m)$ are other IMFs but not modal responses. The overall flowchart of the EMD method with intermittency criteria is shown in Figure 2 and the standard sifting process is shown in the dashed-line box.

Applying the EMD method with intermittency criteria to the sensor measurement data, all the required modal responses can be extracted without leaving the time domain. Those modal responses are then employed to extrapolate the modal responses at the desired location using a transformation equation which is described in next section.

## Transformation Equation

From the above discussion, the sensor measurements in the time domain are decomposed into several IMFs using the EMD method with intermittency criteria. In this section, the dynamic responses other than the sensor locations are extrapolated using transformation equations. Those transformation equations can be derived from the finite element model of the target structure. The extrapolation procedure is discussed below.

Considering an $n$-DOFs dynamical system, its mass, stiffness, and damping matrices are denoted as $\mathbf{M}$, $\mathbf{K}$ and $\mathbf{C}$, respectively. The equation of motion in matrix format can be expressed as,

*Figure 2. Flowchart of the EMD method with intermittency criteria*

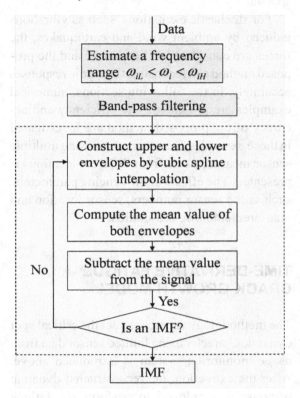

$$\frac{\varphi_{ie}}{\varphi_{iu}} = \frac{\delta_{ie}}{\delta_{iu}}, \tag{5}$$

where the subscript $e$ represents DOFs which can be measured by sensor. $u$ represents the DOFs that are inaccessible for measurements using sensors. $i$ represents the $i$th mode. $\varphi_{ie}$ represents the mode information for the $e$th DOF under $i$th mode and $\delta_{ie}$ represents the modal response for $e$th DOF under $i$th mode. Equation (5) builds a relationship between two coordinates in the structure. Using this equation, the modal response $\delta_{iu}$ at a sensor-inaccessible location can be computed from the modal response $\delta_{ie}$ at a measured location. The transformation equation is derived from finite element analysis of the structure and its simplicity ensures the computation is highly efficient.

Given the mode shape matrix $\Phi$ and $x_i(t)$, the modal response at a desired location can readily be evaluated according to Equation (5). The modal superposition methodology is applied to obtain the dynamic response in time domain after all the modal responses have been calculated. One of the advantages of the proposed method is it is suitable for various dynamic response reconstructions based on the different types of sensor measurements (for example, the proposed method can be used to reconstruct the displacement responses or strain responses if the displacement sensors or strain gages are provided. For the displacement case, the same transformation equation as Equation (5) can be applied; the transformation equation will be different for strain reconstruction case based on the strain-displacement relationship from the mechanics of materials (Kassimali, 1999)). For the displacement or strain reconstruction case, the results provide sufficient local dynamic responses information for further damage prognostics (for example, the local stress for fatigue crack growth analysis can be obtained by classical finite element analysis using the local displacement or strain information).

$$M\ddot{X} + C\dot{X} + KX = F, \tag{3}$$

where $X$ is the response variable vector and $F$ is the load vector. According to structure dynamics, its $n \times n$ mode shape matrix can be obtained by solving the eigen problem of $\left( M^{-1}K \right)$. The resulting mode shape matrix $\Phi$ is expressed as Equation (4)

$$\Phi = \begin{bmatrix} \varphi_{11} & \cdots & \varphi_{1n} \\ \vdots & \ddots & \vdots \\ \varphi_{1n} & \cdots & \varphi_{nn} \end{bmatrix}. \tag{4}$$

Based on the modal analysis, a transformation equation can be written as

Figure 3 presents the overall reconstruction procedure. First, the EMD method with intermittency criteria decompose measured dynamical responses into a set of modal responses. Following this, modal responses at the desired (sensor-inaccessible) location are calculated based on the transformation equation. Finally, modal superposition method transforms modal responses into time domain.

It should be noted that for dynamic responses excited by deterministic and periodic external forces, if the participation factor (amplitude in the Fourier spectra) is strong, it is not suitable for using the proposed method to reconstruct dynamic responses. For the external force which has deterministic frequency (such as periodic excitation), the frequency component from the excitation will be dominant during the excitation duration. The key concept of the proposed method is extrapolating mode shape based on the frequency distribution in Fourier spectra. However, in this situation, the frequency component produced by excitation, which is not associated with any mode, will be dominant and the natural mode frequency

component will be insignificant in the Fourier spectra.

For stochastic excitations, such as vibrations induced by ambient wind and earthquakes, the forces are considered to be random and the proposed method can reconstruct dynamic responses accurately. In the following sections, numerical examples are presented. The efficiency and accuracy of the proposed method are investigated in those examples. Reconstruction using multiple sensor measurements from different locations is presented. The effects of influencing parameters, such as the sensor numbers, sensor location and measurement noise are studied.

## TIME-DERIVATIVE FATIGUE CRACK GROWTH MODEL

The methodology for reconstruct the critical spot dynamic character using limited sensor data from usage monitoring system is introduced above. After the extraction, the reconstructed dynamic response is employed to perform the fatigue analysis. The material fatigue crack growth model used in this chapter has been recently developed in (Lu & Liu, 2010). Detailed description can be found in the referred article. A brief introduction about the fatigue crack growth model employed here is given for the completeness of this chapter. The major new contribution of this chapter is to couple the developed material fatigue damage model with structural sensor measurement and signal reconstruction technique.

Traditional fatigue crack growth analysis uses cycle-based crack growth curves (e.g., well know Paris law), which defines the relationship between the crack growth per cycle and the applied stress intensity factor range(Janssen, Zuidema, & Wanhill, 2004; Paris & Erdogan, 1963; Patankar, Ray, & Lakhtakia, 1998; Porter, 1972). The proposed fatigue crack growth model is based on the incremental crack growth at any arbitrary time instant during a loading cycle (He, Lu, & Liu, 2009, 2011;

*Figure 3. Overall procedure of dynamic response reconstruction using sensor data*

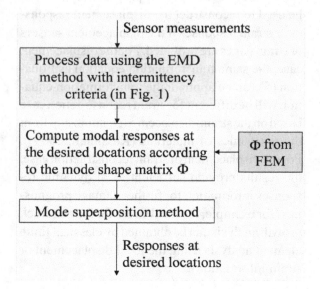

Lu & Liu, 2009). The key concept is to define the fatigue crack kinetics at any arbitrary time instant $(dt)$. The crack will propagate a distance $da$ during the small time scale $dt$. The geometric relationship between the Crack Tip Opening Displacement (CTOD) and the instantaneous crack growth kinetics is shown in Figure 4.

The crack growth rate $da / dt$ is derived based on the geometric relationship shown in Equation (6), where $\theta$ is the crack tip opening angle (CTOA).

$$da = ctg\theta \times d\delta / 2 \qquad (6)$$

Following the derivation in (Lu & Liu, 2010), the instantaneous crack growth rate at an arbitrary time can be expressed as Equation (6).

$$\dot{a} = H\left(\dot{\sigma}\right) \cdot H\left(\sigma - \sigma_{ref}\right) \cdot \frac{2C\lambda}{1 - C\lambda\sigma^2} \cdot \dot{\sigma} \cdot \sigma \cdot a \qquad (7)$$

$H$ is the Heaviside step function. $\sigma_{ref}$ is the reference stress level where the crack begins to grow. The crack length at any arbitrary time can be calculated by the integration of Equation (7). Detailed discussion of the model can be found in (Lu & Liu, 2010). One advantage of the small scale model is that it can be used for fatigue

analysis at variable time and length scales. The fatigue crack growth analysis under random variable amplitude loading can be performed without cycle-counting. This main advantage makes it possible to couple the fatigue crack model (material-level) with dynamic responses (structural-level) for concurrent damage analysis.

## EXAMPLES

In this part, a cantilever beam problem is used as an example to demonstrate the procedure of the dynamic reconstruction and fatigue prognosis. The beam is divided into ten segments with an impact loading at the free end of the beam (shown in Figure 5). A through edge crack is assumed at the fixed end of the beam which is also the critical spot assuming to be inaccessible for sensor. Two displacement sensors are put at the middle and the free end of the beam. More complicate structure and more sensors can be applied based on the methodology proposed in this paper. The impact force and the dynamic response for the free end are shown in Figure 6. The property of the beam is list in Table 1. Only the data from sensor #1 are taken as an example here, the same procedure can be applied to the measurements from sensor #2.

*Figure 4. Crack tip geometry*

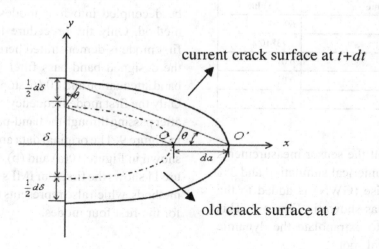

*Figure 5. Illustration diagram of the cantilever beam example*

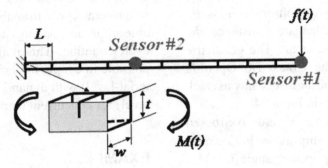

*Figure 6. Applied forces and the displacement responses at the free end*

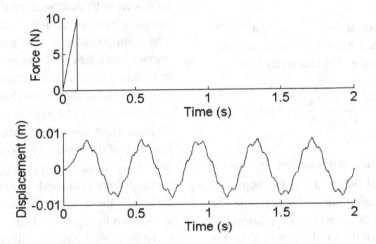

*Table 1. Dimensions and material properties of the beam*

| Property name | Value |
|---|---|
| Young's modulus (*MPa*) | 69600 |
| Density (*kg/m³*) | 2.73×10³ |
| Width of cross section (*m*) | 0.01 |
| Thickness of beam (*m*) | 0.01 |
| Length of beam (*m*) | 1 |

The data represent the sensor measurements are obtained from numerical simulation and 3% Gaussian White Noise (GWN) is added to the deterministic signal as shown in Figure 7. This data are employed to extrapolate the dynamic response at the critical spot.

The first four modes are used to reconstruct the displacement information at the critical spot. The measured displacement signal data need to be decoupled into four modes using the EMD method. Only the procedure for extracting the first mode is demonstrated here. Figure 8 shows the designed band-pass filter with $0.3dB$ pass band attenuation and $10dB$ stop band attenuation. Only the first mode frequency component is kept after passing through the band-pass filter, as shown in Figure 9. The original data and filtered data are shown in Figure 10(a) and (b), respectively. Figure 11 shows the first four IMFs obtained by EMD methods, which also represents the displacements for the first four modes.

*Figure 7. Signal which represents data from sensor #1*

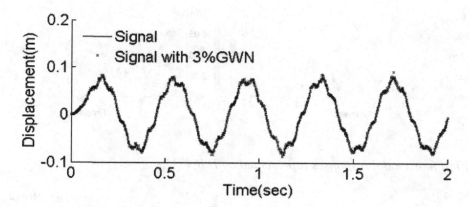

*Figure 8. Frequency response of the band-pass filter*

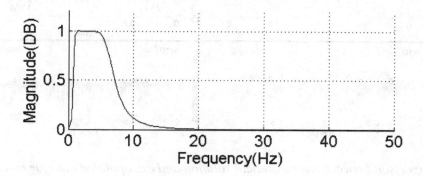

*Figure 9. Fourier spectra of the signal after band-pass filtering*

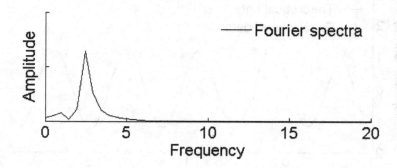

The extrapolation process (Equations (4) - (5)) is then applied to those decoupled data to obtain the mode responses for displacement at the critical spot. The comparison between the theoretical displacement and the extrapolated data for mode one is shown Figure 12. An overall good agree-

ment is observed except in the region near $t = 0$ which is known as the end boundary effect for EMD method wildly discussed in literature (Norden E. Huang, et al., 1998; Yang, et al., 2003). This difference will not affect the fatigue prognosis since the time history for fatigue loading is

*Figure 10. Displacement responses (a) orignial signal (b) band-pass filtered*

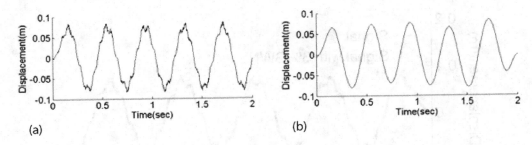

(a)                                             (b)

*Figure 11. First four modal responses represented by IMFs*

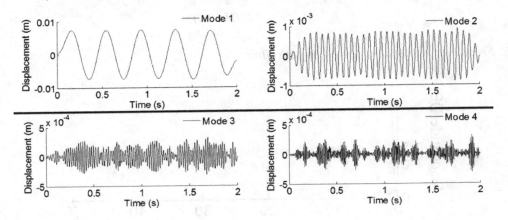

*Figure 12. Comparison between the theoretical solution and extrapolated data for mode one*

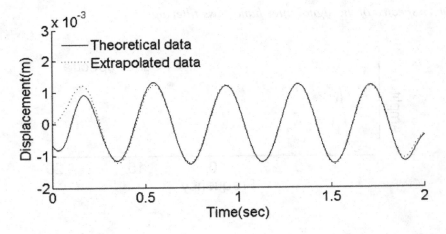

very long. Small difference at the beginning and ending of the signal will not change the RUL prediction much (e.g., 2~3 cycles vs. one million cycles).

The mode superposition method is then used to reconstruct the physical displacement responses in time domain. The same extrapolation and reconstruction procedure is applied to the data for sensor #2. Comparisons between the theoretical

*Figure 13. Comparison between extrapolate data and theoretical solution*

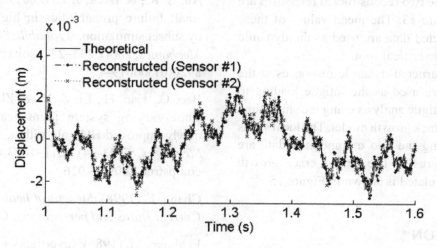

*Figure 14. Stress responses*

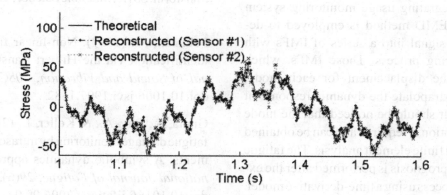

*Figure 15. Crack growth trajectory*

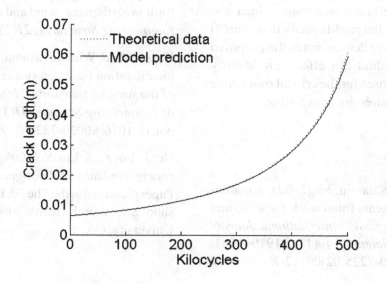

solution and the two reconstructed responses are shown in Figure 13. The mean values of these two reconstructed data are used as the dynamic responses at the critical spot.

The reconstructed dynamic responses at the critical spot are used as the fatigue loading to perform the fatigue analysis using the small time scale fatigue crack growth model. The local stress calculated using the two extrapolated data are shown as Figure 14. The fatigue crack growth trajectory calculated is shown in Figure 15.

## CONCLUSION

In this study, a new methodology for fatigue prognosis integrating usage monitoring system is proposed. EMD method is employed to decompose the signal into a series of IMFs with specific filtering process. Those IMFs, which represented the displacement for each mode, are used to extrapolate the dynamic response at critical spot. It should be noticed that the mode shape information is required and can be obtained from classical finite element analysis. The fatigue crack growth prognosis is performed after the extrapolation process using a time-derivative model. Based on the current study, several conclusions are drawn: (1) The presented study provides a concurrent fatigue crack prognosis, which can be used for on-line fatigue life prediction, and (2) The numerical study demonstrates the proposed reconstruction method can effectively identify the dynamic responses for the critical spot where direct sensor measures are unavailable.

## REFERENCES

Adams, D. E., & Nataraju, M. (2002). A nonlinear dynamical systems framework for structural diagnosis and prognosis. *International Journal of Engineering Science, 40*(17), 1919–1941. doi:10.1016/S0020-7225(02)00112-X

Au, S. K., & Beck, J. L. (2001). Estimation of small failure probabilities in high dimensions by subset simulation. *Probabilistic Engineering Mechanics, 16*(4), 263–277. doi:10.1016/S0266-8920(01)00019-4

Bao, C., Hao, H., Li, Z. X., & Zhu, X. (2009). Time-varying system identification using a newly improved HHT algorithm. *Computers & Structures, 87*(23-24), 1611–1623. doi:10.1016/j.compstruc.2009.08.016

Chang, F. (1998). *Structural health monitoring: Current status and perspectives*. CRC.

Feldman, M. (1985). Invertigation of the natural vibrations of machine elements using the Hilbert transform. *Soviet Machine Science, 2*(0739-8999), 3.

Feldman, M. (1997). Non-linear free vibration identification via the Hilbert transform. *Journal of Sound and Vibration, 208*(3), 475–489. doi:10.1006/jsvi.1997.1182

Gupta, S., Ray, A., & Keller, E. (2007). Online fatigue damage monitoring by ultrasonic measurements: A symbolic dynamics approach. *International Journal of Fatigue, 29*(6), 1100–1114. doi:10.1016/j.ijfatigue.2006.09.011

Gurley, K. (1999). Application of wavelet transform in earthquake, wind and ocean engineering. *Engineering Structures, 21*(2), 149–167.

Haase, M., & Widjajakusuma, J. (2003). Damage identification based on ridges and maxima lines of the wavelet transform. *International Journal of Engineering Science, 41*(13-14), 1423–1443. doi:10.1016/S0020-7225(03)00026-0

He, J., Lu, Z., & Liu, Y. (2009). *A new method for concurrent Multi-scale fatigue damage prognosis*. Paper presented at the The 7th International Workshop Strucutural Health Monitoring, Stanford University, CA.

He, J., Lu, Z., & Liu, Y. (2011). A new method for concurrent dynamic analysis and fatigue damage prognosis of bridges. *Journal of Bridge Engineering, 1*, 196.

Huang, N. E., & Shen, S. S. (2005). *Hilbert-Huang transform and its applications*. World Scientific Pub Co Inc.

Huang, N. E., Shen, Z., & Long, S. R. (1999). A new view of nonlinear water waves: The Hilbert Spectrum 1. *Annual Review of Fluid Mechanics, 31*(1), 417–457. doi:10.1146/annurev.fluid.31.1.417

Huang, N. E., Shen, Z., Long, S. R., Wu, M. C., Shih, H. H., Zheng, Q., et al. (1998). The empirical mode decomposition and the Hilbert spectrum for nonlinear and non-stationary time series analysis. *Proceedings of the Royal Society of London. Series A: Mathematical, Physical and Engineering Sciences, 454*(1971), 903-995.

Janssen, M., Zuidema, J., & Wanhill, R. (2004). *Fracture mechanics* (2nd ed.). London, UK: Taylor & Francis Group.

Kassimali, A. (1999). *Matrix analysis of structures*. Pacific Grove, CA: Brooks, Cole Publishing Company.

Li, C., Wang, X., Tao, Z., Wang, Q., & Du, S. (2011). Extraction of time varying information from noisy signals: An approach based on the empirical mode decomposition. *Mechanical Systems and Signal Processing, 25*(3), 812–820. doi:10.1016/j.ymssp.2010.10.007

Link, M., & Weiland, M. (2009). Damage identification by multi-model updating in the modal and in the time domain. *Mechanical Systems and Signal Processing, 23*(6), 1734–1746. doi:10.1016/j.ymssp.2008.11.009

Lu, Z., & Liu, Y. (2009). *An incremental crack growth model for multi-scale fatigue analysis*. Paper presented at the 50th AIAA/ASME/ASCE/AHS/ASC Structures, Structural Dynamics, and Materials Conference, Palm Springs, CA.

Lu, Z., & Liu, Y. (2010). Small time scale fatigue crack growth analysis. *International Journal of Fatigue, 32*(8), 1306–1321. doi:10.1016/j.ijfatigue.2010.01.010

Luk, R. W. P., & Damper, R. I. (2006). Non-parametric linear time-invariant system identification by discrete wavelet transforms. *Digital Signal Processing, 16*(3), 303–319. doi:10.1016/j.dsp.2005.11.004

Papazian, J. M., Anagnostou, E. L., Engel, S. J., Hoitsma, D., Madsen, J., & Silberstein, R. P. (2009). A structural integrity prognosis system. *Engineering Fracture Mechanics, 76*(5), 620–632. doi:10.1016/j.engfracmech.2008.09.007

Papazian, J. M., Nardiello, J., Silberstein, R. P., Welsh, G., Grundy, D., & Craven, C. (2007). Sensors for monitoring early stage fatigue cracking. *International Journal of Fatigue, 29*(9-11), 1668–1680. doi:10.1016/j.ijfatigue.2007.01.023

Paris, P., & Erdogan, F. (1963). A critical analysis of crack propagation. *Journal of Basic Engineering, 85*, 528–534. doi:10.1115/1.3656900

Patankar, R., Ray, A., & Lakhtakia, A. (1998). A state-space model of fatigue crack growth. *International Journal of Fracture, 90*(3), 235–249. doi:10.1023/A:1007491916925

Pislaru, C., Freeman, J. M., & Ford, D. G. (2003). Modal parameter identification for CNC machine tools using Wavelet Transform. *International Journal of Machine Tools & Manufacture, 43*(10), 987–993. doi:10.1016/S0890-6955(03)00104-4

Poon, C. W., & Chang, C. C. (2007). Identification of nonlinear elastic structures using empirical mode decomposition and nonlinear normal modes. *Smart Structures and Systems, 3*, 14.

Porter, T. (1972). Method of analysis and prediction for variable amplitude fatigue crack growth. *Engineering Fracture Mechanics, 4*(4), 717–736. doi:10.1016/0013-7944(72)90011-2

Tan, J.-B., Liu, Y., Wang, L., & Yang, W.-G. (2008). Identification of modal parameters of a system with high damping and closely spaced modes by combining continuous wavelet transform with pattern search. *Mechanical Systems and Signal Processing, 22*(5), 1055–1060. doi:10.1016/j.ymssp.2007.11.017

Yan, G. (2006). Hilbert-Huang transform-based vibration signal analysis for machine health monitoring. *IEEE Transactions on Instrumentation and Measurement, 55*, 9. doi:10.1109/TIM.2006.887042

Yang, J. N., Lei, Y., Pan, S., & Huang, N. (2003). System identification of linear structures based on Hilbert-Huang spectral analysis. Part 1: Normal modes. *Earthquake Engineering & Structural Dynamics, 32*(9), 1443–1467. doi:10.1002/eqe.287

# Compilation of References

Abernethy, R. B. (1996). *The new Weibull handbook* (2nd ed.). North Palm Beach, FL: Abernethy.

Acevedo, P., Martinez, J., & Gonzalez, M. (2008). Case-based reasoning and system identification for control engineering learning. *IEEE Transactions on Education, 51*(2), 271–281. doi:10.1109/TE.2007.909361

Adams, D. E., & Nataraju, M. (2002). A nonlinear dynamical systems framework for structural diagnosis and prognosis. *International Journal of Engineering Science, 40*(17), 1919–1941. doi:10.1016/S0020-7225(02)00112-X

Aissa, K., & Eddine, K. D. (2009). Vector control using series iron loss model of induction, motors and power loss minimization. *World Academy of Science. Engineering and Technology, 52*, 142–148.

Akagaki, T., Nakamur, M., Monzen, T., & Kawabata, M. (2006). Analysis of the behaviour of rolling bearings in contaminated oil using some condition monitoring techniques. *Proceedings of the Institution of Mechanical Engineers, Part J, Journal of Engineering Tribology, 220*(5), 447–453. doi:10.1243/13506501J00605

An aligned view on shafts? (2006). *Marine Engineers Review, 5,* 44-47.

Anderson, B., & Moore, J. (1971). *Linear optimal control.* Englewood Cliffs, NJ: Prentice Hall.

Andrawus, J. A., Watson, J., Kishk, M., & Adam, A. (2006). The selection of a suitable maintenance strategy for wind turbines. *International Journal of Wind Engineering, 30*(6), 471–486. doi:10.1260/030952406779994141

Andrews, J., Kjørholt, H., & Jøranson, H. (2005, June). *Production enhancement from sand management philosophy: A case study from Statfjord and Gullfaks.* Paper presented at SPE European Formation Damage Conference, Sheveningen, The Netherlands.

Angstenberger, L. (2001). Dynamic fuzzy pattern recognition. In Zimmermann, H. J. (Ed.), *International series in intelligent technologies, 17.* Dordrecht, The Netherlands: Kluwer Academic Publishers.

Arabian-Hoseynabadi, H., Oraee, H., & Tavner, P. (2010). Failure modes and effects analysis (FMEA) for wind turbines. *International Journal of Electrical Power & Energy Systems, 32*(7), 817–824. doi:10.1016/j.ijepes.2010.01.019

Arimoto, S. (1990). Learning control theory for robotic motion. *International Journal of Adaptive Control and Signal Processing, 4,* 543–564. doi:10.1002/acs.4480040610

Arimoto, S., Kawamura, S., & Miyazaki, F. (1984). Bettering operation of robots by learning. *Journal of Robotic Systems, 1*(2), 123–140. doi:10.1002/rob.4620010203

Arulampalam, S. M., Maskell, S., Gordon, N., & Clapp, T. (2002). A tutorial on particle filters for on-line nonlinear/non-Gaussian Bayesian tracking. *IEEE Transactions on Signal Processing, 50*(2), 174–188. doi:10.1109/78.978374

Ashenbaum, B. (2006). *Defining cost reduction and cost avoidance.* CAPS Research.

Astrom, K. (1970). *Introduction to stochastic control theory.* New York, NY: Academic Press.

Atlas, L., Ostendorf, M., & Bernard, G. D. (2000). *Hidden Markov models for monitoring machining tool-wear.* Paper presented at the IEEE International Conference on Acoustics, Speech, and Signal Processing.

Aumeier, S. E., Alpay, B., Lee, J. C., & Akcasu, A. Z. (2006). Probabilistic techniques for diagnosis of multiple component degradations. *Nuclear Science and Engineering, 153*, 101–123.

Au, S. K., & Beck, J. L. (2001). Estimation of small failure probabilities in high dimensions by subset simulation. *Probabilistic Engineering Mechanics, 16*(4), 263–277. doi:10.1016/S0266-8920(01)00019-4

Bachschmid, N., Pennacchi, P., & Tanzi, E. (2010). A sensitivity analysis of vibrations in cracked turbogenerator units versus crack position and depth. *Mechanical Systems and Signal Processing, 24*(3), 844–859. doi:10.1016/j.ymssp.2009.10.001

BAE. (2008). *BAE 61972 BAE annual report.*

Bagajewicz, M. (2000). *Design and upgrade of process plant instrumentation.* Lancanster, PA: Technomic Publishers. doi:10.1002/cjce.5450800101

Bagdonavicius, V., & Nikulin, M. (2000). Estimation in degradation models with explanatory variables. *Lifetime Data Analysis, 7*, 85–103. doi:10.1023/A:1009629311100

Balaban, E., Saxena, A., Narasimhan, S., Roychoudhury, I., & Goebel, K. (2009, September). Experimental *validation of a prognostic health management system for electromechanical actuators.* Paper presented at the Annual Conference of the Prognostics and Health Management Society, San Diego, CA.

Baldi, P. (1995). Gradient descent learning algorithm overview: A general dynamical systems perspective. *IEEE Transactions on Neural Networks, 6*(1), 182–195. doi:10.1109/72.363438

Bankole, O. O., Roy, R., Shehab, E., & Wardle, P. (2009). Affordability assessment of industrial product-service system in the aerospace defense industry. *Proceedings of the CIRP Industrial Product-Service Systems (IPS2) Conference* (p. 230).

Bao, C., Hao, H., Li, Z. X., & Zhu, X. (2009). Time-varying system identification using a newly improved HHT algorithm. *Computers & Structures, 87*(23-24), 1611–1623. doi:10.1016/j.compstruc.2009.08.016

Baraldi, P., Razavi-Far, R., & Zio, E. (2010, November). *A method for estimating the confidence in the identification of nuclear transients by a bagged ensemble of FCM classifiers.* Paper presented at the 7th American Nuclear Society International Topical Meeting on Nuclear Plant Instrumentation, Controls and Human Machine Interface Technology, Las Vegas, Nevada, USA.

Baraldi, P., Zio, E., Compare, M., Rossetti, G., & Despujols, A. (2009, September). *A novel approach to model the degradation of components in electrical production plants.* Paper presented at the European Safety and Reliability Conference, Prague, Czech Republic.

Baraldi, P., Compare, M., Despujols, A., & Zio, E. (2011a). Modelling of the effect of maintenance actions on the degradation of an electric power plant component. *Proceedings of the Institution of Mechanical Engineers, Part O. Journal of Risk and Reliability, 225*(2), 169–184.

Baraldi, P., Compare, M., Rossetti, G., Zio, E., & Despujols, A. (2011b). A modelling framework to assess maintenance policy performance in electrical production plants. In Andrews, J., Berenguer, C., & Jackson, L. (Eds.), *Maintenance modelling and applications* (pp. 263–282). Det Norske Veritas, AS.

Baraldi, P., Razavi-Far, R., & Zio, E. (2011). Bagged ensemble of FCM classifier for nuclear transient identification. *Annals of Nuclear Energy, 38*(5), 1161–1171. doi:10.1016/j.anucene.2010.12.009

Bauer, E., & Kohavi, R. (1999). An empirical comparison of voting classification algorithms: Bagging, boosting and variants. *Machine Learning, 36*, 105–139. doi:10.1023/A:1007515423169

Bayoudh, M., Travé-Massuyès, L., & Olive, X. (2008). Hybrid systems diagnosis by coupling continuous and discrete event techniques, In *Proceedings of the 17th World Congress*, (pp.7265-7270). Seoul, Korea.

Beanum, R. L. (2006). Performance-based logistics and contractor support methods. *Proceedings of the IEEE Systems Readiness Technology Conference (AUTOTESTCON).*

Bechhoefer, E., & Bernhard, A. (2007). *A generalized process for optimal threshold setting in HUMS.* IEEE Aerospace Conference, Big Sky.

Bechhoefer, E., & Kingsley, M. (2009). *A review of time synchronous average algorithms.* Annual Conference of the Prognostics and Health Management Society

Bechhoefer, E., Bernhard, A., & He, D. (2008). *Use of Paris law for prediction of component remaining life.* IEEE Aerospace Conference, Big Sky. 2008

Bechhoefer, E., Duke, A., & Mayhew, E. (2007). *A case for health indicators vs. condition indicators in mechanical diagnostics.* American Helicopter Society Forum 63, Virginia Beach.

Bechhoefer, E., He, D., & Dempsey, P. (2011). *Gear threshold setting based on a probability of false alarm.* Annual Conference of the Prognostics and Health Management Society.

Becker, G., Camarinopoulos, L., & Zioutas, G. (2000). A semi-Markovian model allowing for inhomogenities with respect to process time. *Reliability Engineering & System Safety, 70*(1), 41–48. doi:10.1016/S0951-8320(00)00044-2

Benazera, E., & Travé-Massuyès, L. (2009). Set-theoretic estimation of hybrid system configurations. *IEEE Transactions on Systems, Man, and Cybernetics. Part B, 39*(5), 1277–1291.

Bendell, D., Wightman, D., & Walker, E. (1991). Applying proportional hazard modeling in reliability. *Reliability Engineering & System Safety, 34*, 35–53. doi:10.1016/0951-8320(91)90098-R

Benosman, M. (2010). A survey of some recent results on nonlinear fault tolerant control. *Mathematical Problems in Engineering*, article ID 586169, 1-25.

Bérenguer, C., Grall, A., & Castanier, B. (2000). Simulation and evaluation of condition-based maintenance policies for multi-component continuous-state deteriorating systems. In M. Cottam, D. Harvey, R. Pape, & J. Tait (Eds.), *Proceedings of the ESREL 2000 Foresight and Precaution Conference* (pp. 275-82). Rotterdam, The Netherlands: Balkema Publishers.

Berthold, M., & Hand, J. H. (Eds.). (2003). *Intelligent data analysis.* Berlin, Germany: Springer-Verlag. doi:10.1007/978-3-540-48625-1

Bigerelle, M., & Iost, A. (1999). Bootstrap analysis of FCGR, application to the Paris relationship and to lifetime prediction. *International Journal of Fatigue, 21*, 299–307. doi:10.1016/S0142-1123(98)00076-0

Bilmes, J. (1997). *A gentle tutorial on the EM algorithm and its application to parameter estimation for Gaussian mixture and hidden Markov models.* Berkeley: University of California.

Biswas, G., Cordier, M.-O., Lunze, J., Travé-Massuyès, L., & Staroswiecki, M. (2004). Diagnosis of complex systems: Bridging the methodologies of FDI and DX communities. *IEEE Transactions on Systems, Man and Cybernetics, Part B: Cybernetics. Special section on Diagnosis of Complex Systems. Bridging the Methodologies of FDI and DX Communities, 34*(5), 2159–2162.

Blanke, M., Kinnaert, M., Lunze, J., & Staroswiecki, M. (2006). *Diagnosis and fault-tolerant control* (2nd ed.). Springer Publisher.

Blasi, A., Janssen, J., & Manca, R. (2004). Numerical treatment of homogeneous and non-homogeneous semi-Markov reliability models. *Communications in Statistics Theory and Methods, 33*(3), 697–714. doi:10.1081/STA-120028692

Bocaniala, C. D., & Palade, V. (2006). Computational intelligence methodologies in fault diagnosis: Review and state of the art computational intelligence in fault diagnosis. In Palade, V., Jain, L., & Bocaniala, C. D. (Eds.), *Computational intelligence in fault diagnosis* (pp. 1–36). London, UK: Springer. doi:10.1007/978-1-84628-631-5_1

Bocker, J., & Mathapati, S. (2007). State of the art of induction motor control. *IEEE International Electric Machines & Drives Conference, IEMDC '07*, (pp. 1459–1464).

Bolander, N., Qiu, H., Eklund, N., Hindle, E., & Rosenfeld, T. (2009). *Physics-based remaining useful life prediction for aircraft engine bearing prognosis.* Annula Conference of Prognosis and Health Management Society, San Diego.

Bolotin, V. V., & Shipkov, A. A. (1998). A model of the environmentally affected growth of fatigue cracks. *Journal of Applied Mathematics and Mechanics, 62,* 313. doi:10.1016/S0021-8928(98)00037-9

Bose, K. B. (2002). *Modern power electronics and AC drives.* Prentice Hall.

Bouman, C. A., Shapiro, M., Cook, G. W., Atkins, C. B., & Cheng, H. (1998). *Cluster: An unsupervised algorithm for modeling Gaussian mixtures.* Retrieved from http://dynamo.ecn.purdue.edu/~bouman/software/cluster

Boutros, T., & Liang, M. (2011). Detection and diagnosis of bearing and cutting tool faults using hidden Markov models. *Mechanical Systems and Signal Processing, 25*(6), 2102–2124. doi:10.1016/j.ymssp.2011.01.013

Box, G. E. P., Jenkins, G. M., & Reinsel, G. C. (1994). *Time series analysis: Forecasting and control* (3rd ed.). Upper Saddle River, NJ: Prentice-Hall.

Brauer, D. C., & Brauer, G. D. (2009). Reliability-centered maintenance. *IEEE Transactions on Reliability, 36*(1), 17–24. doi:10.1109/TR.1987.5222285

Breiman, L. (1996). Bagging predictors. *Machine Learning, 24*(2), 123–140. doi:10.1007/BF00058655

Breiman, L. (1999). Combining predictors. In Sharkey, A. J. C. (Ed.), *Combining artificial neural nets.* Springer.

Bringedal, B., Hovda, K., Ujang, P., With, H. M., & Kjørrefjord, G. (2010, November-December). *Using on-line dynamic virtual flow metering and sand erosion monitoring for integrity management and production optimization.* Paper presented at Deep Offshore Technology Conference, Houston, Texas.

Brogan, W. (1991). *Modern control theory.* Upper Saddle River, NJ: Prentice Hall.

Brown, D. W., & Vachtsevanos, G. J. (2011). *A prognostic health management based framework for fault-tolerant control.* Paper presented at the Annual Conference of the Prognostics and Health Management Society, Montreal, Canada.

Brown, D. W., Kalgren, P. W., Byington, C. S., & Roemer, M. J. (2007). Electronic prognostics – A case study using global positioning system (GPS). *Microelectronics and Reliability, 47,* 1874–1881. doi:10.1016/j.microrel.2007.02.020

Bucher, C., & Bourgund, U. (1990). A fast and efficient response surface approach for structural reliability problems. *Structural Safety, 7*(1), 57–66. doi:10.1016/0167-4730(90)90012-E

Butler, S., & Ringwood, J. (2010). Particle filters for remaining useful life estimation of abatement equipment used in semiconductor manufacturing. *Proceedings of the First Conference on Control and Fault-Tolerant Systems,* Nice, France, (pp. 436–441).

Byington, C. S., & Stoelting, P. (2004). A model-based approach to prognostics and health management for flight control actuators. *2004 IEEE Aerospace Conference* (pp. 3551-3562). Big Sky, MT.

Byington, C. S., Roemer, M. J., & Galie, T. (2002). Prognostic enhancements to diagnostic systems for improved condition-based maintenance. *2002 IEEE Aerospace Conference* (pp. 2815-2824). Big Sky, MT, USA.

Byington, C. S., Roemer, M. J., & Kalgren, P. W. (2005, April). *Verification and validation of diagnostic/prognostic algorithms.* Paper presented at Machinery Fault Prevention Technology Conference (MFPT 59). Virginia Beach, Virginia, USA.

Byington, C., Safa-Bakhsh, R., Watson, M., & Kalgren, P. (2003). *Metrics evaluation and tool development for health and usage monitoring system technology.* HUMS 2003 Conference, DSTO-GD-0348.

Cadini, F., Zio, E., & Avram, D. (2009). Monte Carlo-based filtering for fatigue crack growth estimation. *Probabilistic Engineering Mechanics, 24,* 367–373. doi:10.1016/j.probengmech.2008.10.002

Cai, G., & Elishakoff, I. (1994). Refined second-order reliability analysis. *Structural Safety, 14*(4), 267–276. doi:10.1016/0167-4730(94)90015-9

Candy, J. (2009). *Bayesian signal processing: Classical, modern and particle filtering methods.* New Jersey: John Wiley & Sons, Inc.

Carlin, B. P., & Louis, T. A. (2000). *Bayes and empirical Bayes methods for data analysis* (2nd ed.). Boca Raton, LA: Chapman and Hall/CRC.

Casimir, R., Boutleux, E., & Clerc, G. (2003). *Fault diagnosis in an induction motor by pattern recognition methods*. 4th IEEE International Symposium on Power Electronics and Drives, SDEMPED 24-26 August, Atlanta, Georgia, USA.

Cempel, C. (2003). Multidimensional condition monitoring of mechanical system in operation. *Mechanical Systems and Signal Processing*, *17*(6), 1291–1303. doi:10.1006/mssp.2002.1573

Chang, F. (1998). *Structural health monitoring: Current status and perspectives*. CRC.

Chang, J. B. (1981). Round-robin crack growth prediction on center cracked tension specimen under random spectrum loading. In Chang, J. B., & Hudson, C. M. (Eds.), *Methods and models for predicting fatigue crack growth under random loading, ASTM STP 748* (pp. 4–40). doi:10.1520/STP28332S

Chapman, S. (2005). *Electric machinery fundamentals* (4th ed.). McGraw Hill.

Chee, T., Phil, I., & David, M. (2007). A comparative experimental study on the diagnostic and prognostic capabilities of acoustics emission. vibration and spectrometric oil analysis for spur gears. *Mechanical Systems and Signal Processing*, *21*(1), 208–233. doi:10.1016/j.ymssp.2005.09.015

Chen, A., & Wu, G. S. (2007). Real-time health prognosis and dynamic preventive maintenance policy for equipment under aging Markovian deterioration. *International Journal of Production Research*, *45*(15), 3351–3379. doi:10.1080/00207540600677617

Cheng, J., Zhang, J., Cai, C., & Xiao, R. C. (2007). A new approach for solving inverse reliability problems with implicit response functions. *Engineering Structures*, *29*(1), 71–79. doi:10.1016/j.engstruct.2006.04.005

Cheng, S., Azarian, M., & Pecht, M. (2010). Sensor systems for prognostics and health management. *Sensors (Basel, Switzerland)*, *10*, 5774–5797. doi:10.3390/s100605774

Chen, H.-F., & Fang, H.-T. (2004). Output tracking for nonlinear stochastic systems by iterative learning control. *IEEE Transactions on Automatic Control, 49*(4), 583–588. doi:10.1109/TAC.2004.825613

Chen, J., & Patton, R. J. (1999). *Robust model-based fault diagnosis for dynamic systems*. Norwell, MA: Kluwer Academic Publishers.

Chien, C.-J., & Liu, J.-S. (1994). *A P-type iterative learning controller for robust output tracking of nonlinear time-varying systems*. American Control Conference.

Chinnam, R. B. (1999). On-line reliability estimation of individual components, using degradation signals. *IEEE Transactions on Reliability, 48*(4), 403–412. doi:10.1109/24.814523

Chow, A. Y., & Willsky, A. (1984). Analyticaly redundancy and the design of robust failure detection systems. *IEEE Transactions on Automatic Control, 7*(29), 603–614. doi:10.1109/TAC.1984.1103593

Chryssolouris, G., Lee, M., & Ramsey, A. (1996). Confidence interval prediction for neural network models. *IEEE Transactions on Neural Networks, 7*(1), 229–232. doi:10.1109/72.478409

Coble, J. (2010). *Merging data sources to predict remaining useful life – An automated method to identify prognostic parameters* (Doctoral dissertation). Retrieved from http://trace.tennessee.edu/utk_graddiss/683

Coble, J., & Hines, J. W. (2009, April). *Fusing data sources for optimal prognostic parameter selection*. Paper presented at the Sixth American Nuclear Society International Topical Meeting on Nuclear Plant Instrumentation, Control, and Human-Machine Interface Technologies NPIC&HMIT 2009, Knoxville, TN.

Coble, J., & Hines, J. W. (2009b, April). *Analysis of prognostic opportunities in power industry with demonstration*. Paper presented at the 6th American Nuclear Society International Topical Meeting on Nuclear Plant Instrumentation, Controls and Human Machine Interface Technology, Knoxville, Tennessee, USA.

Cocquempot, V., El Mezyani, T., & Staroswiecki, M. (2004). *Fault detection and isolation for hybrid systems using structured parity residuals.* IEEE/IFAC-ASCC'2004, 5th Asian Control Conference, Melbourne, Victoria, Australia, 20-23 June.

Cocquempot, V., Staroswiecki, M., & El Mezyani, T. (2003). Switching time estimation and fault detection for hybrid systems using structured parity residuals. In *Proceedings of the 15th IFAC Symposium on Fault Detection, Supervison and Safety of Technical Processes (SAFEPROCESS '03)*, (pp. 681-686). Washington, DC, USA.

Combet, L., & Gelman, L. (2007). An automated methodology for performing time synchronous averaging of a gearbox signal without seed sensor. *Mechanical Systems and Signal Processing, 21*(6), 2590–2606. doi:10.1016/j.ymssp.2006.12.006

Cordier, M. O., Dague, P., Dumas, M., Lévy, M., Montmain, J., Staroswiecki, M., & Travé-Massuyès, L. (2004). Conflicts vs. analytical redundancy relations: A comparative analysis of the model-based diagnosis from the artificial intelligence and automatic control perspectives. *IEEE Transactions on Systems, Man and Cybernetics, Part B: Cybernetics. Special Section on Diagnosis of Complex Systems: Bridging the Methodologies of FDI and DX Communities, 34*(5), 2163–2177.

Cox, D. (1972). Regression models and life tables. *Journal of the Royal Statistics B, 34*, 187–202.

Cox, D. R., & Oakes, D. (1984). *Analysis of survival data.* USA: Chapman and Hall.

Crowley, J. T., & Choi, Y. K. (1998). Experimental studies on optimal molecular weight distribution control in a batch-free radical polymerization process. *Chemical Engineering Science, 53*(15), 2769–2790. doi:10.1016/S0009-2509(98)00095-5

Dadhich, V., & Roy, S. (2010). Availability-based optimization of preventive maintenance schedule: A parametric study. *Chemical Engineering Transactions, 21*, 949–954.

D'Amico, G., Janssen, J., & Manca, R. (2011). Duration dependent semi-Markov models. *Applied Mathematical Sciences, 5*(42), 2097–2108.

Darwiche, A. (2009). *Modeling and reasoning with Bayesian networks.* Cambridge University Press. doi:10.1017/CBO9780511811357

De Castro, H., & Cavalca, K. (2003). Availability optimization with genetic algorithm. *International Journal of Quality & Reliability Management, 20*(7), 847–863. doi:10.1108/02656710310491258

de Jong, P., & Penzer, J. (2004). The ARMA model in state space form. *Statistics & Probability Letters, 40*(1), 119–125. doi:10.1016/j.spl.2004.08.006

de Kleer, J., & Williams, B. C. (1987). Diagnosing multiple faults. *Artificial Intelligence, 32*(1), 97–130. doi:10.1016/0004-3702(87)90063-4

De Rango, F., Veltri, F., & Marano, S. (2011). Channel modeling approach based on the concept of degradation level Discrete-Time Markov chain: UWB system case study. *IEEE Transactions on Wireless Communications, 10*(4), 1098–1107. doi:10.1109/TWC.2011.012411.091590

Deloux, E., Castanier, B., & Bérenguer, C. (2008, September). *Condition-based maintenance approaches for deteriorating system influenced by environmental conditions.* Paper presented at the European Safety and Reliability Conference, Valencia, Spain.

Dempsey, P., & Afjeh, A. (2002). *Integrating oil debris and vibration gear damage detection technologies using fuzzy logic.* NASA Technical Memorandum 2002-211126.

Dempsey, P., Handschuh, R., & Afjeh, A. (2002). *Spiral bevel gear damage detection using decision fusion analysis.* NASA/TM-2002-211814.

Dempster, A. P., Laird, N. M., & Rubin, D. B. (1977). Maximum likelihood from incomplete data via the EM algorithm. *Journal of the Royal Statistical Society. Series B. Methodological, 39*(1), 1–38.

Dempsy, P., & Keller, J. (2008). *Signal detection theory applied to helicopter transmissions diagnostics thresholds.* NASA Technical Memorandum 2008-215262.

Dennis, J. E. Jr, Gay, D. M., & Walsh, R. E. (1981). An adaptive nonlinear least-squares algorithm. [TOMS]. *ACM Transactions on Mathematical Software, 7*(3), 348–368. doi:10.1145/355958.355965

Der Kiureghian, A., & Dakessian, T. (1998). Multiple design points in first and second-order reliability. *Structural Safety, 20*(1), 37–49. doi:10.1016/S0167-4730(97)00026-X

Der Kiureghian, A., Lin, H. Z., & Hwang, S. J. (1987). Second-order reliability approximations. *Journal of Engineering Mechanics, 113*(8), 1208–1225. doi:10.1061/(ASCE)0733-9399(1987)113:8(1208)

Der Kiureghian, A., Zhang, Y., & Li, C. C. (1994). Inverse reliability problem. *Journal of Engineering Mechanics, 120*(5), 1154–1159. doi:10.1061/(ASCE)0733-9399(1994)120:5(1154)

Dhillon, B. S. (1999). *Engineering maintainability.* Houston, TX: Gulf Publishing Company.

Ding, S., Zhang, P., Ding, E., Naik, A., Deng, P., & Gui, W. (2010). On the application of PCA technique to fault diagnosis. *Tsinghua Science and Technology, 15*(2), 138–144. doi:10.1016/S1007-0214(10)70043-2

Ditlevsen, O., & Madsen, H. O. (1996). *Structural reliability methods* (*Vol. 315*). Citeseer.

Dixon, S. J., & Brereton, R. G. (2009). Comparison of performance of five common classifiers represented as boundary methods: Euclidean distance to centroids, linear discriminant analysis, quadratic discriminant analysis, learning vector quantization and support vector machines, as dependent on data structure. *Chemometrics and Intelligent Laboratory Systems, 95*, 1–17. doi:10.1016/j.chemolab.2008.07.010

Domlan, E. A., Ragot, J., & Maquin, D. (2007). *Active mode estimation for switching systems.* 26th American Control Conference, New York City, NY, USA, ACC'2007, 9-13 July.

Dong, M., & He, D. (2007). A segmental hidden semi-Markov model(HSMM)-based diagnostics and prognostics framework and methodology. *Mechanical Systems and Signal Processing, 21*, 2248–2266. doi:10.1016/j.ymssp.2006.10.001

Dong, M., & He, D. (2007). Hidden semi-Markov model-based methodology for multi-sensor equipment health diagnosis and prognosis. *European Journal of Operational Research, 178*(3), 858–878. doi:10.1016/j.ejor.2006.01.041

Dong, M., He, D., Banerjee, P., & Keller, J. (2006). Equipment health diagnosis and prognosis using hidden semi-Markov models. *International Journal of Advanced Manufacturing Technology, 30*(7-8), 738–749. doi:10.1007/s00170-005-0111-0

Dong, M., & Peng, Y. (2011). Equipment PHM using non-stationary segmental hidden semi-Markov model. *Robotics and Computer-integrated Manufacturing, 27*(3), 581–590. doi:10.1016/j.rcim.2010.10.005

Doucet, A., de Freitas, J. F. G., & Gordon, N. J. (2001). An introduction to sequential Monte Carlo methods. In Doucet, A., de Freitas, J. F. G., & Gordon, N. J. (Eds.), *Sequential Monte Carlo in practice.* New York, NY: Springer-Verlag.

Doucet, A., Godsill, S., & Andrieu, C. (2000). On sequential Monte Carlo sampling methods for Bayesian filtering. *Statistics and Computing, 10*(3), 197–208. doi:10.1023/A:1008935410038

Dubois, D., Prade, H., & Testemale, C. (1988). Weighted fuzzy pattern matching. *Fuzzy Sets and Systems, 28*, 313–331. doi:10.1016/0165-0114(88)90038-3

Duda, R. O., Hart, P. E., & Stork, D. G. (Eds.). (2001). *Pattern classification* (2nd ed.). Wiley-Interscience.

Dunbar, C. A., & Hickey, A. J. (2000). Evaluation of probability density functions to approximate particle size distributions of representative pharmaceutical aerosols. *Journal of Aerosol Science, 31*(7), 813–831. doi:10.1016/S0021-8502(99)00557-1

Du, X., Sudjianto, A., & Chen, W. (2004). An integrated framework for optimization under uncertainty using inverse reliability strategy. *Journal of Mechanical Design, 126*, 562. doi:10.1115/1.1759358

Dzakowic, J. E., & Valentine, G. S. (2007, April). *Advanced techniques for the verification and validation of prognostics and health management capabilities.* Paper presented at Machinery Failure Prevention Technology (MFPT 60). Virginia Beach, Virginia, USA.

Ebert-Uphoff, I. (2009). *A probability-based approach to soft discretization for Bayesian networks.* Georgia Institute of Technology.

Efron, B. (1979). Bootstrap methods: Another look at the Jacknife. *Annals of Statistics, 7*, 1–26. doi:10.1214/aos/1176344552

Efron, B., & Tibshirani, R. J. (1993). *An introduction to the bootstrap*. New York, NY: Chapman and Hall.

Engel, S., Gilmartin, B., Bongort, K., & Hess, A. (2000). Prognostics, the real issues involved with predicting life remaining. *Proceedings of the 2000 IEEE Aerospace Conference*, (pp. 457-469).

Escobet, A., Nebot, À., & Cellier, F. (2011). Fault diagnosis system based on fuzzy logic: Application to a valve actuator benchmark. *Journal of Intelligent and Fuzzy Systems, 22*(2), 155–171.

Espinoza-Trejo, D., Campos-Delgado, D., & Loredo-Flores, A. (2010). *A novel fault diagnosis scheme for FOC induction motor drives by using variable structure observers. Industrial Electronics* (pp. 2601–2606). ISIE.

Fan, X., & Zuo, M. (2006). Fault diagnosis of machines based on D-S evidence theory. Part 2: Application of the improved D-S evidence theory in gearbox fault diagnosis. *Pattern Recognition Letters, 27*(5), 377–385. doi:10.1016/j.patrec.2005.08.024

Fan, X., & Zuo, M. J. (2008). Machine fault feature extraction based on intrinsic mode functions. *Measurement Science & Technology, 19*(4), 1–13. doi:10.1088/0957-0233/19/4/045105

Farrell, J., & Polycarpou, M. (2005). *Adaptive approximation based control, unifying neural, fuzzy and traditional adaptive approximation approaches*. Wiley-Interscience.

Fatemi, A., & Yang, L. (1998). Cumulative fatigue damage and life prediction theories: A survey of the state of the art for homogeneous materials. *International Journal of Fatigue, 20*, 9–34. doi:10.1016/S0142-1123(97)00081-9

Feldman, M. (1985). Invertigation of the natural vibrations of machine elements using the Hilbert transform. *Soviet Machine Science, 2*(0739-8999), 3.

Feldman, K., Jazouli, T., & Sandborn, P. (2009). A methodology for determining the return on investment associated with prognostics and health management. *IEEE Transactions on Reliability, 58*(2), 305–316. doi:10.1109/TR.2009.2020133

Feldman, M. (1997). Non-linear free vibration identification via the Hilbert transform. *Journal of Sound and Vibration, 208*(3), 475–489. doi:10.1006/jsvi.1997.1182

Feng, W. C. (2003). Making a case for efficient supercomputing. *Queue,* Oct, 54 – 64.

Feng, Z. P., Zuo, M. J., & Chu, F. L. (2010). Application of regularization dimension to gear damage assessment. *Mechanical Systems and Signal Processing, 24*(4), 1081–1098. doi:10.1016/j.ymssp.2009.08.006

Flandrin, P., Rilling, G., & Goncalves, P. (2004). Empirical mode decomposition as a filter bank. *IEEE Signal Processing Letters, 11*(2), 112–114. doi:10.1109/LSP.2003.821662

Frank, P. M., & Ding, X. (1997). Survey of robust residual generation and evaluation methods in observer-based fault detection systems. *Journal of Process Control, 7*(6), 403–424. doi:10.1016/S0959-1524(97)00016-4

Frost, N., March, K., & Pook, L. (1999). *Metal fatigue* (pp. 228–244). Mineola, NY: Dover Publications.

Fukunaga, K. (1990). *Introduction to statistical pattern recognition* (p. 75). London, UK: Academic Press.

Fussel, D., & Balle, P. (1997). *Combining neuro-fuzzy and machine learning for fault diagnosis of a DC motor.* American Control Conference, Albuquerque, NM, USA, 4-6 June.

García, C., Escobet, T., & Quevedo, J. (2010, October). *PHM techniques for condition-based maintenance based on hybrid system model representation.* Paper presented at Annual Conference of the Prognostics and Health Management Society, Portland, US.

Gebraeel, N., Elwany, A., & Pan, J. (2009). Residual life predictions in the absence of prior degradation knowledge. *IEEE Transactions on Reliability, 58*(1), 106–117. doi:10.1109/TR.2008.2011659

Gelman, A. JCarlin, J., Stern, H., & Rubin, D. (2004). *Bayesian data analysis,* 2nd ed. Boca Raton, LA: Chapman and Hall/CRC.

Gelman, A., & Meng, X. L. (1998). Simulating normalizing constants: From importance sampling to bridge sampling to path sampling. *Statistical Science, 13*(2), 163–185. doi:10.1214/ss/1028905934

Geramifard, O., Xu, J. X., Zhou, J. H., & Li, X. (2010, 7-10 Dec. 2010). *Continuous health assessment using a single hidden Markov model.* Paper presented at the Control Automation Robotics & Vision (ICARCV), 2010 11th International Conference on.

Gertler, J. J. (1988). *Survey of model-based failure detection and isolation in complex plants* (pp. 3–11). IEEE Control Systems. doi:10.1109/37.9163

Gertler, J. J. (1998). *Fault detection and diagnosis in engineering systems.* New York, NY: Marcel Dekker.

Ghasemi, A., Yacout, S., & Ouali, M.-S. (2010). Parameter estimation methods for condition-based maintenance with indirect observations. *IEEE Transactions on Reliability, 59*(2), 426–439. doi:10.1109/TR.2010.2048736

Ginart, A., Barlas, I., Dorrity, J. L., Kalgren, P., & Roemer, M. J. (2006). Self-healing from a PHM perspective. In *Proceedings of the IEEE Systems Readiness Technology Conference.* New York, NY: IEEE.

Goebe, K. L., Saha, B., & Saxena, A. (2008). A comparison of three data-driven techniques for prognostics. *62nd Meeting of the Society for Machinery Failure Prevention Technology (MFPT)* (pp. 119-131). VA, USA.

Goebel, K., & Bonissone, P. (2005). Prognostic information fusion for constant load systems. In *Proceedings of the 7th Annual Conference on Information Fusion,* Vol. 2 (pp. 1247-1255).

Gola, G., & Nystad, B. H. (2011, May-June). *Comparison of time- and state-space non-stationary Gamma processes for estimating the remaining useful life of choke valves undergoing erosion.* Paper presented at COMADEM Conference, Stavanger, Norway.

Goldman, P., & Muszynska, A. (1999). Application of full spectrum to rotating machinery diagnostics. *Orbit,* First Quarter, 17-21.

Gooijer, J. G. D., & Hyndman, R. J. (2006). 25 years of time series forecasting. *International Journal of Forecasting, 22,* 443–473. doi:10.1016/j.ijforecast.2006.01.001

Grall, A., Bérenguer, C., & Chu, C. (1998). Optimal dynamic inspection/replacement planning in condition-based maintenance. In S. Lydersen, G. Hansen, & H. Sandtorv (Eds.), *Proceedings of the European Safety and Reliability Conference ESREL'98* (pp. 381-388). Rotterdam, The Netherlands: Balkema Publishers.

Grall, A., Bérenguer, C., & Dieulle, L. (2002). A condition-based maintenance policy for stochastically deteriorating systems. *Reliability Engineering & System Safety, 76*(2), 167–180. doi:10.1016/S0951-8320(01)00148-X

Graves, T., Hamada, M., Klamann, R., Koehler, A., & Martz, H. (2008). Using simultaneous higher-level and partial lower-level data in reliability assessments. *Reliability Engineering & System Safety, 93*(8), 1273–1279. doi:10.1016/j.ress.2007.07.002

Greco, S., Inuiguchi, M., & Slowiński, R. (2002). In Alpigini, J., Peters, J., Skowron, A., & Zhong, N. (Eds.), *Dominance-based rough set approach using possibility and necessity measures rough sets and current trends in computing* (pp. 84–85). Berlin, Germany: Springer.

Gregory, P. C. (2005). *Bayesian logical data analysis for the physical sciences: A comparative approach with Mathematica support.* Cambridge, UK: Cambridge Univ Press. doi:10.1017/CBO9780511791277

Groer, P. G. (2000). Analysis of time-to-failure with a Weibull model. In *Proceedings of the Maintenance and Reliability Conference MARCON,* Knoxville, TN, USA.

Guan, X., Giffin, A., Jha, R., & Liu, Y. (2011). (in press). Maximum relative entropy-based probabilistic inference in fatigue crack damage prognostics. *Probabilistic Engineering Mechanics.* doi:doi:10.1016/j.probengmech.2011.11.006

Guan, X., Jha, R., & Liu, Y. (2011). Model selection, updating, and averaging for probabilistic fatigue damage prognosis. *Structural Safety, 33*(3), 242–249. doi:10.1016/j.strusafe.2011.03.006

Gucik-Derigny, D., Outbib, R., & Ouladsine, M. (2009). Estimation of damage behaviour for model-based prognostic. *7th IFAC Symposium on Fault Detection and Safety of Technical Processes,* (pp. 1444-1449). Barcelona.

Guenab, F., Weber, P., Theilliol, D., & Zhang, Y. (2011). Design of a fault tolerant control system incorporating reliability analysis under dynamic behaviour constraints. *International Journal of Systems Science, 42*(1), 219–223. doi:10.1080/00207720903513319

Gulich, J. F. (2010). *Centrifugal pumps*. New York, NY: Spinger. doi:10.1007/978-3-642-12824-0

Guo, L., & Wang, H. (2003). Pseudo-PID tracking control for a class of output PDFs of general non-Gaussian stochastic systems. *Proceedings of the 2003 American Control Conference,* Denver, Colorado, USA, (pp. 362–367).

Guo, L., & Wang, H. (2004). Applying constrained nonlinear generalized PI strategy to PDF tracking control through square root b-spline models. *International Journal of Control, 77*(17), 1481–1492. doi:10.1080/00207170412331326972

Guo, L., & Wang, H. (2005). Fault detection and diagnosis for general stochastic systems using b-spline expansions and nonlinear filters. *IEEE Transactions on Circuits and Systems, 52*(8), 1644–1652. doi:10.1109/TCSI.2005.851686

Gupta, S., Ray, A., & Keller, E. (2007). Online fatigue damage monitoring by ultrasonic measurements: A symbolic dynamics approach. *International Journal of Fatigue, 29*(6), 1100–1114. doi:10.1016/j.ijfatigue.2006.09.011

Gurley, K. (1999). Application of wavelet transform in earthquake, wind and ocean engineering. *Engineering Structures, 21*(2), 149–167.

Guyon, I. (2008). Practical feature selection: From correlation to causality. In *Mining massive data sets for security*. IOS Press.

Haase, M., & Widjajakusuma, J. (2003). Damage identification based on ridges and maxima lines of the wavelet transform. *International Journal of Engineering Science, 41*(13-14), 1423–1443. doi:10.1016/S0020-7225(03)00026-0

Haddad, G., Sandborn, P. A., & Pecht, M. G. (2012a). An options approach for decision support of systems with prognostic capabilities. Submitted to *IEEE Transactions on Reliability*.

Haddad, G., Sandborn, P. A., Eklund, N. H., & Pecht, M. G. (2012b). Using maintenance options to measure the benefits of prognostics for wind farms. Submitted to *Wind Energy*.

Haddad, G., Sandborn, P. A., Jazouli, T., Pecht, M. G., Foucher, B., & Rouet, V. (2011). Guaranteeing high availability of wind turbines. *Proceedings of the European Safety and Reliability Conference (ESREL)*, Troyes, France.

Hakem, A., Pekpe, K. M., & Cocquempot, V. (2011). *Sensor fault diagnosis for bilinear systems using data-based residuals*. IEEE 50th International Conference on Decision and Control and European Control Conference (CDC-ECC), Orlando, FL, USA, December 12-15.

Haldar, A., & Mahadevan, S. (2000). *Probability, reliability, and statistical methods in engineering design*. John Wiley & Sons, Inc.

Han, J., & Kamber, M. (2001). *Data mining: Concepts and techniques*. London, UK: Academic Press.

Hansen, L. K., & Salamon, P. (1990). Neural network ensembles. *IEEE Transactions on Pattern Analysis and Machine Intelligence, 12*(10), 993–1001. doi:10.1109/34.58871

Harkat, M. F., Mourot, G., & Ragot, J. (2003). *Nonlinear PCA combining principal curves and RBF-networks for process monitoring*. 42th Conference of Decision and Control IEEE CDC, Lewis, FL, USA, 9-12 December.

Hasofer, A. M., & Lind, N. C. (1974). Exact and invariant second-moment code format. *Journal of the Engineering Mechanics Division, 100*(1), 111–121.

Hastie, T., & Tibshirani, R. (1990). *Classification by pairwise coupling, Technical report*. Stanford University and University of Toronto.

Hastie, T., Tibshirani, R., & Friedman, J. (2009). *The elements of statistical learning: Data mining, inference, and prediction* (2nd ed.). Springer. doi:10.1007/BF02985802

Hastings, W. K. (1970). Monte Carlo sampling methods using Markov chains and their applications. *Biometrika, 57*(1), 97. doi:10.1093/biomet/57.1.97

Haugen, K., Kvernvold, O., Ronold, A., & Sandberg, R. (1995). Sand erosion of wear resistant materials: Erosion in choke valves. *Wear, 186-187*, 179–188. doi:10.1016/0043-1648(95)07158-X

Haupt, R. L., & Haupt, S. E. (2004). *Practical genetic algorithms*. Hoboken, NJ: John Wiley & Sons Ltd.

He, J., Lu, Z., & Liu, Y. (2009). *A new method for concurrent Multi-scale fatigue damage prognosis*. Paper presented at the The 7th International Workshop Strucutural Health Monitoring, Stanford University, CA.

He, J., Lu, Z., & Liu, Y. (2011). A new method for concurrent dynamic analysis and fatigue damage prognosis of bridges. *Journal of Bridge Engineering, 1*, 196.

Heng, A., Tan, A., Mathew, J., Montgomery, N., Banjevic, D., & Jardine, A. (2009). Intelligent condition-based prediction of machinery reliability. *Mechanical Systems and Signal Processing, 23*(5), 1600–1614. doi:10.1016/j.ymssp.2008.12.006

Heng, A., Zhang, S., Tan, A. C. C., & Mathew, J. (2009). Rotating machinery prognostics: state of the art, challenges and opportunities. *Mechanical Systems and Signal Processing, 23*, 724–739. doi:10.1016/j.ymssp.2008.06.009

Hess, P. E., Bruchman, D., Assakkaf, I., & Ayyub, B. M. (2002). Uncertainties in material strength, geometric, and load variables. *ASNE Naval Engineers Journal, 114*(2), 139–165. doi:10.1111/j.1559-3584.2002.tb00128.x

Hines, J. W., & Garvey, D. R. (2007, October). *Data based fault detection, diagnosis and prognosis of oil drill steering systems*. Paper presented at the Maintenance and Reliability Conference. Knoxville, Tennessee, USA.

Hines, J. W., Garvey, J., Preston, J., & Usynin, A. (2008, January). *Empirical methods for process and equipment prognostics*. Paper presented at 53rd Annual Reliability and Maintainability Symposium. Las Vegas, Nevada, USA.

Hines, J. W., Garvey, D., Seibert, R., & Usynin, A. (2008). Technical review of on-line monitoring techniques for performance assessment: *Vol. 2. Theoretical issues (NUREG/CR-6895)*. Rockville, MD: Nuclear Regulatory Commission.

Hines, J. W., & Usynin, A. (2008). Current computational trends in equipment prognostics. *International Journal of Computational Intelligence Systems, 1*(1), 94–102. doi:10.2991/ijcis.2008.1.1.7

Hong, H. P. (1997). Reliability analysis with nondestructive inspection. *Structural Safety, 19*(4), 383–395. doi:10.1016/S0167-4730(97)00018-0

Hongzhi, T., Jianmin, Zh., Xisheng, J., Yunxian, J., Xinghui, Zh., & Liying, C. (2011). Experimental study on gearbox prognosis using total life vibration analysis. *Prognostics and System Health Management Conference* (PHM-Shenzhen), (pp. 1-6).

Hontelez, J. A. M., Burger, H. H., & Wijnmalen, D. J. D. (1996). Optimum condition-based maintenance policies for deteriorating systems with partial information. *Reliability Engineering & System Safety, 51*, 267–274. doi:10.1016/0951-8320(95)00087-9

Horst, R., & Hoang, T. (1996). *Global optimization: Deterministic approaches*. New York, NY: Springer.

Hovda, K., & Andrews, J. S. (2007, May). *Using Cv models to detect choke erosion - A case study on choke erosion at Statfjord C-39*. Paper presented at SPE Applied Technology Workshop on Sound Control, Phuket, Thailand.

Hovda, K., & Lejon, K. (2010, March). *Effective sand erosion management in chokes and pipelines - Case studies from Statoil*. Paper presented at 4th European Sand management Forum, Aberdeen, UK.

Huang, N. E., Shen, Z., Long, S. R., Wu, M. C., Shih, H. H., Zheng, Q., et al. (1998). The empirical mode decomposition and the Hilbert spectrum for nonlinear and non-stationary time series analysis. *Proceedings of the Royal Society of London. Series A: Mathematical, Physical and Engineering Sciences, 454*(1971), 903-995.

Huang, N. E., Wu, M.-L. C., Long, S. R., Shen, S. S. P., Qu, W., Gloersen, P., & Fan, K. L. (2003). A confidence limit for the empirical mode decomposition and Hilbert spectral analysis. *Proceedings of the Royal Society of London. Series A: Mathematical, Physical and Engineering Sciences, 459*(2037), 2317-2345.

Huang, N. E., & Shen, S. S. (2005). *Hilbert-Huang transform and its applications*. World Scientific Pub Co Inc.

Huang, N. E., Shen, Z., & Long, S. R. (1999). A new view of nonlinear water waves: The Hilbert Spectrum 1. *Annual Review of Fluid Mechanics*, *31*(1), 417–457. doi:10.1146/annurev.fluid.31.1.417

Hu, Q., Yu, D., & Guo, M. (2010). Fuzzy preference based rough sets. *Information Science*, *180*(10), 2003–2022. doi:10.1016/j.ins.2010.01.015

Hurson, A. R., Pakzad, S., & Lin, B. (1994). Automated knowledge acquisition in a neural network based decision support system for incomplete database systems. *Microcomputers in Civil Engineering*, *9*(2), 129–143. doi:10.1111/j.1467-8667.1994.tb00368.x

Hussey, A., Lu, B., & Bickford, R. (2009, April). *Performance enhancement of on-line monitoring through plant data classification*. Paper presented at the 6th American Nuclear Society International Topical Meeting on Nuclear Plant Instrumentation, Controls and Human Machine Interface Technology, Knoxville, TN, US.

Hyman, W. A. (2009). Performance-based contracting for maintenance. *NCHRP Synthesis 389*.

IASC-ASCE SHM Task Group. (1999). ASCE structural health monitoring (SHM) benchmark problem. Retrieved from http://bc029049.cityu.edu.hk/asce.shm/phase1/ascebenchmark.asp

IEC 61508. (1998). *Functional safety of electrical/electronic/programmable electronic safety-related systems*. Geneva, Switzerland: International Electrotechnical Commission.

IEC 61511. (2003). *Functional safety - safety instrumented systems for the process industry*. Geneva, Switzerland: International Electrotechnical Commission.

Investopedia (2011). Real option. *Investopedica*. Retrieved September 1, 2011, from http://www.investopedia.com/terms/r/realoption.asp#axzz1WiKKDwgK

Isermann, R., & Balle, P. (1996). Trends in the application of model based fault detection and diagnosis of technical process. *Proceedings of the IFAC World Congress*, (pp. 1–12).

Isermann, R. (1993). Fault diagnosis of machines via parameter estimation and knowledge processing, tutorial paper. *Automatica*, *29*(4), 815–836. doi:10.1016/0005-1098(93)90088-B

Iserman, R. (2006). *Fault-diagnosis systems: An introduction from fault detection to fault tolerance*. Berlin, Germany: Springer-Verlag.

Ishikawa, A., Amagasa, M., Tomizawa, G., Tatsuta, R., & Mieno, H. (1993). The max-min Delphi method and fuzzy Delphi method via fuzzy integration. *Fuzzy Sets and Systems*, *55*(3), 241–253. doi:10.1016/0165-0114(93)90251-C

ISO 10816-3:2009. (2009). *Mechanical vibration – Evaluation of machine vibration by measurement on non-rotating parts*.

ISO 10825. (2007) *Gears -- Wear and damage to gear teeth – Terminology*.

ISO 13374-1. (2003). *Condition monitoring and diagnostics of machines – data processing, communication and presentation – Part 1: General guidelines*. ISO, Switzerland.

ISO 13374-2. (2007). *Condition monitoring and diagnostics of machines – data processing, communication and presentation – Part 2: Data processing*. ISO, Switzerland.

ISO 13379. (2003). *Condition monitoring and diagnostics of machines — General guidelines on data interpretation and diagnostic techniques*.

ISO 13381-1. (2004). *Condition monitoring and diagnostics of machines – prognostics – Part 1: General guidelines*. ISO, Switzerland.

Jack, L. B., & Nandi, A. K. (2002). Fault detection using support vector machines and artificial neural network, augmented by genetic algorithms. *Mechanical Systems and Signal Processing*, *16*, 373–390. doi:10.1006/mssp.2001.1454

Janssen, J., Blasi, A., & Manca, R. (2002). Discrete time homogeneous and non-homogeneous semi-Markov reliability models. *Proceedings of the Third International Conference on Mathematical Methods in Reliability*, Trondheim, Norway.

Janssen, J., & Manca, R. (2001). Numerical solution of non homogeneous semi Markov processes in transient case. *Methodology and Computing in Applied Probability*, *3*(3), 271–293. doi:10.1023/A:1013719007075

Janssen, M., Zuidema, J., & Wanhill, R. (2004). *Fracture mechanics* (2nd ed.). London, UK: Taylor & Francis Group.

Jardine, A. K. S., Lin, D., & Banjevic, D. (2006). A review on machinery diagnostics and prognostics implementing condition-based maintenance. *Mechanical Systems and Signal Processing*, *20*(7), 1483–1510. doi:10.1016/j.ymssp.2005.09.012

Jarrell, D. B., Sisk, D. R., & Bond, L. J. (2004). Prognostics and condition-based maintenance: A new approach to precursive metrics. *Nuclear Technology*, *145*, 275–286.

Jata, K. V., & Parthasarathy, T. A. (2005). Physics of failure. In *Proceedings of the 1st International Forum on Integrated System Health Engineering and Management in Aerospace*, Napa Valley, CA, USA.

Jay, L. (1996). Measurement of machine performance degradation using a neural network model. *Computers in Industry*, *30*(3), 193–209. doi:10.1016/0166-3615(96)00013-9

Jazouli, T., & Sandborn, P. (2011). Using PHM to meet availability-based contracting requirements. *Proceedings of the IEEE International Conference on Prognostics and Health Management*, Denver, CO.

Jennions, I. K. (2011). *IVHM Centre.* Cranfield University, Retrieved September 28, 2011, from http://www.cranfield.ac.uk/ivhm

Jiang, B., & Chowdhury, F. N. (2005). Fault estimation and accommodation for linear MIMO discrete-time systems. *IEEE Transactions on Control Systems Technology*, *13*(3), 493–499. doi:10.1109/TCST.2004.839569

Jiang, D. X., & Liu, C. (2011). Machine condition classification using deterioration feature extraction and anomaly determination. *IEEE Transactions on Reliability*, *60*(1), 41–48. doi:10.1109/TR.2011.2104433

Jie, S., Hong, G. S., Rahman, M., & Wong, Y. S. (2002). *Feature extraction and selection in tool condition monitoring system.* Paper presented at the 15th Australian Joint Conference on Artificial Intelligence: Advances in Artificial Intelligence.

Jin, X., Li, L., & Ni, J. (2009). Options model for joint production and preventive maintenance system. *International Journal of Production Economics*, *119*(2), 347–353. doi:10.1016/j.ijpe.2009.03.005

Joentgen, A., Mikenina, L., Weber, R., & Zimmermann, H. J. (1999). Dynamic fuzzy data analysis based on similarity between functions. *Fuzzy Sets and Systems*, *105*(1), 81–90. doi:10.1016/S0165-0114(98)00337-6

Jolliffe, I. T. (2002). *Principal component analysis.* New York, NY: Springer-Verlag.

Jones, N., & Li, Y. (2000). A review of condition monitoring and fault diagnosis for diesel engines. *Tribotest*, *6*(3), 267–291. doi:10.1002/tt.3020060305

Jun-Hong, Z., Chee Khiang, P., Lewis, F. L., & Zhao-Wei, Z. (2009). Intelligent diagnosis and prognosis of tool wear using dominant feature identification. *IEEE Transactions on Industrial Informatics*, *5*(4), 454–464. doi:10.1109/TII.2009.2023318

Kadry, S., & Younes, R. (2004). Etude probabiliste d'un systeme mecanique a parametres incertains par une technique basee sur la methode de transformation. *Proceeding of CanCam*, Canada.

Kadry, S., & Samili, K. (2007). One dimensional transformation method in reliability analysis. *Journal of Basic and Applied Sciences*, *3*(2).

Kalgren, P., Byington, C., & Watson, M. (2006). *Defining PHM, a lexical evolution of maintenance and logistics* (pp. 353–358). Anaheim, California: IEEE AUTOTESTCON. doi:10.1109/AUTEST.2006.283685

Kalos, M. H., & Whitlock, P. A. (2008). *Monte carlo methods.* Wiley-VCH. doi:10.1002/9783527626212

Kanninen, M. F., & Popelar, C. H. (1985). *Advanced fracture mechanics.* New York, NY: Oxford University Press.

Karny, M., Nagy, I., & Novovicova, J. (2002). Mixed-data multi-modelling for fault detection and isolation. *International Journal of Adaptive Control and Signal Processing*, *16*, 61–83. doi:10.1002/acs.672

Kassimali, A. (1999). *Matrix analysis of structures.* Pacific Grove, CA: Brooks, Cole Publishing Company.

Khalid, Y. A., & Sapuan, S. M. (2007). Wear analysis of centrifugal slurry pump impellers. *Industrial Lubrication and Tribology*, *59*(1), 18–28. doi:10.1108/00368790710723106

Kiddy, J. S. (2003, June). *Remaining useful life prediction based on known usage data*. Paper presented at the SPIE Conference, Santa Fe, NM.

Kim, M. J., Makis, V., & Jiang, R. (2010). Parameter estimation in a condition-based maintenance model. *Statistics & Probability Letters*, *80*(21-22), 1633–1639. doi:10.1016/j.spl.2010.07.002

Kiureghiana, A., Ditlevsenb, D., & Song, J. (2007). Availability, reliability and downtime of systems with repairable components. *Reliability Engineering & System Safety*, *92*, 231–242. doi:10.1016/j.ress.2005.12.003

Klein, J., & Moeschberger, M. L. (1997). *Survival analysis: Techniques for censored and truncated data*. New York, NY: Springer.

Kliger, S., Yemini, S., Yemini, Y., Oshie, D., & Stolfo, S. (1995). A coding approach to event correlation. In *Proceedings of the 4th International Symposium on Intelligent Network Management* (pp. 266 – 277). London, UK: Chapman and Hall.

Kodukula, P., & Papudesu, C. (2006). *Project valuation using real options: A practitioner's guide*. Fort Lauderdale, FL: J. Ross Publishing.

Kohonen, T. (1986). *Learning vector quantization. Technical Report*. Otaniemi, Finland: Helsinki Univ. of Tech.

Kohonen, T. (2001). *Self-organizing maps*. Berlin, Germany: Springer-Verlag.

Kohonen, T., Hynninen, J., Kangas, J., Laaksonen, J., & Torkkola, K. (1995). *LVQ Pak: The learning vector quantization program package*. Helsinki, Finland: Helsinki Univ. of Tech.

Koide, Y., Kaito, K., & Abe, M. (2001). Life cycle cost analysis of bridges where the real options are considered. In Das, P. C., Frangopol, D. M., & Nowak, A. S. (Eds.), *Current and future trends in bridge design, construction and maintenance* (Vol. 2, pp. 387–394). London, UK: Thomas Telford Publishing.

Kopnov, V. A. (1999). Optimal degradation process control by two-level policies. *Reliability Engineering & System Safety*, *66*, 1–11. doi:10.1016/S0951-8320(99)00006-X

Kozin, F., & Bogdanoff, J. L. (1989). Probabilistic models of fatigue crack growth: Results and speculations. *Nuclear Engineering and Design*, *115*, 143–171. doi:10.1016/0029-5493(89)90267-7

Kraskov, A., Stogbauer, H., & Grassberger, P. (2004). Estimating mutual information. *Physical Review E: Statistical, Nonlinear, and Soft Matter Physics*, *69*(6), 1–16. doi:10.1103/PhysRevE.69.066138

Kressel, U. (1999). *Pairwise classification and support vector machines*. Cambridge, MA: MIT Press.

Krysander, M., Åslund, J., & Nyberg, M. (2008). An efficient algorithm for finding minimal overconstrained subsystems for model-based diagnosis systems. *IEEE Transactions on Man and Cybernetics. Part A*, *38*(1), 197–206.

Kuipers, B. (1994). *Qualitative reasoning – Modelling and simulation with incomplete knowledge*. Cambridge, MA: MIT Press. doi:10.1016/0005-1098(89)90099-X

Lam, C., & Yeh, R. (1994). Optimal maintenance policies for deteriorating systems under various maintenance strategies. *IEEE Transactions on Reliability*, *43*, 423–430. doi:10.1109/24.326439

Lam, H. (2004). Phase I IASC-ASCE structural health monitoring benchmark problem using simulated data. *Journal of Engineering Mechanics*, *130*(3).

Lang, M., & Marci, G. (1999). Influence of single and multiple overloads on fatigue crack propagation. *Fatigue of Engineering Materials and Structures*, *22*, 257–271. doi:10.1046/j.1460-2695.1999.00165.x

Lanktree, C., & Briere, J. (1991). *Early data on the trauma symptom checklist for children (TSC-C)*. Paper presented at the meeting of the American Professional Society on the Abuse of Children, San Diego, CA.

Laser, M. (2000). Recent safety and environmental legislation. *Transactions of the Institution of Chemical Engineers*, *78*(B), 419-422.

Lee, C. W., & Han, Y. S. (1998). The directional Wigner distribution and its applications. *Journal of Sound and Vibration, 216*(4), 585–600. doi:10.1006/jsvi.1998.1715

Lee, I., Choi, K., Du, L., & Gorsich, D. (2008). Inverse analysis method using MPP-based dimension reduction for reliability-based design optimization of nonlinear and multi-dimensional systems. *Computer Methods in Applied Mechanics and Engineering, 198*(1), 14–27. doi:10.1016/j.cma.2008.03.004

Lee, J., Ghaffari, M., & Elmeligy, S. (2011). Self-maintenance and engineering immune systems: Towards smarter machines and manufacturing systems. *Annual Reviews in Control, 35*(1), 111–122. doi:10.1016/j.arcontrol.2011.03.007

Lehmann, A. (2006). Joint modeling of degradation and failure time data. *Journal of Statistical Planning and Inference, 139*(5), 1693–1706. doi:10.1016/j.jspi.2008.05.027

Lei, Y., & Zuo, M. J. (2009a). Fault diagnosis of rotating machinery using an improved HHT based on EEMD and sensitive IMFs. *Measurement Science & Technology, 20*(12), 1–12. doi:10.1088/0957-0233/20/12/125701

Lei, Y., & Zuo, M. J. (2009b). Gear crack level identification based on weighted K nearest neighbor classification algorithm. *Mechanical Systems and Signal Processing, 23*(5), 1535–1547. doi:10.1016/j.ymssp.2009.01.009

Lei, Y., Zuo, M. J., He, Z. J., & Zi, Y. Y. (2010). A multidimensional hybrid intelligent method for gear fault diagnosis. *Expert Systems with Applications, 37*(2), 1419–1430. doi:10.1016/j.eswa.2009.06.060

Lensu, A. (2002). *Computationally intelligent methods for qualitative data analysis*. Jyväskylä, Finland: Jyväskylä University Printing House.

Leontopoulos, C., Dovies, P., & Park, K. (2005). Shaft alignment analysis: Solving the reverse problem. *Proceedings of the Institute of Marine Engineering, Science and Technology Part B. Journal of Marine Design and Operations, 8B*, 3–12.

Liao, F., Wang, J. L., & Yang, G.-H. (2002). Reliable robust flight tracking control: An LMI approach. *IEEE Transactions on Control Systems Technology, 10*(1), 76–89. doi:10.1109/87.974340

Liberatore, M. T., & Stylianou, A. C. (1994). Using knowledge-based systems for strategic market assessment. *Information & Management, 27*(4), 221–232. doi:10.1016/0378-7206(94)90050-7

Li, C. J., & Limmer, J. D. (2000). Model-based condition index for tracking gear wear and fatigue damage. *Wear, 241*(1), 26–32. doi:10.1016/S0043-1648(00)00356-2

Li, C., Wang, X., Tao, Z., Wang, Q., & Du, S. (2011). Extraction of time varying information from noisy signals: An approach based on the empirical mode decomposition. *Mechanical Systems and Signal Processing, 25*(3), 812–820. doi:10.1016/j.ymssp.2010.10.007

Li, H., & Foschi, R. O. (1998). An inverse reliability method and its application. *Structural Safety, 20*(3), 257–270. doi:10.1016/S0167-4730(98)00010-1

Lin, C. C., Liu, P. L., & Yeh, P. L. (2009). Application of empirical mode decomposition in the impact-echo test. *NDT & E International, 42*(7), 589–598. doi:10.1016/j.ndteint.2009.03.003

Lin, D., & Makis, V. (2004). On-line parameter estimation for a failure-prone system subject to condition monitoring. *Journal of Applied Probability, 41*(1), 211–220. doi:10.1239/jap/1077134679

Lindely, D. V., & Smith, A. F. (1972). Bayes estimates for linear models. *Journal of the Royal Statistical Society. Series B. Methodological, 34*(1), 1–41.

Link, M., & Weiland, M. (2009). Damage identification by multi-model updating in the modal and in the time domain. *Mechanical Systems and Signal Processing, 23*(6), 1734–1746. doi:10.1016/j.ymssp.2008.11.009

Lisnianski, A., & Levitin, G. (2003). *Multi-state system reliability: Assessment, optimization and applications*. Singapore: World Scientific.

Liu, J., & Yan, X. (2010). Study of diagnosis system framework using remote knowledge service. *Proceedings of the 2010 Prognostics and System Health Management Conference*, Macau, China, January 12, 2010 - January 14.

Liu, J. S. (1996). Metropolized independent sampling with comparisons to rejection sampling and importance sampling. *Statistics and Computing, 6*(2), 113–119. doi:10.1007/BF00162521

Liu, J., Djurdjanovic, D., Marko, K. A., & Ni, J. (2009). A divide and conquer approach to anomaly detection, localization and diagnosis. *Mechanical Systems and Signal Processing*, *23*(8), 2488–2499. doi:10.1016/j.ymssp.2009.05.016

Liu, J., Zuo, B., Zeng, X., Vroman, P., & Rabensolo, B. (2010). Nonwoven uniformity identification using wavelet texture analysis and LVQ neural network. *Expert Systems with Applications*, *37*, 2241–2246. doi:10.1016/j.eswa.2009.07.049

Li, W., Gu, F., Ball, A., Leung, A., & Phipps, C. (2001). A study of the noise from diesel engines using the independent component analysis. *Mechanical Systems and Signal Processing*, *15*(6), 1165–1184. doi:10.1006/mssp.2000.1366

Li, Z., & Yan, X. (2011a). Independent component analysis and manifold learning with applications to fault diagnosis of VSC-HVDC systems. *Hsi-An Chiao Tung Ta Hsueh*, *45*(1), 44–48.

Li, Z., Yan, X., Wu, J., & Peng, Z. (in press). Study on the fusion of multi-dimensional sensors for health monitoring of rolling bearings. [in press]. *Sensor Letters*.

Li, Z., Yan, X., Yuan, C., Peng, Z., & Li, L. (2011b). Virtual prototype and experimental research gear multi-fault diagnosis using wavelet-autoregressive model and principal component analysis method. *Mechanical Systems and Signal Processing*, *25*(7), 2589–2607. doi:10.1016/j.ymssp.2011.02.017

Li, Z., Yan, X., Yuan, C., Zhao, J., & Peng, Z. (2010a). A new method of nonlinear feature extraction for multi-fault diagnosis of rotor systems. *Noise and Vibration Worldwide*, *41*(10), 29–37. doi:10.1260/0957-4565.41.10.29

Li, Z., Yan, X., Yuan, C., Zhao, J., & Peng, Z. (2010b). The fault diagnosis approach for gears using multidimensional features and intelligent classifier. *Noise and Vibration Worldwide*, *41*(10), 76–86. doi:10.1260/0957-4565.41.10.76

Li, Z., Yan, X., Yuan, C., Zhao, J., & Peng, Z. (2011c). Fault detection and diagnosis of the gearbox in marine propulsion system based on bispectrum analysis and artificial neural networks. *Journal of Marine Science and Application*, *10*(1), 17–24. doi:10.1007/s11804-011-1036-7

Long, Y., Shi, T., & Xiong, L. (2010). Excimer laser electrochemical etching n-Si in the KOH solution. *Optics and Lasers in Engineering*, *48*(5), 570–574. doi:10.1016/j.optlaseng.2009.12.001

Looney, D., & Mandic, D. P. (2009). Multiscale image fusion using complex extensions of EMD. *IEEE Transactions on Signal Processing*, *57*(4), 1626–1630. doi:10.1109/TSP.2008.2011836

López Droguett, E., das Chagas Moura, M., Magno Jacinto, C., & Feliciano Silva Jr., M. (2008). A semi-Markov model with Bayesian belief network based human error probability for availability assessment of down hole optical monitoring systems. *Simulation Modelling Practice and Theory*, *16*(10), 1713–1727. doi:10.1016/j.simpat.2008.08.011

Loutridis, S. J. (2006). Instantaneous energy density as a feature for gear fault detection. *Mechanical Systems and Signal Processing*, *20*(5), 1239–1253. doi:10.1016/j.ymssp.2004.12.001

Low, K., & Lim, S. (2004). Propulsion shaft alignment method and analysis for surface crafts. *Advances in Engineering Software*, *35*, 45–58. doi:10.1016/S0965-9978(03)00082-6

Lu, Z., & Liu, Y. (2009). *An incremental crack growth model for multi-scale fatigue analysis.* Paper presented at the 50th AIAA/ASME/ASCE/AHS/ASC Structures, Structural Dynamics, and Materials Conference, Palm Springs, CA.

Lu, C. J., & Meeker, W. (1993). Using degradation measures to estimate a time-to-failure distribution. *Technometrics*, *35*(2), 161–174.

Luk, R. W. P., & Damper, R. I. (2006). Non-parametric linear time-invariant system identification by discrete wavelet transforms. *Digital Signal Processing*, *16*(3), 303–319. doi:10.1016/j.dsp.2005.11.004

Lumme, V. (1996). *Self learning method in the condition monitoring and diagnostics of rotating machines.* FI102857, Helsinki, Finland: Finnish Patent Office

Lumme, V. (2005). *Method in handing of samples.* FI115486. Helsinki, Finland: Finnish Patent Office

Lumme, V. (2011). *Diagnosis of multi-descriptor condition monitoring data.* Paper presented at the International Conference on Prognostics and Heath Management, Denver, CO.

Lumme, V., & Seppä, J. (2001). *Method in monitoring the condition of machines.* EP1292812. Munich, Germany: European Patent Office.

Lunze, J., & Supavatanakul, P. (2002). Diagnosis of discrete event systems described by timed automata. In *Proceeding of IFAC Congress,* Barcelona, Spain.

Luo, J., Pattipati, K. R., Qiao, L., & Chigusa, S. (2005). Integrated model-based and data-driven diagnostic strategies applied to an anti-lock brake system. In *Proceedings IEEE Aerospace Conference* (pp. 3702-2308). Big Sky, MT.

Luo, J., Pattipati, K., Qiao, L., & Chigusa, S. (2008). Model-based prognostic techniques applied to a suspension system. *IEEE Transactions on Systems, Man, and Cybernetics. Part A, Systems and Humans,* 38(5), 1156–1168. doi:10.1109/TSMCA.2008.2001055

Lu, Z., & Liu, Y. (2010). Small time scale fatigue crack growth analysis. *International Journal of Fatigue,* 32(8), 1306–1321. doi:10.1016/j.ijfatigue.2010.01.010

Ly, C., Tmo, K., Byington, C. S., Patrick, R., & Vachtsevanos, G. J. (2009). Fault diagnosis and failure prognosis for engineering systems: A global perspective. In *Proceedings Annual IEEE Conference on Automation Science and Engineering* (pp. 108-115). Bangole, India.

Madsen, H. (1977). *Some experience with the Rackwitz-Fiessler algorithm for the calculation of structural reliability under combined loading,* (pp. 73-98). DIALOG-77. Lyngby, Denmark: Danish Engineering Academy.

Madsen, H., Krenk, S., & Lind, N. (1986). *Methods of structural safety.* Englewood Cliffs, NJ: Prentice-Hall, Inc.

Mahulkar, V., Adams, D. E., & Derriso, M. (2009). Minimization of degradation through prognosis based control for a damaged aircraft actuator. In *Prooceedings of Dynamic Systems and Control Conference* (pp. 669-676). Hollywood, California, USA.

Mallat, S. (2008). *A wavelet tour of signal processing, The sparse way* (3rd ed.). Academic Press.

Marino, R., Tomei, P., & Verrelli, C. (2010). *Induction motor control design.* London, UK: Springer.

Marseguerra, M., Zio, E., & Podofillini, L. (2002). Condition-based maintenance optimization by means of genetic algorithms and Monte Carlo simulation. *Reliability Engineering & System Safety,* 77, 151–165. doi:10.1016/S0951-8320(02)00043-1

Maru, M., Castillo, R., & Padovese, L. (2007). Study of solid contamination in ball bearings through vibration and wear analyses. *Tribology International,* 40(3), 433–440. doi:10.1016/j.triboint.2006.04.007

Mathew, J., & Stecki, J. (1987). Comparison of vibration and direct reading ferrographic techniques in application to high-speed gears operating under steady and varying load conditions. *Lubrication Engineers,* 43(8), 646–653.

Maxwell, H., & Johnson, B. (1997). Vibration and lube oil analysis in an integrated predictive maintenance program. In *Proceedings of the 21st Annual Meeting of the Vibration Institute,* USA.

McFadden, P. (1987). A revised model for the extraction of periodic waveforms by time domain averaging. *Mechanical Systems and Signal Processing,* 1(1), 83–95. doi:10.1016/0888-3270(87)90085-9

McFadden, P., & Smith, J. (1985). A signal processing technique for detecting local defects in a gear from a signal average of the vibration. *Proceedings - Institution of Mechanical Engineers,* 199(4).

Mehra, R., & Peschon, J. (1971). An innovations approach to fault detection and diagnosis in dynamic systems. *Automatica,* 7(5), 637–640. doi:10.1016/0005-1098(71)90028-8

Meland, E. (2011). *Condition monitoring of safety critical valves.* PhD thesis, Department of Marine Engineering, Norwegian University of Science and Technology, Trondheim, Norway.

Melchers, R. E. (1999). *Structural reliability analysis and prediction.* John Wiley & Son Ltd.

Meseguer, J., Puig, V., & Escobet, T. (2010). Fault diagnosis using a timed discrete-event approach based on interval observers: Application to sewer networks. *IEEE Transactions on Systems, Man, and Cybernetics. Part A, Systems and Humans,* 40(5), 900–916. doi:10.1109/TSMCA.2010.2052036

Metso Automation. (2005). *Flow control manual* (4th ed.).

Meyer, R. M., Ramuhalli, P., Bond, L. J., & Cumblidge, S. E. (2011). Developing effective continuous on-line monitoring technologies to manage service degradation of nuclear power plants. In *Proceedings of IEEE Conference on Prognostics and Health Management* (PHM), Denver, CO, USA.

Miller, L. T., & Park, C. S. (2004). Economic analysis in the maintenance, repair, and overhaul industry: an options approach. *The Engineering Economist, 49*, 21–41. doi:10.1080/00137910490432593

Milne, R. (1996). Knowledge based systems for maintenance management. In Rao, B. K. N. (Ed.), *Handbook of condition monitoring* (pp. 377–393). Oxford, UK: Elsevier Advanced Technology.

Mitchell, J., Bond, T., Bever, K., & Manning, N. (1998). Mimosa - Four years later. *Sound and Vibration*, 12–21.

Moghaddass, R., & Zuo, M. J. (2011a). A parameter estimation method for a multi-state deteriorating system with incomplete information. *Proceedings of the 7th International Conference on Mathematical Methods in Reliability: Theory, Methods, and Applications (MMR 2011)* (pp. 34-32), Beijing, China.

Moghaddass, R., & Zuo, M. J. (2011b). A parameter estimation method for condition monitored equipment under multi-state deterioration. Manuscript submitted for publication in *Reliability Engineering and System Safety*.

Montes de Oca, S., & Puig, V. (2010). Fault-tolerant control design using a virtual sensor for LPV systems. In *Proceedings of Conference on Control and Fault Tolerant Systems* (pp. 88-93). Nice, France.

Morcous, G., Lounis, Z., & Mirza, M. S. (2003). Identification of environmental categories for Markovian deterioration models of bridge decks. *Journal of Bridge Engineering, 8*(6), 353–361. doi:10.1061/(ASCE)1084-0702(2003)8:6(353)

More, J. (1978). The Levenberg-Marquardt algorithm: Implementation and theory. *Numerical Analysis*, 105-116.

Moura, M. C., & Droguett, E. L. (2009). Mathematical formulation and numerical treatment based on transition frequency densities and quadrature methods for non-homogeneous semi-Markov processes. *Reliability Engineering & System Safety, 94*(2), 342–349. doi:10.1016/j.ress.2008.03.032

Muller, A., Crespo Marquez, A., & Iung, B. (2008). On the concept of e-maintenance: Review and current research. *Reliability Engineering & System Safety, 93*(8), 1165–1187. doi:10.1016/j.ress.2007.08.006

Muller, A., Suhner, M., & Iung, B. (2008). Formalisation of a new prognosis model for supporting proactive maintenance implementation on industrial system. *Reliability Engineering & System Safety, 93*(2), 234–253. doi:10.1016/j.ress.2006.12.004

Myotyri, E., Pulkkinen, U., & Simola, K. (2006). Application of stochastic filtering for lifetime prediction. *Reliability Engineering & System Safety, 91*, 200–208. doi:10.1016/j.ress.2005.01.002

Naikan, V., & Rao, P. (2005). Maintenance cost benefit analysis of production shops – An availability based approach. *Proceedings of Reliability and Maintainability Symposium*, (pp. 404-409). January 24-27.

Natke, H. G., & Cempel, C. (2001). The symptom observation matrix for monitoring and diagnostics. *Journal of Sound and Vibration, 248*(4), 597–620. doi:10.1006/jsvi.2001.3800

Neapolitan, R. (2003). *Learning Bayesian networks*. Prentice Hall.

Nejjari, F., Pérez, R., Escobet, T., & Travé-Massuyès, L. (2006). Fault diagnosability utilizing quasi-static and structural modeling. *Mathematical and Computer Modelling, 45*, 606–616. doi:10.1016/j.mcm.2006.06.008

Ng, I. C. L., Maull, R., & Yip, N. (2009). Outcome-based contracts as a driver for systems thinking and service-dominant logic in service science: Evidence from the defence industry. *European Management Journal, 27*(6), 377–387. doi:10.1016/j.emj.2009.05.002

Ngkleberg, L., & Sontvedt, T. (1995). Erosion in choke valves-oil and gas industry applications. *Wear, 186-187*(2), 401–412. doi:10.1016/0043-1648(95)07138-5

Nimmo, I. (1995). Adequately address abnormal situation operations. *Chemical Engineering Progress, 91*(9), 36–45.

Nomack, M. (2011, January). *BP's Deepwater Horizon accident investigation report*. Paper presented at NCSE 11th National conference on Science, Policy, and the Environment, Washington, DC.

Nystad, B. H., Gola, G., Hulsund, J. E., & Roverso, D. (2010, October). *Technical condition assessment and remaining useful life estimation of choke valves subject to erosion.* Paper presented at the Annual Conference of the Prognostics and Health Management Society, Portland, Oregon.

Ocak, H., Loparo, K. A., & Discenzo, F. M. (2007). Online tracking of bearing wear using wavelet packet decomposition and probabilistic modeling: A method for bearing prognostics. *Journal of Sound and Vibration, 302*(4-5), 951–961. doi:10.1016/j.jsv.2007.01.001

Ofsthun, S. (2002). Integrated vehicle health management for aerospace platforms. *IEEE Instrumentation and Measurement Magazine*, September.

OLF. (2001). *Application of IEC 61508 and IEC 61511 in the Norwegian petroleum industry.* Retrieved from http://www.itk.ntnu.no/sil/OLF-070-Rev2.pdf

Ong, C.-M. (1998). *Dynamic simulations of electric machinery using MATLAB Simulink.* Prentice Hall.

Opitz, D. W., & Shavlik, J. W. (1996). Actively searching for an effective neural network ensemble. *Connection Science, 8*(3-4), 337–354. doi:10.1080/095400996116802

Orchard, E., & Vachtsevanos, G. (2007). *A particle filtering-based framework for real-time fault diagnosis and failure prognosis in a turbine engine.* 15th Mediterranean Conference on Control & Automation, Athens, Greece.

Orchard, M., Kacprzynski, G., Goebel, K., Saha, B., & Vachtsevanos, G. (2008, October). *Advances in uncertainty representation and management for particle filtering applied to prognostics.* Paper presented at International Conference on Prognostics and Health Management, Denver, USA.

Orchard, M. E., & Vachtsevanos, G. J. (2011). A particle-filtering approach for on-line fault diagnosis and failure prognosis. *Transactions of the Institute of Measurement and Control, 31*(3-4), 221–246. doi:10.1177/0142331208092026

Orchard, M., & Vachtsevanos, G. (2007). A particle filtering approach for on-line failure prognosis in a planetary carrier plate. *International Journal of Fuzzy Logic and Intelligent Systems, 7*(4), 221–227. doi:10.5391/IJFIS.2007.7.4.221

Oswald, G. F., & Schueller, G. I. (1984). Reliability of deteriorating structures. *Engineering Fracture Mechanics, 20*(1), 479–488. doi:10.1016/0013-7944(84)90053-5

Palma, L. B., Coito, F. J., & Silva, R. N. (2002). Adaptive observer based fault diagnosis approach applied to a thermal plant. *Proceedings of the 10th MED. Conference on Control and Automation*, Lisbon, Portugal, (pp. 115-123). 9-12 July.

Pan, Y., Chen, J., & Li, X. (2010). Bearing performance degradation assessment based on lifting wavelet packet decomposition and fuzzy c-means. *Mechanical Systems and Signal Processing, 24*(2), 559–566. doi:10.1016/j.ymssp.2009.07.012

Papadimitriou, C., Beck, J. L., & Katafygiotis, L. (2001). Updating robust reliability using structural test data. *Probabilistic Engineering Mechanics, 16*(2), 103–113. doi:10.1016/S0266-8920(00)00012-6

Papazian, J. M., Anagnostou, E. L., Engel, S. J., Hoitsma, D., Madsen, J., & Silberstein, R. P. (2009). A structural integrity prognosis system. *Engineering Fracture Mechanics, 76*(5), 620–632. doi:10.1016/j.engfracmech.2008.09.007

Papazian, J. M., Nardiello, J., Silberstein, R. P., Welsh, G., Grundy, D., & Craven, C. (2007). Sensors for monitoring early stage fatigue cracking. *International Journal of Fatigue, 29*(9-11), 1668–1680. doi:10.1016/j.ijfatigue.2007.01.023

Papoulis, A. (2002). *Probability, random variables and stochastic processes* (4th ed.). Boston, MA: McGraw-Hill.

Paris, P., & Erdogan, F. (1963). A critical analysis of crack propagation. *Journal of Basic Engineering, 85*, 528–534. doi:10.1115/1.3656900

Patankar, R., Ray, A., & Lakhtakia, A. (1998). A state-space model of fatigue crack growth. *International Journal of Fracture, 90*(3), 235–249. doi:10.1023/A:1007491916925

Patel, T. H., & Darpe, A. K. (2008). Vibration response of a cracked rotor in presence of rotor–stator rub. *Journal of Sound and Vibration, 317*(3-5), 841–865. doi:10.1016/j.jsv.2008.03.032

Patel, T. H., & Darpe, A. K. (2009). Use of full spectrum cascade for rotor rub identification. *Advances in Vibration Engineering, 8*(2), 139–151.

Patel, T. H., & Zuo, M. J. (2010). *The pump loop, its modifications, and experiments conducted during phase II of the slurry pump project* (p. 25). Edmonton, Canada: University of Alberta.

Patton, R. (1994). Robust model-based fault diagnosis: The state of the art. *IFAC Symposium on Fault Detection, Supervision and Safety for Technical Processes SAFEPROCESS '94* (pp. 1-24). Helsinki, Finland, 13-15 June.

Patton, R. J. (1997). Fault-tolerant control: The 1997 situation. In *Proceedings of IFAC Symposium on Fault Detection Supervision and Safety for Technical Processes* (pp. 1033-1054). Hull, UK.

Patton, R., Frank, P., & Clark, R. (1989). *Fault diagnosis in dynamic systems: Theory and application.* Englewood Cliffs, NJ: Prentice Hall.

PCT/NO2008/00293. (2008). *System and method for empirical ensemble-based virtual sensing.*

Pearl, J. (1988). *Probabilistic reasoning in intelligent systems.* San Mateo, CA: Morgan Kaufmann.

Pecht, M. (2008). *Prognostics and health management of electronics.* New Jersey: John Wiley. doi:10.1002/9780470385845

Pedregal, D. J., & Carnero, M. C. (2006). State space models for condition monitoring: A case study. *Reliability Engineering & System Safety, 91*, 171–180. doi:10.1016/j.ress.2004.12.001

Peel, L. (2008, October). *Data driven prognostics using a Kalman filter ensemble of neural network models.* Paper presented at the International Conference on Prognostics and Health Management, Denver, CO.

Pekpe, K. M., Mourot, G., & Ragot, J. (2004). *Subspace method for sensor fault detection and isolation-application to grinding circuit monitoring.* 11th IFAC Symposium on automation in Mining, Mineral and Metal processing, (MMM 2004), Nancy, France, 8-10 septembre.

Peng, Z., & Kessissoglou, N. (2003). An integrated approach to fault diagnosis of machinery using wear debris and vibration analysis. *Wear, 255*(7-12), 1221-1232.

Peng, H. C., Long, F., & Ding, C. (2005). Feature selection based on mutual information: criteria of max-dependency, max-relevance, and min-redundancy. *IEEE Transactions on Pattern Analysis and Machine Intelligence, 27*(8), 1226–1238. doi:10.1109/TPAMI.2005.159

Peng, Y., & Dong, M. (2010). A prognosis method using age-dependent hidden semi-Markov model for equipment health prediction. *Mechanical Systems and Signal Processing, 25*(1), 237–252. doi:10.1016/j.ymssp.2010.04.002

Peng, Y., & Dong, M. (2011). A hybrid approach of HMM and grey model for age-dependent health prediction of engineering assets. *Expert Systems with Applications, 38*(10), 12946–12953. doi:10.1016/j.eswa.2011.04.091

Peng, Y., Zhang, S., & Pan, R. (2010). Bayesian network reasoning with uncertain evidences. *International Journal of Uncertainty. Fuzziness and Knowledge-Based Systems, 18*(5), 539–564. doi:10.1142/S0218488510006696

Peng, Z. K., Tse, P. W., & Chu, F. L. (2005). An improved Hilbert-Huang transform and its application in vibration signal analysis. *Journal of Sound and Vibration, 286*(1-2), 187–205. doi:10.1016/j.jsv.2004.10.005

Peng, Z., Kessissoglou, N., & Cox, M. (2005). A study of the effect of contaminant particles in lubricants using wear debris and vibration condition monitoring techniques. *Wear, 258*(11-12), 1651–1662. doi:10.1016/j.wear.2004.11.020

Penha, R. L., & Hines, J. W. (2002). Hybrid system modeling for process diagnostics. *Proceedings of the Maintenance and Reliability Conference MARCON*, Knoxville, USA.

Petrovic, G., Marinkovic, Z., & Marinkovic, D. (2011). Optimal preventive maintenance model of complex degraded systems: A real life case study. *Journal of Scientific and Industrial Research, 70*(6), 412–420.

Pires, V. F., Martins, J. F., & Pires, A. J. (2010). Eigenvector/eigenvalue analysis of a 3D current referential fault detection and diagnosis of an induction motor. *Energy Conversion and Management, 51*(5), 901–907. doi:10.1016/j.enconman.2009.11.028

Pislaru, C., Freeman, J. M., & Ford, D. G. (2003). Modal parameter identification for CNC machine tools using Wavelet Transform. *International Journal of Machine Tools & Manufacture, 43*(10), 987–993. doi:10.1016/S0890-6955(03)00104-4

Platt, J. C., Cfisfianini, N., & Shawe, T. J. (2000). *Large margin DAGs for multi-class classification.* Cambridge, MA: MIT Press.

Polikar, R. (2006). Ensemble based systems in decision making. *IEEE Circuits and Systems Magazine, 6*(3), 21–45. doi:10.1109/MCAS.2006.1688199

Pommier, S., & Freitas, M. D. (2002). Effect on fatigue crack growth of interactions between overloads. *Fatigue of Engineering Materials and Structures, 25*, 709–722. doi:10.1046/j.1460-2695.2002.00531.x

Poon, C. W., & Chang, C. C. (2007). Identification of nonlinear elastic structures using empirical mode decomposition and nonlinear normal modes. *Smart Structures and Systems, 3*, 14.

Porter, T. (1972). Method of analysis and prediction for variable amplitude fatigue crack growth. *Engineering Fracture Mechanics, 4*(4), 717–736. doi:10.1016/0013-7944(72)90011-2

Powell, M. (1970). A FORTRAN subroutine for solving systems of nonlinear algebraic equations. *Numerical Methods for Nonlinear Algebraic Equations, 115*.

Provan, J. W. (1987). *Probabilistic fracture mechanics and reliability.* Dordrecht, The Netherlands: Martinus Nijhoff.

Puig, V., & Quevedo, J. (2001). Fault-tolerant PID controllers using a passive robust fault diagnosis approach. *Control Engineering Practice, 9*(11), 1221–1234. doi:10.1016/S0967-0661(01)00068-5

Pulido, B., & Alonso, C. (2004). Possible conflicts: A compilation technique for consistency-based diagnosis. *IEEE Transactions on Systems, Man and Cybernetics, Part B: Cybernetics. Special Section on Diagnosis of Complex Systems: Bridging the Methodologies of FDI and DX Communities, 34*(5), 2192–2206.

Pulkkinen, U. (1991). A stochastic model for wear prediction through condition monitoring. In Holmberg, K., & Folkeson, A. (Eds.), *Operational reliability and systematic maintenance* (pp. 223–243). London, UK: Elsevier.

Qiu, H., Lee, J., Lin, J., & Yu, G. (2003). Robust performance degradation assessment methods for enhanced rolling element bearing prognostics. *Advanced Engineering Informatics, 17*(3-4), 127–140. doi:10.1016/j.aei.2004.08.001

Rabiner, L. R. (1989). A tutorial on hidden Markov models and selected applications in speech recognition. *Proceedings of the IEEE, 77*(2), 257–286. doi:10.1109/5.18626

Rackwitz, R. (2001). Reliability analysis-a review and some perspectives. *Structural Safety, 23*(4), 365–395. doi:10.1016/S0167-4730(02)00009-7

Rackwitz, R., & Flessler, B. (1978). Structural reliability under combined random load sequences. *Computers & Structures, 9*(5), 489–494. doi:10.1016/0045-7949(78)90046-9

Randal, R. B. (2011). *Vibration-based condition monitoring.* West Sussex, UK: John Wiley & Sons. doi:10.1002/9780470977668

Rao, B. K. N. (1996). *Handbook of condition monitoring.* Elsevier Advanced Technology.

Rehman, N., & Mandic, D. P. (2010b). Multivariate empirical mode decomposition. *Proceedings of the Royal Society A: Mathematical, Physical and Engineering Science, 466*(2117), 1291-1302.

Rehman, N., & Mandic, D. P. (2010a). Empirical mode decomposition for trivariate signals. *IEEE Transactions on Signal Processing, 58*(3), 1059–1068. doi:10.1109/TSP.2009.2033730

Reineke, D., Murdock, W., Pohl, E., & Rehmert, I. (1999). Improving availability and cost performance for complex systems with preventive maintenance. *Proceedings of Reliability and Maintainability Symposium*, (pp. 383-388).

Reiter, R. (1987). A theory of diagnosis from first principles. *Artificial Intelligence, 32*, 57–95. doi:10.1016/0004-3702(87)90062-2

Renewables, G. L. (2007). *Guidelines for the certification of condition monitoring systems for wind turbines.* Retrieved from http://www.gl-group.com/en/certification/renewables/CertificationGuidelines.php

Rilling, G., Flandrin, P., Goncalves, P., & Lilly, J. M. (2007). Bivariate empirical mode decomposition. *IEEE Signal Processing Letters, 14*(12), 936–939. doi:10.1109/LSP.2007.904710

Rish, I., Brodie, M., & Ma, S. (2002). Accuracy vs. efficiency trade-offs in probabilistic diagnosis. In *Proceedings of the 18th National Conference on Artificial Intelligence,* Edmonton, Canada.

Robinson, M. E., & Crowder, M. T. (2000). Bayesian methods for a growth-curve degradation model with repeated measures. *Lifetime Data Analysis, 6,* 357–374. doi:10.1023/A:1026509432144

Roemer, M. J., Byington, C. S., Kacprzynski, G. J., & Vachtsevanos, G. (2006). An overview of selected prognostic technologies with application to engine health management. *51st ASME Turbo Expo* (pp. 707-715). Barcelona, Spain.

Rosich, A., Sarrate, R., Puig, V., & Escobet, T. (2007). Efficient optimal sensor placement for model-based FDI using an incremental algorithm. In *Proceedings of IEEE Conference Decision and Control* (pp. 2590-2595). New Orleans, USA.

Ross, P. (2008). *The night the sea caught fire: Remembering Piper Alpha.* Scotland on Sunday, June 15, 2008.

Roverso, D. (2009, September). *Empirical ensemble-based virtual sensing – A novel approach to oil-in-water monitoring.* Paper presented at Oil-in-Water Monitoring Workshop, Aberdeen, UK.

Rumelhart, D. E., & McClelland, J. L. (1986). *Parallel distributed processing.* Cambridge, MA: MIT Press.

Russell, S., & Norvig, P. (2009). *Artificial intelligence: A modern approach.* Prentice Hall.

Saha, B., Goebel, K., Poll, S., & Christopherson, J. (2007). An integrated approach to battery health monitoring using Bayesian regression, classification and state estimation. In *Proceedings of IEEE Autotestcon.* New York, NY: IEEE.

Saha, B., Goebel, K., Poll, S., & Christophersen, J. (2009). Prognostics methods for battery health monitoring using a Bayesian framework. *IEEE Transactions on Instrumentation and Measurement, 58*(2), 291–296. doi:10.1109/TIM.2008.2005965

Sait, A. S., & Sharaf-Eldeen, Y. I. (2011). *A review of gearbox condition monitoring based on vibration analysis techniques diagnostics and prognostics,* Vol. 5. Paper presented at the Rotating Machinery, Structural Health Monitoring, Shock and Vibration.

Samanta, P. K., Vesely, W. E., Hsu, F., & Subudly, M. (1991). *Degradation modeling with application to ageing and maintenance effectiveness evaluations. NUREG/CR-5612.* US Nuclear Regulatory Commission.

Sandborn, P. (2012). *Electronic systems cost modeling.* Singapore: World Scientific.

Sandborn, P., & Myers, J. (2008). Designing engineering systems for sustainment. In Misra, K. B. (Ed.), *Handbook of performability engineering* (pp. 81–103). London, UK: Springer. doi:10.1007/978-1-84800-131-2_7

Sandborn, P., & Wilkinson, C. (2007). A maintenance planning and business case development model for the application of prognostics and health management (PHM) to electronic systems. *Microelectronics and Reliability, 47*(12), 1889–1901. doi:10.1016/j.microrel.2007.02.016

Sansom, J., & Thomson, P. (2001). Fitting hidden semi-Markov models to breakpoint rainfall data. *Journal of Applied Probability, 38,* 142–157. doi:10.1239/jap/1085496598

Saranyasoontorn, K., & Manuel, L. (2004). Efficient models for wind turbine extreme loads using inverse reliability. *Journal of Wind Engineering and Industrial Aerodynamics, 92*(10), 789–804. doi:10.1016/j.jweia.2004.04.002

Sarrate, R., Puig, V., Escobet, T., & Rosich, A. (2007). Optimal sensor placement for model-based fault detection and isolation. In *Proceedings of IEEE Conference Decision and Control* (pp. 2584-2598). New Orleans, USA.

Sawyer, S., & Tapia, A. (2005). The sociotechnical nature of mobile computing work: Evidence from a study of policing in the United States. *International Journal of Technology and Human Interaction, 1*(3), 1–14. doi:10.4018/jthi.2005070101

Saxena, A., & Goebel, K. (2008). *C-MAPSS data set.* NASA Ames Prognostics Data Repository. Retrieved from http://ti.arc.nasa.gov/project/prognostic-data-repository

Saxena, A., Celaya, J., Balaban, E., Goebel, K., Saha, B., Saha, S., & Schwabacher, M. (2008, October). *Metrics for evaluating performance of prognostic techniques*. Paper presented at International Conference on Prognostics and Health Management. Denver, USA.

Saxena, A., Goebel, K., Simon, D., & Eklund, N. (2008, October). *Damage propagation modeling for aircraft engine run-to-failure simulation*. Paper presented at International Conference on Prognostics and Health Management, Denver, CO.

Schijve, J., Skorupa, M., Skorupa, A., Machniewicz, T., & Gruszczynsky, P. (2004). Fatigue crack growth in aluminium alloy D16 under constant and variable amplitude loading. *International Journal of Fatigue*, 26(1), 1–15. doi:10.1016/S0142-1123(03)00067-7

Schmidt, B., & Bartlett, F. (2002). Review of resistance factor for steel: Data collection. *Canadian Journal of Civil Engineering*, 29(1), 98–108. doi:10.1139/l01-081

Schneider, J. J. (2006). *Stochastic optimization*. Berlin, Germany: Springer-Verlag.

Schuëller, G., & Pradlwarter, H. (2007). Benchmark study on reliability estimation in higher dimensions of structural systems-An overview. *Structural Safety*, 29(3), 167–182. doi:10.1016/j.strusafe.2006.07.010

Schütz, W. (1989). Standardized stress-time histories—An overview. In Potter, J. M., & Watanabe, R. T. (Eds.), *Development of fatigue loading spectra, ASTM STP 1006* (pp. 3–16). doi:10.1520/STP10346S

Schwabacher, M. (2005, September). *A survey of data driven prognostics*. Paper presented at AIAAA Infotech@ Aerospace Conference, Atlanta, USA.

Schwabacher, M., & Goebel, K. (2007). A survey of artificial intelligence for prognostics. *Working Notes of 2007 AAAI Fall Symposium: AI for Prognostics* (pp. 107-114). Arlington, VA, USA.

Sharma, U., & Misra, K. (1988). Optimal availability design of a maintained system. *Reliability Engineering & System Safety*, 20(2), 147–159. doi:10.1016/0951-8320(88)90094-4

Sheng, C., Xie, H., Zhang, Y., Hui, F., & Wu, J. (in press). Nonlinear blind source separation of multi-sensor signals for marine diesel engine fault diagnosis. [in press]. *Sensor Letters*.

Sheng, C., Yan, X., Wu, T., Zhang, Y., & Wu, J. (in press). Wear properties of marine diesel engine under abnormal operation state using Ferrographic analysis and Spectrum analysis. [in press]. *Mechanika*.

Sheppard, J. W., Butcher, S. G. W., & Ramendra, R. (2007, September). *Electronic systems Bayesian stochastic prognosis: Algorithm development*. Technical Report JHU-NISL-07-002.

Sheppard, J. W., Wilmering, T. J., & Kaufman, M. A. (2008). IEEE standards for prognostics and health management. In *Proceedings of* (pp. 243–248). Anaheim, California: IEEE AUTOTESTCON.

Shibasaki, Y., Araki, T., Nagahata, R., & Ueda, M. (2005). Control of molecular weight distribution in polycondensation polymers 2- poly(ether ke-tone) synthesis. *European Polymer Journal*, 41(10), 2428–2433. doi:10.1016/j.eurpolymj.2005.05.001

Shi, D., El-Farra, N. H., Mhaskar, M., & Li, P., & Christodes, P. D. (2006). Predictive control of particle size distribution in particulate processes. *Chemical Engineering Science*, 61(1), 268–28. doi:10.1016/j.ces.2004.12.059

Sikorska, J. Z., Kelly, P. J., & McGrath, J. (2010, June). *Maximizing the remaining life of cranes*. Paper presented at the ICOMS Asset Management Conference, Adelaide, Australia.

Simani, S., Fantuzzi, C., & Patton, R. J. (2003). *Model-based fault diagnosis in dynamic systems using identification techniques*. New York, NY: Springer-Verlag, Inc.

Simola, K., & Pulkkinen, U. (1998). Models for non-destructive inspection data. *Reliability Engineering & System Safety*, 60, 1–12. doi:10.1016/S0951-8320(97)00087-2

Singpurwalla, N. (1995). Survival in dynamic environments. *Statistical Science*, 10(1), 86–103. doi:10.1214/ss/1177010132

Singpurwalla, N. (2006). *Reliability and risk: A Bayesian perspective*. West Sussex, UK: Wiley & Sons.

Sobezyk, K., & Spencer, B. F. (1992). *Random fatigue: From data to theory*. Boston, MA: Academic Press.

Solmaz, S., Akar, M., & Shorten, R. (2006). Online center of gravity estimation in automotive vehicles using multiple models and switching. In *Proceedings of the 9th International Conference on Control, Automation, Robotics and Vision ICARCV06,* Singapore, Dec. 5-8.

Solmaz, S., Akar, M., & Shorten, R. (2007). *Method for determining the center of gravity for an automotive vehicle*. Irish patent ref: (s2006/0162), European Patent Pending, February.

Solmaz, S., Akar, M., Shorten, R., & Kalkkuhl, J. (2008). Realtime multiple-model estimation of center of gravity position in automotive vehicles. *Vehicle System Dynamics Journal, 46*(9), 763–788. doi:10.1080/00423110701602670

Solo, V. (1990). Stochastic adaptive control and martingale limit theory. *IEEE Transactions on Automatic Control, 35*(1), 66–71. doi:10.1109/9.45146

Sotiris, V., & Pecht, M. (2007, July). *Support vector prognostics analysis of electronic products and systems*. Paper presented at the AAAI Conference on Artificial Intelligence, Vancouver, Canada.

Sotiris, V., Tse, P., & Pecht, M. (2010, June). Anomaly detection through a Bayesian support vector machine. *IEEE Transactions on Reliability, 59*(2), 277–286. doi:10.1109/TR.2010.2048740

Spieler, S., Staudacher, S., & Fiola, R. (2008). Probabilistic engine performance scatter and deterioration modeling. *Journal of Engineering for Gas Turbines and Power, 130,* 1–9. doi:10.1115/1.2800351

Srichander, R., & Walker, K. B. (1993). Stochastic stability analysis for continuous-time fault tolerant control systems. *International Journal of Control, 57,* 433–452. doi:10.1080/00207179308934397

Standards. (2006). *Electric motors and generators MG 1-2006.*

Staroswiecki, M., Cassar, J. P., & Declerk, P. (2000). A structural framework for the design of FDI system in large scale industrial plants. In Patton, R. J., Frank, P. M., & Clark, R. N. (Eds.), *Issues of fault diagnosis for dynamic systems*. Springer.

Steinder, M., & Sethi, A. S. (2002). End-to-end service failure diagnosis using belief networks, *Proceedings of Network Operations and Management Symposium*. IEEE Conference Publications.

Stengel, R. F. (1991). Intelligent failure-tolerant control. *IEEE Control Systems Magazine, 11*(4), 14–23. doi:10.1109/37.88586

Sudjianto, A., Du, X., & Chen, W. (2005).. *SAE Transactions, 114*(6), 267–276.

Sunder, R., & Prakash, R. V. (1996). A study of fatigue crack growth in lugs under spectrum loading. In Mitchell, M. R., & Landgraf, R. W. (Eds.), *Advances in fatigue lifetime predictive techniques* (*Vol. 3,* p. 1292). ASTM STP. doi:10.1520/STP16141S

Sun, X., Yue, H., & Wang, H. (2006). Modelling and control of the flame temperature distribution using probability density function shaping. *Transactions of the Institute of Measurement and Control, 28*(5), 401–428. doi:10.1177/0142331206073124

Tamhane, A. C., & Dunlop, D. D. (1999). *Statistics and data analysis: From elementary to intermediate*. Upper Saddle River, NJ: Prentice Hall.

Tandon, N., & Choudhury, A. (1999). A review of vibration and acoustic measurement methods for the detection of defects in rolling element bearings. *Tribology International, 32*(8), 469–480. doi:10.1016/S0301-679X(99)00077-8

Tang, L., Kacprzynski, G. J., Goebel, K., Saxena, A., Saha, B., & Vachtsevanos, G. (2008). Prognostics-enhanced automated contingency management for advanced autonomous systems. In Proceeding of *International Conference on Prognostics and Health Management*, Denver, CO.

Tan, J.-B., Liu, Y., Wang, L., & Yang, W.-G. (2008). Identification of modal parameters of a system with high damping and closely spaced modes by combining continuous wavelet transform with pattern search. *Mechanical Systems and Signal Processing, 22*(5), 1055–1060. doi:10.1016/j.ymssp.2007.11.017

Tao, Y., Huang, S., & Wei, G. (2004). Research on remote network monitoring and fault diagnosis system of steam turbine sets. *Power Engineering, 24,* 840–844.

Tay, F., & Shen, L. (2003). Fault diagnosis based on rough set theory. *Engineering Applications of Artificial Intelligence*, *16*(1), 39–43. doi:10.1016/S0952-1976(03)00022-8

Thurston, M. G. (2001). An open standard for Web-based condition-based maintenance systems. In *Proceedings of IEEE Systems Readiness Technology Conference* (pp. 401 – 415). Valley Forge, PA, USA.

Tierney, L., & Kadane, J. B. (1986). Accurate approximations for posterior moments and marginal densities. *Journal of the American Statistical Association*, *81*(393), 82–86.

Tran, V., Yang, B., & Tan, A. (2009). Multi-step ahead direct prediction for machine condition prognosis using regression trees and neuro-fuzzy systems. *Expert Systems with Applications*, *36*, 9378–9387. doi:10.1016/j.eswa.2009.01.007

Travé-Massuyès, L., Escobet, T., & Olive, X. (2006). Diagnosability analysis based on component supported analytical redundancy relations. *IEEE Transactions on Systems, Man. Cybernetics: Part A*, *36*(6), 1146–1160. doi:10.1109/TSMCA.2006.878984

Tsang, A. H. C. (1995). Condition-based maintenance: Tools and decision making. *Journal of Quality in Maintenance Engineering*, *1*(3), 3–17. doi:10.1108/13552519510096350

Tse, P. W., Peng, Y. H., & Yam, R. (2001). Wavelet analysis and envelope detection for rolling element bearing fault diagnosis---Their effectiveness and flexibilities. *Journal of Vibration and Acoustics*, *123*(3), 303–310. doi:10.1115/1.1379745

Turhan-Sayan, G. (2005). Real time electromagnetic target classification using a novel feature extraction technique with PCA-based fusion. *IEEE Transactions on Antennas and Propagation*, *53*(2), 766–776. doi:10.1109/TAP.2004.841326

Uckun, S., Goebel, K., & Lucas, P. (2008). Standardizing research methods for prognostics. *PHM 2008, International Conference on Prognostics and Health Management*, (pp. 1-10).

Union of Concerned Scientists. (n.d.). *Davis-Besse outage report*. Retrieved from http://www.ucsusa.org/assets/documents/nuclear_power/davis-besse-ii.pdf

Upadhyaya, B. R., Naghedolfeizi, M., & Raychaudhuri, B. (1994). Residual life estimation of plant components. *P/PM Technology*, June, 22 – 29.

US Department of Energy. (2002, May). *Instrumentation, controls and human-machine interface*. Presented at the IC&HMI Technology Workshop. Gaithersburg, Maryland, PA.

Vachtsevanos, G. (2003). Performance metrics for fault prognosis of complex systems. In *AUTOTESTCON 2003, IEEE Systems Readiness Technology Conference* (pp. 341-345). IEEE Conference Publications.

Vachtsevanos, G., & Wang, P. (2001). Fault prognosis using dynamic wavelet neural networks. In *AUTOTESTCON Proceedings, IEEE Systems Readiness Technology Conference* (pp.857-870). IEEE Conference Publications.

Vachtsevanos, G., Lewis, F. L., Roemer, M., Hess, A., & Wu, B. (2006). *Intelligent fault diagnosis and prognosis for engineering systems* (1st ed.). Hoboken, NJ: John Wiley & Sons. doi:10.1002/9780470117842

Van Dijk, G. M. (1975). *Introduction to a fighter aircraft loading standard for fatigue evaluation*. Amsterdam, The Netherlands: National Aerospace Lab.

van Noortwijk, J. M., & Pandey, M. D. (2003, November). *A stochastic deterioration process for time-dependent reliability analysis*. Paper presented at IFIP WG 7.5 Working Conference on Reliability and Optimization of Structural Systems, Banff, Canada.

Vanier, D. J. (2001). Why industry needs management tools. *Journal of Computing in Civil Engineering*, *15*(1), 35–43. doi:10.1061/(ASCE)0887-3801(2001)15:1(35)

Vapnik, V. N. (1998). *Statistical learning theory*. New York, NY: John Wiley & Sons.

Vapnik, V. N. (1999). An overview of statistical learning theory. *IEEE Transactions on Neural Networks*, *10*(5), 988–999. doi:10.1109/72.788640

Vaurio, J. K. (1997). Reliability characteristics of components and systems with tolerable repair times. *Reliability Engineering & System Safety, 56*(1), 43–52. doi:10.1016/S0951-8320(96)00133-0

Vecer, P., Kreidl, M., & Smid, R. (2005). Condition indicators for gearbox condition monitoring systems. *Acta Polytechnica, 45*(6).

Veillette, R. J. (1995). Reliable linear-quadratic state-feedback control. *Automatica, 31*(1), 137–143. doi:10.1016/0005-1098(94)E0045-J

Veillette, R. J., Medanic, J. V., & Perkins, W. R. (1992). Design of reliable control systems. *IEEE Transactions on Automatic Control, 37*(3), 290–304. doi:10.1109/9.119629

Venkatasubramanian, V., Rengaswamy, R., Yin, K., & Kavuri, S. N. (2003). A review of process fault detection and diagnosis: Part I and III. *Computers & Chemical Engineering, 27*, 293–311, 327–346. doi:10.1016/S0098-1354(02)00160-6

Vento, J., Puig, V., & Sarrate, R. (2010). Fault detection and isolation of hybrid systems using diagnosers that combine discrete and continuous dynamics. In *Proceedings of Conference on Control and Fault Tolerant Systems* (pp. 149-154). Nice, France.

Vidyasagar, M., & Viswanadham, N. (1985). Reliable stabilization using a multi-controller configuration. *Automatica, 21*(5), 599–602. doi:10.1016/0005-1098(85)90008-1

Vlok, P. J., Coetzee, J. L., Banjevic, D., Jardine, A. K. S., & Makis, V. (2002). An application of vibration monitoring in proportional hazards models for optimal component replacement decisions. *The Journal of the Operational Research Society, 53*(2), 193–202. doi:10.1057/palgrave.jors.2601261

Vrouwenvelder, A. (2002). Developments towards full probabilistic design codes. *Structural Safety, 24*(2-4), 417–432. doi:10.1016/S0167-4730(02)00035-8

Wackerly, D., Mendenhall, W., & Scheaffer, R. (1996). *Mathematical statistics with applications*. Belmont, CA: Buxbury Press.

Wald, A. (1945). Sequential tests of statistical hypotheses. *Annals of Mathematical Statistics, 16*(2), 117–186. doi:10.1214/aoms/1177731118

Walford, C., & Roberts, D. (2006). Condition monitoring of wind turbines: Technology overview, seeded-fault testing, and cost benefit analysis. Tech. Rep. 1010419. Palo Alto, CA EPRI.

Wallace, M. S., Dempster, W. M., Scanlon, T., Peters, J., & McCulloch, S. (2004). Prediction of impact erosion in valve geometries. *Wear, 256*, 927–936. doi:10.1016/j.wear.2003.06.004

Wang, A., Wang, H., & Guo, L. (2009). Recent advances on stochastic distribution control: Probability density function control. *Control and Decision Conference, CCDC'09* (pp. xxxv–xli).

Wang, H. (2003). Multivariable output probability density function control for non-Gaussian stochastic systems using simple MLP neural networks. *Proceedings of the IFAC International Conference on Intelligent Control Systems and Signal Processing,* Algarve, Portugal, (pp. 84– 89).

Wang, H., & Afshar, P. (2006). Radial basis function based iterative learning control for stochastic distribution systems. *Proceedings of the IEEE International Symposium on Intelligent Control,* (pp. 100–105).

Wang, H., Afshar, P., & Yue, H. (2006). ILC-based generalised PI control for output PDF of stochastic systems using LMI and RBF neural net-works. *Proceedings of the IEEE Conference on Decision and Control,* (pp. 5048–5053).

Wang, T., Yu, J., Siegel, D., & Lee, J. (2008, October). *A similarity based prognostic approach for remaining useful life estimation of engineered systems*. Paper presented at the International Conference on Prognostics and Health Management. Denver, USA.

Wang, X., Yu, G., & Lee, J. (2002, September). *Wavelet neural network for machining performance assessment and its implication to machinery prognostic*. Paper presented at 5th International Conference on Managing Innovations in Manufacturing (MIM), Milwaukee, Wisconsin, USA.

Wang, A., Afshar, P., & Wang, H. (2008). Complex stochastic systems modelling and control via iterative machine learning. *Neurocomputing, 71*(13-15), 2685–2692. doi:10.1016/j.neucom.2007.06.018

Wang, H. (1999). Robust control of the output probability density functions for multivariable stochastic systems with guaranteed stability. *IEEE Transactions on Automatic Control, 44*(11), 2103–2107. doi:10.1109/9.802925

Wang, H. (2000). *Bounded dynamic stochastic systems: Modelling and control.* London, UK: Springer-Verlag.

Wang, H., & Afshar, P. (2009). ILC-based fixed-structure controller design for output PDF shaping in stochastic systems using LMI techniques. *IEEE Transactions on Automatic Control, 54*(4), 760–773. doi:10.1109/TAC.2009.2014934

Wang, H., & Lin, W. (2000). Applying observer based FDI techniques to detect faults in dynamic and bounded stochastic distributions. *International Journal of Control, 73*, 1424–1436. doi:10.1080/002071700445433

Wang, H., & Wang, C. (1997). Intelligent agents in the nuclear industry. *IEEE Computer Magazine, 30*(11), 28–34.

Wang, W. Q., Goldnaraghi, M. F., & Ismail, F. (2004). Prognosis of Machine health condition using neuro-fuzzy systems. *Mechanical Systems and Signal Processing, 18*, 813–831. doi:10.1016/S0888-3270(03)00079-7

Wang, W.-J., & Wang, C.-C. (1998). A new composite adaptive speed controller for induction motor based on feedback linearization. *IEEE Transactions on Energy Conversion*, 1–6. doi:10.1109/60.658196

Wang, W., Tse, P., & Lee, J. (2007). Remote machine maintenance system through Internet and mobile communication. *International Journal of Advanced Manufacturing Technology, 31*(5), 783–789.

Wang, X., Rabiei, M., Hurtado, J., Modarres, M., & Hoffman, P. (2009). A probabilistic-based airframe integrity management model. *Reliability Engineering & System Safety, 94*(5), 932–941. doi:10.1016/j.ress.2008.10.010

Wasynczuk, O., Sudhoff, S. D., Corzine, K. A., Tichenor, J. L., Krause, P. C., & Hamen, I. G. (1998). A maximum torque per ampere control strategy for induction motor drivers. *IEEE Transactions on Energy Conversion, 13*(2), 163–169. doi:10.1109/60.678980

Wei, P. R. (2010). *Fracture mechanics: Integration of mechanics, materials science & chemistry.* Cambridge, UK: Cambridge University press. doi:10.1017/CBO9780511806865

Wemhoff, E., Chin, H., & Begin, M. (2007). Gearbox diagnostics development using dynamic modeling. *Proceedings of the AHS 63ʳᵈ Annual Forum*, Virginia Beach, VA, 2007.

Weston, J., & Watkins, C. (1999). *Multi-class support vector machines.* Paper presented at the ESANN99, Brussels, Belgium.

Wheatley, G., Wu, X. Z., & Estrin, Y. (1999). Effects of a single tensile overload on fatigue crack growth in a 316L steel. *Fatigue of Engineering Materials and Structures, 22*, 1041–1051. doi:10.1046/j.1460-2695.1999.00225.x

Widodo, A., & Yang, B.-S. (2007). Application of nonlinear feature extraction and support vector machines for fault diagnosis of induction motors. *Expert Systems with Applications, 33*(1), 241–250. doi:10.1016/j.eswa.2006.04.020

Wikipedia. (2008). *Low-discrepancy sequence.* Retrieved from http://en.wikipedia.org/wiki/Low-discrepancy_sequence

Williams, J. H., Davies, A., & Drake, P. R. (1994). *Condition-based maintenance and machine diagnostics.* London, UK: Chapman & Hall.

Worden, K., & Manson, G. (2007). Damage identification using multivariate statistics: Kernel discriminant analysis. *Inverse Problems in Engineering, 8*(1), 25–46. doi:10.1080/174159700088027717

World Nuclear News. (2008). *Cook 1 restart September at the earliest.* Retrieved from http://www.world-nuclear-news.org/C-Cook_1_restart_September_at_the_earliest-0212088.html

Wu, F. Q., & Meng, G. (2006). Compound rub malfunctions feature extraction based on full-spectrum cascade analysis and SVM. *Mechanical Systems and Signal Processing, 20*(8), 2007–2021. doi:10.1016/j.ymssp.2005.10.004

Wu, J., Ng, T., Xie, M., & Huang, H. Z. (2010). Analysis of maintenance policies for finite life-cycle multi-state systems. *Computers & Industrial Engineering, 59*(4), 638–646. doi:10.1016/j.cie.2010.07.013

Wu, T. F., Lin, C. J., & Weng, R. C. (2004). Probability estimates for multi-class classification by pairwise coupling. *Journal of Machine Learning Research, 5,* 975–1005.

Wu, T., Mao, J., Wang, J., Wu, J., & Xie, Y. (2009). A new on-line visual ferrograph. *Tribology Transactions, 52*(5), 623–631. doi:10.1080/10402000902825762

Wu, T., Wang, J., Wu, J., Mao, J., & Xie, Y. (2011). Wear characterization by on-line ferrograph image. *Proceedings of the Institution of Mechanical Engineers, Part J, Journal of Engineering Tribology, 225*(1), 23–24.

Xia, Y., Shi, P., Liu, G., & Rees, D. (2006). Sliding mode control for stochastic jump systems with time-delay. *The Sixth World Congress on Intelligent Control and Automation, WCICA 2006,* Vol. 1, (pp. 354–358).

Xuan, J., Shi, T., Liao, G., & Liu, S. (2009). Statistical analysis of frequency estimation methods of vibration signal. *Key Engineering Materials, 413-414,* 195–200. doi:10.4028/www.scientific.net/KEM.413-414.195

Xu, Z., Xuan, J., & Shi, T. (2008). Fault diagnosis of bearings based on the wavelet grey moment vector and CHMM. *China Mechanical Engineering, 19*(15), 1858–1862.

Xu, Z., Xuan, J., & Shi, T. (2009). Application of a modified fuzzy ARTMAP with feature-weight learning for the fault diagnosis of bearing. *Expert Systems with Applications, 36*(6), 9961–9968. doi:10.1016/j.eswa.2009.01.063

Yan, J., & Lee, J. (2008, October). *A hybrid method for on-line performance assessment and life prediction in drilling operations.* Paper presented at International Conference on Prognostics and Health Management, Denver, USA.

Yan, X., Zhang, Y., Sheng, C., & Yuan, C. (2010). Study on remote fault diagnosis system using multi monitoring methods on dredger. In *Proceedings of the 2010 Prognostics and System Health Management Conference,* Macau, China, January 12, 2010 - January 14.

Yan, G. (2006). Hilbert-Huang transform-based vibration signal analysis for machine health monitoring. *IEEE Transactions on Instrumentation and Measurement, 55,* 9. doi:10.1109/TIM.2006.887042

Yang, B. S. (2005). Condition classification of small reciprocating compressor for refrigerators using artificial neural networks and support vector machines. *Mechanical Systems and Signal Processing, 19,* 371–390. doi:10.1016/j.ymssp.2004.06.002

Yang, J. N., Lei, Y., Pan, S., & Huang, N. (2003). System identification of linear structures based on Hilbert-Huang spectral analysis. Part 1: Normal modes. *Earthquake Engineering & Structural Dynamics, 32*(9), 1443–1467. doi:10.1002/eqe.287

Yan, J., Koç, M., & Lee, J. (2004). A prognostic algorithm for machine performance assessment and its application. *Production Planning and Control, 15*(8), 796–801. doi:10.1080/09537280412331309208

Yan, X., Zhao, C., Lu, Z., Zhou, X., & Xiao, H. (2005a). A study of information technology used in oil monitoring. *Tribology International, 38*(10), 879–886. doi:10.1016/j.triboint.2005.03.012

Yao, H., Bin, L., & Liu, G. (2007). Plan and design of remote service center for steam turbine. *Turbine Technology, 49,* 81–84.

Yao, J. T. P., Kozin, F., Wen, Y. K., Yang, J. N., Schueller, G. I., & Ditlevsen, O. (1986). Stochastic fatigue, fracture and damage analysis. *Structural Safety, 3,* 231–267. doi:10.1016/0167-4730(86)90005-6

Yao, L.-N., Wang, A., & Wang, H. (2008). Fault detection, diagnosis and tolerant control for non-Gaussian stochastic distribution systems using a rational square-root approximation model. *International Journal of Modelling. Identification and Control, 3*(2), 162–172. doi:10.1504/IJMIC.2008.019355

Yeh, R. H. (1997). State-age-dependent maintenance policies for deteriorating systems with Erlang sojourn time distributions. *Reliability Engineering & System Safety, 58,* 55–60. doi:10.1016/S0951-8320(97)00049-5

Yeh, R. H., & Chang, W. L. (2007). Optimal threshold value of failure-rate for leased products with preventive maintenance actions. *Mathematical and Computer Modelling, 46,* 730–737. doi:10.1016/j.mcm.2006.12.001

Yesilyurt, I., & Ozturk, H. (2006). Tool condition monitoring in milling using vibration analysis. *International Journal of Production Research*, *45*(4), 1013–1028. doi:10.1080/00207540600677781

Yuan, C., Peng, Z., Zhou, X., & Yan, X. (2005a). The surface roughness evolutions of wear particles and wear components under lubricated rolling wear condition. *Wear, 259*(1-6), 512-518.

Yuan, C., Yan, X., & Peng, Z. (2007). Prediction of surface features of wear components based on surface characteristics of wear debris. *Wear, 263*(7-12), 1513-1517.

Yuan, C., Jin, Z., Tipper, J., & Yan, X. (2009). Numerical surface characterization of wear debris from artificial joints using atomic force microscopy. *Chinese Science Bulletin, 54*(24), 4583–4588. doi:10.1007/s11434-009-0588-2

Yuan, C., Peng, Z., & Yan, X. (2005b). Surface characterisation using wavelet theory and confocal laser scanning microscopy. *Journal of Tribology, ASME, 127*(2), 394–404. doi:10.1115/1.1866161

Yuan, C., Peng, Z., Yan, X., & Zhou, X. (2008b). Surface roughness evolutions in sliding wear process. *Wear, 265*(3-4), 341–348. doi:10.1016/j.wear.2007.11.002

Yuan, C., Yan, X., & Zhou, X. (2008a). Effects of Fe powder on sliding wear process under lubricants with different viscosity. *Proceedings of the Institution of Mechanical Engineers, Part J, Journal of Engineering Tribology, 222*(4), 611–616. doi:10.1243/13506501JET316

Yuan, S., & Chu, F. L. (2006). Support vector machines-based fault diagnosis for turbo-pump rotor. *Mechanical Systems and Signal Processing, 20*, 939–952. doi:10.1016/j.ymssp.2005.09.006

Yuen, K. V. (2010). *Bayesian methods for structural dynamics and civil engineering*. Wiley. doi:10.1002/9780470824566

Yura, H. T., & Hanson, S. G. (2011). Digital simulation of an arbitrary stationary stochastic process by spectral representation. *Journal of the Optical Society of America. A, Optics, Image Science, and Vision, 28*(4), 675–685. doi:10.1364/JOSAA.28.000675

Yu, S. Z. (2010). Hidden semi Markov models. *Artificial Intelligence, 174*(2), 215–243. doi:10.1016/j.artint.2009.11.011

Yu, S. Z., & Kobayashi, H. (2003, January). An efficient forward-backward algorithm for an explicit duration hidden Markov model. *IEEE Signal Processing Letters, 10*(1), 11–14. doi:10.1109/LSP.2002.806705

Zaita, V., Buley, G., & Karlsons, G. (1998). Performance deterioration modeling in aircraft gas turbine engines. *Journal of Engineering for Gas Turbines and Power, 120*, 344–349. doi:10.1115/1.2818128

Zakrajsek, J., Townsend, D., & Decker, H. (1993). *An analysis of gear fault detection method as applied to pitting fatigue failure damage*. NASA Technical Memorandum 105950.

Zhang, C., Liu, S., Shi, T., & Tang, Z. (2009). Improved model-based infrared reflectrometry for measuring deep trench structures. *Journal of the Optical Society of America. A, Optics, Image Science, and Vision, 26*(11), 2327–2335. doi:10.1364/JOSAA.26.002327

Zhang, D., Li, W., Xiong, X., & Liao, R. (2011). Evaluating condition index and its probability distribution using monitored data of circuit breaker. *Electric Power Components and Systems, 39*(10), 965–978. doi:10.1080/15325008.2011.552091

Zhang, S., Hodkiewicz, M., Ma, L., & Mathew, J. (2006). *Machinery condition prognosis using multivariate analysis engineering asset management* (pp. 847–854). London, UK: Springer.

Zhang, Y. M., Guo, L., & Wang, H. (2006). Filter-based fault detection and diagnosis using output PDFs for stochastic systems with time delays. *International Journal of Adaptive Control and Signal Processing, 20*(4), 175–194. doi:10.1002/acs.894

Zhang, Y. M., & Jiang, J. (2003). Fault tolerant control system design with explicit consideration of performance degradation. *IEEE Transactions on Aerospace and Electronic Systems, 39*(3), 838–848. doi:10.1109/TAES.2003.1238740

Zhang, Y., & Jiang, J. (2008). Bibliographical review on reconfigurable fault-tolerant control systems. *Annual Reviews in Control, 32*(2), 229–252. doi:10.1016/j.arcontrol.2008.03.008

Zhao, X., Fouladirad, M., Berenguer, C., & Bordes, L. (2008, September). *Optimal periodic inspection/replacement policy of deteriorating systems with explanatory variables*. Paper presented at the European Safety and Reliability Conference, Valencia, Spain.

Zhao, X., Zuo, M. J., & Liu, Z. (2011b). *Diagnosis of pitting damage levels of planet gears based on ordinal ranking*. Paper presented at the IEEE International Conference on Prognostics and Health management, Denver, U.S.

Zhao, F. (Ed.). (2006). *Maximize business profits through e-partnerships*. Hershey, PA: IRM Press.

Zhao, X., Hu, Q., Lei, Y., & Zuo, M. J. (2010a). Vibration-based fault diagnosis of slurry pump impellers using neighbourhood rough set models. *Proceedings of the Institution of Mechanical Engineers. Part C, Journal of Mechanical Engineering Science, 224*(4), 995–1006. doi:10.1243/09544062JMES1777

Zhao, X., Patel, T. H., Sahoo, A., & Zuo, M. J. (2010b). *Numerical simulation of slurry pumps with leading or trailing edge damage on the impeller* (p. 34). Edmonton, Canada: University of Alberta.

Zhao, X., Patel, T. H., & Zuo, M. J. (2011a). (in press). Multivariate EMD and full spectrum based condition monitoring for rotating machinery. *Mechanical Systems and Signal Processing*.

Zhao, X., Zuo, M. J., & Patel, T. H. (2011c). (Manuscript submitted for publication). Generating an indicator for pump impeller damage using half and full spectra, fuzzy preference based rough sets, and PCA. *Measurement Science & Technology*.

Zhao, Y. G., & Ono, T. (1999). A general procedure for first/second-order reliability method (form/sorm). *Structural Safety, 21*(2), 95–112. doi:10.1016/S0167-4730(99)00008-9

Zhou, K., & Ren, Z. (2001). A new controller architecture for high performance, robust, and fault-tolerant control. *IEEE Transactions on Automatic Control, 46*(10), 1613–1618. doi:10.1109/9.956059

Zhu, K. P., Hong, G. S., & Wong, Y. S. (2008). A comparative study of feature selection for hidden Markov model-based micro-milling tool wear monitoring. *Mining Science and Technology, 12*(3), 348–369.

Zille, V., Despujols, A., Baraldi, P., Rossetti, G., & Zio, E. (2009, September). *A framework for the Monte Carlo simulation of degradation and failure processes in the assessment of maintenance programs performance*. Paper presented at the European Safety and Reliability Conference, Prague, Czech Republic.

Zio, E. (2006). A study of the bootstrap method for estimating the accuracy of artificial neural networks in predicting nuclear transient processes. *IEEE Transactions on Nuclear Science, 53*(3), 1460–1478. doi:10.1109/TNS.2006.871662

Zio, E., & Di Maio, F. (2010). A data-driven fuzzy approach for predicting the remaining useful life in dynamic failure scenarios of a nuclear power plant. *Reliability Engineering & System Safety, 95*(1), 49–57. doi:10.1016/j.ress.2009.08.001

Zio, E., & Peloni, G. (2011). Particle filtering prognostic estimation of the remaining useful life of nonlinear components. *Reliability Engineering & System Safety, 96*(3), 403–409. doi:10.1016/j.ress.2010.08.009

# About the Contributors

**Seifedine Kadry** is an Associate Professor of Applied Mathematics in the American University of the Middle East Kuwait. He received his Masters' degree in Modelling and Intensive Computing (2001) from the Lebanese University – EPFL -INRIA. He did his Doctoral research (2003-2007) in Applied Mathematics from Blaise Pascal University-Clermont Ferrand II, France. He worked as Head of Software Support and Analysis Unit of First National Bank where he designed and implement the data warehouse and business intelligence; he has published one book and more than 50 papers on applied math, computer science, and stochastic systems in peer-reviewed journals.

* * *

**Eric Bechhoefer** received his B.S. in Biology from the University of Michigan, his M.S. in Operations Research from the Naval Postgraduate School, and a Ph.D. in General Engineering from Kennedy Western University. His is a former Naval Aviator who has worked extensively on condition based mainteance, rotor track and balance, vibration analysis of rotating machinary, and fault detection in electronic systems. Dr. Bechhoefer is a board member of the Prognostics Health Mangement Society, and a member of the IEEE Reliability Sociciety. Dr. Bechnoefer is currently working as a Chief Engineer for NRG Systems.

**Jamie Baalis Coble** is currently a research scientist at Pacific Northwest National Laboratory where her research focuses on monitoring and prognostic needs for Light Water Reactor Sustainability. She received a B.S. in Nuclear Engineering and Mathematics from the University of Tennessee in May, 2005. She earned an M.S. in Nuclear Engineering in 2006 and an M.S. in Reliability and Maintenance Engineering in 2009. She completed her PhD in Nuclear Engineering in May, 2010 with a dissertation entitled, "Merging Data Sources to Predict Remaining Useful Life – An Automated Method to Identify Prognostic Parameters." Her research interests include empirical methods for monitoring complex engineering systems, fault detection and diagnostics, and system prognostics. She is a member of the PHM Society, the American Nuclear Society, and Women in Nuclear, the IEEE Reliability Society, and the IEEE Women in engineering society.

**Vincent Cocquempot** received the Ph.D. degree in Automatic Control from the Lille University of Sciences and Technologies, in 1993. He is currently a full Professor in Automatic Control and Computer Science at the "Institut Universitaire de Technologies de Lille," France. He is a researcher of the LAGIS-CNRS UMR 8219: Automatic Control, Computer Science and Signal Processing Laboratory from Lille 1 University and Head of the team Fault Tolerant Systems in this laboratory. His research interests include robust on-line Fault Detection and Isolation (FDI) for uncertain dynamical nonlinear systems and Fault Tolerant Control (FTC) for Hybrid Dynamical Systems (HDS).

**Teresa Escobet** received a degree in Industrial Engineering at UPC in 1989 and PhD at the same University in 1997. She began to work at UPC as an Assistant Professor in 1986 and she became Associate Professor in 2001. Her teaching activities are related to issues of Automatic Control. She is a member of the research group Advanced Control Systems (SAC). Her main research interests are in dynamic system modelling and identification applied to fault detection, isolation, fault-tolerant control, and condition-based maintenance. She has been involved in several international and national research projects and networks. Recent publication includes 13 articles in journals and 100 papers at international conferences.

**Claudia Maria García** was born in Terrassa, Barcelona, Spain in 1984. She holds a Technical Engineering degree specialized in Industrial Electronics, at EUETIT, Universitat Politècnica de Catalunya (UPC), Barcelona. Final project developed in Karel de Grote-Hogeschool, Antwerp, Belgium. She did an Automatic Control and Industrial Electronics Master's in Engineering, at ETSEIAT, graduated with honors for finalizing as first of the promotion UPC (2008). An Automotive Electronic Technology Bachelor degree was developed as European Program (2007-2008) at Glyndwr Univeristy, Whales, UK, with a First award. As a Ph.D student she was a participant at the PHM Society Doctoral Consortium'10 with *PHM Strategies Applying Hybrid System Representation for Advanced Maintenance*, and in January 2010 finishes a Master in Automatic Control & Robotics, cursing Diagnosis and Fault Tolerant Control lectures and Hybrid System between others. She is a PhD student on the theme of PHM applied to maintenance in SAC research group in Terrassa, Barcelona, Spain.

**Omid Geramifard** received his Bachelor's degree in Computer Engineering from Isfahan University of Technology, Isfahan, Iran in 2008. He was awarded Singapore International Graduate Award (SINGA) and joined National University of Singapore as a PhD student in 2009. Currently, He is a PhD candidate in Electrical and Computer Engineering Department. His research interests lie primarily in machine learning and its applications in condition based maintenance, as well as diagnostics and prognostics in the machinery systems.

**Giulio Gola** has an MSc in Nuclear Engineering, PhD in Nuclear Engineering, from Polytechnic of Milan, Italy. He is currently working as a Research Scientist at the Institute for Energy Technology (IFE) and OECD Halden Reactor Project (HRP) within the Computerized Operations and Support Systems department. His research topics deal with the development of physical and empirical models based on artificial intelligence for data fusion, data reconciliation, and data mining with application to the nuclear and oil and gas industries. In particular, his research focuses on methods for condition monitoring, system diagnostics and prognostics, parameter estimation, and large-scale sensor monitoring and signal validation. He works also for the consulting company FirstSensing AS on a variety of projects of condition monitoring with applications to air quality monitoring, waste-water treatment and particle emission monitoring. He is author of more than twenty publications on international journals.

**Xuefei Guan** is a research scientist at Siemens Corporate Research, Siemens Corporation. He obtained his PhD degree in Mechanical Engineering from Clarkson University in 2011, and received his B.S. degree in Reliability Engineering and M.S. degree in Aeronautical Engineering from Beihang University in China in 2005 and 2008, respectively. His research interests include probabilistic modelling and analysis for engineering structures, Bayesian and entropy-based inference methods, Markov Chain Monte Carlo simulations, reliability analysis, and combinatorial optimizations. He is currently working on non-destructive inspection and damage assessment methods for energy generation infrastructures.

**Gilbert Haddad** received his BE in Mechanical Engineering from the American University of Beirut in 2007, and his MS degree from the University of Wisconsin at Madison in 2009. He completed his Ph.D. at the University of Maryland, College Park at the Centre for Advanced Life Cycle Engineering (CALCE) in 2011. His research interests include prognostics and health management, data mining, analytics, and decision support systems. Gilbert most recently worked at GE Global Research at the Machine Learning Lab (summers of 2010 and 2011), and is now a Mechanical Engineer at Schlumberger in Sugar Land, Texas working on prognostics and health management for oil and gas applications.

**Assia Hakem** is currently PhD student in the Engineering School Polytech'Lille at the University of Lille 1. Her research interests are in model and data based system monitoring and dynamical system identification. She holds her Master's degree in intelligent systems for transportation from University of Compiègne in France.

**David He** is a Professor and Director of the Intelligent Systems Modelling & Development Laboratory in the Department of Mechanical and Industrial Engineering, The University of Illinois-Chicago. He got his Ph.D. degree in Industrial Engineering from The University of Iowa in 1994. Dr. He's research areas include: machinery health monitoring, diagnosis and prognosis, system failure modelling and analysis, quality, and reliability engineering.

**Jingjing He** is a graduate Research Assistant in department of Civil Engineering at Clarkson University. Her research interests are fatigue analysis of metals, structure health monitoring system, and sensor networks. She received her B.S. degree in Reliability Engineering and M.S. degree in Aerospace System Engineering from Beihang University in China in 2005 and 2008, respectively.

**J. Wesley Hines** is a Postelle Professor and Head of the Nuclear Engineering Department at the University of Tennessee, Knoxville, as well as the Director of the Reliability and Maintainability Engineering program. He received the BS degree in Electrical Engineering from Ohio University in 1985, and then served as a nuclear qualified submarine officer in the US Navy. He later received both an MBA and an MS in Nuclear Engineering from The Ohio State University in 1992 and a Ph.D. in Nuclear Engineering from The Ohio State University in 1994. Dr. Hines teaches and conducts research in artificial intelligence and advanced statistical techniques applied to process diagnostics, condition based maintenance, and prognostics. Much of his research program involves the development of algorithms and methods to monitor high value equipment, detect abnormalities, and predict time to failure. He has authored over 250 papers and holds several patents in the area of advanced process monitoring and prognostics techniques.

**Taoufik Jazouli** received the B.S. degree in Mechanical Engineering and the M.S. degree in Mechanical Engineering: Design and Mechanical Manufacturing, both from the National Higher School of Electricity and Mechanics – University Hassan II Ain Chock – Morocco, in 2005. After two years of professional experience in engineering consulting, he attended the University of Maryland College Park, and received M.S. degree and Ph.D. degree in Mechanical Engineering: Systems Design and Support, in 2011. Currently, he is a Lead Systems Engineer at CSSI INC, supporting the Configuration Management Automation (CMA) program, in compliance with the Federal Aviation Administration (FAA) strategic transition to the Next Generation Air Transportation System (NextGen).

**Ratneshwar Jha** has conducted research in composite and smart structures, structural health monitoring, UAV, active flow control, and multidisciplinary design optimization. He has established Smart Structures Laboratory at Clarkson University for vibro-acoustic research. Dr. Jha's contributions include both theoretical and experimental research resulting in over 90 publications in archival journals and conferences. He has served as advisor to 25 graduate students and three undergraduate honors students. Dr. Jha is an Associate Fellow of AIAA and a member of ASME and ASEE. He is a member of the AIAA Adaptive Structures Technical Committee and has served as Session Chair/Co-Chair for Adaptive Structures Conferences. Dr. Jha has received research grants from NSF, AFOSR, NASA, ARO, ARDEC, New York State Energy Research and Development Authority, and several industries. Prior to joining Clarkson, Dr. Jha worked in the aerospace industry from 1983 to 1995 where he led a team of engineers working on conceptual and preliminary designs of combat aircraft.

**Dongxiang Jiang** is a Professor at the Department of Thermal Engineering, Tsinghua University, Beijing, China. He received the B. Sc. degree in Electronic Engineering from the Shenyang University of Technology in 1983, the M. Sc. degree in Electrical Engineering from Harbin Institute of Technology in 1989, and the PhD degree in Astronautics and Mechanics from Harbin Institute of Technology in 1994. He worked as an Assistant Engineer and an Engineer at Harbin Research Institute of Electrical Instrumentation for six years. He was a post doctor at the Department of Thermal Engineering, Tsinghua University, from 1994 to 1996. His research interests include condition monitoring, and diagnostics for rotating machinery, and wind power. He is a member of ASME, Machinery Fault Diagnostic Division of the Chinese Society for Vibration Engineering.

**Zhixiong Li** holds an MD and is PhD candidate in school of Energy and Power Engineering, Wuhan University of Technology, China. He is a reviewer for the *Journal of Mechanical Science and Technology* and a program Chairman of the International Conference on Green Power, Materials and Manufacturing Technology and Applications (GPMMTA), in Chongqing, China. He has published more than 10 technical journal papers and international conference papers. His research interests include: (1) condition monitoring and fault diagnosis of marine power machinery; (2) dynamical modeling of ship propulsion systems, and their adaptive control.

**Chao Liu** is a Ph.D. candidate in the Department of Thermal Engineering at Tsinghua University, Beijing, China. He received the B. Sc. degree in Thermal and Power Engineering from Huazhong University of Science and Technology in 2008. His research interests include structural dynamics, fault diagnostics, and prognostics of rotating machinery.

**Yongming Liu** is an Assistant Professor in the department of Civil and Environmental Engineering. He completed his PhD at Vanderbilt University in 2006, and obtained his Bachelors' and Masters' degrees from Tongji University in China in 1999 and 2002, respectively. His research interests include fatigue and fracture analysis of engineering materials and structures, probabilistic methods, diagnostics and prognostics, and risk management. Dr. Liu's group is currently working on several projects related to structural damage assessment and risk evaluation from various federal agencies, including NSF, NASA, AFOSR, and FAA. He is the recipient of the Air Force Young Investigator Award in 2011.

**Veli Lumme** is a PhD student in the Tampere University of Technology and is also involved in the training and certification of condition monitoring analysts. He has an extensive background in condition monitoring since receiving his M.Sc. in Mechanical Engineering from the Helsinki University of Technology in 1980. Veli was the first to introduce the Finnish industry to the computer-aided vibration analysis. Later he has studied various methods in the remote and automated diagnosis of vibration data. This work has resulted in a four patented inventions. As a recognition of his work, Veli has been granted the the Grand Prize in productive maintenance in 2005 and the Golden Badge of Merit in 2004 both awarded by the Finnish Maintenance Society.

**Jinghua Ma** received her B.S. and M.S. degrees in Systems Engineering from Beijing University of Aeronautics and Astronautics 2002 and 2008, respectively. She is a Research Assistant in the Intelligent Systems Modelling & Development Laboratory and a Ph.D. Candidate in the Department of Mechanical and Industrial Engineering, The University of Illinois-Chicago.

**Komi Midzodzi** is currently assistant Professor in the Engineering School Polytech'Lille at the University of Lille 1. His research interests are in model and data based system monitoring and dynamical system identification. He teaches automatic control and data analysis. Doctor Komi Midzodzi PEKPE obtained his Bachelor's in Applied Mathematics from University of Lomé in Togo, and holds a Master's degree in Signal Control and Communication from INPL (National Polytechnic Institute of Lorraine), France. He obtained his PhD in the field of Automatic and Signal Procession is from INPL. He was temporarily Researcher and Professor in September 2004 at IUT de Saint Dié des Vosges and he has been an Assistant Professor at University of Lille 1 since September 2005. He is a member of French research groups S3 (Sûreté, Surveillance, Supervision) and GT Identification (Groupe de Travail Identification). He works on model and data based method for system monitoring using identified or first principle models of nonlinear dynamical systems. He applies the methods to biological systems. He is interested in vibration based faults detection technics. He applies these methods to helicopter monitoring and energy production systems monitoring.

**Ramin Moghaddass** is a PhD candidate in the department of Mechanical Engineering at University of Alberta. He received both his BSc and MSc in Industrial Engineering from Sharif University of Technology, Tehran, Iran. He is currently a member of reliability research lab in the department of Mechanical Engineering at the University of Alberta. His research activities are in Multi-state reliability modeling, prognostic health management (PHM), and condition-based maintenance.

**Fatiha Nejjari** received her Master degree in Physics from Hassan II University and the Ph.D. degree in Automatic Control from Cadi Ayyad University of Marrakech in 1993 and 1997, respectively. She is currently an Associate Professor at the Automatic Control Department of the Technical University of Catalonia (UPC). She is also a member of the Aeronautical and Space Science Research Centre (CRAE) of UPC. Her active research areas include nonlinear control, parameter and state estimation, model-based fault diagnosis, adaptive and fault tolerant control. Her expertise is specifically in the field of joint state and parameter estimation, fault detection and isolation, nonlinear and fault tolerant control with application to bioprocesses, environmental systems, and electrical systems. She has published several papers in journals and international events and participated in several projects and networks related with these topics.

**Bent Helge Nystad** was awarded an MSc in Cybernetics by RWTH, Aachen in Germany, 1993, and a PhD in Marine Technology by the University of Trondheim (NTNU), Norway, 2008. He has work experience as a condition monitoring expert from Raufoss ASA (a Norwegian missile and ammunition producer) and he has been a Principal Research Scientist at the Institute for Energy Technology (IFE) and OECD Halden Reactor Project (HRP) since 1998. He is the author of 15 publications in international journals and conference proceedings. His experience and research interests have ranged from data-driven algorithms and first principle models for prognostics, performance evaluation of prognostic algorithms, requirement specification for prognostics, technical health assessment, and prognostics in control applications.

**Vicenç Puig** has a degree in Engineering in Telecommunications and a PhD degree in Automatic Control. He is an Associate Professor in the Automatic Control Department of UPC since 2001. Prof. Puig is leader of the UPC research group —Advanced Control Systems (SAC). Research interests cover themes in advanced control systems, such as model identification, model predictive control, and supervision (fault diagnosis and tolerant control) of complex systems. He has participated in more than 15 international & national research projects. He was responsible at UPC for —Decentralized and Wireless Control of Large Scale Systems (WIDE) project (FP-7-ICT-224168). Recent publications: 50 articles in journals and more than 250 papers in international conferences.

**Joseba Quevedo** received the Master's degree in Electrical, Electronic and Control Engineering and the PhD in Control Engineering from the University Paul Sabatier of Toulouse (France) and the PhD in Computer Engineering from the Technical University of Catalonia. He is full Professor since 1990. Joseba Quevedo has published more than 250 journal and conference papers in the areas of advanced control, identification and parameter estimation, fault detection and diagnosis, and fault tolerant control and their applications to large scale systems (water distribution systems and sewer networks) and to industrial processes.

**Peter Sandborn** is a Professor in the CALCE Electronic Products and Systems Center at the University of Maryland. Dr. Sandborn's group develops life-cycle cost models and business case support for long field life systems. This work includes: obsolescence forecasting algorithms, strategic design refresh planning, lifetime buy quantity optimization, and return on investment models for maintenance planning (including the application of PHM to systems). Dr. Sandborn is an Associate Editor of the IEEE Transactions on Electronics Packaging Manufacturing, the North American Editor of the *International Journal of Performability Engineering*, and on the Board of Directors of the International PHM Society. He has authored over 150 technical publications and several books on electronic packaging and electronic systems cost analysis; and was the winner of the 2004 SOLE Proceedings Award and the 2006 Eugene L. Grant Award for best paper in the journal *Engineering Economics*. He has a B.S. degree in Engineering

Physics from the University of Colorado, Boulder, in 1982, and the M.S. degree in Electrical Science and Ph.D. degree in Electrical Engineering, both from the University of Michigan, Ann Arbor, in 1983 and 1987, respectively.

**Chenxing Sheng**, PhD, associate professor in school of Energy and Power Engineering, Wuhan University of Technology, China. He received his PhD degree from Wuhan University of Technology, China, in 2009. He has published more than 10 technical journal papers and international conference papers. His research interests include condition monitoring and fault diagnosis of marine power machinery.

**Zakwan Skaf** was born in Hama, Syria, in 1978. He received the B.S. degree from power department of the faculty of Mechanical Engineering in Aleppo University, Aleppo, Syria, in 2001, and the M.S. and Ph.D. degrees in advanced control systems from Control Systems Centre, University of Manchester, Manchester, U.K., in 2006 and 2011, respectively. His research interests are nonlinear control, stochastic distribution control, iterative learning control, fault detection and diagnosis, active fault tolerant control, and minimum entropy control.

**Jian-Xin Xu** received the Bachelor's degree from Zhejiang University, China, the Master's and PhD degrees from the University of Tokyo, Japan. All his degrees are in Electrical Engineering. He worked for one year in Hitachi Research Laboratory, Japan, and one year in Ohio State University as a visiting scholar, and 6 months at Yale University as a visiting research fellow. In 1991, he joined the National University of Singapore, and is currently a Professor in the Department of Electrical and Computer Engineering, and the area director for control, intelligent systems and robotics. His research interests include learning theory, system control, and robotics. He has 400 publications including 143 peer re-viewed journal papers and 5 books. He has served as the general co-chair for 2007 IEEE Conference on Evolutionary Computation, and program chairs for a number of international conferences. He served as the IEEE Singapore EMBS Chapter chair.

**Xinping Yan**, PhD, Professor in school of Energy and Power Engineering, Wuhan University of Technology, China. He received his BSc in Marine Mechanical Engineering from Wuhan University of Water Transportation Engineering, China in 1982 and his Master from the same University in 1987, and his PhD in Mechanical Engineering from Xi'an Jiaotong University, China in 1997. His research interests include: tribology and its industrial application, condition monitoring and fault diagnosis for marine power system, intelligent transport system, and maritime education. He is the member of ISO/TC108/SC5 Committee and member of STLE. He is VP of Chinese Tribology Institute (CTI). He is the editorial member of *Journal of COMADEM* (U.K.) and *Journal of Maritime Environment* (U.K.). Now he is a Professor of Vehicle Operational Engineering and head of Reliability Engineering Institute in Wuhan University of Technology, China.

**Chengqing Yuan**, PhD, is Professor in school of Energy and Power Engineering, Wuhan University of Technology, China. He received his PhD degree from Wuhan University of Technology, China, in 2005. He was a full-time research officer of an ARC project in Australia from July 2002 to January 2005 and a postdoctoral research fellow in Leeds University, UK from February 2007 to November 2007 supported by Royal Society KC Wong Fellowship. He has published more than 50 journal papers and won 4 provincial level awards for science and technology. His research interests include: (1) tribology problems under marine rigorous condition; (2) condition monitoring and fault diagnosis of marine power machinery; (3) maintenance theory and approaches for mechanical system; (4) and study and applications of clean energy or new energy in ship.

**Yuelei Zhang**, PhD, squadron leader in Unit 94270 of Chinese People's Liberation Army (PLA), China. He received his PhD degree from Wuhan University of Technology, China, in 2011. He has published more than 10 technical journal papers and international conference papers. His research interests include condition monitoring, fault diagnosis, and aviation equipment maintenance.

**Jiangbin Zhao**, PhD, Lecturer in school of Energy and Power Engineering, Wuhan University of Technology, China. He received his PhD degree from Huazhong University of Science and Technology, China, in 2007. He was a postdoctoral research fellow in Huazhong University of Science and Technology, China from February 2007 to November 2009. He has published more than 10 technical journal papers and international conference papers. His research interests include (1) condition monitoring, fault diagnosis, and aviation equipment maintenance; (2) and sensor network and virtual instrument.

**Xiaomin Zhao** is a PhD candidate in the department of Mechanical Engineering at University of Alberta. She received both her BSc and MSc from Harbin Institute of Technology, Harbin, China. She is currently a research assistant at Reliability Research Lab of University of Alberta. Her research activities are in fault diagnosis, condition monitoring, and prognostic health management (PHM).

**Junhong Zhou** received the B.Eng. degree from the Automation Department, Tsinghua University, Beijing, China, in 1987 and the M. Eng. degree from the Department of Electrical and Computer Engineering, National University of Singapore (NUS) in 1995. She is currently working towards the Ph.D. degree at the School of Mechanical and Aerospace Engineering, Nanyang Technological University, Singapore. Since 1996, she worked as a Research Fellow at the Singapore Institute of Manufacturing Technology, A*STAR, Singapore. Her current research interests include equipment health monitoring, intelligent condition-based maintenance, fault diagnosis and failure prognosis, multimodal sensing and feature extraction, performance and serviceability optimization, artificial intelligence, and statistical analysis.

**Junda Zhu** received his B.S. degree in Mechanical Engineering from North-eastern University, Shenyang, China and M.S. degree in Mechanical Engineering from The University of Illinois-Chicago, 2009. He is currently working toward the Ph.D. degree in Industrial Engineering in the Department of Mechanical and Industrial Engineering at The University of Illinois-Chicago. His research interests include: lubrication oil condition monitoring and degradation analysis, physics based and data mining based full ceramic bearing diagnostics and prognostics, and AutoCAD/FE and design analysis.

**Enrico Zio** (BS in Nuclear Engng., Politecnico di Milano, 1991; MSc in Mechanical Engng., UCLA, 1995; PhD, in Nuclear Engng., Politecnico di Milano, 1995; PhD, in Nuclear Engng., MIT, 1998) is Director of the Chair in Complex Systems and the Energetic Challenge of Ecole Centrale Paris and Supelec, full Professor, Rector's delegate for the Alumni Association and past-Director of the Graduate School at Politecnico di Milano, adjunct Professor at University of Stavanger. He is the Chairman of the European Safety and Reliability Association ESRA, member of the Korean Nuclear society and China Prognostics and Health Management society, and past-Chairman of the Italian Chapter of the IEEE Reliability Society. He is serving as Associate Editor of *IEEE Transactions on Reliability* and as editorial board member in various international scientific journals. He has functioned as Scientific Chairman of three International Conferences and as Associate General Chairman of two others. His research topics are: analysis of the reliability, safety, and security of complex systems under stationary and dynamic conditions, particularly by Monte Carlo simulation methods; development of soft computing techniques for safety, reliability, and maintenance applications, system monitoring, fault diagnosis, and prognosis. He is author or co-author of five international books and more than 170 papers on international journals.

**Ming J. Zuo** received an MSc in 1986 and a PhD in 1989, both in Industrial Engineering from Iowa State University, Ames, Iowa. He is a Professor in the Department of Mechanical Engineering at the University of Alberta. His research interests include system reliability analysis, maintenance planning and optimization, and fault diagnosis. He is Associate Editor of *IEEE Transactions on Reliability and* Department Editor of *IIE Transactions*. He is a senior member of IEEE and IIE.

# Index